COMBINATORICS
OF COMPOSITIONS
AND WORDS

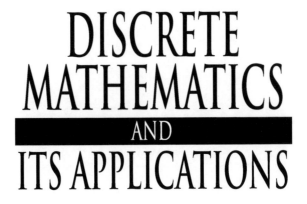

DISCRETE MATHEMATICS
AND
ITS APPLICATIONS

Series Editor
Kenneth H. Rosen, Ph.D.

Continued Titles

DISCRETE MATHEMATICS AND ITS APPLICATIONS

Series Editor KENNETH H. ROSEN

COMBINATORICS OF COMPOSITIONS AND WORDS

SILVIA HEUBACH
CALIFORNIA STATE UNIVERSITY
LOS ANGELES, CALIFORNIA, U.S.A.

TOUFIK MANSOUR
UNIVERSITY OF HAIFA
ISRAEL

CRC Press
Taylor & Francis Group
Boca Raton   London   New York

CRC Press is an imprint of the
Taylor & Francis Group, an **informa** business

A CHAPMAN & HALL BOOK

Chapman & Hall/CRC
Taylor & Francis Group
6000 Broken Sound Parkway NW, Suite 300
Boca Raton, FL 33487-2742

First issued in paperback 2017

© 2010 by Taylor and Francis Group, LLC
Chapman & Hall/CRC is an imprint of Taylor & Francis Group, an Informa business

No claim to original U.S. Government works

ISBN 13: 978-1-138-11667-2 (pbk)
ISBN 13: 978-1-4200-7267-9 (hbk)

Library of Congress Cataloging-in-Publication Data

Heubach, Silvia.
    Combinatorics of compositions and words / Silvia Heubach, Toufik Mansour.
      p. cm.
    Includes bibliographical references and index.
    ISBN 978-1-4200-7267-9 (hardcover : alk. paper)
    1. Combinatorial analysis. 2. Word problems (Mathematics) 3. Combinatorial enumeration problems. 4. Combinatorial probabilities. 5. Composition (Language arts)--Mathematics. I. Mansour, Toufik. II. Title.

QA164.H48 2010
511'.6--dc22
                                                                   2009020908

Visit the Taylor & Francis Web site at
http://www.taylorandfrancis.com

and the CRC Press Web site at
http://www.crcpress.com

Für meine Eltern - ihr habt mir die Freiheit gelassen meinen eigenen Weg zu gehen.

Silvia

لِأُمِّي - لِكَوْنِها قُدْوَةً لنا بِصُمُودِها في وَجْهِ ٱلصِّعَابِ.

توفيق

Contents

List of Tables

List of Figures

Preface

This text gives an introduction to and an overview of the methods used in the combinatorics of pattern avoidance and pattern enumeration in compositions and words, a very active area of research in the last decade. The earliest results on enumeration of compositions appeared in a paper by P.A. MacMahon in 1893, while Axel Thue is credited with starting research on the combinatorics of words in a paper in 1906. MacMahon considered words in the context of partitions and permutations, while Thue approached words from a number theoretic background. In the decades that followed the main focus of research in enumerative combinatorics was on partitions and permutations, driven in part by MacMahon's prolific writing. Until the late 1960s, individual articles on various aspects of compositions appeared, but there was no concentrated research interest.

This changed in the 1970s, when several groups of authors developed new research directions. They studied compositions and words that were restricted in some way, and not only enumerated the total number of these objects, but also certain of their characteristics (statistics). Another focus was the study of random compositions to obtain asymptotic results. Most publications on compositions and words have been within the last decade. Authors have studied various aspects of compositions and words, generalizing previous results and introducing new concepts. In particular, research on pattern avoidance in words and compositions has followed earlier very active research on pattern avoidance in permutations.

We wrote this book to provide a comprehensive resource for anybody interested in this new area of research. The text aims to:

- provide a self-contained, broadly accessible introduction to this research area;

- introduce the reader to a variety of tools and approaches that are also applicable to other areas of enumerative combinatorics;

- give an overview of the history of research on enumeration of and pattern avoidance in compositions and words;

- showcase known and new results; and

- provide a comprehensive and extensive bibliography.

Our book is based on our own research, and on that of our collaborators and other researchers in the field. We present these results with consistent notation and have modified some proofs to relate to other results in the book.

As a general rule, theorems listed without specific references give results from articles by the authors and their collaborators while results from other authors are given with specific references.

Audience

This book can be used as the primary text for a second course in discrete mathematics at the undergraduate level or for a beginning graduate level course in the subject, with a focus on compositions and words. In addition the book serves as a one-stop reference for a bibliography of papers on the subject, known results, and research directions for any researcher who is interested in entering this new field.

Outline

In Chapter 1 we give a historical perspective of research on compositions and words and present basic definitions and early results. This is followed by an overview of the six major themes: compositions with restricted arrangement, compositions with restricted parts, statistics on compositions, random compositions, pattern avoidance in compositions and words, and asymptotics. We also provide a time line for the articles in these major areas of research.

In Chapter 2 we introduce techniques to solve recurrence relations, which arise naturally when enumerating compositions. This section contains many examples of important integer sequences, such as the Fibonacci and Lucas sequences, to illustrate the techniques of setting up and of solving recurrence relations. Methods for solving recurrence relations include guess and check, iteration, characteristic polynomial, and generating functions, including the Lagrange Inversion Formula.

Chapter 3 focuses on results for compositions. We start by giving basic results for compositions that are restricted with regard to arrangement (palindromic compositions and Carlitz compositions) and with regard to parts (compositions with 1s and 2s or with odd parts). This is followed by results for compositions with parts in a general set A, which are then applied to several specific sets. We also discuss the use of computer algebra systems (Maple[1] and $Mathematica$[2]) to perform computations when explicit formulas cannot be derived. Finally, we connect compositions to tilings (another very active research area) and discuss some special compositions, namely n-colored and cracked compositions.

Chapter 4 deals with statistics on compositions. A statistic is a characteristic such as the number of times two adjacent terms in a composition are the same. We connect the early research on the statistics rises, levels and falls with enumeration of patterns and pattern avoidance, presenting specific results for avoidance of the subword patterns of length three and for families of longer subword patterns.

---

[1] Maple[TM] is a registered trademark of Waterloo Maple Software.

[2] $Mathematica$® is a registered trademark of Wolfram Research, Inc.

In Chapter 5 we present results on pattern avoidance for subsequence, generalized, and partially ordered patterns. We provide a complete classification with respect to Wilf-equivalence for subsequence patterns of lengths three, four, and five, and provide generating functions for patterns of length three. In addition, we present new results on Wilf classification of generalized patterns of type $(2, 1)$ and on the respective generating functions. We conclude the chapter with results on avoidance of partially ordered patterns.

In Chapter 6 we apply the results for compositions from the previous chapters to words. We also discuss specialized techniques used for enumeration of words and for pattern avoidance in words. In particular, we compare and contrast three approaches: the Noonan-Zeilberger algorithm, the block decomposition method, and the scanning-element algorithm, by applying them to the problem of enumerating words that avoid the subsequence pattern 123.

One tool that has been successfully used for pattern avoidance in permutations and words is automata. In Chapter 7 we provide graph theory basics to develop the notion of an automaton and its corresponding transfer matrix. This matrix is very useful to describe the asymptotic behavior of the generating functions. We give results for increasing patterns using automata and discuss the closely related ECO method and generating trees. We also derive new results for compositions using automata.

As in many other areas of combinatorics, randomization is useful in the study of compositions. In Chapter 8 we focus on random compositions, after first providing tools from probability theory. We then describe asymptotic results for various statistics on random compositions. Finally, we derive asymptotic results using tools from complex analysis.

The chapters build upon each other, with the exception of Chapter 8 which can be read independently of Chapter 7. Most chapters start with a section describing the history of the particular topic and its relation to other chapters. New methods and definitions are illustrated with worked examples and we provide Maple and *Mathematica* code whenever applicable. Both these computer algebra systems have powerful built-in functions that can assist in solving recurrence relations or in finding generating functions. Each chapter ends with a list of exercises that range in level from easy to difficult (marked with a *), followed by research directions that extend the results given in the chapter. Questions posed in the research directions can be used for projects, thesis topics, or research.

The appendix gives basic background from linear algebra, algebra, probability theory, complex analysis, and Chebychev polynomials. We also provide a list of useful generating functions and combinatorial formulas and a number of C++ programs that are used in the classification of patterns according to Wilf-equivalence. We give examples of their use and indicate how they can be modified for exercises and questions posed in the research directions. In addition, we provide a detailed description of the relevant *Mathematica* and Maple functions.

Support Features

We provide an instructor's solutions manual with worked-out solutions. In addition, the C++ programs and in some instances a Java version are available for download from Toufik Mansour's webpage `http://math.haifa.ac.il/toufik/manbook.html`. This webpage also has links to tables containing lists of values of the number of compositions avoiding a specific subsequence pattern of length four or five, and similar tables for the number of words avoiding specific subsequence patterns of length four, five or six.

We will regularly update Toufik's website to post references to new papers in this research area as well as any corrections that might be necessary. (No matter how hard we try, we know that there will be typos.) We invite readers to give us feedback at either

`sheubac@calstatela.edu` or `toufik@math.haifa.il`

July 2009

Acknowledgment

Every book has a story of how it came into being and the people that supported the author(s) along the way. Ours is no exception. The idea for this book came when the authors met for the first time in person at an FPSAC conference in Vancouver, after collaborating via the Internet on two papers. Silvia had been working on enumeration of rises, levels, and drops in restricted compositions, while Toufik had a background in pattern avoidance in permutations and words. Combining the two lines of inquiry we started to work on pattern avoidance in compositions.

Several years later, after a mostly long-distance collaboration that has made full use of modern electronic technology (email and Skype[1]), this book has taken its final shape. Along the way we have received encouragement for our endeavor from Martin Golumbic, Einar Steingrímsson, and Doron Zeilberger. We appreciate the financial support of the Department of Mathematics and the Caesarea Rothschild Institute at the University of Haifa which made visits to Haifa possible. We also would like to thank Bill Chen at the Center for Combinatorics at Nankai University, Tianjin, China for hosting Silvia for a visit while Toufik was a visiting professor at the center.

And then there are the people in our lives who supported us on a daily basis by giving us the time and the space to write this book. Their moral support has been very important. Silvia thanks Ken, who has been supportive throughout and has been very understanding when the book needed time and took precedence. In addition, his guidance on the intricacies of the English language is very much appreciated. Toufik thanks his wife Ronit for her support and understanding when the work on the book took him away from spending time with her and from playing with their children Itar, Atil, and Hadil. We both thank our larger families for cheering us on and supporting us even though they do not understand the mathematics that we describe.

Finally, we thank the staff at Chapman & Hall/CRC who have assisted us as we navigated the book production process.

---

[1]Skype[TM] is a registered trademark of Skype Limited.

Author Biographies

Silvia Heubach obtained a Masters degree in Wirtschaftsmathematik (Mathematics and Economics) from the University of Ulm in 1986, and Masters and Ph.D. degrees in Applied Mathematics from the University of Southern California in 1992. Her area of specialty originally was probability theory, but starting in 1998 she published several papers concerning the enumeration of compositions, with a number of different co-authors. In 2003 she started collaborating with Toufik Mansour who had been very active in research on pattern avoidance in permutations and words.

Silvia has been a faculty member at California State University Los Angeles since 1994 and was promoted to Full Professor in 2004. She received the 1999/2000 CSULA Outstanding Professor Award and has been the chair of the Department of Mathematics since 2007. She has given numerous talks at national and international conferences and has been a reviewer for *Discrete Mathematics*, *Integers*, *Ars Combinatoria* and the *Journal of Difference Equations and Applications*.

Toufik Mansour obtained his Ph.D. degree in Mathematics from the University of Haifa in 2001. He spent one year as a postdoctoral researcher at the University of Bordeaux (France) supported by a Bourse Chateaubriand scholarship, and a second year at the Chalmers Institute in Gothenburg (Sweden) supported by a European Research Training Network grant. He has also received a prestigious MAOF grant from the Israeli Council for Education. Toufik has been a permanent faculty member at the University of Haifa since 2003 and was promoted to Associate Professor in 2008. He spends his summers as a visitor at institutions around the globe, for example, at the Center for Combinatorics at Nankai University (China) where he was a faculty member from 2004 to 2007.

Toufik's area of specialty is enumerative combinatorics and more generally, discrete mathematics and its applications to physics, biology, and chemistry. Originally focusing on pattern avoidance in permutations he has extended his interest to colored permutations, partitions of a set, words and compositions. Toufik has authored or co-authored more than 60 papers in this area, many of them concerning the enumeration of words and compositions. He has given talks at national and international conferences and is very active as a reviewer for several journals, incuding *Discrete Mathematics*, *Discrete Applied Mathematics*, *Advances in Applied Mathematics*, the *Journal of Integer Sequences*, the *Journal of Combinatorial Theory Series A*, *Annals of Combinatorics*, and the *Electronic Journal of Combinatorics*.

Chapter 1

Introduction

1.1 Historical overview – Compositions

P.A. MacMahon

The study of compositions has a long and rich history. The earliest publication on compositions is by Percy Alexander MacMahon[1](1854 – 1929) in 1893, entitled *Memoir on the Theory of Compositions of a Number* [136], which appeared in the Philosophical Transactions of the Royal Society of London. At the time, articles were submitted and read before the society prior to being published. MacMahon begins the memoir by stating that "Compositions are merely partitions in which the order of occurrence of the parts is essential," and then proceeds to give the partitions and compositions of $n = 3$ as an example.

This more informal style of writing was quite common in the papers at the time, where definitions, examples, statements and proofs followed each other without much of the formal structure one sees in today's papers. Therefore, when quoting results given by MacMahon, we will indicate the page on which the statement appears.

Definition 1.1 *A* partition *of a positive integer n is a nonincreasing sequence $p = p_1 p_2 \ldots p_m$ of positive integers such that $\sum_{i=1}^{m} p_i = n$. A* composition *of a positive integer n is any sequence $\sigma = \sigma_1 \sigma_2 \ldots \sigma_m$ of positive integers such that $\sum_{i=1}^{m} \sigma_i = n$. The p_i and σ_i are called the* parts, *and $\mathrm{par}(p) = m$ and $\mathrm{par}(\sigma) = m$ denote the number of parts of the partition and composition, respectively. We refer to n as the* order *of the partition or composition and denote it by $\mathrm{ord}(p)$ and $\mathrm{ord}(\sigma)$.*

---

[1] http://www-history.mcs.st-andrews.ac.uk/Mathematicians/MacMahon.html
[1] http://en.wikipedia.org/wiki/Percy_Alexander_MacMahon

Note that the difference between partitions and compositions is a matter of order – for partitions we do not care about order, while for compositions, two differently ordered sequences result in two different compositions of n. Thus for every partition with at least two different parts, there are more compositions than partitions, as we can reorder the partition in all possible ways to get the associated compositions.

Example 1.2 *The partitions of 4 are 4, 31, 22, 211 and 1111, while the compositions of 4 are 4, 31, 13, 22, 211, 121, 112 and 1111. The partition 211 can be reordered in three different ways and results in the compositions 211, 121, 112. However, the partition 22 results in only one composition 22.*

MacMahon derived a number of results, for example the total number of compositions and the number of compositions with a given number of parts, using generating functions. He also provided a representation of a composition as a graph, which allowed for the derivation of those results using a simple combinatorial argument. As defined by MacMahon, "the graph of a number n is taken to be a straight line divided at $n - 1$ points into equal segments. The graph of a composition of the number n is obtained by placing nodes at certain of the $n - 1$ points of division." Figure 1.1 illustrates the graph of the composition 215 of eight in the format given in [136]. We refer to this representation as the *line graph of a composition* to distinguish it from other representations that will be introduced in Chapter 3.

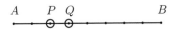

FIGURE 1.1: Line graph of the composition 215.

The individual parts are read off as the number of segments between nodes or between the first and the last nodes and the endpoints A and B, respectively. Using this representation, MacMahon gave a combinatorial proof of the following result.

Theorem 1.3 *[136, page 835] The number of compositions of n with exactly m parts is given by $\binom{n-1}{m-1}$, and the total number of compositions of n is 2^{n-1}.*

Proof Since the number of parts of a composition is one more than the number of nodes, and the number of division points is $n - 1$, we merely select at which of the $n - 1$ division points we place the $m - 1$ nodes. Furthermore, since each of the $n - 1$ division points is either a node or not a node in each composition, the total number of compositions is 2^{n-1}. □

MacMahon also looked at relationships between compositions, for example conjugate and inverse compositions.

Definition 1.4 *Two compositions of n are* conjugate *if in their respective line graphs, one has nodes exactly at the positions where the other one does not have nodes and vice versa. We denote the conjugate of σ by* conj(σ).

Figure 1.2 shows the line graphs of two conjugate compositions. Note that it is very easy to derive the conjugate composition using the line graph, but that an algorithmic description of how to compute a conjugate composition without the line graph is quite lengthy (see [136, pages 837-838]) and much less obvious.

FIGURE 1.2:   Line graphs of conjugate compositions 215 and 131111.

Definition 1.5 *If σ* = σ₁σ₂...σₘ, *then its* inverse *or* reverse *composition is given by R(σ)* = σₘσₘ₋₁...σ₁. *A composition is called* self-inverse *or* palindromic *if it reads the same from left to right as from right to left. A composition is called* reverse conjugate *if the reverse of σ is also the conjugate of σ.*

Example 1.6 *The palindromic compositions of 4 are 4, 22, 121, and 1111, none of which are reverse conjugate.*

MacMahon enumerated palindromic compositions and gave several results involving conjugate compositions. The remainder of his memoir is devoted to connections of generalized compositions and partitions, including line graphs for these more general compositions.

In the years that followed the publication of his memoir [136], the main focus of research in enumerative combinatorics was on partitions and permutations, driven in part by MacMahon's prolific writing. Until the late 1960s, individual articles on various aspects of compositions appeared, but there was no concentrated research interest. This was followed by roughly a decade in which several groups of authors developed new research directions in compositions. They studied compositions that were restricted in some way, and not only enumerated the total number of these compositions, but also certain characteristics, or *statistics*, for example rises, levels, and drops, of these compositions. However, the majority of publications on compositions has been within the last decade, when many different authors have studied various aspects of compositions, generalizing papers from the 1970s, introducing new types of compositions, or extending research questions and methods for pattern avoidance in permutations to compositions. Another focus was the study of random compositions to obtain asymptotic results, that is, the rate of growth for the quantities of interest as n tends to infinity. Tables 1.1, 1.2 and 1.3 give an overview of research activity for compositions over time in five major areas: restrictions on the set from which the parts can be chosen, restrictions on the arrangements of parts within the composition, enumeration of statistics of compositions in $\mathbb{N}$, pattern avoidance in compositions, and asymptotic results based on methods from complex analysis and probability theory. Papers that do not fall into one of these areas are listed under variations on compositions, and we will briefly indicate the areas studied here. Research in the five major areas will be described in more detail in Section 1.3. Note that we use † to denote articles that have multiple authors, and that we have abbreviated a few authors' names to fit the information into the table. In particular, Knopfmacher is shortened to Knop., Srini. refers to Srinivase Rao, and Serv. refers to Servedio.

Table 1.1 shows the time line from 1893 to 1982, which includes the decade when Hoggatt, Carlitz and their co-authors were very active.

TABLE 1.1: Time line of research in major areas for compositions

Year	Restricted sets	Restricted arrangements	Statistics	Variations
1893			MacMahon [136]	
1955				Narayana [161]
1958				Narayana† [162]
1964				Gould [74]
1965				Mohanty [155]
1967				Mohanty [156]
1968			Hoggatt† [99]	
1969			Hoggatt† [100]	
1975	Alladi† [7]	Hoggatt† [98]	Abramson [1]	
1976		Carlitz† [38]	Abramson [2]	
1977			Carlitz [39]	
1978			Carlitz [40]²	
1982				Cerasoli [45] Charalambides [46]

Of the papers listed under variations, there was a cluster in the 1950s and 1960s. Narayana and Mohanty independently studied domination of compositions [155, 156, 161], Narayana and Fulton investigated the structure of compositions as a distributive lattice [162], and Gould found an identity that involves binomial coefficients, the bracket function and compositions with rela-

---

²Even though title or paper refers to partitions, it is compositions that are studied

tively prime parts [74]. In 1982, Cerasoli [45] and Charalambides [46] provided connections between compositions and important number sequences.

TABLE 1.2: Time line of research in major areas for compositions

Year	Restricted sets	Restricted arrangements	Statistics	Asymptotics	Variations
1995				Richmond[171]	Serv.[178] Zanten [188]
1997				Hwang[102]	
1998		Knop.[119]			
1999			Knop.[118] Knop.[117]	Hitczenko[95]	
2000	Grimaldi [76] Srini.[181]		Rawlings [168]	Hitczenko[97]	Agarwal [3]
2001	Grimaldi [77]			Hitczenko[93]	
2002				Goh[73] Hitczenko[94] Louchard[134]	
2003	Chinn[50] Chinn[51] Chinn[52]		Chinn[48] Heubach[87]	Louchard[133] Knop.[120]	Agarwal [4]
2004			Heubach[89] Merlini[154]	Bender[20] Hitczenko[96]	Corteel[56]
2005			Heubach[90]	Knop.[121]	Bender[19] Corteel[57]

In 1995, the first papers on asymptotics for compositions appeared. Tables 1.2 and 1.3 give the time line from 1995 to the present. Here, the variations on compositions mostly result from new types of restrictions on compositions, with the exception of the papers by van Zanten [188], who connected compositions with Gray codes, and Servedio and Yeh [178], who studied the coloring of circular-arc graphs which leads to circular compositions. Corteel and her co-authors [56, 57] studied compositions where the i-th part satisfies a system of linear inequalities in the parts that follow the i-th part. The authors gave generating functions that can be computed from the coefficients of the linear inequalities. Bender and Canfield [19] considered compositions of n in which the first part, the last part, and the differences between adjacent parts all lie in prescribed sets. Under quite general conditions, the number of such compositions is determined asymptotically. In addition, the number of parts, rises, falls, and various other statistics are shown to satisfy a central limit theorem. Finally, Jackson and Ruskey [103] found interesting connections between meta-Fibonacci sequences and compositions in which the i-th coefficient is in a specified set.

Other types of compositions do not stem from restricting the compositions, but from introducing characteristics that can be translated into a coloring of the individual parts. Examples are the n-color compositions introduced by Agarwal, where the part n occurs in n colors [3, 4]. Narang and Agarwal [159, 160] studied n-color self-inverse compositions and connections between lattice paths and n-color compositions. Another variation is the cracked compositions discussed in Chapter 3, or the compositions that have two types of 1s, but only one type of any other part studied by Grimaldi [79]. He enumerated for example the total number of summands and derived connections with the alternate Fibonacci sequence.

TABLE 1.3:   Time line of research in major areas for compositions

Year	Statistics	Asymptotics	Pattern avoidance	Variations
2006	Kitaev[†][111] Mansour[†][145]		Heubach[†][91] Savage[†][176]	Narang[†][159] Jackson[†][103]
2007	Heubach[†][92]		Heubach[†][88]	Grimaldi [79]
2008			Jelínek[†][105] Baril[†][17]	Narang[†][160]
2009		Brennan[†][28] Knop.[†][115]		

1.2   Historical overview – Words

We now turn our attention to words. The origins of research activity for words are a little harder to pin down because they arise naturally in many different settings, for example in group theory to describe algebras, in probability theory to describe coin-tossing experiments and in computer science in the context of automata (see Chapter 7) and formal languages.

Definition 1.7 *Let* $[k] = \{1, 2, \ldots, k\}$ *be a (totally ordered) alphabet on k letters. A* word w *of length n on the alphabet $[k]$ is an element of $[k]^n$ and is also called* word of length n on k letters *or* k-ary word of length n*. Words with letters from the set $\{0, 1\}$ are called* binary words *or* binary strings*.

Example 1.8 *The words of length three on two letters are* 111, 112, 121, 122, 211, 212, 221 *and* 222*, while the binary strings of length two are given by* 00, 01, 10, *and* 11*.

Basic results such as the total number of words were part of well-known mathematical folklore long before a systematic study of their properties was undertaken.

Lemma 1.9 *The number of k-ary words of length n is given by k^n.*

Proof There are k choices for the first letter, and for any choice of the first letter, there are k choices for the second letter, for a total of k^2 choices for the first two letters. By the same argument, there are k^3 words of length 3, and in general, k^n words of length n over the alphabet $[k]$. □

Axel Thue

The first systematic study of words or sequences of numbers ("Zeichenreihen") seems to have appeared in three papers of Axel Thue[3] (1863-1922) in 1906, 1912, and 1914 [184, 185, 186], and in a paper by MacMahon in 1913 [137]. Thue's work is on infinite words (sequences of symbols) and he investigated words from a number-theoretic viewpoint. In the first paper, he asked the following question: Given any finite word $w = w_1 \cdots w_m$ on $P = \{p_1, \ldots, p_k\}$ and any infinite word w' on $Q = \{q_1, \ldots, q_n\}$, is there a map Φ from elements in P to finite words in Q, such that the word $\Phi(w_1) \cdots \Phi(w_k)$ has to occur in Q?

Thue showed that the answer is negative. In the subsequent papers [185, 186], he investigated equivalence of sequences that are transformed by certain rules. As an introduction to the 1912 paper, he wrote:

> "Für die Entwicklung der logischen Wissenschaften wird es, ohne Rücksicht auf etwaige Anwendungen, von Bedeutung sein, ausgedehnte Felder für die Spekulation über schwierige Probleme zu finden." (It will be necessary for the development of the logical sciences to find vast areas for research on difficult problems, without regard to potential applications.)

By contrast, MacMahon approached words, which he called *assemblage of objects*, in the context of partitions and permutations. His viewpoint and questions are the first instance of one of the main topics covered in this text, namely enumerating statistics on words. We will state his definitions and results in the modern terminology that will be used throughout the book,

---

[3]http://www-history.mcs.st-andrews.ac.uk/Mathematicians/Thue.html

but mention how he referred to those objects. Coming from the study of permutations, MacMahon started with a list of parts, the *assemblage*, and then considered any word that can be created from the parts in this set, that is, all the permutations of the assemblage.

Definition 1.10 *A permutation of* $[n] = \{1, 2, \ldots, n\}$ *is a one-to-one function from* $[n]$ *to itself. We denote permutations of* $[n]$ *by* $\pi = \pi_1 \pi_2 \cdots \pi_n$, *and the set of all permutations of* $[n]$ *by* $\mathcal{S}_n$ ($\mathcal{S}$ *stands for symmetric group).*

The definition easily generalizes to any set of n elements, as any such set can be associated with the set $[n]$.

Example 1.11 *There are six permutations of the assemblage 123, namely 123, 132, 213, 231, 312 and 321, and three permutations of the assemblage 112, namely 112, 121, and 211.*

The statistics MacMahon investigated are based on the following relations between consecutive letters in a sequence.

Definition 1.12 *For any sequence* $s_1 s_2 \cdots s_k$ *(permutation, word or composition), a* rise *consists of a part followed by a larger part, a* level *or* straight *occurs when a part is followed by itself, and a* drop *or* fall *consists of a part followed by a smaller part. A rise, level, or drop occurs at position* i *if* s_i *is less than, equal to, or bigger than* s_{i+1}.

Note that MacMahon referred to a rise as a *minor contact*, a level as an *equal contact*, and a drop as a *major contact*.

Definition 1.13 *For a given assemblage* $a_1^{i_1} a_2^{i_2} \cdots a_k^{i_k}$ *(where* $a_j^{i_j}$ *indicates that there are* i_j *copies of the letter* a_j), *let* w *be a permutation of these letters. Then the* greater index p *of* w *is defined to be the sum of the positions* i *in* w *such that a drop occurs at position* i. *Likewise, the* equal index q *of* w *is defined to be the sum of the positions where a level occurs, and the* lesser index r *is the sum of the positions where a rise occurs. If there are* m *rises in* w, *then* w *belongs to the class* m *with regard to rises. Likewise, we define the class* m *with regard to levels and drops.*

MacMahon then gave the generating function (see Definition 2.36) for the greater index p. Using the symmetry of this function, he derived the following interesting result.

Theorem 1.14 *[138, item 106] The number of words associated with the assemblage*

$$a_1^{i_1} a_2^{i_2} \cdots a_k^{i_k}$$

with m *rises is the same as the number of words with* m *rises associated with any assemblage* $a_1^{i'_1} a_2^{i'_2} \cdots a_k^{i'_k}$, *where* $i'_1 i'_2 \cdots i'_k$ *is a permutation of* $i_1 i_2 \cdots i_k$.

Example 1.15 *There are exactly six permutations that have a greater index of $p = 3$ in each of the assemblages*

$$1^3 2^2 3, \quad 1^3 2 3^2, \quad 1^2 2^3 3, \quad 1^2 2 3^3, \quad 1 2^3 3^2, \quad 1 2^2 3^3.$$

For the assemblage 111223, the six words with a greater index of 3 are

$$112123, \quad 122113, \quad 321112, \quad 113122, \quad 123112, \quad 223111;$$

for the assemblage 112333, the six words with a greater index of 3 are

$$113233, \quad 321133, \quad 123133, \quad 133123, \quad 233113, \quad 333112.$$

From this result, he derived a second, related result.

Theorem 1.16 *[138, item 107] Let p denote the greater index, and q denote the lesser index. Then for any assemblage, $\sum x^p = \sum x^r$, where the sum is over all permutations of the given assemblage.*

Proof We can map any word w of a given assemblage to a word w' by mapping the largest part to the smallest part and vice versa, the second largest part to the second smallest part, etc. Then each rise becomes a drop and vice versa, so the lesser index of w matches the greater index of w' and vice versa. Now we use Theorem 1.14 because the assemblage of w' is different from the assemblage of w (the powers have been mapped accordingly), and the result follows. □

MacMahon also investigated the special case of words with just two letters with regard to the difference of the greater and the lesser index and derived averages of the statistics of interest, a topic that we will discuss in Chapter 8.

Theorem 1.17 *[138, item 109] For any word created from the assemblage $1^i 2^j$, the difference between the greater and the lesser index, $p - r$, equals $-i$ or j, depending on whether the word terminates with a 2 or a 1.*

Proof The truth of the statement is easily checked for words of length one or two. Assume the hypothesis is true for words associated with the assemblage $1^i 2^j$ and let $(1^i 2^j)_1$ and $(1^i 2^j)_2$ indicate that the word ends in 1 or 2, respectively. Then the hypothesis follows for words of length $i + j + 1$ by checking the four different cases and computing the value of $p - r$ in each case:

$$(1^i 2^j)_1 1 : p - r = j,$$
$$(1^i 2^j)_1 2 : p - r = j - (i + j) = -i,$$
$$(1^i 2^j)_2 1 : p - r = -i + (i + j) = j,$$
$$(1^i 2^j)_2 2 : p - r = -i,$$

so the result follows. □

After these initial results on words, the literature became quiet for a while, with the exception of a few individual papers. Schensted [177] in 1961 derived results on the number of longest increasing and decreasing subsequences in words, and Carlitz and his collaborators enumerated statistics on words as well as the number of restricted words in a number of papers from the 1970s. In 1983, the first of two volumes on the combinatorics of words from an algebraic and group theoretic point of view appeared, authored by M. Lothaire, the nom du plume for a group of authors that initially consisted of students of M. P. Schützenberger. These volumes [130, 131] treat words from a different perspective, and the reader interested in algebraic and group theoretic results is invited to study these texts and the references therein. In the algebraic context, the term "avoidable word" often occurs, but this is different from the pattern avoidance considered here. Note that we will not give any further references to this branch of study of words, except in cases where there is an overlap.

Table 1.4 shows the time line for research on words in the early years, while Table 1.5 shows the more recent developments. We once more have to abbreviate an author's name – Mac. refers to MacMahon in Table 1.4 and Prod. refers to Prodinger in Table 1.5.

TABLE 1.4:　　Time line of research in major areas for words

Year	Statistics	Restricted arrangements	Pattern avoidance	Variations
1906				Thue [184]
1912				Thue [185]
1913	Mac. [137]			
1914				Thue [186]
1961			Schensted [177]	
1972		Carlitz[†][41]	Carlitz [36]	
1973		Carlitz [37]		
1974			Carlitz[†][43]	
1976	Carlitz[†][42]			
1983				Lothaire[130]

Major activity in the statistics of words started at the end of the last decade of the 20th century. Evdokimov and Nyu [62] derived results on the smallest length of a binary word w such that every word with exactly k ones and $n-k$ zeros can be obtained from w by deleting some letters. In 1997, incidentally at the same time that the *Combinatorics of Words* [130] was reprinted, Foata, who was one of the authors of that book, published two papers concerning statistics on words. He and Han [70] showed that a certain transformation gives an inverse for words which preserves the number of external inversions, but not the number of internal inversions, and also showed existence of an involution which preserves both types of inversions. The second paper by the same authors [71] in 1998 established that certain transformations that show the equality of some statistics can be obtained as special cases of a more general procedure.

TABLE 1.5: Time line of research in major areas for words

Year	Statistics	Restricted arrangements	Pattern avoidance	Variations
1992				Evdokimov[†][62]
1997				Foata[†][70]
1998			Burstein[†][31]	Foata[†][71]
			Regev [169]	Régnier[†][170]
2001				Tracy [187]
2002				Lothaire [131]
2003	Burstein[†][34]		Burstein[†][33]	Burstein[†][32]
	Burstein[†][35]		Kitaev[†][109]	
2004				Dollhopf[†][61]
2005			Brändén[†][27]	Lothaire [132]
2006			Mansour [141]	
2007			Bernini[†][23]	
2008	Kitaev[†][110]	Knop.[†] [116]	Jelínek[†][105]	Mansour [142]
	Prod. [166]	Mansour [144]		Firro[†][65]
	Myers[†][158]			Mansour [143]

In the same year, the study of subsequence patterns in words was initiated by Regev [169], who found the asymptotic behavior for the number of k-ary words avoiding an increasing subsequence of length ℓ. Subsequence patterns were the first type of patterns to be studied in permutations and subsequently in words, by generalizing proof techniques developed for permutations. Tracy and Widom [187] considered the asymptotic behavior of a related statistic, namely the length of the longest increasing subsequence, by considering random words. However, the foundation of research on pattern avoidance in words was laid in Burstein's thesis [31], where he gave results for generating functions for the number of k-ary words of length n that avoid a set of patterns $T \subset S_3$. Various generalizations followed, to pattern avoidance in words for multi-permutation subsequence patterns [33, 105] or pairs of more general patterns of varying length. Mansour [141] considered the special case of words that avoid both the pattern 132 and some arbitrary pattern, which leads to generating functions that are expressed in terms of continued fractions and Chebyshev polynomials of the second kind. Many of these papers generalize methods used for permutations avoiding permutation patterns, but several authors utilized different methods, such as automata theory [27], the ECO method [23], and the scanning-element algorithm [65]. Another variation was considered by Dollhopf et al. [61], who enumerated words that avoid a set of patterns of length two that describes a reflexive, acyclic relation.

Several authors considered other types of patterns besides subsequence patterns and also studied enumeration instead of avoidance. Burstein and Mansour [34] considered the enumeration of words containing a subword pattern of length ℓ exactly r times, while Kitaev and Mansour [109] studied partially ordered generalized patterns. Burstein and Mansour [35] also enumerated all generalized patterns of length three. A somewhat related paper by Burstein et al. [32] investigated the packing density for subword patterns, generalized patterns and superpatterns.

The most recent papers investigated a wide range of topics. Kitaev et al. [110] enumerate the number of rises, levels and falls in words that have a prescribed first element, while Prodinger [166] enumerated words according to the number of records (= sum of the positions of the rises, or in the terminology of MacMahon, the lesser index). Myers and Wilf [158] investigated left-to-right maxima in words and multiset permutations. Knopfmacher et al. [116] and Mansour [144] looked at smooth words and smooth partitions, respectively, and obtained connections with Chebyshev polynomials of the second kind. Finally, Mansour enumerated words [143] and words that avoid certain patterns [142] according to the longest alternating subsequence.

Another variation is avoidance or occurrence of substrings in words. Régnier and Szpankowski [170] used a combinatorial approach to study the frequency of occurrences of strings from a given set in a random word, when overlapping copies of the strings are counted separately (see [170, Theorem 2.1]). Various examples on enumeration of substrings in words are included in general texts on combinatorics (see for example [75, 182]). This research area is also treated in the third volume of the Lothaire series [132] which focuses on applied combinatorics of words, for example genetics, where words are used to describe DNA fragments (see Example 1.23). The reader who is interested in the algorithmic aspects and the statistics connected to substrings in words will find this third volume a good starting point. We will restrict our focus to patterns that encode relations as opposed to a specific string of letters.

We now give a more detailed overview of the major research directions for both compositions and words.

1.3   A more detailed look

In this section we will have a closer look at the five major research areas for compositions and those for words, by giving relevant definitions and basic examples, as well as an overview of the structure of this text.

1.3.1   Restricted compositions and words

Restrictions come in many flavors, one of which is to restrict the parts of the composition or word to a subset A of $\mathbb{N}$. In the case of words this is not interesting at all – all that matters is the total number of letters. Any set of n elements can be mapped bijectively to the set $[n]$. However, for compositions, this type of restriction leads to interesting results. This was the direction taken by Alladi and Hoggatt [7], who considered $A = \{1, 2\}$, and enumerated, among other things, the number of compositions, number of parts, and number of rises, levels and drops in compositions of n, exhibiting connections to the Fibonacci sequence (see Chapter 3). Interest in the connection between compositions and the Fibonacci sequence came from earlier studies by Hoggatt and Lind of weighted compositions [99, 100], who found connections to identities for Fibonacci numbers.

Chinn, Grimaldi and Heubach [50, 51, 52, 76, 77] have recently revived this research direction. They have generalized to $A = \{1, k\}$, $A = \mathbb{N}\setminus\{1\}$, $A = \mathbb{N}\setminus\{2\}$, $A = \mathbb{N}\setminus\{k\}$, and $A = \{1, 3, 5, \ldots\}$ and have counted the number of compositions, rises, levels and drops. For $A = \mathbb{N}\setminus\{1\}$ and $A = \mathbb{N}\setminus\{2\}$ they have found combinatorial proofs for connections between the various sequences of interest. At the same time, Srinivasa Rao and Agarwal [181] enumerated compositions in which each part is bounded above and below, but did not investigate the statistics considered by the other authors.

A second type of restriction is to place limitations on the arrangement of the parts within the word or composition, the direction taken by Carlitz and his various co-authors. It started with a paper by Carlitz [36] in which he enumerated words according to the number of rises and falls, a refinement of the Simon Newcomb problem [60]. A natural extension is the study of up-down sequences (words in which rises and falls alternate) by Carlitz [37] and by Carlitz and Scoville [41], and, more generally, to the enumeration of words or compositions that do not have levels [38]. This restriction on the placement of the parts produces compositions that are "closest" to permutations. Carlitz and his co-authors referred to these words or compositions as *waves*, while other authors have called them *Smirnov sequences* (see for example [75]). Knopfmacher and Prodinger [119] introduced the name *Carlitz compositions* in honor of Carlitz, which has become the name commonly used.

Definition 1.18 *A* Carlitz composition *of n is a composition of n in which no adjacent parts are the same.*

Example 1.19 *The Carlitz compositions of* 5 *are* 131, 212, 32, 23, 41, 14 *and* 5.

Another example of a restriction on the arrangement of parts are palindromic compositions. Hoggatt and Bicknell [98] investigated palindromic compositions in $\mathbb{N}$ according to rises and falls, and various other authors who studied compositions from restricted sets also considered palindromic compositions from those restricted sets.

We will study restricted compositions, as well as other variations of compositions in Chapter 3.

1.3.2 Statistics on compositions in $\mathbb{N}$

At the same time as Alladi and Hoggatt, Carlitz [38, 39, 40] and other authors studied rises, levels and falls for compositions with parts from $\mathbb{N}$, as well as other statistics. For example, Abramson [1, 2] enumerated compositions according to longest runs and rises of a specific size. Carlitz and Vaughan [43] derived the generating functions for the number of compositions according to specification (a list of counts for the occurrences of each integer), rises, falls and maxima, while Carlitz et al. [42] enumerated pairs of sequences according to rises, levels and falls. More recently, Rawlings [168] enumerated compositions according to weak rises and falls (equality allowed) in connection with restricted words by adjacencies. He also introduced the notion of *ascent variation* (the sum of the increases of the rises within a composition), which is motivated by a connection to the perimeter of directed vertically convex polyominoes. Knopfmacher and Mays enumerated compositions according to the number of distinct parts [117] and gave results on the sum of distinct parts in compositions and partitions [118]. Grimaldi, Chinn and Heubach enumerated rises, falls, levels, "+" signs, as well as the number of parts in all compositions [48, 87]. These results were generalized by Heubach and Mansour [89, 90] and Merlini et al. [154], who provided a general framework for compositions with parts from a set A, and derived generating functions for the number of parts, rises, levels and falls.

Results on the statistics rise, level and fall can be extended by thinking of them as two-letter subword patterns. To define what we mean by a pattern, we need to introduce some terminology.

Definition 1.20 *The* reduced form *of a sequence $\sigma = \sigma_1\sigma_2\ldots\sigma_m$ is given by the sequence $s_1s_2\cdots s_m$, where $s_i = \ell$ if σ_i is ℓ-th smallest term.*

Example 1.21 *The reduced form of the sequence* 35237 *is* 23124, *as the terms of the sequence are in order* $2 \leq 3 \leq 5 \leq 7$, *and therefore,* 2 *is the smallest element,* 3 *is the second smallest element,* 5 *is the third smallest element, and* 7 *is the fourth smallest element. The reduced form takes into account only the relative size of the sequence terms and maps the sequence to the set* $[k]$, *where* k *is the number of distinct terms in the sequence.*

Definition 1.22 *A sequence (permutation, word, composition or partition)* σ contains a substring w *if the word* w *occurs in* σ. *Otherwise, we say that* σ avoids the substring w *or is* w-*avoiding.*

Substring patterns are particularly relevant when the sequence consists of letters or symbols that do not have an order relation associated with them. A prime example is a DNA sequence, which can be considered a word from the alphabet of nucleotides $\{A, C, G, T\}$. Since there is biological relevance to the order in which the nucleotides occur, statistics on the number of substrings in a DNA sequence has become an important area of research.

Example 1.23 *The DNA sequence* $ACTGGATCTGAGA$ *contains the substring* CT *twice, the substring* GA *three times, and avoids the substring* GGG.

Definition 1.24 *A sequence (permutation, word, composition or partition)* σ contains a subword pattern $\tau = \tau_1\tau_2\ldots\tau_k$ *if the reduced form of any subsequence of* k *consecutive terms of* σ *equals* τ. *Otherwise we say that* σ *avoids the subword pattern* τ *or is* τ-*avoiding. A sequence avoids a set* T *of subword patterns if it simultaneously avoids all subword patterns in* T. *We denote the set of sequences with parts in a set* $\mathcal{F}$ *that avoid subword patterns from the set* T *by* $\mathcal{F}(T)$.

Example 1.25 *The sequence* 1413364 *avoids the substring* 12, *but has three occurrences of the subword pattern* 12 *(namely* 14, 13 *and* 36*). It also avoids the subword pattern* 123.

For longer sequences, it is often useful to write the sequence in standard short-hand notation as follows.

Definition 1.26 *The notation* k^j *in a sequence of nonnegative integers (permutation, word, composition or partition) stands for a sequence of* j *consecutive occurrences of* k.

Example 1.27 *The sequence* 1111234415333333 *is written as* $1^4234^2153^6$ *in short-hand notation.*

Now that we have defined the notion of a subword pattern, it is easy to see that a level corresponds to the subword pattern 11, while a rise corresponds to the subword pattern 12. Heubach and Mansour [92] enumerated compositions according to subword patterns of length three. Some of their results were further generalized by Mansour and Sirhan [145] to enumerating compositions according to patterns from families of ℓ-letter patterns, for example the pattern $1^{\ell-1}2$.

Having obtained very general results on the statistics rises, levels, parts and order of compositions, we can look at special cases. Since partitions can be thought of as the nonincreasing compositions, they can be enumerated as compositions without rises. This allows us to obtain results for partitions from our results for compositions. We will present these results in Chapter 4.

1.3.3 Pattern avoidance in compositions

In Chapter 5, we will change our focus from enumeration of compositions and words according to the number of times a certain pattern occurs to the avoidance of such patterns. Clearly, we can obtain results for pattern avoidance of subword patterns from the previous chapter by looking at zero occurrences. In this chapter, we will look at more general patterns, namely subsequence, generalized, and partially ordered patterns. Subsequence patterns are those where the individual parts of the pattern do not have adjacency requirements, while generalized patterns have some adjacency requirements. Partially ordered patterns is a further generalization in the sense that some of the letters of the pattern are not comparable.

Pattern avoidance was first studied for $\mathcal{S}_n$, the set of permutations of $[n]$, avoiding a pattern $\tau \in \mathcal{S}_3$. Knuth [122] found that, for any $\tau \in \mathcal{S}_3$, the number of permutations of $[n]$ avoiding τ is given by the n-th Catalan number. Later, Simion and Schmidt [179] determined $|\mathcal{S}_n(T)|$, the number of permutations of $[n]$ simultaneously avoiding any given set of patterns $T \subseteq \mathcal{S}_3$. Burstein [31] extended this to words of length n on the alphabet $[k]$, determining the number of words that avoid a set of patterns $T \subseteq \mathcal{S}_3$. Burstein and Mansour [33, 34] considered forbidden patterns with repeated letters. Recently, pattern avoidance has been studied for compositions. Savage and Wilf [176] considered pattern avoidance in compositions for a single subsequence pattern $\tau \in \mathcal{S}_3$, and showed that the number of compositions of n with parts in $\mathbb{N}$ avoiding $\tau \in \mathcal{S}_3$ is independent of τ. Heubach and Mansour answered the same questions for subsequence patterns of length three with repeated letters [91]. For longer subsequence patterns, the results by Jelínek and Mansour [105] allow for a complete classification according to avoidance in both words and compositions.

We will also provide new results on the avoidance of generalized patterns of length three in Section 5.3, both for permutation patterns and those that have repeated letters, thereby giving results on the only three-letter patterns not yet studied. Finally, Section 5.4 contains results on the number of compositions that contain or avoid partially ordered patterns. Kitaev [108] originally studied avoidance of partially ordered patterns in permutations, and he and

Mansour [109] generalized to words. This was followed by a paper by Heubach et al. [88] who gave the corresponding results for avoidance of partially ordered patterns in compositions. A somewhat different approach was used by Kitaev et al. [111] to enumerate compositions that contain partially ordered patterns.

1.3.4 Pattern avoidance and enumeration in words

Historically, pattern avoidance in words has preceded pattern avoidance in compositions. However, one can consider words as special cases of compositions (when only looking at a fixed number of parts, not the order of the composition) and therefore all results derived for compositions also apply for words and we recover results for pattern avoidance in words [33, 35]. However, there are some results on pattern avoidance in words that do not easily extend to compositions. In Chapter 6 we will present the results on pattern avoidance in words, both by stating the results that can be obtained via the results for compositions, as well as approaches that are specific to words. In particular, we will present the result on avoidance of the subsequence pattern 132, one of the patterns studied in Burstein's thesis [31] in 1998. The original result was derived using the Noonan-Zeilberger algorithm. Since then, two different approaches, the block decomposition method and the scanning-element algorithm, have been used to re-derive Burstein's result in easier ways.

The block decomposition method was first used by Mansour and Vainshtein [147] to study the structure of 132-avoiding permutations, and permutations containing a given number of occurrences of 132 [149]. Mansour [141] extended the use of the block decomposition approach to pattern avoidance in words. Besides providing a different and simpler proof of Burstein's result, the block decomposition method reveals more of the structure of the 132-avoiding words, and can also be used to derive generating functions for words that avoid a set of patterns $\{132, \tau\}$ for families of patterns. Likewise, the scanning-element algorithm was first used for pattern avoidance and enumeration in permutations by Firro and Mansour [64], and then extended to study the number of k-ary words of length n that satisfy a certain set of conditions [65]. Not only does the approach work equally well for the patterns 123 and 132, it can also be used more easily for the enumeration of patterns, especially for the pattern 123.

1.3.5 Automata, generating trees and the ECO method

In Chapter 7 we will describe three approaches that are variations on the same theme, namely defining sequences recursively and then expressing the recursive structures in an associated graph. The first approach we discuss is a finite automaton. Bränden and Mansour [27] were the first to use this approach for pattern avoidance in words, using the transfer matrix method of the automaton to count the number of words that avoid a subsequence pattern.

Not only did they obtain exact enumeration results, but also the asymptotics for the number of words of length n on k letters avoiding a pattern τ. In addition, they gave the first combinatorial proof of a formula for the number of words of length n on k letters avoiding the pattern 123. This approach can be easily extended to pattern avoidance in compositions.

The second approach, generating trees, consists of defining recursive rules that induce a tree, as the name suggests. Generating trees were introduced by West [191] to enumerate permutations avoiding certain subsequence patterns. Even though the two approaches look rather different at first, each generating tree (which usually has infinitely many vertices) can be mapped to a graph with a finite number of vertices in the form of an automaton. Using several examples, we describe how a recursive creation for the sequence translates into recursive rules and labels for the vertices of the tree.

We conclude this chapter by discussing the ECO method, which is a general framework that describes a process to obtain recurrences for enumerating combinatorial objects. In essence, it is a road map that lays out the task to be done, but does not give a formulaic description of the steps. The general idea (but not with the name ECO method) was introduced in 1978 by Chung et al. [54] who enumerated permutations avoiding a pair of generalized patterns using a generating tree. In 1995, Barcucci et al. [14] enumerated Motzkin paths and various other combinatorial objects using the ECO method. Soon thereafter, these authors established a general methodology for plane tree enumeration [15], followed by a survey on the general theory [16] in which many enumerative examples are considered in great detail. We refer the interested reader to [16] and [63] for further details and results.

At roughly the same time, West [191, 192] used what is in principle the ECO method in the form of generating trees and the associated directed graphs for enumerating restricted permutations. In fact, most applications of the ECO method to date reduce to generating trees, or equivalently, automata. Recently, the ECO method has been applied by Baril and Do [17] to compositions and by Bernini et al. [23] to words. We will present the general framework of the ECO method and the results of Bernini et al. [23] for pattern avoidance of pairs of generalized patterns in words.

1.3.6　Asymptotics

In the study of statistics on compositions, sometimes one is not able to get explicit results or nice generating functions. Then the question becomes what can be said about the quantity of interest as n tends to infinity. Tools to obtain such asymptotic results come from probability theory and complex analysis, or a combination of the two. In order to utilize the probability approach, one considers the compositions to be random, with each composition of n occurring equally likely. With this approach, several authors have derived interesting asymptotic results.

Hitczenko and Savage [94, 95, 96] studied the probability that a randomly chosen part size in a random composition is unrepeated, and more generally, gave results on the multiplicity of parts. Hitczenko and Stengle [97] presented an asymptotic result for the expectation of the number of distinct part sizes in a random composition of an integer, a result that has been obtained independently and by an entirely different method by Hwang and Yeh [102]. Gho, Hitczenko, Louchard and Prodinger [73, 93, 133, 134] studied the distinctness of a random composition, that is, the number of distinct parts, a natural and important parameter in theoretical models for many applications. They also studied a variant of distinctness, namely the number of distinct parts of multiplicity m and other characteristics of random compositions and random Carlitz compositions.

Knopfmacher and Robbins investigated asymptotics for compositions whose parts are either powers of two or Fibonacci numbers [120], and for compositions with parts that are constrained by the leading summand [121]. Bender et al. [20] considered irreducible pairs of compositions, where the subsums of the first terms are all different from each other. They enumerated the number of irreducible ordered pairs of compositions of n into k parts, and approximated the probability of choosing an irreducible ordered composition of n. Most recently, Knopfmacher and Mansour have investigated the asymptotic behavior for record statistics in random compositions [115], while Brennan and Knopfmacher [28] have investigated the distribution of the ascents of size d or more in compositions. We will give an introduction to the tools from complex analysis and probability theory and then present some of these results in Chapter 8.

1.4 Exercises

Exercise 1.1

(1) *List all partitions of 5.*

(2) *List all compositions of 5.*

(3) *Identify which of the compositions are palindromes.*

(4) *Associate each composition with its conjugate.*

(5) *Which of the compositions, if any, are reverse conjugate?*

Exercise 1.2 *List the reduced form of the following sequences:*

(1) 3751293.

(2) 342615.

(3) 34567.

(4) 12345.

(5) 333333.

Exercise 1.3 (1) *List all words of length three on the alphabet* [3].

(2) *List the words of length three on the alphabet* [3] *that avoid the substring* 12.

(3) *List the words of length three on the alphabet* [3] *that avoid the subword pattern* 12.

Exercise 1.4 *For the sequence* 1342573264, *determine*

(1) *whether it avoids the subword pattern* 123;

(2) *the number of occurrences of the subword pattern* 21.

Exercise 1.5 *A word* $w = w_1 \cdots w_n$ *is said to be a* partition word *if the first occurrence of* i *precedes the first occurrence of* j *for any* $i < j$ *(and therefore, $w_i \leq i$). For example, the partition words of length 3 are* 111, 112, 121, 122 *and* 123. *Partition words of length* n *provide an alternative representation for the set partitions of* $[n]$, *a well known combinatorial object enumerated by the Bell (exponential) numbers (see Definition 2.59 and [180, A000110]). List all the partition words of length four.*

Exercise 1.6 *Which of the following types of compositions are closed under conjugation (that is, the conjugate compositions are of the same type)? Either give a proof that the conjugates are of the same type, or exhibit a counter example.*

(1) *Compositions without* 1*s.*

(2) *Compositions with odd parts.*

(3) *Carlitz compositions.*

Exercise 1.7

(1) *Show that if a reverse conjugate composition of* n *has* p *parts, then* $n = 2p - 1$. *Thus, the reverse conjugate compositions of* $2m + 1$ *have* $m + 1$ *parts.*

(2) *Use the result of Part* (1) *to give a combinatorial argument that there are* 2^m *reverse conjugate compositions of* $2m + 1$.

Exercise 1.8 *It can be shown that the total number of palindromic composi-tions of $2m+1$ is also 2^m (see Theorem 3.4). Derive a bijection (see Definition 2.4) between the palindromic and the reverse conjugate compositions of $2m+1$. Show the correspondence between the reverse conjugate and the palindromic compositions of 7 according to your bijection.*

Exercise 1.9 *The number of compositions of n in $\mathbb{N}$ is 2^{n-1}, which is also the number of subsets of $[n-1]$. Find a bijection (see Definition 2.4) between the compositions of n and the subsets of $[n-1]$.*

Exercise 1.10 *Are there any proper subsets A of $\mathbb{N}$ such that the set of all compositions of n with parts in A, denoted by $C_A(n)$, are closed under conjuga-tion for $n \geq n_0(A)$, that is, does there exist an $n_0(A)$ such that for $n \geq n_0(A)$, $\sigma \in C_A(n)$ implies $\mathrm{conj}(\sigma) \in C_A(n)$?*

Exercise 1.11

(1) *Find a composition with the fewest parts (in reduced form) that con-tains all the subword patterns of length two on two letters, that is, a composition that contains 11, 12 and 21. Is this answer unique?*

(2) *Find a composition with the fewest parts (in reduced form) that contains all the subword patterns of length three on two letters, that is, a compo-sition that contains 111, 112, 121, 211, 221, 212 and 122. Is this answer unique?*

(3) *Are the respective compositions also the ones of smallest order? Is the composition of smallest order unique?*

Exercise 1.12 *A composition of n is called smooth if the difference between any two adjacent parts is either -1, 0, or 1, and strictly smooth if the dif-ferences are -1 or 1. List the smooth compositions and the strictly smooth compositions of 5.*

Exercise 1.13 *A zero-composition of n with m parts is a composition of n with m parts for which zero is an allowed part. List all the zero-compositions of 4 with three parts.*

Exercise 1.14 *A k-ary circular word $w = w_1 w_2 \cdots w_n$ of length n is a word on n circular positions around a disk (but still written in linear form). For example, as circular words, 123, 231 and 312 all denote the same word. List all the circular words of lengths one, two, three, and four on the alphabet $[2]$.*

Exercise 1.15 *A circular composition $\sigma = \sigma_1 \sigma_2 \cdots \sigma_m$ of n is a composition of n on m circular positions around a disk (but still written in linear form).*

(1) *List all the circular compositions of 1, 2, 3, and 4.*

(2) *Give an example of a circular composition that is not a partition.*

Chapter 2

Basic Tools of the Trade

Combinatorics is the art of counting. Usually, we are interested in counting objects that depend on a parameter (or two, or three...), for example the number of compositions of n with only 1s and 2s. (Here the parameter is n.) It is pretty straightforward to enumerate the number of such compositions for small values of n by exhibiting all possibilities, but as n increases, the number of compositions grows exponentially, and so we need to be smarter about counting. In this chapter we will give an overview of some basic techniques and ideas of combinatorics. We will illustrate important techniques such as the use of recurrence relations and generating functions with basic examples, which we will build upon and expand on in subsequent chapters.

2.1   Sequences

When we count combinatorial objects that depend on a parameter n, then we obtain different answers for each value of n, that is, we obtain a sequence (in this case of nonnegative terms).

Definition 2.1 *A sequence with values in B, in short a sequence in B, can be expressed formally as a function $a : I \rightarrow B$, where $I \subseteq \mathbb{Z}$, the set of integers, and, for the objects we are interested in, $B \subseteq \mathbb{N}_0$, the set of nonnegative integers. The set I is called the* index set, *and the set B consists of the values of the sequence. If $I = [k]$, then a is called a* finite sequence of length k.

We will use a_n and $a(n)$ interchangeably to denote the value associated with index n. The full sequence is usually given as an ordered list. For example, when $I = \mathbb{N}$, then $a = \{a_1, a_2, \ldots\}$, or $a = \{a_n\}_{n=1}^{\infty} = \{a_n\}_{n \geq 1}$ for short. If we select a subset of the terms of a sequence and preserve the order in which they occur, then we obtain a subsequence of the original sequence.

Definition 2.2 *A sequence $\{b_n\}_{n \in I'}$ is a subsequence of the sequence $\{a_n\}_{n \in I}$ if for each index n there exists an index i_n such that $b_n = a_{i_n}$, where $i_n > i_{n-1}$ for $n > 1$.*

Example 2.3 *The sequence $\{2n\}_{n\in\mathbb{N}}$ of even natural numbers is a subsequence of the sequence $\{n\}_{n\in\mathbb{N}}$ of natural numbers.*

In combinatorics, the notion of a one-to-one correspondence is especially important, as it allows us to establish that two sets have the same cardinality, that is, the same number of elements. This idea will be used repeatedly when we discuss Wilf-equivalence in Chapters 4 and 5.

Definition 2.4 *A function $f : A \rightarrow B$ is* injective *or* one-to-one *if for all $x, y \in A$, $x \neq y$ implies that $f(x) \neq f(y)$, that is, every element of the codomain B is mapped to by at most one element of the domain A. A function $f : A \rightarrow B$ is* surjective *or* onto *if for all $y \in B$ there is an element $x \in A$ such that $f(x) = y$, that is, the range is equal to the codomain. A function is* bijective *or* one-to-one and onto *if and only if it is both injective and surjective, that is, every element of the codomain is mapped to exactly one element of the domain. A bijective function is a* bijection *or a* one-to-one correspondence.

Example 2.5 *We will use a bijection to establish that the number of words on the alphabet $[3]$ that avoid the pattern 12–3 equals the number of words on $[3]$ that avoid the pattern 21–3. Avoiding the pattern 12–3 means that there is no subsequence of the form $w_i w_{i+1} w_j$ such that $w_i < w_{i+1} < w_j$ with $j > i + 1$. These two patterns are examples of generalized patterns, which we will discuss in Chapter 5. Since the alphabet is $[3]$, we have two cases: If w does not contain the letter 3, then it automatically avoids both 12–3 and 21–3. If w contains at least one letter 3, the pattern 12–3 is avoided if $w = w^{(1)}3w^{(2)}3\cdots 3w^{(s)}$, where $s \geq 2$ and each of the $w^{(i)}$ for $i < s$ is a word on the alphabet $[2]$ that avoids the pattern 12, and $w^{(s)}$ is any word on $[2]$. Thus, each $w^{(i)}$ for $i < s$ is of the form $2^j 1^\ell$, with $j, \ell \geq 0$. On the other hand, a word w avoids 21–3 if either the word w does not contain the letter 3, or $w = w^{(1)}3w^{(2)}3\cdots 3w^{(s)}$, where $s \geq 2$ and each of the $w^{(i)}$ is a word on the alphabet $[2]$ that avoids the pattern 21, and therefore is of the form $1^j 2^\ell$, with $j, \ell \geq 0$. So what is the bijection? Here it is: For a word $w = w^{(1)}3w^{(2)}3\cdots w^{(s-1)}3w^{(s)}$ that avoids 12–3, there is exactly one corresponding word $w' = f(w)$ that avoids 21–3 where*

$$f(w) = R(w^{(1)})3R(w^{(2)})3\cdots R(w^{(s-1)})3w^{(s)}$$

and R denotes the reversal map (see Definition 1.5). Likewise, starting with a word

$$\bar{w} = \bar{w}^{(1)}3\bar{w}^{(2)}3\cdots \bar{w}^{(s-1)}3\bar{w}^{(s)}$$

that avoids 21–3, we can map it to exactly one word w that avoids 12–3 using the function f^{-1} defined by

$$f^{-1}(\bar{w}) = R(\bar{w}^{(1)})3R(\bar{w}^{(2)})3\cdots R(\bar{w}^{(s-1)})3w^{(s)}.$$

Clearly, $f(f^{-1}\bar{w}) = \bar{w}$ and $f^{-1}f(w) = w$, so we have established a bijection or one-to-one correspondence between the set of words on [3] that avoid 12–3 and those that avoid 21–3.

Now we return to sequences. They can be defined either by an explicit formula or via a recurrence relation. An explicit formula for a sequence allows for direct computation of any term of the sequence by knowing just the value of n, without the need to compute any other term(s) of the sequence.

Example 2.6 (*Permutations*) *Given a set of n symbols, we want to count the number of ways these n symbols can be arranged, which equals the number of bijections from $[n]$ to $[n]$ or the number of permutations of $[n]$ (see Definition 1.10). There are n choices for the first element in the arrangement, $n - 1$ choices for the next element, $n - 2$ choices for the third element, ... , one choice for the last element. Therefore,*

$$S_n = |\mathcal{S}_n| = n \cdot (n - 1) \cdot (n - 2) \cdots 2 \cdot 1 = n!.$$

Note that $n!$ reads "n factorial", and $0! = 1$ by definition. With this explicit formula it is easy to compute the number of permutations of $[10]$ as $10! = 3,628,800$.

A recurrence relation, on the other hand, defines the value of a_n in terms of the preceding value(s) of the sequence, together with an initial condition or a set of initial conditions. The initial conditions are necessary to ensure a uniquely defined sequence, and consist of the first k values of the sequence, where k equals the difference between the highest and the lowest indices that occur in the recurrence relation.

Example 2.7 (*Recursion for the Fibonacci sequence*) *The Fibonacci sequence is defined by*

$$F_n = F_{n-1} + F_{n-2}, \quad F_0 = 0, F_1 = 1,$$

that is, each term is the sum of the two preceding terms, except for the first two values, which are needed to get the recursion started. Hence, the recurrence relation only applies for $n \geq 2$. From the initial conditions we can easily compute $F_2 = F_1 + F_0 = 1$, $F_3 = 2$, ... The first sixteen terms of the sequence are given by 0, 1, 1, 2, 3, 5, 8, 13, 21, 34, 55, 89, 144, 233, 377 and 610 (see A000045 in [180]). To obtain these (and, with appropriate modification, more) values we can use the following Maple code:

```
aseq:=proc(n)
  if n<0 then return "seq not defined for negative indices";
  elif n=0 then return 1;
    elif n=1 then return 1;
      else aseq(n-1)+aseq(n-2);
  end if;
end proc:
seq(aseq(n),n=0..15);
```

The corresponding Mathematica *code is*

```
F[0] = 0; F[1] = 1;
F[n_] := F[n] = F[n - 1] + F[n - 2];
Table[F[n], {n, 0, 15}]
```

Note that the delayed assignment `F[n_] := F[n]` `=` *is a way to prompt* Mathematica *to retain already computed values, which is crucial in recursive computations.*

If we change the initial conditions for the Fibonacci recurrence we obtain another famous sequence.

Example 2.8 (*Recursion for the Lucas sequence*) *The Lucas sequence is defined by*

$$L_n = L_{n-1} + L_{n-2}, \quad L_0 = 2, L_1 = 1,$$

which leads to the sequence whose terms for $n = 0, \ldots, 15$ *are given by* 2, 1, 3, 4, 7, 11, 18, 29, 47, 76, 123, 199, 322, 521, 843 *and* 1364 (*see [180], A000032 and A000204*). *Modifying the code from Example 2.7, we can compute the sequence of values as*

```
aseq:=proc(n)
  if n<0 then return "seq not defined for negative indices";
  elif n=0 then return 2;
    elif n=1 then return 1;
      else aseq(n-1)+aseq(n-2);
  end if;
end proc:
seq(aseq(n),n=0..15);
```

in Maple; *in* Mathematica, *we can use*

```
L[0] = 2; L[1] = 1;
L[n_] := L[n] = L[n - 1] + L[n - 2];
Table[L[n], {n, 0, 15}]
```

For a history of these two sequences, as well as applications where these sequences occur, see [125]. It is customary to use the letters F and L for these two sequences. A third sequence that is closely related to these two is the shifted Fibonacci sequence, defined as $f_n = F_{n+1}$ for $n \geq 0$, which occurs naturally in many different contexts. We will now look at additional examples, namely compositions in $\{1, 2\}$ and words on [3]; the first of these will give rise to the shifted Fibonacci sequence.

Example 2.9 (*Compositions in* $\{1, 2\}$) *We want to count the number of compositions of* n *that contain only 1s and 2s. For the first few values of* n, *we can easily make a list of such compositions, as shown in Table 2.1, and count their number directly.*

TABLE 2.1: Compositions of n with 1s and 2s

n	# of compositions	compositions
1	1	1
2	2	11, 2
3	3	111, 12, 21
4	5	1111, 112, 121, 211, 22
5	8	11111, 1112, 1121, 1211, 122, 2111, 212, 221

The resulting sequence for the number of compositions looks very much like the Fibonacci sequence, except it does not start correctly. We also have to check whether this Fibonacci pattern continues beyond the first few values. Let's think about a systematic way to create the compositions of n in $\{1,2\}$ recursively. Each such composition of n (for $n \geq 2$) either starts with a 1 or a 2. We can create those starting with a 1 by appending a composition of $n-1$, and those starting with a 2 by appending a composition of $n-2$. It is customary to define the number of compositions of 0 to be 1 (for the empty composition). If we denote the number of compositions of n in $\{1,2\}$ by C_n, then we obtain the following recursion:

$$C_n = C_{n-1} + C_{n-2}, \quad C_0 = C_1 = 1,$$

which results in the shifted Fibonacci sequence.

Example 2.10 (*Words avoiding given substrings*) *Let's determine a_n, the number of words of length n on the alphabet $\{1,2,3\}$ which do not start with 3 and do not contain the substrings 13, 22, 31, and 32. If the word starts with a 1, we can append any of the a_{n-1} such words of length $n-1$. If the word starts with a 2, then the second letter has to be a 1 or a 3, that is, the word starts with 21 or with 23. In the first case, there are no restrictions on how the word can continue, so we can append any word of length $n-2$ without the forbidden substrings, for a total of a_{n-2} such words. If the word starts with 23, then the only possible words are of the form $233\cdots 3$, and there is exactly one. The initial conditions are $a_0 = 1$ and $a_1 = 2$. Overall,*

$$a_n = a_{n-1} + a_{n-2} + 1, \qquad a_0 = 1, a_1 = 2.$$

In these two examples, the recurrence relation was very easy to derive, whereas an explicit formula is hard to obtain directly. Very often, a recursive relation can be obtained naturally, as it expresses how a sequence evolves from one step to the next. Notice how we divided the objects to be counted into classes or categories, each of which is counted separately. In the example above, we focused on the first part of the composition or word. In the case of the compositions, we could have just as easily focused on the last part, as each composition either ends in a 1 or a 2. Focusing on the first and last parts are common ways to categorize, but we will see other possibilities in the examples throughout the book.

The drawback of a recurrence relation is that in order to obtain a specific term of the sequence, say C_{100}, all preceding values $C_1, C_2, \ldots, C_{99}$ have to be computed, unless an explicit formula can be derived. We will discuss ways of deriving an explicit formula from a recurrence relation in Section 2.2.

Sometimes we are interested in more than one parameter, for example, in addition to the number of compositions we may want to keep track of the number of parts in the compositions of n, or the number of times the part 1 occurs in all compositions of n. In this case, we obtain a sequence with two indices.

Definition 2.11 *A sequence with k indices is a function $a: I^k \to B$, denoted by $\{a_{n_1,\ldots,n_k}\}_{n_1,\ldots,n_k \in I}$ or $\{a_{\overrightarrow{n}}\}_{\overrightarrow{n} \in I^k}$, where $I \subseteq \mathbb{Z}$. The element $a_{\overrightarrow{n}}$ of a sequence $\{a_{\overrightarrow{n}}\}_{\overrightarrow{n} \in I}$ is called the $\overrightarrow{n}$-th term, and the vector $\overrightarrow{n}$ of integers is the sequence vector of indices.*

If a sequence has only two indices, it is easy to list the sequence values in a two-dimensional table.

Example 2.12 (*Continuation of Example 2.9*) *Let $a_{n,m}$ denote the number of compositions of n in $\{1, 2\}$ with exactly m parts. From Table 2.1, we can read off that $a_{4,1} = 0$, $a_{4,2} = 1$, $a_{4,3} = 3$, and $a_{4,4} = 1$. To obtain the recurrence relation for this sequence, we now also keep track of what happens to the number of parts when we create the compositions recursively. Whenever we either prepend a 1 or a 2 to a composition of $n-1$ or $n-2$, respectively, the number of parts increases by one. Thus,*

$$a_{n,m} = a_{n-1,m-1} + a_{n-2,m-1}, \quad a_{0,0} = 1, a_{1,1} = 1, a_{n,0} = 0 \text{ if } n \neq 0.$$

From the definition, $a_{n,m} = 0$ if $m > n$, and we therefore present the values as a triangular array in Table 2.2, where only the values corresponding to $m \leq n$ are listed. Note that summing the elements across a row gives the number of compositions of the respective n.

TABLE 2.2:   Compositions of n in $\{1, 2\}$ with m parts

$n \backslash m$	0	1	2	3	$\cdots$
0	$a_{0,0} = 1$				
1	$a_{1,0} = 0$	$a_{1,1} = 1$			
2	$a_{2,0} = 0$	$a_{2,1} = 1$	$a_{2,2} = 1$		
3	$a_{3,0} = 0$	$a_{3,1} = 0$	$a_{3,2} = 2$	$a_{3,3} = 1$	$\cdots$
4	$a_{4,0} = 0$	$a_{4,1} = 0$	$a_{4,2} = 1$	$a_{4,3} = 3$	$\cdots$
$\vdots$	$\vdots$	$\vdots$	$\vdots$	$\vdots$	$\ddots$

To compute the values for $a_{i,j}$ given in Table 2.2, we can use the following Mathematica *code:*

```
a[n_, 0] := Piecewise[{{1, n == 0}, {0, n != 0}}];
a[n_, m_] := a[n, m] = a[n - 1, m - 1] + a[n - 2, m - 1];
Table[a[n, m], {n, 0, 4}, {m, 0, 3}]
```

The Maple code is given by

```
a:=Matrix(5,5):
a[1,1]:=1:
 for m from 2 to 5 do
   a[1,m]:=0:
od:
for n from 2 to 5 do
  a[n,1]:=0;
  for m from 2 to 5 do
    if n=m then a[n,m]:=1;
    elif n<m then a[n,m]:=0;
      else a[n,m]:=a[n-1,m-1]+a[n-2,m-1];
    end if;
  od:
od:
"print/rtable"(a);
```

Other examples of sequences with two indices are *k-element permutations* and *k-element combinations* of n objects.

Example 2.13 (*Permutations revisited*) *In Example 2.6 we looked at all possible arrangements of a set of n objects. What if we only want to arrange $k \leq n$ of the objects? In how many ways can this be accomplished? Arguing as before, there are n choices for the first element in the arrangement, $n-1$ choices for the second element, ..., $n-k+1$ choices for the k-th element. Overall, if $S_{n,k}$ denotes the* number *of k-element permutations of n objects, then*

$$S_{n,k} = n \cdot (n-1) \cdots (n-k+1) = \frac{n!}{(n-k)!}.$$

Note that S_n in Example 2.6 is given by $S_{n,n} = n!/0! = n!$.

For permutations, order matters. However, if we look at all possible k-element subsets of a set with n elements, then order does not matter. What matters is only whether a particular element is contained in the subset or not. Selections without order are called *combinations*.

Example 2.14 (*Combinations*) *How can we count the number of k-element combinations of n objects? Rather than counting the k-combinations directly, we make a connection with k-element permutations of n objects by creating the k-element permutations in two steps. First we select the k elements, and* *then we arrange them in all possible orders. If we let $C_{n,k}$ denote the number of k-element combinations of n objects, then the first task can be done in $C_{n,k}$ ways, and each such selection can be arranged in $S_{k,k} = k!$ ways. Therefore, $S_{n,k} = C_{n,k} \cdot k!$, which implies that*

$$C_{n,k} = \frac{S_{n,k}}{k!} = \frac{n!}{(n-k)!k!} = \binom{n}{k}.$$

The values $C_{n,k}$ occur in the Binomial theorem, and $\binom{n}{k}$ reads "n choose k". These values also occur in the famous *Pascal's Triangle*, which shows yet another way to display a sequence with two indices in tabular form.

Example 2.15 (*Pascal's Triangle*) *Pascal's Triangle is named after Blaise Pascal[1], but was known about 500 years earlier both in the Middle East by Al-Karaji[2] and in China by Jia Xian[3].*

It was named after Pascal (at least in Europe) due to Pascal's important Treatise on the Arithmetical Triangle, *in which the binomial coefficients of the form $\binom{n}{k}$ for $0 \leq k \leq n$ are given in a triangular array.*

TABLE 2.3:   Pascal's triangle.

$$
\begin{array}{ccccccc}
 & & & \binom{0}{0} & & & \\
 & & \binom{1}{0} & & \binom{1}{1} & & \\
 & \binom{2}{0} & & \binom{2}{1} & & \binom{2}{2} & \\
\binom{3}{0} & & \binom{3}{1} & & \binom{3}{2} & & \binom{3}{3} \\
 & & & \vdots & & &
\end{array}
$$

$$
\begin{array}{ccccccc}
 & & & 1 & & & \\
 & & 1 & & 1 & & \\
 & 1 & & 2 & & 1 & \\
1 & & 3 & & 3 & & 1 \\
 & & & \vdots & & &
\end{array}
$$

When displayed in the form of Table 2.3, where rows correspond to the values of n, and the diagonals from lower left to upper right correspond to the values

[1]http://www-history.mcs.st-andrews.ac.uk/Mathematicians/Pascal.html
[2]http://turnbull.mcs.st-and.ac.uk/~history/Mathematicians/Al-Karaji.html
[3]http://www-history.mcs.st-andrews.ac.uk/Mathematicians/Jia_Xian.html

of k, many interesting properties can be derived. One very useful property is the fact that the two diagonals that "enclose" the triangle on the sides consist of all 1s, and all other entries can be computed by adding the two values directly above to the left and right of the entry. This property can be verified by showing that

$$\binom{n}{k} = \binom{n-1}{k-1} + \binom{n-1}{k}, \tag{2.1}$$

either by using the formula for $\binom{n}{k}$, or by giving a combinatorial argument. Recall that $\binom{n}{k}$ counts the number of ways to select a subset of k elements from a set of n elements, say the set $[n]$. Now any subset either does or does not contain the element n. If the subset contains the element n, then the remaining $k-1$ elements are chosen from the set $[n-1]$ in $\binom{n-1}{k-1}$ ways. If the subset does not contain the element n, then all k elements are chosen from the set $[n-1]$, which can be done in $\binom{n-1}{k}$ ways. Since these are the only possible cases, and the two cases are disjoint, we can add the two counts to obtain the result of (2.1). The entries of the enclosing diagonals are of the form $\binom{n}{0}$ or $\binom{n}{n}$, and there is exactly one way to either select no or all elements from a set of n objects.

2.2 Solving recurrence relations

We will now discuss methods to obtain explicit formulas from recurrence relations. Different approaches can be used, some of which apply only to certain types of recurrence relations. Therefore, we start by classifying recurrence relations.

Definition 2.16 *A linear (in the a_i) recurrence relation of order r is of the form*

$$P_0(n)a_n + P_1(n)a_{n-1} + \cdots + P_r(n)a_{n-r} = q(n)$$

for all $n \geq r$, where P_r is nonzero. The recurrence relation is called P-recursive if $P_0, P_1, \ldots, P_r$ and q are polynomials in n. In the special case where all the polynomials P_i are constant, the recurrence relation reduces to a linear recurrence relation of order r with constant coefficients. If $q(n) = 0$, then the recurrence relation is called homogeneous, *otherwise it is called* nonhomogeneous.

The order of the recurrence relation is the number of prior terms needed to compute the next one, and also equals the number of initial values a_{n_0}, $a_{n_0+1}, \ldots, a_{n_0+r-1}$ needed to obtain a unique sequence. The order can be computed as the difference between the highest and lowest indices in the recurrence relation. Note that the order may differ from the number of terms in the recurrence relation.

Example 2.17 (*Compositions in $\{1,2\}$ avoiding the substring 11*) *Let a_n be the number of compositions of n in $\{1,2\}$ that avoid the substring 11. Then $a_0 = 1$ (the empty composition), $a_1 = 1$ (the composition 1), and $a_2 = 1$ (the composition 2). In general, any such composition either starts with a 1 or with a 2. A composition of n that starts with a 2 can be created by combining a 2 with a composition of $n-2$ that avoids the substring 11. If the composition starts with a 1, then it has to start with 12 (to avoid the substring 11), so we can combine a composition of $n - 3$ with 12 to obtain those compositions of n. All together,*

$$a_n = a_{n-2} + a_{n-3}, \quad a_0 = a_1 = a_2 = 1.$$

In this case, the order of the recurrence is $n - (n - 3) = 3$.

We are now ready to discuss several methods of solving recurrence relations:

- Guess and check.
- Iteration (repeated substitution).
- Characteristic polynomial.
- Generating functions.

Note that with the exception of linear recurrence relations with constant coefficients, there are no general methods for obtaining explicit formulas from recurrences relations.

2.2.1   Guess and check

As the name indicates, the first method consists of guessing a solution and then proving by induction that the guess is correct. Obtaining the right guess is a matter of either smart observation or just plain luck.

Example 2.18 *Consider the recurrence relation*

$$a_n = n \cdot a_{n-1}, \quad n \geq 1, \quad a_0 = 1.$$

Computing the first few values gives $a_0 = 1$, $a_1 = 1$, $a_2 = 2$, $a_3 = 6$, $a_4 = 24$, and $a_5 = 120$. A reasonable guess (both from the values and the recurrence relation) is that $a_n = n!$ for all $n \geq 0$, which is easy to prove by induction:

$$a_{n+1} = (n + 1) \cdot a_n = (n + 1) \cdot n! = (n + 1)!.$$

Note that this recurrence relation is P-recursive. We can also find the explicit formula for a_n by using the Maple function `rsolve` *or the Mathematica function* `RSolve`:

```
rsolve({a(n)=n*a(n-1),a(0)=1},a(n));
```

and

```
RSolve[{a[n]==n*a[n - 1], a[0]==1}, a[n], n]
```

2.2.2 Iteration

The second method, iteration or repeated substitution, is most useful when the recurrence relation is of order one. The method consists of applying the recurrence relation repeatedly, simplifying the result and then recognizing a pattern.

Example 2.19 *Consider the recurrence relation*

$$a_n = \frac{2n+1}{n+1} \cdot a_{n-1}, \quad n \geq 1, \quad a_0 = 1.$$

Repeated substitution gives

$$
\begin{aligned}
a_n &= \frac{2n+1}{n+1} \cdot a_{n-1} = \frac{(2n+1)(2n-1)}{(n+1)n} \cdot a_{n-2} \\
&= \frac{(2n+1)(2n-1)(2n-3)}{(n+1)n(n-1)} \cdot a_{n-3} \\
&= \cdots \\
&= \frac{(2n+1)(2n-1)\cdots 5 \cdot 3}{(n+1)n\cdots 2} \cdot a_0 \\
&= \frac{(2n+1)!}{(n+1)! \cdot 2n(2n-2)\cdots 2} \\
&= \frac{(2n+1)!}{2^n n!(n+1)!} = \frac{1}{2^n}\binom{2n+1}{n}.
\end{aligned}
$$

The explicit formula for a_n *can be obtained by modifying the Maple and* Mathematica *codes of Example 2.18:*

```
rsolve({a(n)=(2*n+1)/(n+1)*a(n-1),a(0)=1},a(n));
```

and

```
RSolve[{a[n]==(2n+1)/(n+1)a[n - 1], a[0]==1}, a[n], n]
```

Note that the answer given by Mathematica *is in terms of Gamma functions, which is no surprise, as the Gamma functions are a generalization of the binomial coefficients for noninteger values.*

More generally, the explicit formula for the recurrence relation

$$a_n = q(n) \cdot a_{n-1}, \quad n \geq 1$$

with initial condition a_0 is given by $a_n = a_0 \prod_{j=1}^{n} q(j)$. In particular, if $q(j) = b$ for all j, then the solution is of the form $a_0 b^n$.

Iteration can also work nicely for nonlinear recurrence relations of order one.

Example 2.20 *Consider the recurrence relation*

$$a_n = 2\,a_{n-1}^2, \quad n \geq 1, \quad a_0 = 3,$$

which is nonlinear because of the squared term. Repeated substitution gives

$$a_n = 2 \cdot a_{n-1}^2 = 2(2a_{n-2}^2)^2 = 2 \cdot 2^2 \cdot (2a_{n-3}^2)^4$$

$$= \cdots$$

$$= 2 \cdot 2^2 \cdot 2^4 \cdots 2^{n-1} a_0^{2^n} = 2^{2^n - 1} \cdot 3^{2^n}$$

The explicit formula for a_n can be computed using the Maple code

```
rsolve({a(n)=2*a(n-1)^2,a(0)=3},a(n));
```

or the Mathematica *code*

```
RSolve[{a[n]==2 a[n - 1]^2, a[0]==3}, a[n], n]
```

2.2.3   Characteristic polynomial

We now describe how to derive an explicit solution for a linear recurrence relation with constant coefficients of order r of the form

$$a_n + \alpha_1 a_{n-1} + \alpha_2 a_{n-2} + \cdots + \alpha_r a_{n-r} = q(n) \qquad \text{for all } n \geq 0, \qquad (2.2)$$

where $q(n)$ is a polynomial in n, not depending on a_n.

Solving a nonhomogeneous recurrence relation, when possible, requires solving the corresponding homogeneous recurrence relation as part of the process, so we will start by solving linear homogeneous recurrence relations with constant coefficients of the form

$$a_n + \alpha_1 a_{n-1} + \alpha_2 a_{n-2} + \cdots + \alpha_r a_{n-r} = 0 \qquad \text{for all } n \geq 0. \qquad (2.3)$$

Like many results in mathematics, Theorem 2.22 below arose from an educated guess and then working out the algebraic details. Taking inspiration from Example 2.19, where the solution is of the form $a_0 \cdot b^n$, one might guess that for linear recurrence relations with constant coefficients, the solution may be of the form ξ^n, give or take a multiplicative constant. Checking the guess by substituting it into the recurrence relation and taking out common factors reduces to finding the roots of the polynomial

$$\Delta(x) = x^r + \alpha_1 x^{r-1} + \alpha_2 x^{r-2} + \cdots + \alpha_r$$

in order to determine the value of ξ.

Definition 2.21 *The polynomial $\Delta(x) = x^r + \alpha_1 x^{r-1} + \alpha_2 x^{r-2} + \cdots + \alpha_r$ is called the* characteristic polynomial *of the recurrence relation*

$$a_n + \alpha_1 a_{n-1} + \alpha_2 a_{n-2} + \cdots + \alpha_r a_{n-r} = 0.$$

By the Fundamental Theorem of Algebra, the characteristic polynomial has r (possibly complex) roots, counting multiplicities. The following theorem tells us how the solutions look in the case of roots with multiplicity bigger than one, and ensures that indeed there are r independent solutions for a linear recurrence relation with constant coefficients of order r.

Theorem 2.22 *Let $\xi \in \mathbb{C}$ be any root of*

$$\Delta(x) = x^r + \alpha_1 x^{r-1} + \alpha_2 x^{r-2} + \cdots + \alpha_r$$

with multiplicity m, then the basic solutions $n^i \xi^n$, $i = 0, 1, \ldots, m-1$, satisfy (2.3).

Proof See Exercise 2.10. □

We now form a linear combination of the r different basic solutions given in Theorem 2.22. It is easy to show that the resulting sequence is a solution to (2.3) because of the linearity of the recurrence relation.

Theorem 2.23 *If $a_1(n), \ldots, a_r(n)$ are different sequences satisfying (2.3), then for any constants $c_1, \ldots, c_r \in \mathbb{C}$, the sequence $a_n = \sum_{j=1}^{n} c_j \cdot a_j(n)$, called the* general solution, *satisfies (2.3).*

Now, for a given linear homogeneous recurrence relation as described in (2.3) with r initial conditions, we can obtain an explicit formula using the following steps:

1. Find all the roots of the characteristic polynomial $\Delta(x)$ and their multiplicity.

2. Determine the general solution as described in Theorem 2.23.

3. Use the initial conditions to obtain a linear system of r equations in r unknowns, then solve it to obtain a specific solution.

Example 2.24 *Consider the recurrence relation $a_n = 4a_{n-1} - 5a_{n-2} + 2a_{n-3}$ with initial conditions $a_0 = 1$, $a_1 = 1$, and $a_2 = 2$. The characteristic polynomial of this recurrence relation is*

$$\Delta(x) = x^3 - 4x^2 + 5x - 2 = (x - 2)(x^2 - 2x + 1) = (x - 2)(x - 1)^2.$$

The characteristic polynomial has two roots: $\xi = 2$ with multiplicity one, and $\xi = 1$ with multiplicity two. From Theorem 2.23, we obtain that the general solution for this recurrence relation is given by

$$a_n = c_1 \cdot 2^n + c_2 \cdot 1^n + c_3 \cdot n \cdot 1^n = c_1 \cdot 2^n + c_2 + c_3 \cdot n.$$

Using the initial conditions results in the equations $c_1 + c_2 = 1$, $2\,c_1 + c_2 + c_3 = 1$, and $4\,c_1 + c_2 + 2\,c_3 = 2$. The values of c_1, c_2, c_3 can be computed by using the Maple code

```
solve({c_1*2^0+c_2+c_3*0=1,c_1*2^1+c_2+c_3*1=1,
c_1*2^2+c_2+c_3*2=2},{c_1,c_2,c_3});
```

or the Mathematica *code*

```
Solve[{c[1] 2^0+c[2]+c[3] 0==1, c[1] 2^1+c[2]+c[3] 1==1,
c[1] 2^2+c[2]+c[3] 2==2},{c[1],c[2],c[3]}]
```

We obtain that $c_1 = 1$, $c_2 = 0$, and $c_3 = -1$, which implies that the explicit formula is given by $a_n = 2^n - n$.

Note that a homogeneous linear recurrence relation with constant coefficients of order two can always be solved (as we can compute the exact roots using the quadratic formula). For higher orders, the Fundamental Theorem of Algebra guarantees that there are as many roots as the order of the recurrence relation, but we do not have formulas to find the exact roots unless we can factor the characteristic polynomial nicely.

Example 2.25 (*Fibonacci and Lucas Sequences*) *We can now find explicit solutions for the Fibonacci and the Lucas sequence. Recall that the two sequences have the same recurrence relation $a_n = a_{n-1} + a_{n-2}$, with initial conditions $F_0 = 0$, $F_1 = 1$ for the Fibonacci sequence and $L_0 = 2$, $L_1 = 1$ for the Lucas sequence. Since the recurrence relation is the same, they both have the same characteristic polynomial $\Delta(x) = x^2 - x - 1$, and hence the same general solution. Using the quadratic formula, we obtain that $\xi_{1,2} = \frac{1 \pm \sqrt{5}}{2}$, and thus, the general solution is given by*

$$a_n = c_1 \cdot \left(\frac{1 + \sqrt{5}}{2} \right)^n + c_2 \cdot \left(\frac{1 - \sqrt{5}}{2} \right)^n.$$

Note that it is customary in the context of the Fibonacci sequence to define $\alpha = \frac{1+\sqrt{5}}{2}$, the Golden Ratio[4] Φ, *and $\beta = \frac{1-\sqrt{5}}{2}$. The initial conditions for the Fibonacci sequence give $0 = c_1 + c_2$ and $1 = c_1 \cdot \alpha + c_2 \cdot \beta$. Using any method*

---

[4]http://mathworld.wolfram.com/GoldenRatio.html

to solve a system of two equations, we obtain $c_1 = 1/\sqrt{5}$ and $c_2 = -1/\sqrt{5}$, and thus

$$F_n = \frac{1}{\sqrt{5}} \left[\left(\frac{1+\sqrt{5}}{2} \right)^n - \left(\frac{1-\sqrt{5}}{2} \right)^n \right]. \tag{2.4}$$

This formula for the Fibonacci sequence is somewhat surprising, as the powers of irrational numbers balance in just the correct way to give an integer result for any value of n (see Exercise 2.9). This formula was first derived by Abraham De Moivre[5], and independently by Daniel Bernoulli[6]. For the derivation of the explicit formula for the Lucas sequence see Exercise 2.8. We can find the explicit formula for F_n by using the following Maple code:

```
rsolve({a(n)=a(n-1)+a(n-2),a(0)=0,a(1)=1},a(n));
```

Using the corresponding Mathematica *code*

```
RSolve[{a[n]==a[n - 1] + a[n - 2], a[0]==0, a[1]==1},a[n],n]
```

does not result in an explicit formula, but rather in the answer that a_n is the n-th Fibonacci number, a function that is defined in Mathematica*:*

```
{{a[n] -> Fibonacci[n]}}.
```

Now we are ready to indicate how to solve a linear nonhomogeneous recurrence relation with constant coefficients. Theorem 2.26 gives the form of the general solution, which consists of two parts.

Theorem 2.26 *The general solution of the nonhomogeneous linear recurrence relation with constant coefficients of order r as given in (2.2) is of the form $h(n) + p(n)$, where $h(n)$ is the general solution of the associated homogeneous recurrence relation (2.3), and $p(n)$ is any solution of (2.2).*

The solution $p(n)$ in Theorem 2.26 is called the *particular solution*. To find the general solution for a nonhomogeneous recurrence relation, we need to use the following steps:

1. Find the general solution for the associated homogeneous recurrence relation.

2. Guess a particular solution.

3. Use the initial conditions to determine the specific solution.

The difficult part is to find the particular solution, which often' involves guess and check. There is no approach that works for all types of functions that may occur on the right-hand side of (2.2). However, for certain common functions, the form of the particular function is known. If $p(n)$ is a constant

---

[5]http://www-groups.dcs.st-and.ac.uk/~history/Mathematician/De_Moivre.html

[6]http://www-groups.dcs.st-and.ac.uk/~history/Mathematicians/Bernoulli_Daniel.html

multiple of a function listed on the left-hand side of Table 2.4, then the function given on the right-hand side is a particular solution (if it is not also a solution of the homogeneous recurrence relation). The constants d, r, and θ are given, while the constants c_i have to be determined from the given recurrence relation. If $q(n)$ is a linear combination of functions given on the left-hand side of Table 2.4, and none of these functions is a solution of the homogeneous recurrence relation, then the particular solution is a linear combination of the respective functions given on the right-hand side of Table 2.4. For example, if $q(n) = d^n + \cos(\theta n)$, then $p(n) = c_1 \cdot d^n + c_2 \sin(\theta n) + c_3 \cos(\theta n)$. For a more detailed discussion of how to solve nonhomogeneous linear recurrence relations with constant coefficients, specifically what to do when the candidate particular solution also solves the homogeneous recurrence relation, see [78, 182, 189].

TABLE 2.4: Particular solutions for certain $q(n)$

$q(n)$	$p(n)$
d	c_0
d^n	$c \cdot d^n$
$n^r, r \in \mathbb{N}$	$c_r n^r + \cdots + c_1 n + c_0$
$n^r d^n, r \in \mathbb{N}$	$d^n (c_r n^r + \cdots + c_1 n + c_0)$
$\sin(\theta n)$ or $\cos(\theta n)$	$c_1 \sin(\theta n) + c_2 \cos(\theta n)$
$d^n \sin(\theta n)$ or $d^n \cos(\theta n)$	$c_1 d^n \sin(\theta n) + c_2 d^n \cos(\theta n)$

To demonstrate how to solve a nonhomogeneous recurrence relation, we revisit Example 2.10.

Example 2.27 (*Words avoiding substrings*) *The recurrence relation for the number of words on $\{1, 2, 3\}$ that do not start with 3 and do not contain the substrings 22, 31, and 32 is given by*

$$a_n = a_{n-1} + a_{n-2} + 1, \qquad (2.5)$$

with initial conditions $a_0 = 1$ and $a_1 = 2$. The associated homogeneous recurrence relation is the same as the one for the Fibonacci sequence, and therefore, the general solution is given by $a_n = c_1 \alpha^n + c_2 \beta^n$ (see Example 2.25) with $\alpha = (1 + \sqrt{5})/2$ and $\beta = (1 - \sqrt{5})/2$. The next step is to find a particular solution. In this case, the polynomial $q(n)$ is a constant, so the particular solution is constant too, say $p(n) = c$. Substituting this particular solution into (2.5) leads to $c = c + c + 1$ or $c = -1$. Thus, the general solution for the nonhomogeneous recurrence relation is given by

$$a_n = c_1 \alpha^n + c_2 \beta^n - 1 \qquad (2.6)$$

and substituting the initial conditions results in the system of equations

$$1 = c_1 + c_2 - 1 \quad and \quad 2 = c_1\alpha + c_2\beta - 1.$$

Using standard methods to solve a system of two equations gives $c_1 = \frac{\sqrt{5}+2}{\sqrt{5}}$ *and* $c_2 = \frac{\sqrt{5}-2}{\sqrt{5}}$. *The sequence of values for* a_n *for* $n = 0, \ldots, 10$ *is given by* 1, 2, 4, 7, 12, 20, 33, 54, 88, 143 *and* 232. *These first few values satisfy* $a_n = F_{n+3} - 1$. *Comparing this expression with* (2.6), *the pattern would work in general if* $c_1 = \frac{1}{\sqrt{5}}\alpha^3$ *and* $c_2 = -\frac{1}{\sqrt{5}}\beta^3$, *which can be shown to be true using some algebra. Substituting these coefficients into the specific solution then gives* $a_n = \frac{1}{\sqrt{5}}\left(\alpha^3 - \beta^3\right) - 1 = F_{n+3} - 1$. *Once more we can find the explicit formula for* $a(n)$ *by using the Maple code*

```
rsolve({a(n)=a(n-1)+a(n-2)+1,a(0)=1,a(1)=2},a(n));
```

In Mathematica, *we use*

```
RSolve[{a[n]==a[n-1]+a[n-2]+1,a[0]==1,a[1]==2},a[n],n]
```

which returns an answer in terms of the Fibonacci and Lucas sequences:

```
{{a[n] -> -1 + 2 Fibonacci[n] + LucasL[n]}}
```

LucasL *is a built-in* Mathematica *function which computes the n-th Lucas number. The two expressions for* a_n, *when set equal to each other, yield that* $2F_n + L_n = F_{n+3}$.

2.3 Generating functions

We will now discuss another important tool to obtain explicit formulas from recurrence relations, namely generating functions. Generating functions are formal power series, and we will discuss the theory of formal power series after stating the definitions and some basic examples. There are two main types of generating functions used in combinatorics, ordinary and exponential generating functions.

Definition 2.28 *The* ordinary generating function *for the sequence* $\{a_n\}_{n\in\mathbb{N}_0}$ *is given by* $A(x) = \sum_{n\geq 0} a_n x^n$, *and the* exponential generating function *for the sequence* $\{a_n\}_{n\in\mathbb{N}_0}$ *is given by* $E(x) = \sum_{n\geq 0} a_n \frac{x^n}{n!}$.

The name generating function comes from the fact that the power series expansion of the functions $A(x)$ and $E(x)$ "generates" the values of the sequence as the coefficients of the terms x^n and $\frac{x^n}{n!}$, respectively. With tools such as Maple and *Mathematica*, knowing the generating function for a sequence allows for easy computation of any value, even if we cannot find an explicit formula for the coefficients. Since we will primarily deal with ordinary generating functions, we will omit "ordinary" in the remainder of the book. We start by computing the generating functions for some well known sequences.

Example 2.29 (*Fibonacci Sequence*) *We will derive the generating function for the Fibonacci sequence and from it the explicit formula given in Example 2.25. We start with the recurrence relation for the Fibonacci numbers, $F_n = F_{n-1} + F_{n-2}$ for $n \geq 2$, with initial conditions $F_0 = 0, F_1 = 1$. In order to connect this recurrence relation to the generating function, we multiply the recurrence relation by x^n, and then sum over the values of n for which the recurrence is valid:*

$$\sum_{n \geq 2} F_n \, x^n = \sum_{n \geq 2} F_{n-1} \, x^n + \sum_{n \geq 2} F_{n-2} \, x^n.$$

Let $F(x) = \sum_{n \geq 0} F_n x^n$ be the generating function of the Fibonacci sequence. We express each of the series in terms of $F(x)$ (this is where the initial terms come in), factor out appropriate powers of x so that the remaining powers match the sequence index, and re-index the series on the right-hand side of the equation to obtain

$$F(x) - F_0 - F_1 x = x \sum_{n \geq 2} F_{n-1} \, x^{n-1} + x^2 \sum_{n \geq 2} F_{n-2} \, x^{n-2}$$

$$= x \left(F(x) - F_0 \right) + x^2 \, F(x). \tag{2.7}$$

Solving for $F(x)$ gives

$$F(x) = \frac{x}{1 - x - x^2} = \frac{x}{-(x + \alpha)(x + \beta)}, \tag{2.8}$$

where $\alpha = \frac{1+\sqrt{5}}{2}$ and $\beta = \frac{1-\sqrt{5}}{2}$. To obtain the coefficients of x^n, we need the Taylor series expansion of $F(x)$. Instead of using the definition of the Taylor series expansion which involves derivatives of the function, we make use of functions whose Taylor series we know. One of these is the geometric series *$\sum_{n \geq 0} t^n = \frac{1}{1-t}$, which will be used repeatedly in computing generating functions. Since $\alpha \beta = -1$, we can rewrite the denominator as product of factors in the correct form for the geometric series:*

$$\frac{1}{-(x + \alpha)(x + \beta)} = \frac{1}{\frac{1}{\beta}(-\beta x - \beta \alpha)\frac{-1}{\alpha}(-\alpha x - \alpha \beta)} = \frac{1}{(1 - \beta x)(1 - \alpha x)}.$$

We then compute the partial fraction decomposition of the function, that is, we determine the constants c_1 and c_2 satisfying

$$\frac{1}{(1 - \beta x)(1 - \alpha x)} = \frac{c_1}{1 - \alpha x} + \frac{c_2}{1 - \beta x}.$$

Solving for the constants in the numerator gives $c_1 = \frac{\alpha}{\alpha - \beta} = \frac{\alpha}{\sqrt{5}}$ *and* $c_2 = \frac{-\beta}{\alpha - \beta} = \frac{-\beta}{\sqrt{5}}$. *Thus,*

$$F(x) = \frac{x}{1 - x - x^2} = \frac{x}{\sqrt{5}} \left(\frac{\alpha}{1 - \alpha x} - \frac{\beta}{1 - \beta x} \right).$$

Using the geometric series, this gives

$$F(x) = \frac{x}{\sqrt{5}} \left(\alpha \sum_{n \geq 0} (\alpha x)^n - \beta \sum_{n \geq 0} (\beta x)^n \right) = \sum_{n \geq 1} \frac{1}{\sqrt{5}} (\alpha^n - \beta^n) x^n.$$

Now we can read off F_n *as the coefficient of* x^n *to obtain* $F_n = \frac{1}{\sqrt{5}} (\alpha^n - \beta^n)$ *as in Example 2.25. In Maple, the generating function can be obtained from the recurrence relation by giving an optional argument to* `rsolve`:

```
rsolve({F(n)=F(n-1)+F(n-2),F(0)=0,F(1)=1},F,'genfunc'(x));
```

The Mathematica *function* RSolve *does not have this option, but we can use* RSolve *in conjunction with* GeneratingFunction *(Mathematica Version 7.0) to obtain the generating function.*

```
sol = RSolve[{F[n]==F[n-1]+F[n-2], F[0]==0, F[1]==1}, F[n], n];
GeneratingFunction[sol[[1, 1, 2]], n, x]
```

In Mathematica *Version 6.0, we can use the following code to arrive at the same answer:*

```
sol = RSolve[{F[n]==F[n-1]+F[n-2], F[0]==0, F[1]==1}, F[n], n];
Sum[sol[[1, 1, 2]] x^n, {n,0,Infinity}] // FullSimplify
```

We will now give some examples of exponential generating functions. In most cases, the exponential generating function will result in logarithmic or exponential functions, unless the sequence terms contain factorials.

Example 2.30 *To compute the exponential generating function for* S_n, *the number of permutations of* $[n]$ *(of which there are* $n!$), *we use the definition and obtain*

$$E(x) = \sum_{n \geq 0} S_n \frac{x^n}{n!} = \sum_{n \geq 0} x^n = \frac{1}{1 - x}.$$

Example 2.31 *Modifying Example 2.30 slightly, we derive the exponential generating function for the sequence* $a_n = (n-1)!$. *Then*

$$E(x) = \sum_{n \geq 1} a_n \frac{x^n}{n!} = \sum_{n \geq 1} \frac{(n-1)!}{n!} x^n = \sum_{n \geq 1} \frac{x^n}{n} = -\log(1 - x),$$

where we make use of the known Taylor series expansion for the natural logarithm $\log(x)$. *We can find the generating function* $E(x)$ *by using the Maple function* sum *or the Mathematica function* Sum, *respectively:*

```
sum((n-1)!/n!*x^n,n=1..infinity);
```

```
Sum[(n - 1)!/n! x^n, {n, 1, Infinity}]
```

The last two examples show that it is convenient to use the exponential generating function rather than the ordinary generating function when the sequence terms or the recurrence relation involve factorial terms. Note also that we have used the generating function in two different ways: in Example 2.29 we started from the recurrence relation, multiplying the relation by x^n and summing over n, then solving for the generating function. If we already have an explicit formula, then we just use the definition.

In Example 2.29, we derived an explicit formula for the Fibonacci sequence from the recurrence relation using the generating function approach. Previously (see Example 2.25) we derived the same formula from the solution procedure for linear recurrence relations with constant coefficients. Do we always have these two choices? The answer is yes, that is, any linear recurrence relation with constant coefficients can also be solved using the generating function approach, as we will demonstrate now. Note that the generating function approach has the advantage that we can solve the nonhomogeneous recurrence relation directly, without first solving the associated homogeneous relation.

Lemma 2.32 *Given a linear recurrence relation with constant coefficients of the form*

$$a_n + \alpha_1 a_{n-1} + \alpha_2 a_{n-2} + \cdots + \alpha_r a_{n-r} = f(n) \qquad \text{for all } n \geq r,$$

where $\alpha_1, \ldots, \alpha_r$ are constants and $f(n)$ is any function, the associated generating function is given by

$$A(x) = \frac{\sum_{j=0}^{r} \alpha_j x^j \sum_{i=0}^{r-j-1} a_i x^i + \sum_{n \geq r} f(n) x^n}{\sum_{j=0}^{r} \alpha_j x^j}. \qquad (2.9)$$

Proof Given a linear recurrence relation with constant coefficients, we proceed as in Example 2.25, multiplying the recurrence relation by x^n and summing over $n \geq r$ to obtain

$$\sum_{n \geq r} \left(\sum_{j=0}^{r} \alpha_j a_{n-j} \right) x^n = \sum_{n \geq r} f(n) x^n.$$

Separating terms that are independent of n, and splitting off x^j gives

$$\sum_{j=0}^{r} \alpha_j x^j \sum_{n \geq r} a_{n-j} x^{n-j} = \sum_{n \geq r} f(n) x^n,$$

or equivalently, after re-indexing the inner sum and writing it in terms of the generating function,

$$\sum_{j=0}^{r} \alpha_j x^j \left(A(x) - \sum_{i=0}^{r-j-1} a_i x^i \right) = \sum_{n \geq r} f(n) x^n.$$

Solving for $A(x)$ gives (2.9). $\qquad\square$

We can apply this formula to the nonhomogeneous recurrence relation of Example 2.27.

Example 2.33 *Let $a_n = a_{n-1} + a_{n-2} + 1$ with initial conditions $a_0 = 1$ and $a_1 = 2$. The relevant parameters are $r = 2$, $\alpha_0 = 1$, $\alpha_1 = \alpha_2 = -1$, $f(n) = 1$ for all n, and therefore, $\sum_{n \geq 2} f(n) x^n = x^2 \sum_{n \geq 0} x^n = \frac{x^2}{1-x}$. Substituting this expression and the parameter values into (2.9), we obtain that*

$$A(x) = \frac{\alpha_0 (a_0 + a_1 x) + \alpha_1 x a_0 + \frac{x^2}{1-x}}{\alpha_0 + \alpha_1 x + \alpha_2 x^2} = \frac{1 + x + \frac{x^2}{1-x}}{1 - x - x^2}$$

$$= \frac{1}{(1-x)(1 - x - x^2)} = \frac{1}{1 - 2x + x^3}.$$

Maple is able to find the generating function also in this case:

```
rsolve({a(n)=a(n-1)+a(n-2)+1,a(0)=1,a(1)=2},a,'genfunc'(x));
```

In Mathematica 7.0, we use once more RSolve and GeneratingFunction. Note that RSolve expresses the answer as a linear combination of the Fibonacci and Lucas sequences.

```
sol = RSolve[{a[n]==a[n-1]+a[n-2]+1, a[0]==1, a[1]==2}, a[n], n]
GeneratingFunction[sol[[1, 1, 2]], n, x] // Simplify
```

We will return to this example once we have developed a few more tools to see how we can read off the explicit formula for $\{a_n\}_{n \geq 0}$ from this generating function.

In order for generating functions to be useful, we need to be able to add and multiply them. However, we will not be concerned with questions of convergence at this point, but rather work with *formal power series*. (We will change our point of view in Chapter 8 to obtain results on the asymptotic behavior of the coefficients.) Let $\{a_{\vec{n}}\}_{\vec{n} \in \mathbb{N}^k}$ be a sequence with k indices, $\mathbb{L}$ be a ring (see Definition B.27), and $\mathbb{L}[x_1, \ldots, x_k] = \mathbb{L}[\vec{x}]$ be the set of all polynomials in k indeterminates $\vec{x} = (x_1, \ldots, x_k)$ with coefficients in $\mathbb{L}$.

Definition 2.34 *The ordinary generating function and the exponential generating function for the sequence* $\{a_{\overrightarrow{n}}\}_{\overrightarrow{n} \in \mathbb{N}^k}$ *are given by*

$$A(\overrightarrow{x}) = \sum_{\overrightarrow{n} \in \mathbb{N}^k} a_{\overrightarrow{n}} \overrightarrow{x}^{\overrightarrow{n}} \qquad and \qquad E(\overrightarrow{x}) = \sum_{\overrightarrow{n} \in \mathbb{N}^k} a_{\overrightarrow{n}} \frac{\overrightarrow{x}^{\overrightarrow{n}}}{\overrightarrow{n}!},$$

respectively, where $\overrightarrow{x}^{\overrightarrow{n}} = \prod_{j=1}^k x_j^{n_j}$ *and* $\overrightarrow{n}! = \prod_{j=1}^k n_j!$. *Two formal power series* $A(\overrightarrow{x})$ *and* $B(\overrightarrow{x})$ *are equal if* $a_{\overrightarrow{n}} = b_{\overrightarrow{n}}$ *for all* $\overrightarrow{n} \in \mathbb{N}^k$ *and we write* $A(\overrightarrow{x}) = B(\overrightarrow{x})$. *The set of formal power series or generating functions in* $\overrightarrow{x} = (x_1, \ldots, x_k)$ *is denoted by* $\mathbb{L}[[\overrightarrow{x}]]$. *On the set* $\mathbb{L}[[\overrightarrow{x}]]$ *we define addition and subtraction of two power series by*

$$A(\overrightarrow{x}) \pm B(\overrightarrow{x}) = \sum_{\overrightarrow{n} \in \mathbb{N}^k} (a_{\overrightarrow{n}} \pm b_{\overrightarrow{n}}) \overrightarrow{x}^{\overrightarrow{n}}.$$

Multiplication of two power series is defined using the Cauchy product rule

$$C(\overrightarrow{x}) = A(\overrightarrow{x}) \cdot B(\overrightarrow{x}) = \sum_{\overrightarrow{n} \in \mathbb{N}^k} \sum_{\overrightarrow{m}' + \overrightarrow{m}'' = \overrightarrow{n}} a_{\overrightarrow{m}'} b_{\overrightarrow{m}''} \overrightarrow{x}^{\overrightarrow{n}},$$

where $\overrightarrow{m}', \overrightarrow{m}'' \in \mathbb{N}^k$ *and* $\overrightarrow{m}' + \overrightarrow{m}'' = (m_1' + m_1'', \ldots, m_k' + m_k'')$. *The function* C *is called the* convolution *of* A *and* B. *For generating functions in one variable this reduces to*

$$\sum_{n \geq 0} a_n x^n \sum_{n \geq 0} b_n x^n = \sum_{n \geq 0} \sum_{j=0}^n a_j b_{n-j} x^n.$$

It can be shown that the set $\mathbb{L}[[\overrightarrow{x}]]$ of all formal power series in $\overrightarrow{x}$ over $\mathbb{L}$ is a ring with the addition and multiplication as defined above. The multiplicative operation might seem complicated at first, but it turns out to be exactly what we need. Often, when counting combinatorial objects, we will break those of size n into two parts of size k and $n - k$, where the two parts may have different characteristics or types, and then sum over all possible values of k. The resulting sum has the form of the Cauchy product, and the generating function for the sequence in question can be obtained by multiplying the generating functions for the individual sequences that count the objects of the two different types. Here is an example of this approach.

Example 2.35 *To obtain the generating function for the number of compositions of n with parts in $\mathbb{N}$, we can count the compositions according to the value with which they start, as in Example 2.9. If the composition starts with* k, $k = 1, 2, \ldots, n$, *then it is followed by any composition of $n - k$. Let C_n denote the number of compositions of n with parts in $\mathbb{N}$. Then we get that*

$$C_n = \sum_{k=1}^n C_{n-k} = \sum_{k=1}^n 1 \cdot C_{n-k} \qquad for \ n \geq 1,$$

that is, we obtain the number of compositions as a convolution of the sequence $\{b_n\}_{n\geq 1}$ *with* $b_n = 1$, *and the sequence* $\{C_n\}_{n\geq 0}$. *The sequence* $\{b_n\}_{n\geq 1}$ *has generating function* $B(x) = \frac{x}{1-x}$ *(why?), and therefore,*

$$C(x) - 1 = \frac{x}{1-x}C(x) \quad \Rightarrow \quad C(x) = \frac{1-x}{1-2x}.$$

(It is important not to forget to adjust for the term C_0 *on the left-hand side since the recurrence only holds for* $n \geq 1$.)

We now provide an alternative definition of the generating function which is useful for certain questions.

Definition 2.36 *Let* $A_{\overrightarrow{n}} = A_{n_1,n_2,\ldots,n_k}$ *be a finite set parameterized by* n_1, n_2, $\ldots$, n_k *and let* A *denote the disjoint union of the* $A_{\overrightarrow{n}}$. *We say that* $\sigma \in A_{\overrightarrow{n}}$ *has order* $\overrightarrow{n}$ *and write* $\mathrm{ord}(\sigma) = \overrightarrow{n}$. *For* $\sigma \in A$, *we define a statistic* $\overrightarrow{s}_\sigma : A \to (\mathbb{N}_0)^k$. *Then the ordinary and exponential generating function for the number of elements of* A *according to the order of* σ *and the statistic* $\overrightarrow{s}_\sigma$ *are given by*

$$A(\overrightarrow{x}, \overrightarrow{y}) = \sum_{\sigma \in A} \overrightarrow{x}^{\,\mathrm{ord}(\sigma)} \overrightarrow{y}^{\,\overrightarrow{s}_\sigma} \quad and \quad E(\overrightarrow{x}, \overrightarrow{y}) = \sum_{\sigma \in A} \frac{\overrightarrow{x}^{\,\mathrm{ord}(\sigma)}}{\mathrm{ord}(\sigma)!} \overrightarrow{y}^{\,\overrightarrow{s}_\sigma},$$

respectively.

Example 2.37 *Say we want to find an explicit formula for the number of compositions in* $\{1,2\}$ *of order* n *with* m_1 *ones and* m_2 *parts. Then, the statistic of interest is* $\overrightarrow{s}_\sigma = (\mathrm{ones}(\sigma), \mathrm{par}(\sigma))$, *where* $\mathrm{ones}(\sigma)$ *is the number of 1s in* σ. *We let* A *denote the set of all compositions with parts in* $\{1,2\}$. *Then the generating function is given by*

$$A(x,y) = \sum_{\sigma \in A} x^{\mathrm{ord}(\sigma)} \overrightarrow{y}^{\,\overrightarrow{s}_\sigma}$$

$$= \sum_{n \geq 0} x^n \sum_{m_1, m_2 \geq 0} y_1^{m_1} y_2^{m_2} \left| \{\sigma \in A \mid \mathrm{ord}(\sigma) = n, \overrightarrow{s}_\sigma = (m_1, m_2)\} \right|.$$

We now need to derive an expression for the generating function. Since any composition $\sigma \in A$ *is either empty or starts with a 1 or a 2, it is not too hard to see that* $A(x,y) = 1 + (xy_1 y_2 + x^2 y_2)A(x,y)$, *which is equivalent to*

$$A(x,y) = \frac{1}{1 - xy_2(y_1 + x)} = \sum_{m_2 \geq 0} y_2^{m_2} x^{m_2} (y_1 + x)^{m_2}$$

$$= \sum_{m_2 \geq 0} \sum_{m_1 = 0}^{m_2} \binom{m_2}{m_1} y_1^{m_1} y_2^{m_2} x^{2m_2 - m_1}.$$

This implies that the number of compositions in $\{1,2\}$ of order n with m_1 ones and $m_2 = (n + m_1)/2$ parts is given by $\binom{m_2}{m_1}$, which can be explained combinatorially by choosing m_1 parts to be the ones among the m_2 parts in the composition.

Since we often want to refer to the coefficient of x^n, or more generally, of $\overrightarrow{x}^{\overrightarrow{n}}$, we define this useful notation.

Definition 2.38 *Let $A(x)$ be a generating function. Then $[x^n]A(x)$ denotes the coefficient of x^n in $A(x)$, and more generally, $[\overrightarrow{x}^{\overrightarrow{n}}]$ denotes the coefficient of $\overrightarrow{x}^{\overrightarrow{n}}$ in $A(\overrightarrow{x})$.*

Example 2.39 *Using this notation for the generating function of Example 2.37, we obtain*

$$[x^n y_1^{m_1} y_2^{m_2}]A(x,y) = \begin{cases} \binom{m_2}{m_1} & \text{if } n = 2m_2 - m_1 \\ 0 & \text{if } n \neq 2m_2 - m_1 \end{cases}.$$

Definition 2.40 *A generating function $A(\overrightarrow{x}) \in \mathbb{L}[[x]]$ is called* rational *if there exist two polynomials $p(\overrightarrow{x}), q(\overrightarrow{x}) \in \mathbb{L}[\overrightarrow{x}]$ such that $A(\overrightarrow{x}) = \frac{p(\overrightarrow{x})}{q(\overrightarrow{x})}$. It is* algebraic *if there exists a nontrivial polynomial $R \in \mathbb{L}[\overrightarrow{x}]$ such that $A(\overrightarrow{x})$ satisfies a polynomial equation $R(A; x_1, \ldots, x_k) = 0$.*

Note that every rational generating function is algebraic.

Example 2.41 *The generating function $F(x) = \frac{x}{1-x-x^2}$ for the Fibonacci sequence is rational, whereas the generating function $f(x)$ satisfying the polynomial equation $f(x) = 1 + x\, f(x)^3$ is algebraic, but not rational.*

We can also define two other operations on the ring $\mathbb{L}[[\overrightarrow{x}]]$ which will assist us greatly in manipulating generating functions.

Definition 2.42 *The* derivative *and the* integral *of $A(\overrightarrow{x})$ with respect to x_i are defined by*

$$\frac{\partial}{\partial x_i} A(\overrightarrow{x}) = \sum_{\overrightarrow{n} \in \mathbb{N}^k} a_{\overrightarrow{n}} n_i x_i^{n_i - 1} \prod_{j \neq i} x_j^{n_j}$$

and

$$\int A(\overrightarrow{x}) dx_i = \sum_{\overrightarrow{n} \in \mathbb{N}^k} a_{\overrightarrow{n}} \frac{1}{n_i + 1} x_i^{n_i + 1} \prod_{j \neq i} x_j^{n_j},$$

respectively.

Example 2.43 *The generating function for the sequence* $\{n \cdot m\}_{n,m \geq 0}$ *is given by*

$$\sum_{n,m \geq 0} n\, m\, x^n y^m = xy \sum_{n,m \geq 0} n\, m\, x^{n-1} y^{m-1} = xy \frac{\partial^2}{\partial x \partial y} \sum_{n,m \geq 0} x^n y^m$$

$$= xy \frac{\partial^2}{\partial x \partial y} \left(\frac{1}{(1-x)(1-y)} \right) = \frac{xy}{(1-x)^2 (1-y)^2}.$$

Now that we are on solid ground with regard to the allowed operations on formal power series, we will look at rules that let us easily compute the generating functions for new sequences from known generating functions. These rules will allow us to go directly from the recurrence relation to a functional equation for the generating function without having to apply the definition of the generating function. In addition, those rules will make it easy (with some practice) to do the opposite, namely read off the coefficients of the sequence from the generating function by expressing it in terms of known generating functions (see Example 2.46).

Definition 2.44 *We use* $A \overset{ops}{\leftrightarrow} \{a_n\}_{n \geq 0}$ *to indicate that* $A(x) = \sum_{n \geq 0} a_n x^n$ *is the ordinary power series (ops) generating function of the sequence* $\{a_n\}_{n \geq 0}$.

Let us see how we can get the generating function of a new sequence from the generating function of a related series. For example, if $A \overset{ops}{\leftrightarrow} \{a_n\}_{n \geq 0}$, what is the generating function of the sequence $\{a_{n+1}\}_{n \geq 0}$, that is, the sequence that is shifted over by one? We use the definition of the generating function to obtain

$$\sum_{n \geq 0} a_{n+1} x^n = \frac{1}{x} \sum_{n \geq 0} a_{n+1} x^{n+1} = \frac{A(x) - a_0}{x}.$$

This tells us that

$$A \overset{ops}{\leftrightarrow} \{a_n\}_{n \geq 0} \Rightarrow \frac{A - a_0}{x} \overset{ops}{\leftrightarrow} \{a_{n+1}\}_{n \geq 0}.$$

Applying this rule repeatedly or using the definition of the generating function, we obtain a general rule for computing the generating function of a shifted sequence from the generating function of the original sequence.

Rule 2.45 *[193, Rule 1, Chapter 2] If* $A \overset{ops}{\leftrightarrow} \{a_n\}_{n \geq 0}$, *then for any integer* $k > 0$,

$$\frac{A - a_0 - a_1 x - \ldots - a_{k-1} x^{k-1}}{x^k} \overset{ops}{\leftrightarrow} \{a_{n+k}\}_{n \geq 0}.$$

Proof   See Exercise 2.11.   □

Example 2.46 (*Continuation of Example 2.33*) *We have derived the generating function for the sequence* $\{a_n\}_{n\geq 0}$ *given by*

$$a_n = a_{n-1} + a_{n-2} + 1, \qquad a_0 = 1, a_1 = 2$$

and obtained

$$A(x) = \frac{1}{(1-x)(1-x-x^2)}.$$

We will now use knowledge of generating functions and their properties to easily read off $[x^n]A(x)$. *First we notice that the generating function is a product of the geometric series with a function that is almost the generating function of the Fibonacci sequence. Seeing a product should put us on the alert for a convolution. Next we bring the second function to a convenient form by multiplying with* x/x. *The resulting factor of* $\frac{1}{x}$ *should be translated into: "Aha! A shifted sequence," due to Rule 2.45. We need now only check whether the zero-th term of the sequence is equal to zero which is true for the Fibonacci sequence. All together we have that*

$$A(x) = \frac{1}{(1-x)(1-x-x^2)} = \frac{1}{(1-x)}\frac{1}{x}\frac{x}{(1-x-x^2)} = \frac{1}{1-x}\frac{F(x)-F_0}{x}.$$

Therefore, the sequence $\{a_n\}_{n\geq 0}$ *is a convolution of the sequence* $\{1\}_{n\geq 0}$ *and the shifted Fibonacci sequence* $\{F_{n+1}\}_{n\geq 0}$, *that is*

$$a_n = \sum_{i=0}^{n} 1 \cdot F_{n+1-i} = \sum_{i=1}^{n+1} F_i = F_{n+3} - 1,$$

where the last equality follows from a well-known identity (see Exercise 2.7). Thus, we have derived once more the result of Example 2.27.

Another very common modification of a sequence is multiplication by n, or in general, by a polynomial in n. Let's start again with the simplest case and determine the generating function of $\{n\,a_n\}_{n\geq 0}$ from the generating function A of $\{a_n\}_{n\geq 0}$. Again, we use the definition of the generating function and use algebraic transformations to express it in terms of $A(x)$:

$$\sum_{n\geq 0} n\,a_n x^n = x \sum_{n\geq 0} n\,a_n x^{n-1} = x\,A'(x).$$

Thus, multiplying a sequence by n results in differentiating the generating function and then multiplying by x.

Definition 2.47 *We denote the differential operator by D, and write Df to indicate f' for any function f. The second derivative of f will be denoted by D^2f, and in general, D^kf denotes the k-th derivative of f.*

Therefore, $xDA \overset{ops}{\leftrightarrow} \{n\,a_n\}_{n\geq0}$. Repeatedly applying this fact gives that if $A \overset{ops}{\leftrightarrow} \{a_n\}_{n\geq0}$, then $(xD)^kA \overset{ops}{\leftrightarrow} \{n^k\,a_n\}_{n\geq0}$. Together with the linearity of the generating function we obtain this general rule.

Rule 2.48 *[193, Rule 2, Chapter 2] Let $P(n)$ be a polynomial in n. Then*

$$A \overset{ops}{\leftrightarrow} \{a_n\}_{n\geq0} \Rightarrow P(xD)A \overset{ops}{\leftrightarrow} \{P(n)\cdot a_n\}_{n\geq0}\,.$$

We illustrate this rule with an example.

Example 2.49 *What is the generating function of $\{3n + n^2\}_{n\geq0}$? Looking at this sequence through the generating function lens, we identify this as the sequence $\{1\}$, multiplied by the polynomial $P(n) = 3n + n^2$. Thus,*

$$3(xD)\frac{1}{1-x} + (xD)^2\frac{1}{1-x} \overset{ops}{\leftrightarrow} \{3n + n^2\}_{n\geq0}\,,$$

and we compute the generating function A of the sequence $3n + n^2$ as follows:

$$A(x) = 3(xD)\frac{1}{1-x} + (xD)^2\frac{1}{1-x} = 3x\left(\frac{1}{1-x}\right)' + (xD)\left(x\left(\frac{1}{1-x}\right)'\right)$$

$$= \frac{3x}{(1-x)^2} + x\left(\frac{x}{(1-x)^2}\right)' = \frac{3x}{(1-x)^2} + \frac{x(1+x)}{(1-x)^3} = \frac{2x(2-x)}{(1-x)^3}\,.$$

Example 2.50 *(Dyck words and Catalan numbers) A Dyck word of length $2n$ is a word on the alphabet $\{1,2\}$ consisting of n 1s and n 2s such that no initial segment of the word has more 2s than 1s. For example, the Dyck words of length six are*

$$111222, \quad 112122, \quad 112212, \quad 121122, \quad and \quad 121212.$$

Dyck words of length $2n$ can be used to encode valid arrangements of parentheses in a mathematical expression, with a 1 representing a left parenthesis and a 2 representing a right parenthesis. The above Dyck words correspond to the following parentheses arrangements:

$$((())), \quad (()()), \quad (())(), \quad ()(()), \quad and \quad ()()().$$

We now want to count the number of Dyck words using generating functions. To do so, we need to set up a recurrence relation of some form. Similar to how we proceeded in Example 2.35, we split up the Dyck word into smaller Dyck words as follows: For each position j, compute the difference between the number of 1s and the number of 2s in the first j positions. Since the word is a Dyck word, $j \geq 0$. We condition on the first position j where there is an equal number of 1s and 2s. Then we can write the Dyck word w of length $2n$ as $w = 1\,w'\,2\,w''$, where the 2 is at position $j = 2k + 2$. (Note that w' and w''

may be empty words.) By construction, w' is a Dyck word of length $2k$, and since the number of $1s$ and $2s$ balance at position $2k+2$, the word w'' is also a Dyck word, and it has length $2(n-k-1)$. This representation is unique, and we can therefore count the Dyck words according to this decomposition to get the following recurrence relation, where d_n counts the Dyck words of length $2n$:

$$d_n = \sum_{k=0}^{n-1} d_k\, d_{n-k-1} \quad for \ \ n \geq 1.$$

Adjusting for the fact that the convolution sum has upper index $n-1$ instead of n we get

$$D(x) - 1 = xD(x)^2 \quad \Rightarrow \quad D(x) = \frac{1 - \sqrt{1-4x}}{2x}.$$

To extract the coefficients $[x^n]D(x)$, we make use of the following well-known identity (see Equation (A.2))

$$\sqrt{1+t} = \sum_{m\geq 0} \binom{1/2}{m} t^m = 1 + \sum_{m\geq 1}(-1)^{m-1}\frac{(2m-2)!}{2^{2m-1}m!(m-1)!}t^m.$$

Thus,

$$D(x) = -\frac{1}{2x}\sum_{m\geq 1}(-1)^{m-1}\frac{(2m-2)!}{2^{2m-1}m!(m-1)!}(-4x)^m$$

$$= \sum_{m\geq 1}\frac{1}{m}\binom{2m-2}{m-1}x^{m-1} = \sum_{m\geq 0}\frac{1}{m+1}\binom{2m}{m}x^m. \qquad (2.10)$$

This implies that d_m is given by the m-th Catalan number $\frac{1}{m+1}\binom{2m}{m}$. The Catalan sequence[7] $1, 1, 2, 5, 14, 132, 429, 1430, 4862, 16796, \ldots$ which occurs as sequence A000108 in [180] counts an enormous number of different combinatorial structures (see Stanley [182, Page 219 and Exercise 6.19]). It was first described in the 18th century by Leonard Euler[8], who was interested in the number of different ways of dividing a polygon into triangles. The sequence is named after Eugène Charles Catalan[9], who also worked on the problem and discovered the connection to parenthesized expressions.

    Any Dyck word can be visualized by a *Dyck path*, a lattice path in the plane integer lattice $\mathbb{Z} \times \mathbb{Z}$ consisting of up-steps $(1,1)$ and down-steps $(1,-1)$ which never passes below the x-axis. From the definition of Dyck words it is obvious that a Dyck word on $\{1,2\}$ of length $2n$ can be represented as a Dyck path where a 1 represents an up-step and a 2 represents a down-step. For example, the Dyck path for the Dyck word 112122 is given in Figure 2.1.

---

[7]http://mathworld.wolfram.com/CatalanNumber.html
[8]http://turnbull.mcs.st-and.ac.uk/ history/Mathematicians/Euler.html
[9]http://turnbull.mcs.st-and.ac.uk/ history/Mathematicians/Catalan.html

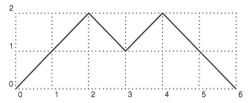

FIGURE 2.1: Dyck path for the Dyck word 112122.

We next state two rules that are an immediate result of the way power series are multiplied.

Rule 2.51 *If* $A_i \overset{ops}{\leftrightarrow} \left\{a_n^{(i)}\right\}_{n \geq 0}$ *for* $i = 1, \ldots, m$ *and* $A \overset{ops}{\leftrightarrow} \{a_n\}_{n \geq 0}$, *then*

$$A_1 A_2 \cdots A_m \overset{ops}{\leftrightarrow} \left\{ \sum_{i_1 + i_2 + \cdots + i_m = n} a_{i_1}^{(1)} a_{i_2}^{(2)} \cdots a_{i_m}^{(m)} \right\}_{n \geq 0},$$

and

$$A^m \overset{ops}{\leftrightarrow} \left\{ \sum_{i_1 + i_2 + \cdots + i_m = n} a_{i_1} a_{i_2} \cdots a_{i_m} \right\}_{n \geq 0}.$$

Finally, we look at the generating function of the partial sum of a sequence.

Rule 2.52 *[193, Rule 5, Chapter 2] If* $A \overset{ops}{\leftrightarrow} \{a_n\}_{n \geq 0}$, *then*

$$\frac{A}{1 - x} \overset{ops}{\leftrightarrow} \left\{ \sum_{i=0}^{n} a_i \right\}_{n \geq 0}.$$

We illustrate this rule by deriving the generating function for the sequence of harmonic numbers.

Definition 2.53 *The* n-*th harmonic number is defined by*

$$H_n = 1 + \frac{1}{2} + \frac{1}{3} + \cdots + \frac{1}{n} \quad \text{for } n \geq 1.$$

Example 2.54 *Let* $H \overset{ops}{\leftrightarrow} \{H_n\}_{n \geq 1}$. *Then by Rule 2.52,* $H = \frac{A}{1-x}$, *where* $A \overset{ops}{\leftrightarrow} \left\{\frac{1}{n}\right\}_{n \geq 1}$ *which means that we have to find* $A(x) = \sum_{n \geq 1} \frac{x^n}{n}$. *Since*

$$(DA)(x) = \sum_{n \geq 1} x^{n-1} = \frac{1}{1 - x},$$

we have that $A(x) = \int \frac{1}{1-x} dx = -\log(1-x)$, and therefore,

$$\frac{1}{1-x} \log \frac{1}{1-x} \overset{ops}{\leftrightarrow} \{H_n\}_{n\geq 1}.$$

Similar rules can be derived for exponential generating functions.

Definition 2.55 *We use $E \overset{egf}{\leftrightarrow} \{a_n\}_{n\geq 0}$ to indicate that $E(x) = \sum_{n\geq 0} a_n \frac{x^n}{n!}$ is the exponential generating function (egf) of the sequence $\{a_n\}_{n\geq 0}$.*

We start by determining the exponential generating function for a shifted sequence.

Rule 2.56 *[193, Rule 1', Chapter 2] If $A \overset{egf}{\leftrightarrow} \{a_n\}_{n\geq 0}$, then for any integer $k > 0$,*

$$D^k A \overset{egf}{\leftrightarrow} \{a_{n+k}\}_{n\geq 0}.$$

Proof   See Exercise 2.11.   □

So in this instance, the rule for the exponential generating function of a shifted sequence is simpler than the corresponding one for the ordinary generating function. If we look at the impact of multiplying a sequence with a polynomial, we obtain exactly the same result as in the case of ordinary generating functions.

Rule 2.57 *[193, Rule 2', Chapter 2] Let $P(n)$ be a polynomial in n and $E \overset{egf}{\leftrightarrow} \{a_n\}_{n\geq 0}$. Then*

$$P(xD)E \overset{ops}{\leftrightarrow} \{P(n) \cdot a_n\}_{n\geq 0}.$$

When multiplying exponential generating functions, the answer is still a convolution, but with a nice twist that results in a very useful formula.

Rule 2.58 *[193, Rule3', Chapter 2] If $E \overset{ops}{\leftrightarrow} \{a_n\}_{n\geq 0}$ and $\tilde{E} \overset{ops}{\leftrightarrow} \{b_n\}_{n\geq 0}$, then*

$$E \cdot \tilde{E} \overset{egf}{\leftrightarrow} \left\{ \sum_{r=0}^{n} \binom{n}{r} a_r b_{n-r} \right\}_{n\geq 0},$$

and more generally,

$$E^k \overset{egf}{\leftrightarrow} \left\{ \sum_{r_1+\cdots+r_k=n} \frac{n!}{r_1! r_2! \cdots r_k!} a_{r-1} a_{r-2} \cdots a_{r-k} \right\}_{n\geq 0}.$$

Proof We have

$$E(x) \cdot \tilde{E}(x) = \left\{ \sum_{i=0}^{\infty} \frac{a_i x^i}{i!} \right\} \left\{ \sum_{j=0}^{\infty} \frac{b_j x^j}{j!} \right\} = \sum_{i,j=0}^{\infty} \frac{a_i b_j x^{i+j}}{i! j!}$$

$$= \sum_{n \geq 0} x^n \left\{ \sum_{i+j=n} \frac{a_i b_j}{i! j!} \right\}.$$

Thus, the coefficient of $x^n/n!$ is given by

$$\left[\frac{x^n}{n!} \right] E \cdot \tilde{E} = \sum_{i+j=n}^{\infty} \frac{n! a_i b_j}{i! j!} = \sum_{r=0}^{n} \binom{n}{r} a_r b_{n-r}.$$

$\square$

We illustrate this rule with an example involving set partitions.

Definition 2.59 *A partition* $\{S_1, S_2, \ldots, S_k\}$ *of a set S is a collection of non-empty sets such that* $S_i \cap S_j = \emptyset$ *for* $i \neq j$ *and* $\bigcup_{i=1}^{k} S_i = S$. *The number of partitions of a set S with n elements is enumerated by the* Bell numbers[10] *(named in honor of Eric Temple Bell[11]). For* $n = 0, \ldots, 10$, *the values of* B_n, *the n-th Bell number, are* 1, 1, 2, 5, 15, 52, 203, 877, 4140, 21147, *and* 115975 *(sequence A000110 in [180]).*

Example 2.60 *We compute the exponential generating function for the Bell numbers. It can be shown (see Exercise 2.16) that the Bell numbers satisfy the recurrence* $B_{n+1} = \sum_{k=0}^{n} \binom{n}{k} B_k$ *for* $n \geq 0$, *with* $B_0 = 1$. *Looking at the right-hand side of the equation, we see that this represents the series obtained by multiplying the exponential generating functions of the Bell numbers and the sequence of all 1s. If we denote the generating function for the Bell numbers by* $B(x)$, *then the exponential generating function of the left-hand side is given by* $(DB)(x)$. *All together,*

$$B' = e^x B,$$

which implies that $B(x) = C \exp(e^x)$ *for some constant* C. *Since* $B_0 = B(0) = 1$, *we have that* $C = e^{-1}$, *and therefore,*

$$B(x) = \exp(e^x - 1).$$

We now return to the generating function for the Catalan sequence derived in Example 2.50. It has a special property, namely that it can expressed as a *continued fraction*. As the name suggests, continued fractions are expressions

---

[10] http://mathworld.wolfram.com/BellNumber.html
[11] http://www-groups.dcs.st-and.ac.uk/ history/Mathematicians/Bell.html

in which the denominator has a term which is a fraction, which in turn has another term that is a fraction, etc. Early traces of continued fractions appear as far back as 306 B.C. Other records have been found that show that the Indian mathematician Aryabhata (475-550) used a continued fraction to solve a linear equation [114, pp. 28-32]. However, he did not develop a general method. Continued fractions were only used in specific examples for more than $1,000$ years, and the use of continued fractions was limited to the area of number theory. Since the beginning of the twentieth century, continued fractions have become more common in various other areas, for example, enumerative combinatorics. The first connection between the pattern avoidance problem (which will be discussed in Chapter 4) and continued fractions was discovered by Robertson, Wilf, and Zeilberger [174], who found the ordinary generating function for the number of 132-avoiding permutations in $\mathcal{S}_n$ with a prescribed number of occurrences of the pattern 123. Recently, connections between generating functions for the enumeration of restricted permutations or words and continued fractions have become more numerous.

Definition 2.61 *A continued fraction is an expression of the form*

$$a_0 + \cfrac{b_1}{a_1 + \cfrac{b_2}{a_2 + \cfrac{b_3}{a_3 + \cdots}}},$$

where the letters $a_0, a_1, b_1, a_2, b_2, \ldots$ denote independent variables, and may be interpreted as real or complex numbers, functions, etc. If $b_i = 1$ for all i, then the expression is called a simple continued fraction. *If the expression contains finitely many terms, then it is called a* finite continued fraction; *otherwise, it is called an* infinite continued fraction. *The numbers a_i are called the* partial quotients. *If the expression is truncated after k partial quotients, then the value of the resulting expression is called the k-th* convergent *of the continued fraction and we will refer to the continued fraction as having k* levels. *Due to the complexity of the expression above, mathematicians have adopted several more convenient notations for simple continued fractions, the most common being $[a_0, a_1, \ldots]$ for the infinite simple continued fraction and $[a_0, a_1, \ldots, a_k]$ for the finite simple continued fraction.*

We will now show that the generating function of the Catalan numbers can be expressed as a continued fraction.

Example 2.62 *Let*

$$f(x) = \cfrac{1}{1 - \cfrac{x}{1 - \cfrac{x}{1 - \cfrac{x}{1 - \cdots}}}},$$

Then it is easy to see that $f(x)$ satisfies $f(x) = \frac{1}{1-x\,f(x)}$, or equivalently, $f(x) = 1 + x\,f^2(x)$, *which is the functional equation for the generating function of the Catalan numbers. This phenomenon is not restricted to the Catalan numbers, in fact, any generating function that has a functional equation of the form*

$$A(x) = a(x) + b(x)\,A(x) + c(x)\,A(x)^2 \qquad (2.11)$$

can be rewritten as

$$A(x) = \cfrac{a(x)}{1 - b(x) - c(x)\,A(x)} = \cfrac{a(x)}{1 - b(x) - \cfrac{c(x)\,a(x)}{1 - b(x) - c(x)\,A(x)}}$$

$$= \cfrac{a(x)}{1 - b(x) - \cfrac{c(x)\,a(x)}{1 - b(x) - \cfrac{c(x)\,a(x)}{1 - b(x) - \cfrac{c(x)\,a(x)}{\cdots}}}}.$$

Finally we introduce the *Lagrange Inversion Formula*, a strong tool for solving certain kinds of functional equations. At its best it can give explicit formulas where other approaches fail. It is also useful for asymptotic analysis (in which case we have to leave the world of formal power series and deal with issues of convergence). The Lagrange Inversion Formula applies to functional equations of the form $u = x\,\phi(u)$, where $\phi(u)$ is a given function of u, when the goal is to determine u as a function of x. More specifically, suppose we are given the power series expansion of the function $\phi = \phi(u)$, convergent in some neighborhood of the origin in the u–ϕ plane. How can we find the power series expansion of the solution of $u = x\,\phi(u)$, say $u = u(x)$, in some neighborhood of the origin in the x–u plane? The answer is surprisingly explicit, and it even allows us to find the expansion of $f(u(x))$ for some function f.

Theorem 2.63 *[193, Theorem 5.1.1] (Lagrange Inversion Formula) Let $f(u)$ and $\phi(u)$ be formal power series in u, with $\phi(0) = 1$. Then there is a unique formal power series $u = u(x)$ that satisfies $u = x\phi(u)$. Furthermore, the value $f(u(x))$ of f at that root $u = u(x)$, when expanded in a power series in x about $x = 0$, satisfies*

$$[x^n]f(u(x)) = \frac{1}{n}[u^{n-1}](f'(u)\phi^n(u)).$$

Note that the function f in the Lagrange Inversion Formula (LIF) allows us to compute with equal ease the coefficients of the unknown function $u(x)$ (in which case $f(u) = u$), or any function of $u(x)$, for example the k-th power (in which case $f(u) = u^k$). Oftentimes, we are just interested in the coefficients of the unknown function itself, and the function f does not really show up at all in the computations. Let's see an example of the LIF in action: we will derive once more that Dyck paths of length $2n$ are counted by the Catalan numbers, this time with the help of the Lagrange Inversion Formula.

Example 2.64 *In Example 2.50 we have derived that the generating function $C(x)$ for the number of Dyck paths of length $2n$ satisfies the functional equation $C(x) = 1 + x\,C^2(x)$. If we define $u(x) = C(x) - 1$, then this is indeed of the form $u = x\,\phi(u)$ with $u(x) = C(x) - 1$, $\phi(u) = (1 + u)^2$ and $\phi(0) = 1$. We choose $f(u) = u$, and apply Theorem 2.63 to obtain*

$$[x^n]C(x) = \frac{1}{n}[u^{n-1}](1 + u)^{2n} = \frac{1}{n}\binom{2n}{n-1} = \frac{1}{n+1}\binom{2n}{n}.$$

Here we have used the binomial expansion for the term $(1 + u)^{2n}$, and then determined the index i such that the power of u is $n - 1$, followed by some algebraic manipulations. We have once more shown that the number of Dyck paths of length $2n$ for $n \geq 1$ is given by the n-th Catalan number $\frac{1}{n+1}\binom{2n}{n}$.

This example shows that the Lagrange Inversion Formula can give results very easily; earlier, we had to work a little harder and use a special identity to obtain the coefficients in Example 2.50.

We close the section with an example where we obtain the generating function not from a recurrence relation, but by breaking the combinatorial object of interest into parts for which we know the generating functions. By deriving the generating function in two different ways, we obtain an interesting (well-known) combinatorial identity. In fact, enumerating a combinatorial object in two different ways can give rise to new identities, or provide combinatorial explanations for known identities. Nice examples of this technique can be found in [22, 193].

Example 2.65 (*Partitions*) *Recall that a partition of n is a decreasing sequence of integers that sum to n. We will derive the generating function for the number of partitions in two different ways, which leads to an interesting identity.*

Since the parts of the partition are written in decreasing order, the only thing that matters is which parts are included and how many of each. Each part of size $i > 0$ contributes a factor of x^i to the generating function. If we look at the parts of size i, then there can be any number of them, thus their contribution is given by

$$\sum_{k=0}^{\infty}(x^i)^k = \frac{1}{1 - x^i}.$$

Since the generating function for the number of partitions is the convolution of all these generating functions for parts of size i, we obtain

$$A(x) = \prod_{i=1}^{\infty}\frac{1}{1 - x^i}.$$

On the other hand, if we separate the partitions into different classes, we obtain $A(x)$ as the sum of the respective generating functions. A partition can either be empty, consist of only 1s (and at least one 1), consist of at least one 2 and any number (or no) 1s, consist of at least one 3 and any number (or no) 1s or 2s, etc. If there is at least one k, then the generating function is given by $\frac{x^k}{1-x^k}$, and therefore

$$A(x) = 1 + \frac{x}{1-x} + \frac{x^2}{1-x^2} \cdot \frac{1}{1-x} + \frac{x^3}{1-x^3} \cdot \frac{1}{1-x^2} \cdot \frac{1}{1-x} + \cdots$$

$$= \sum_{i=0}^{\infty} \frac{x^i}{\prod_{j=1}^{i}(1-x^i)}.$$

This yields the identity

$$\prod_{i=1}^{\infty} \frac{1}{1-x^i} = \sum_{i=0}^{\infty} \frac{x^i}{\prod_{j=1}^{i}(1-x^i)}.$$

The fact that there is no easy recursion for partitions gives an indication that partitions are harder to analyze than compositions. There is a large body of research on partitions, and a good reference is [10].

2.4 Exercises

Exercise 2.1 *Find a recurrence relation*

(1) *(of order 2) for the number of compositions of n with parts in $\mathbb{N}$;*

(2) *for the number of words of length n on $[k]$;*

(3) *for the number of permutations of $[n]$;*

(4) *for the reverse conjugate compositions of n.*

Use the iteration method to find explicit formulas for the above recurrence relations.

Exercise 2.2 *Use Maple or* Mathematica *to find the first 15 terms of the sequences*

(1) $a_n = a_{n-1} + 3a_{n-2}$ *with $a_0 = 1$ and $a_1 = 2$.*

(2) $b_n = \frac{n}{n-1}b_{n-1} + 1$ *with $b_1 = 2$.*

Exercise 2.3 *Solve the recurrence relations of Exercise 2.2 using Maple or Mathematica.*

Exercise 2.4 *Determine the number of words of length n on the alphabet $[3]$, and derive the generating function for the number of words of length $n - 1$ on the alphabet $[3]$.*

Exercise 2.5 *Find recurrence relations for the following counting problems:*

(1) *The number of words $w = w_1 w_2 \cdots w_n$ of length n on $\{1, 2\}$ satisfying $w_i \geq w_{i+2}$ for all i.*

(2) *The number of words $w = w_1 w_2 \cdots w_n$ of length n on $\{1, 2, 3\}$ satisfying $w_i \geq w_{i+2}$ for all i.*

Use the iteration method to find explicit formulas for the above recurrence relations.

Exercise 2.6

(1) *Find an explicit formula for the number of Carlitz compositions of n in $\{1, 2\}$.*

(2) *A word is called* Carlitz *if it does not contain two consecutive letters that are the same. For example, 121 and 212 are the only Carlitz words on $\{1, 2\}$ of length three. Find an explicit formula for the number of Carlitz words on $\{1, 2\}$ of length n.*

Exercise 2.7 *Prove the identity $\sum_{i=1}^{n+1} F_i = F_{n+3} - 1$ for $n \geq 0$ (where F_n is the n-th Fibonacci number) by using rules for generating functions.*

Exercise 2.8 *Derive the explicit formula for the Lucas sequence.*

Exercise 2.9 *Prove by induction that the explicit formulas for the Fibonacci and Lucas sequences produce integer values.*

Exercise 2.10 *Prove Theorem 2.22.*

Exercise 2.11 *Prove Rules 2.45 and 2.56 either by induction on k or by using the definition of the generating function.*

Exercise 2.12 *Compute the generating function for the sequence $\{a_n\}_{n \geq 0}$, where $a_0 = 1$ and $a_n = F_{2n}$ for $n \geq 1$.*

Exercise 2.13 *Let $C(m; n)$ denote the number of compositions of n with m parts in $\mathbb{N}$. Prove that for fixed $m \geq 1$, $\sum_{n \geq 0} C(m; n) x^n = \dfrac{x^m}{(1 - x)^m}$.*

Exercise 2.14

(1) *Derive the generating function for the number of compositions of n in $\{1,2\}$ that avoid the substring* 11. *The recurrence relation for this sequence was derived in Example 2.17.*

(2) *The first few values of the sequence are given by 1, 1, 1, 2, 2, 3, 4, 5, 7, 9, 12, 16 and 21. Checking this sequence in the Online Encyclopedia of Integer Sequences [180] produces a match with the Padovan sequence A000931. However, the sequence for the number of compositions is shifted in relation to the Padovan sequence. Use the generating function given for the Padovan sequence and apply Rule 2.45 to derive the generating function for the number of compositions in $\{1,2\}$ that avoid the substring* 11. *Compare your answer to Part* (1).

Exercise 2.15 *Find the recurrence relation and the generating function for the number of compositions of n in $\{1,2\}$ without d consecutive 1s. (Hint: condition on the position of the first 2.)*

Exercise 2.16 *Derive a recurrence relation for the Bell numbers B_n (see Definition 2.59).*

Exercise 2.17 *A* smooth *word is a word in which the difference between any two adjacent letters is either -1, 0 or 1. Find an explicit formula for the generating function for the number of smooth words of length n on the alphabet* [3]. *Use either Maple or* Mathematica *to find an explicit formula for the number of smooth words of length n on the alphabet* [3].

Exercise 2.18 *A* strictly smooth *word is a word in which the difference between any two adjacent letters is either -1 or 1. Find an explicit formula for the generating function for the number of strictly smooth words of length n on the alphabet* [3]. *Use either* Mathematica *or Maple to find an explicit formula for the number of strictly smooth words of length n on the alphabet* [3].

Exercise 2.19 *Find an explicit formula for the generating function for the number of smooth compositions (see Exercise 1.12) of n with parts in* [3].

Exercise 2.20 *Find an explicit formula for the generating function for the number of strictly smooth compositions (see Exercise 1.12) of n with parts in* [3].

Exercise 2.21 *A k-ary tree is a plane tree in which each vertex has either out-degree 0 or k (see Definitions 7.2 and 7.20). A vertex is said to be* internal *if its out-degree is k. Use the LIF to find an explicit formula for the number of k-ary trees on n internal vertices.*

Exercise 2.22 *Go to the On-Line Encyclopedia of Integer Sequences [180] and look up the functional equation for the large Schröder numbers (sequence A006318). Verify that it has the form of (2.11) and derive its continued fraction form.*

Exercise 2.23* *Let* $f_{m,l}(x,y) = \left(\dfrac{(1-x-y)^m}{(1-2x-(1-x)y)^{m+1}} \right)$. *Show that*

$$[y^\ell] f_{m,l}(x,y) = \frac{(1-x)^{\ell-m}}{(1-2x)^{\ell+1}} \sum_{i \geq 0} \binom{\ell+i}{i} \binom{m}{i} \frac{x^{2i}}{(1-2x)^i}.$$

Hint: $[y^l] f_{m,l}(x,y) = \dfrac{(1-x)^l}{(1-2x)^l} [y^l] f_{m,l}\left(x, \tfrac{(1-2x)}{(1-x)} y\right).$

Chapter 3

Compositions

In this section we will derive basic results on compositions, and also look at some variations of compositions, by restricting either the placement of the parts within the composition or the set from which the parts are taken. Examples of restriction in terms of placement of the parts are Carlitz compositions and palindromic compositions. In the area of restricted sets, we will present the results of Alladi and Hoggatt [7] and Grimaldi [76], who studied compositions in $\{1,2\}$ and compositions with odd parts. We will then derive results for a general set A which includes their results as well as results on the sets $A = \{1,k\}$ [50], $A = \mathbb{N}\backslash\{1\}$ [77], $A = \mathbb{N}\backslash\{2\}$ [52], and more generally, $A = \mathbb{N}\backslash\{k\}$ [51] as special cases.

To provide alternative viewpoints which often simplify proofs, we will give two additional graphical representations of compositions, namely bar graphs and tilings. Both bar graphs and tilings are combinatorial objects studied in their own right (see for example [26] for bar graphs and [29, 44, 66, 81, 82, 84, 85, 86, 107, 106, 151, 152, 153, 167] for a small sampling of tiling related articles). Framing questions about compositions in terms of bar graphs or tilings has the potential to lead to new questions and insights. We conclude with colored compositions, especially the n-color compositions introduced by Agarwal [3], and a variation of compositions that arises from the bar graph representation of a composition.

3.1 Definitions and basic results (one variable)

In the examples given in Chapter 2, we looked at compositions in $\mathbb{N}$ and also at compositions in restricted sets, such as $A = \{1,2\}$ or the set of odd integers. We will use the following notation to refer to compositions with parts in a set A.

Definition 3.1 *Let $A = \{a_1, a_2, \ldots\} \subseteq \mathbb{N}$ be an ordered set. We denote the number of compositions of n with parts in A (respectively with m parts in A) by $C_A(n)$ (respectively $C_A(m; n)$). The corresponding generating functions are* *given by*

$$C_A(x) = \sum_{n \geq 0} C_A(n)x^n \quad and \quad C_A(m; x) = \sum_{n \geq 0} C_A(m; n)x^n,$$

with $C_A(0) = C_A(0; 0) = 1$, $C_A(m; 0) = 0$ for $m \geq 1$, and $C_A(m; n) = 0$ for $n < 0$. We will omit the reference to A if $A = \mathbb{N}$, that is, $C(n)$ denotes the number of compositions in $\mathbb{N}$.

Note that

$$\begin{aligned} C_A(x) &= \sum_{n \geq 0} \left(\sum_{m \geq 0} C_A(m; n) \right) x^n = \sum_{m \geq 0} \sum_{n \geq 0} C_A(m; n)x^n \\ &= \sum_{m \geq 0} C_A(m; x). \end{aligned} \tag{3.1}$$

We start by introducing a visualization of a composition as a bar graph that will be useful for deriving recurrence relations for the generating functions.

Definition 3.2 *A bar graph of a composition $\sigma = \sigma_1\sigma_2 \cdots \sigma_m$ is a sequence of columns composed of cells, where column j has σ_j cells.*

Figure 3.1 shows the bar graphs corresponding to the compositions of 4.

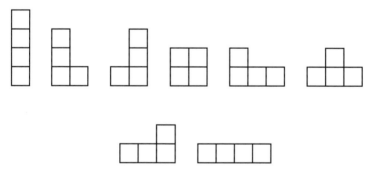

FIGURE 3.1:   Bar graphs associated with compositions of 4.

Using the bar graph of a composition to set up a recurrence relation, we give a different proof of Theorem 1.3. In addition, we derive the respective generating functions.

Theorem 3.3 *For $1 \leq m \leq n$, $C(n) = 2^{n-1}$ and $C(m; n) = \binom{n-1}{m-1}$. The respective generating functions are $C(x) = \dfrac{1 - x}{1 - 2x}$ and $C(m; x) = \dfrac{x^m}{(1 - x)^m}$.*

Proof We give two different proofs to illustrate the techniques of Chapter 2, namely iteration and use of generating functions. For both approaches we will rely on the fact that each composition σ of n can be written in the form $\sigma = \sigma_1 \sigma'$, where $\sigma_1 \in [n]$ and σ' is a composition of $n - \sigma_1$, as illustrated in Figure 3.2.

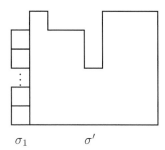

σ_1 $\qquad\qquad$ σ'

FIGURE 3.2: Recursive structure of compositions.

Thus, for all $n \geq 1$,

$$C(n) = \sum_{j=1}^{n} C(n-j) = \sum_{j=0}^{n-1} C(j). \tag{3.2}$$

This implies that $C(n) - C(n-1) = C(n-1)$, or $C(n) = 2C(n-1)$. Iterating this equation and using the initial condition $C(0) = 1$ we get that $C(n) = 2^{n-1}$ for all $n \geq 1$.

We now use the generating function approach. One way to do this is to multiply (3.2) by x^n, sum over $n \geq 1$ and then express everything in terms of the generating function $C(x)$. However, with a little practice, we can go directly from the recurrence for $C(n)$ to a recurrence for the generating function without these intermediate steps. The structure given in Figure 3.2 indicates that a composition (translates into $C(x)$) consists of a part σ_1 (which contributes a factor of x^{σ_1}) combined with another composition (contributes the factor $C(x)$). Taking into account the empty composition (which contributes 1), we arrive at

$$C(x) = 1 + \sum_{\sigma_1 \geq 1} x^{\sigma_1} C(x) = 1 + \frac{x}{1-x} C(x).$$

This implies that

$$C(x) = \frac{1-x}{1-2x} = (1-x) \sum_{n \geq 0} 2^n x^n = 1 + \sum_{n \geq 1} 2^{n-1} x^n,$$

and $C(n) = [x^n]C(x) = 2^{n-1}$, as expected.

To compute $C(m; n)$, notice that every part of a composition has generating function $x/(1-x)$, because the possible sizes for a part are $m \geq 1$. Since the parts add to the total, by Rule 2.51, $C(m; n)$ is the convolution of the generating functions of the individual parts, that is,

$$
C(m; x) = \left(\frac{x}{1-x} \right)^m = \frac{x^m}{(1-x)^m} = x^m \sum_{j \geq 0} \binom{m-1+j}{j} x^j
$$

$$
= \sum_{n \geq m} \binom{n-1}{m-1} x^n, \tag{3.3}
$$

where we have used (A.3) for the third equality. Reading off the coefficient of x^n, we obtain that $C(m; n) = [x^n]C(m; x) = \binom{n-1}{m-1}$. □

3.2 Restricted compositions

In this section we consider compositions that have certain restrictions. Restrictions can occur in at least two ways: restricting the way the parts are arranged within the composition, or restricting the set from which the parts of the composition are taken. We start by looking at compositions that are restricted in the way the parts are arranged.

3.2.1 Palindromic compositions

Recall that palindromic or self-inverse compositions are those that read the same from left to right as from right to left. The results on the number of such compositions and those that have a given number of parts were already known to MacMahon [136, Statement 5, pp. 838–839]. The generating function in the case of a general set A was given by Hoggatt and Bicknell [98, Theorem 1.2]. We will use the notation introduced in Definition 3.1 for compositions by replacing C with P when referring to palindromic compositions. For example, $P_A(n)$ denotes the number of palindromic compositions of n with parts in A.

Theorem 3.4 *The number of palindromic compositions of n is given by*

$$
P(n) = 2^{\lfloor n/2 \rfloor},
$$

where $\lfloor x \rfloor$ denotes the largest integer less than or equal to x, with generating function $P(x) = \dfrac{1+x}{1-2x^2}$.

Proof Using the same argument as in the case of compositions, but taking into account that any part that occurs at the beginning also has to occur at the end of a palindromic composition, is not hard to see that $P(n)$ satisfies the recurrence relation

$$P(n) = 1 + P(n-2) + P(n-4) + P(n-6) + \cdots, \quad n \geq 2,$$

with initial conditions $P(0) = P(1) = 1$, and $P(n) = 0$ for all $n < 0$. This implies that $P(n) = 2P(n-2)$ and iteration gives that $P(n) = 2^{\lfloor n/2 \rfloor}$, as required. From the recurrence, we obtain (by multiplying by x^n and summing over $n \geq 2$) that

$$P(x) - x - 1 = \frac{x^2}{1-x} + \frac{x^2}{1-x^2} P(x).$$

Alternatively, using an appropriate modification of Figure 3.2, we obtain the recursion for the generating function directly as

$$P(x) = \frac{1}{1-x} + \sum_{\sigma_1 \geq 1} x^{2\sigma_1} P(x) = \frac{1}{1-x} + \frac{x^2}{1-x^2} P(x),$$

where the factor $\frac{1}{1-x}$ accounts for the empty palindromic compositions and those consisting of a single part. Solving either equation for $P(x)$ gives the stated result. $\qquad \square$

If we count palindromic compositions according to their number of parts, we get a result that depends on the parity of the number of parts.

Theorem 3.5 *The number of palindromic compositions*

(1) *of $2n$ with $2k$ parts equals $\binom{n-1}{k-1}$,*

(2) *of either $2n$ or $2n-1$ with $2k+1$ parts equals $\binom{n-1}{k}$.*

The respective generating functions are

$$P(2k; x) = \left(\frac{x^2}{1-x^2} \right)^k$$

and

$$P(2k+1; x) = \left(\frac{x^2}{1-x^2} \right)^k \left(\frac{x}{1-x} \right).$$

Proof First we look at palindromic compositions with an even number of parts $m = 2k$, which always result in a composition of an even n. Such a palindromic composition consists of a composition with k parts and its reverse. Since we get each part in duplicate, it follows from (3.3) that the generating function is given by

$$P(2k; x) = C(k; x^2) = \sum_{n \geq k} \binom{n-1}{k-1} x^{2n},$$

which implies that $P(2k; 2n) = \binom{n-1}{k-1}$. If there are an odd number of parts $m = 2k + 1$, then we obtain a composition of either odd or even n. In this case, the palindromic composition consists of a nonzero center part that occurs once, and a nonzero composition with $k \geq 1$ parts and its reverse. Hence,

$$P(2k + 1; x) = \left(\frac{x}{1 - x}\right) C(k; x^2) = \sum_{i \geq 1} x^i \sum_{j \geq k} \binom{j-1}{k-1} x^{2j}.$$

(Note that the formula also covers the case $k = 0$, the composition consisting of just one part.) Expanding the double sum and collecting terms according to the power of x we obtain that

$$P(2k + 1; 2n) = P(2k + 1; 2n - 1) = \sum_{i=1}^{n-k} \binom{n-i-1}{k-1} = \binom{n-1}{k},$$

where the last equality follows from the well known identity (A.6) for $a = k-1$ and $b = n - 2$. $\square$

3.2.2   Carlitz compositions

Recall that a Carlitz composition of n is a composition in which no adjacent parts are the same (see Definition 1.18).

Definition 3.6 *We denote the number of Carlitz compositions of n with parts in A by $C\!C_A(n)$, with generating function $C\!C_A(x) = \sum_{n \geq 0} C\!C_A(n) x^n$, where we define $C\!C_A(0) = 1$ as usual.*

Theorem 3.7 *[38, Section 2] The generating function for the number of Carlitz compositions of n with parts in $\mathbb{N}$ is given by*

$$C\!C(x) = \frac{1}{1 - \sum_{j \geq 1} \dfrac{x^j}{1 + x^j}} = \frac{1}{1 + \sum_{j \geq 1} (-1)^j \dfrac{x^j}{1 - x^j}}.$$

Proof As in the proof of Theorem 3.3, we set up a recursion for the number of Carlitz compositions of n in terms of Carlitz compositions of $k < n$ by splitting off the first part. However, we now need to keep track of the size of this part σ_1 as the remainder of the composition cannot start with that part. Let $\mathit{CC}(\sigma_1 \cdots \sigma_\ell | x)$ be the generating function for the number of Carlitz compositions that start with $\sigma_1 \cdots \sigma_\ell$. Since each Carlitz composition is either the empty composition or starts with some $j \in \mathbb{N}$, we have that

$$\mathit{CC}(x) = 1 + \sum_{j \in \mathbb{N}} \mathit{CC}(j | x). \tag{3.4}$$

Looking at the first two parts of σ, we have that a composition starting with j either consists of just the part j or starts with ji, where $i \neq j$. Thus,

$$\mathit{CC}(j | x) = x^j + \sum_{j \neq i} \mathit{CC}(ji | x) = x^j + x^j \sum_{j \neq i} \mathit{CC}(i | x)$$
$$= x^j + x^j (\mathit{CC}(x) - \mathit{CC}(j | x) - 1) = x^j (\mathit{CC}(x) - \mathit{CC}(j | x)),$$

which gives that $\mathit{CC}(j | x) = \frac{x^j}{1+x^j} \mathit{CC}(x)$. Substituting this result into (3.4) leads to $\mathit{CC}(x) = 1 + \left(\sum_{j \in \mathbb{N}} \frac{x^j}{1+x^j} \right) \mathit{CC}(x)$. Solving for $\mathit{CC}(x)$ gives the first representation for $\mathit{CC}(x)$. The second form is the one given originally by Carlitz [38], and equality of the two forms is a matter of simple algebraic manipulations (see (8.25), where we will treat a more general case). $\qquad \square$

Combining the two restrictions discussed so far leads to *palindromic Carlitz compositions* of n. Note that a palindromic Carlitz composition with an even number of parts cannot occur, as the two middle parts would have to be the same.

Definition 3.8 *We denote the number of palindromic Carlitz compositions of n with parts in A by $\mathit{CP}_A(n)$ with $\mathit{CP}_A(0) = 1$. The corresponding generating function is denoted by $\mathit{CP}_A(x) = \sum_{n \geq 0} \mathit{CP}_A(n) x^n$.*

Theorem 3.9 *The generating function for the number of palindromic Carlitz compositions of n is given by*

$$\mathit{CP}(x) = 1 + \frac{\sum_{j \geq 1} \dfrac{x^j}{1+x^{2j}}}{1 - \sum_{j \geq 1} \dfrac{x^{2j}}{1+x^{2j}}}.$$

Proof We proceed as in the proof of Theorem 3.7. Let $\mathit{CP}(\sigma_1 \cdots \sigma_\ell | x)$ be the generating function for the number of palindromic Carlitz compositions that start with $\sigma_1 \cdots \sigma_\ell$. As before, $\mathit{CP}(x) = 1 + \sum_{j \in \mathbb{N}} \mathit{CP}(j | x)$. The only difference in the argument is that we split off both the first and the last part for the recursion, resulting in a factor of x^{2j} rather than x^j, except for the composition $\sigma = j$. Thus,

$$CP(j|x) = x^j + x^{2j} \sum_{j \neq i} CP(i|x) = x^j + x^{2j}(CP(x) - CP(j|x) - 1),$$

which gives (after simplification) that

$$CP(j|x) = \frac{x^j(1 - x^j)}{1 + x^{2j}} + \frac{x^{2j}}{1 + x^{2j}} CP(x).$$

Substituting this result into the equation for $CP(x)$ and solving for $CP(x)$ gives the desired result. □

3.3 Compositions with restricted parts

We now look at compositions in which the parts are restricted, that is, A is a strict subset of $\mathbb{N}$. Several of these restrictions lead to nice connections with the Fibonacci sequences. The simplest set from which to form compositions is the set $A = \{1, 2\}$. This set was investigated by Alladi and Hoggatt [7] who proved results on the number of compositions and palindromic compositions with parts in A.

Theorem 3.10 *[7, Theorems 1 and 6][1] For $A = \{1, 2\}$ and $n \geq 0$,*

(1) $C_A(n) = F_{n+1}$.

(2) $P_A(2n + 1) = F_{n+1}$ and $P_A(2n) = F_{n+2}$.

Proof (1) Since each composition starts either with a 1 or a 2 followed by a composition of $n - 1$ or $n - 2$, respectively, we get that

$$C_A(n) = C_A(n - 1) + C_A(n - 2).$$

This is the recursion for the Fibonacci sequence, so we just need to check the initial conditions. Since $C_A(0) = C_A(1) = 1$, we obtain that $C_A(n) = F_{n+1}$.

(2) For palindromic compositions of $2n + 1$, the middle part must be odd (and therefore a 1), with a composition of n on the left, and its reverse on the right. Thus,

$$P_A(2n + 1) = C_A(n) = F_{n+1}.$$

---

[1]The formula for $P_A(2n + 1)$ in [7] contains a typo.

For palindromic compositions of $2n$, there are two possibilities. If the palindromic composition has an odd number of parts, then the middle part has to be a 2 combined with a composition of $n-1$ and its reverse. If there are an even number of parts, then the palindromic composition is made up from a composition of n and its reverse. Thus,

$$P_A(2n) = C_A(n-1) + C_A(n) = F_n + F_{n+1} = F_{n+2},$$

as claimed. □

Alladi and Hoggatt also enumerated these compositions according to rises, levels and drops (see Definition 1.12). In the case $A = \{1, 2\}$, a rise corresponds to a 1 followed by a 2, a drop occurs whenever a 2 is followed by a 1, while 11 or 22 create a level. It is easy to see that the number of rises in all compositions is equal to the number of drops in all compositions, as each rise in a composition σ is matched by a drop in the reverse composition $R(\sigma)$.

Theorem 3.11 *[7, Theorems 9 and 10] Let $r(n)$ and $l(n)$ denote the numbers of rises and levels in all compositions of n with parts in $A = \{1, 2\}$. Then*

(1) $r(n+1) = r(n) + r(n-1) + F_{n-1}$,

(2) $l(n+1) = l(n) + l(n-1) + L_{n-1}$,

where F_n and L_n are the Fibonacci and Lucas numbers, respectively. The associated generating functions are given by

$$A_r(x) = \frac{x^3}{(1-(x+x^2))^2} \quad and \quad A_l(x) = \frac{x(x+x^3)}{(1-(x+x^2))^2}.$$

Note that $r(n)$ is a shifted convolution of two Fibonacci sequences, and that $l(n)$ is the convolution of the Fibonacci sequence and the Lucas sequence (with $L_0 := 1$).

Proof (1) The compositions of $n+1$ are obtained by adding either a 1 to a composition of n or a 2 to a composition of $n-1$. In each case we get all the rises in those compositions, plus an additional rise if a 2 is added to a composition of $n-1$ ending in 1, of which there are $C(n-2) = F_{n-1}$. This establishes the recurrence. The derivation of the generating function is left as Exercise 3.9.

(2) By similar reasoning, we have that $l(n+1) = l(n) + F_n + l(n-1) + F_{n-2}$. Since $L_n = F_{n+1} + F_{n-1}$ (which can be easily established by induction), we obtain the recurrence relation. The generating function for $l(n)$ is derived by first proving that $l(n) = r(n+1) + r(n-1)$. Since $l(2) = r(3) = 1$ and $r(1) = 0$, the formula holds for $n = 2$. Assume it holds true for n. Then

$$\begin{aligned} l(n+1) &= l(n) + l(n-1) + F_n + F_{n-2} \\ &= (r(n+1) + r(n-1)) + (r(n) + r(n-2)) + F_n + F_{n-2} \\ &= r(n+2) + r(n). \end{aligned}$$

Thus,

$$A_l(x) = \sum_{n=1}^{\infty} (r(n+1) + r(n-1))x^n$$

$$= \frac{1}{x} A_r(x) + x A_r(x) = \frac{x(x+x^3)}{(1-(x+x^2))^2},$$

which completes the proof.   □

We now look at a different set A, namely the set of odd integers. Compositions with parts in $A = \{2k + 1, k \geq 0\}$ were studied by Grimaldi [76], who found interesting connections with the Fibonacci and Lucas sequences. We give one of his results.

Theorem 3.12 *[76, Section 1] Let A be the set of positive odd integers and let $a(n, 2k+1)$ denote the number of occurrences of $2k+1$ in the compositions of n. Then*

(1) $C_A(n) = F_n$ *for $n \geq 1$.*

(2) $a(1, 1) = 1$ *and* $a(n, 1) = \frac{9}{5} F_{n-1} + \frac{n-1}{5} L_{n-1}$ *for $n \geq 2$.*

(3) $a(n, 2k + 1) = a(n - 2k, 1)$ *for $k \geq 0$, $n \geq 2k + 1$.*

Proof (1) The compositions of n with odd parts can be recursively created by either appending a 1 to the compositions of $n - 1$ or by increasing the last part of a composition of $n - 2$ by 2. Thus, $C_A(n) = C_A(n - 1) + C_A(n - 2)$. Since the initial conditions are $C_A(1) = C_A(2) = 1$, we have $C_A(n) = F_n$ for $n \geq 1$.

(2) Using the same construction as in part (1), but now keeping track of the number of 1s, we obtain a recurrence relation for the sequence $a(n, 1)$. Those compositions that end with 1 contribute an additional 1 to the total count. However, those compositions of $n - 1$ which had a 1 at the end that was increased to a 3 contribute a loss of one 1 per composition. Those are exactly the compositions of $n - 2$. Therefore,

$$a(n + 1, 1) = a(n, 1) + C_A(n) + a(n - 1, 1) - C_A(n - 2)$$
$$= a(n, 1) + a(n - 1, 1) + F_{n-1}.$$

This is a nonhomogeneous second order linear recurrence relation. Let $a_n = a(n, 1)$. Then the general solution is of the form $a_n = h(n) + p(n)$, where $h(n) = c_1 \alpha^n + c_2 \beta^n$ (since the homogeneous recurrence relation is the Fibonacci recurrence relation) and $p(n) = c_3 n \alpha^n + c_4 n \beta^n$ (since the nonhomogeneous part consists of the Fibonacci sequence; see for example [78]).

Substituting first $p(n)$ and the explicit formula for F_{n-1} into the recurrence relation and using that $\alpha^2 - \alpha - 1 = \beta^2 - \beta - 1 = 0$ results in equations that can be solved for c_3 and c_4, and we obtain that $c_3 = (-1 + \sqrt{5})/10$ and $c_4 = (-1 - \sqrt{5})/10$. Using the initial conditions $a_2 = 2$ and $a_3 = 3$, we then obtain $c_1 = 1 - (7\sqrt{5}/25)$ and $c_2 = 1 + (7\sqrt{5}/25)$. This gives an expression for a_n which can be transformed into a combination of Fibonacci and Lucas numbers by splitting off a factor of α and β from each of the terms α^n and β^n and then collecting terms according to $\alpha^{n-1} - \beta^{n-1}$ and $\alpha^{n-1} + \beta^{n-1}$ (see Exercise 3.10).

(3) To illustrate the general case, look at the following example. The compositions of 3 into odd parts are given by 111 and 3. The compositions of 5 into odd parts are 11111, 113, 131, 311, and 5. The only compositions involving a 3 are 113, 131, and 311. We can obtain each of these if we increase each of the 1s in the composition 111 (one at a time) to 3. In general, each part of size $2k - 1$ in a composition of $n - 2$ can be increased by 2 to account for one of the parts of size $2k + 1$ in the compositions of n. Therefore, $a(n, 2k + 1) = a(n - 2, 2k - 1)$, and iteration will give the stated result for $k \geq 0$, $n \geq 2k + 1$. $\square$

Note that in this case, starting with the recurrence equation for $a(1, 1)$ has led to the connection with the Fibonacci and Lucas sequence, which would not have been so easy to see from the generating functions. The moral of the story is that each approach has its own advantages and disadvantages.

So far we have looked at specific sets A, and utilized the structure of the particular set to derive recursions and generating functions. We will now derive structural results for a general set A, from which we can then obtain specialized results by using a particular set A.

Theorem 3.13 *Let $A \subseteq \mathbb{N}$ and $m \geq 0$. Then the generating function for the number of compositions of n with m parts in A is given by*

$$C_A(m; x) = \left(\sum_{a \in A} x^a \right)^m ,$$

and the generating function for the total number of compositions of n with parts in A is given by

$$C_A(x) = \frac{1}{1 - \sum_{a \in A} x^a}.$$

Proof Each part of a composition can consist of any of the elements in A, thus the generating function for a single part is given by $\sum_{a \in A} x^a$. A composition of n with m parts is the sum of these parts, hence the generating functions multiply, which gives the first result. From (3.1), we have that

$$C_A(x) = \sum_{m \geq 0} \left(\sum_{a \in A} x^a \right)^m = \frac{1}{1 - \sum_{a \in A} x^a},$$

which completes the proof.   □

Example 3.14 *Chinn and Heubach [51] studied $(1, k)$ compositions (that is, compositions in $\{1, k\}$), a generalization of the $(1, 2)$ compositions studied by Alladi and Hoggatt [7]. Using Theorem 3.13 with $A = \{1, k\}$, we can easily recover their result that $C_A(x) = \frac{1}{1-x-x^k}$ (see [51, Lemma 1, Part 1]). For $k = 2$, this reduces to the generating function of the shifted Fibonacci sequence (see (A.1)), and we obtain the result given in Theorem 3.10.*

Example 3.15 *Applying Theorem 3.13 to the set $A = \{1, 2, ..., \ell\}$, we obtain the result of Hogatt and Bicknell [98]:*

$$C_A(x) = \frac{1}{1 - x - x^2 - \cdots - x^\ell} = \frac{1}{x^{\ell-1}} \cdot \frac{x^{\ell-1}}{1 - x - x^2 - \cdots - x^\ell}.$$

Thus, the number of compositions of n with parts that are less than or equal to ℓ is given by the shifted ℓ-generalized Fibonacci number. Specifically,

$$C_{\{1,2,...,\ell\}}(n) = F(\ell; n + \ell - 1)$$

for $n \geq 0$, where $F(\ell; n) = \sum_{i=1}^{\ell} F(\ell; n - i)$ with $F(\ell; \ell - 1) = 1$ and $F(\ell; 0) = \cdots = F(\ell; \ell - 2) = 0$ [69]. Combinatorially, this can be seen using the following recursive method: to create the compositions of n, add i to the right end of the compositions of $n - i$, for $i = 1, 2, \ldots, \ell$. The ℓ-generalized Fibonacci sequences occur in [180] as sequence A000073 for $\ell = 3$, A000078 for $\ell = 4$, A001591 for $\ell = 5$, and A001592 for $\ell = 6$.

We can also easily obtain results on odd and even compositions.

Example 3.16 *Theorem 3.13 when applied to the odd or even positive integers gives*

$$C_{\{1,3,5,...\}}(j; x) = \left(\frac{x}{1 - x^2}\right)^j \text{ and } C_{\{2,4,6,...\}}(j; x) = \left(\frac{x^2}{1 - x^2}\right)^j,$$

that is, $C_{\{2,4,6,...\}}(j; x) = x^j C_{\{1,3,5,...\}}(j; x)$. This equality follows easily, since every odd composition of n with j parts uniquely corresponds to an even composition of $n + j$ with j parts (increase each odd part by 1). So now let $A = \{2, 4, 6, \ldots\}$. To obtain $C_A(j; n) = [x^n]C_A(j; x)$, the number of even compositions, we need to expand $C_A(j; x)$ into a series. Using (A.3), we get that

$$C_A(j; x) = \frac{x^{2j}}{(1 - x^2)^j} = x^{2j} \sum_{i \geq 0} \binom{i + (j-1)}{j-1} x^{2i} = \sum_{\ell \geq j-1} \binom{\ell}{j-1} x^{2\ell+2},$$

after appropriate re-indexing. Thus, the number of compositions of $2n$ with j parts in $\{2, 4, 6, \ldots\}$ is given by $C_{\{2,4,6,\ldots\}}(j; 2n) = \binom{n-1}{j-1}$, which is also the number of compositions of $2n - j$ with j odd parts.

Turning our attention to the total number of compositions with odd parts, Theorem 3.13 gives

$$C_{\{1,3,5,\ldots\}}(x) = \frac{1}{1 - \dfrac{x}{1 - x^2}} = \frac{1 - x^2}{1 - x - x^2}.$$

This is the generating function of the modified Fibonacci sequence ($\hat{F}_0 = 1$, $\hat{F}_n = F_n$ for $n \geq 1$), therefore, the number of compositions of n with parts in $\{1, 3, 5, \ldots\}$ is given by F_n for $n \geq 1$, which was proved in Theorem 3.12 using the recurrence relation. More generally, let $\mathbb{N}_{d,e}$ be the set of all numbers of the form $k \cdot d + e$ where $k \geq 0$, $d, e \in \mathbb{N}$. Then Theorem 3.13 yields

$$C_{\mathbb{N}_{d,e}}(x) = \frac{1}{1 - \dfrac{x^e}{1 - x^d}} = \frac{1 - x^d}{1 - x^e - x^d}.$$

If $d = 2e$, then the set A consists of all odd multiples of e, and

$$C_{\mathbb{N}_{2e,e}}(x) = C_{\{e,3e,5e,\ldots\}}(x) = \frac{1 - x^{2e}}{1 - x^e - x^{2e}} = g_{\hat{F}}(x^e).$$

Thus, the number of compositions of $n = k \cdot e$ with parts in $\{e, 3e, 5e, \ldots\}$ is given by F_k for $k \geq 1$. This result can be readily explained using a combinatorial argument: multiply each part of a composition of n with parts in $\{1, 3, 5, \ldots\}$ by e to create a composition of $n \cdot e$ with parts in $\{e, 3e, 5e, \ldots\}$.

In the examples discussed so far, the generating function we obtain when applying Theorem 3.13 to a specific set A can be expanded into a series that allows us to obtain an explicit expression for $C_A(n)$. This is not always the case, as will be seen in the next example.

Example 3.17 *Let $A = \{2, 3, 5, 7, 11, 13, \ldots\}$ be the set of prime numbers. Then*

$$C_A(x) = \frac{1}{1 - \sum_{p \in A} x^p}.$$

Since powers of x^n for $n \leq K$ cannot involve powers of x^p with $p > K$ for some constant K, we can use the function $1/(1 - \sum_{p \in A, p \leq K} x^p)$ to approximate $C_A(x)$ for powers up to K. Using Maple or Mathematica, *we expand the approximate function into its Taylor series. Since the 11th prime number is 31, we can use the finite sum with 11 terms to get correct values for $C_A(n)$ for $n \leq 31$. Here is the* Mathematica *code for series expansion:*

```
Series[1/(1 - Sum[x^Prime[n], {n, 1, 11}]), {x, 0, 30}]
```

The corresponding Maple code for the series expansion is given by

```
taylor(1/(1- sum(x^ithprime(n), 'n'=1..20)), x=0, 20);
```

Mathematica *also provides a command to get just the list of coefficients, rather than the series:*

```
CoefficientList[Series[1/(1-Sum[x^Prime[n],{n,1,11}]),{x,0,30}],x]
```

We can read off the values for $n = 0, 1, \ldots, 30$ as 1, 0, 1, 1, 1, 3, 2, 6, 6, 10, 16, 20, 35, 46, 72, 105, 152, 232, 332, 501, 732, 1081, 1604, 2352, 3493, 5136, 7595, 11212, 16534, 24442, *and* 36039, *which occur as sequence A023360 in* [180].

3.4   Connection between compositions and tilings

We have already seen two graphical representations of a composition, namely the line graph introduced by MacMahon (see Chapter 1) and the bar graph of a composition. Having different ways to think about a composition allows one to pick the viewpoint that is most useful for a given question. For example, the line graph representation results in an easy combinatorial proof for the total number of compositions and provides a quick way to obtain the conjugate composition, whereas a description of the algorithm without the line graph representation is very cumbersome. Last but not least, certain questions arising naturally in one representation may be reframed into another representation, with the potential of giving new insights or raising new questions.

We will now look at the representation of a composition as a tiling of a $1 \times n$ board with tiles of size $1 \times k$. The tiles correspond to the parts of the composition in the obvious way. The tiling can also be connected to the line graph as follows: A node in the line graph corresponds to the end of a tile, and the lengths of the tiles correspond to the number of segments between nodes. Figure 3.3 shows the tiling and line graph corresponding to the composition 2231. (Arranging the tiles in columns results in the bar graph.)

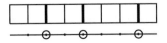

FIGURE 3.3: Correspondence between line graph and tiling.

The simplest tiling problem is to enumerate the number of ways one can tile a $1 \times n$ board with 1×1 tiles, called *squares*, and 1×2 tiles, called *dominoes*. This tiling problem is identical to the question on the total number of compositions in $\{1, 2\}$ studied by Alladi and Hoggatt [7]. We give the derivation in terms of tilings.

Example 3.18 (*Tiling with squares and dominoes*) *Suppose you are given a large number of squares and dominoes. In how many different ways can a $1 \times n$ board be tiled using these two types of tiles? Let T_n denote the number of tilings of a $1 \times n$ board, with $T_0 = 1$. There is exactly one way to tile the 1×1 board. In general, for $n \geq 2$, each tiling of a $1 \times n$ board either starts with a square followed by a tiling of a $1 \times (n-1)$ board (of which there T_{n-1}) or a domino, followed by a tiling of a $1 \times (n-2)$ board (of which there T_{n-2}). Thus we obtain the following recurrence relation:*

$$T_n = T_{n-1} + T_{n-2}, \quad T_0 = T_1 = 1.$$

This recursive relation is identical to that of the Fibonacci numbers, except that the initial conditions are shifted. The resulting sequence is often denoted by f_n, with $f_0 = 1$. Using Rule 2.45, the generating function for this shifted sequence is given by

$$\frac{F(x) - F_0}{x} = \frac{1}{x} \cdot \frac{x}{1 - x - x^2} = \frac{1}{1 - x - x^2}.$$

Tiling questions have been studied independently of compositions. Questions on square and domino tilings of larger boards arise for example from questions in physics. There they are known as the monomer-dimer or dimer problems, where monomers correspond to squares and dimers correspond to dominoes (see for example [44, 66, 107, 106, 151, 167] and references therein).

3.5 Colored compositions and other variations

When looking at compositions in their representation as tilings of a $1 \times n$ board, it is natural to ask about the total number of tilings when tiles of a certain size come in several colors. Let $A = \{a_1, a_2, \ldots, a_d\}$ be the set of possible sizes for the tiles, and let c_i indicate the number of possible colors for the tile of size a_i. Then we can easily establish the following result.

Theorem 3.19 *The number of tilings of a $1 \times n$ board with tiles of size a_i that come in c_i colors is given by*

$$T_n = \sum_{i=1}^{d} c_i T_{n-a_i} \quad \text{for } n \geq a_d,$$

with initial conditions $T_0 = 1, T_1, \ldots, T_{a_d}$.

This general recursion may be solved for specific values of A and the number of colors c_i.

Example 3.20 *Let $A = \{1, 2, 3\}$, and $c_1 = 1$, $c_2 = 4$, and $c_3 = 2$. The recurrence relation becomes*

$$T_n = T_{n-1} + 4 T_{n-2} + 2 T_{n-3},$$

which expresses the sequence $\{T_n\}$ as a convolution with the sequence $\{1, 4, 2\}$. Thus we have

$$T(x) = 1 + c(x)T(x) \quad \Rightarrow \quad T(x) = \frac{1}{1 - c(x)}, \qquad (3.5)$$

where $c(x)$ is the generating function for the (finite) sequence of colors. By the definition of the generating function,

$$c(x) = x + 4x^2 + 2x^3,$$

and therefore, $T(x) = 1/(1 - x - 4x^2 - 2x^3)$. (This example arose in the context of tilings of larger boards with L-shaped tiles and squares investigated by Chinn et al. [49], which can be mapped to this particular tiling question on the $1 \times n$ board.)

Another example of colored compositions was introduced by Agarwal [3], which leads to a nice connection with Fibonacci numbers.

Definition 3.21 *An n-color composition is a composition with parts in $\mathbb{N}$ in which the part of size i exists in i different colors.*

Example 3.22 *Distinguishing different colors by a subscript for $i \geq 2$, we can write down the eight 3-color compositions of 3, namely*

$$3_1 \quad 3_2 \quad 3_3 \quad 2_1 1 \quad 2_2 1 \quad 1 2_1 \quad 1 2_2 \quad 1 1 1.$$

We use (3.5) with $c(x) = \sum_{i=1}^{\infty} i x^i = x/(1-x)^2$ to obtain the generating function for the n-colored compositions as

$$T(x) = \frac{1}{1 - \dfrac{x}{(1-x)^2}} = \frac{(1-x)^2}{1 - 3x + x^2}.$$

Note that this generating function differs from the generating function given in [3], as Agarwal uses $T_0 = 0$ instead of the convention we have followed, defining $T_0 = 1$.

Theorem 3.23 *[3, Theorem 1] Let $\bar{C}(n)$ $(\bar{C}(m;n))$ be the number of n-color compositions of n (with m parts), and let $\bar{C}(x)$ $(\bar{C}(m;x))$ be the respective generating functions, where $\bar{C}(0) = \bar{C}(m;0) = 0$ for all m. Then*

$$\bar{C}(m;x) = \frac{x^m}{(1-x)^{2m}} \quad \text{and} \quad \bar{C}(x) = \frac{x}{1-3x+x^2}.$$

The corresponding sequences are given by

$$\bar{C}(m;n) = \binom{n+m-1}{2m-1} \quad \text{and} \quad \bar{C}(n) = F_{2n},$$

where F_{2n} is the $2n$-th Fibonacci number.

Agarwal [3] also presents a more general result in which the parts and the colors for a given part can be restricted. He obtains recursions for the generating functions and the number of n-color compositions of n that have exactly m parts from the set $[k]$ and which use only the first r colors for each of the parts.

Example 3.24 *There are 21 n-color compositions of 6 with exactly two parts from $\{1, 2, 3, 4\}$, using up to 3 colors for each of the parts: six of the form $4_i 2_j$, six of the form $2_j 4_i$, and nine of the form $3_i 3_j$.*

We now look at a variation of compositions, based on the representation of the compositions as a bar graph. *Cracked compositions* are compositions that have parts of size k with potential "cracks" between the cells in the bar graph representation. In particular, cracks do not occur at the beginning or at the end of the column. These cracks do not increase the size of the part, but the different possible crack patterns can be thought of as each giving rise to a different color. We indicate the occurrence of a crack by placing an "×" on the border between two successive cells. Figure 3.4 shows the bar graphs of the cracked compositions for $n = 3$.

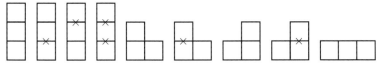

FIGURE 3.4: Bar graphs of cracked compositions for $n = 3$.

We will now answer a few questions about cracked compositions:

- How many compositions have exactly one crack?

- How many compositions have at most one crack in each part?

- How many compositions have at most m cracks in each part?

To answer the first question, we condition on where the part with the crack occurs. If the cracked part occurs as the i-th part, then the preceding composition is just a regular composition (that is, with no cracks, and potentially empty), and the remainder of the composition is also of that form.

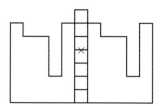

FIGURE 3.5:   Cracked compositions with exactly one crack.

We already know the generating function for compositions, so we just need to compute the generating function for the cracked part. If there is exactly one crack in the column, then there can be any number $k_1 \geq 1$ of cells below the crack, and likewise any number $k_2 \geq 1$ above the crack. The generating function counting the number of cells above or below the crack is given by $x/(1-x)$, and the number of these cells have to add to the height of the column. By Rule 2.51, the generating function for the number of parts with exactly one crack is the convolution of $x/(1-x)$ with itself. Thus, the generating function $A(x)$ for the number of cracked compositions with exactly one crack is given by

$$A(x) = C(x)\left(\frac{x}{1-x}\right)^2 C(x) = \left(\frac{1-x}{1-2x}\right)^2 \left(\frac{x}{1-x}\right)^2 = \frac{x^2}{(1-2x)^2}.$$

Note that this generating function is a convolution of $\tilde{C}(x)$ with itself, where $\tilde{C}(x) = x/(1-2x) = C(x) - 1$ is the generating function for the nonempty compositions.

Such an equality of two different generating functions indicates that there exists a bijection between the items counted by the two generating functions. In this case, we are looking for a bijection between the cracked compositions of n with exactly one crack and sequences that are created by concatenating two nonempty compositions of sizes k and $n - k$, for $k = 1, \ldots, n - 1$. Note that a given concatenated sequence may be obtained in several different ways. For example, the concatenated sequence 111 is obtained as 11|1 or 1|11, where we place a | between the two compositions to clearly mark the concatenation point.

We now describe the bijection. To create a cracked composition with exactly one crack, proceed as follows. Let $\sigma_1 \cdots \sigma_k$ denote the first composition, and $\bar{\sigma}_1 \cdots \bar{\sigma}_\ell$ the second one. For each value σ_i, create a column of height σ_i. Place a crack on top of the last column of height σ_k, then add $\bar{\sigma}_1$ cells above the crack. This produces the piece with exactly one crack. Then create columns of heights $\bar{\sigma}_2, \ldots, \bar{\sigma}_\ell$. Clearly, this process is reversible. Also, because both compositions contain at least one part, we always create a cracked part, and have therefore established the desired bijection. Figure 3.6 shows the cracked compositions of 3 with exactly one crack and the corresponding concatenation of two nonempty compositions.

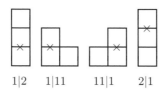

$$1|2 \qquad 1|11 \qquad 11|1 \qquad 2|1$$

FIGURE 3.6: Cracked compositions and associated sequences.

To answer the second question, we again break up the composition according to the parts that have a crack. We can visualize the cracked composition with at most one crack in each part as a sequence of ordinary compositions interspersed with a part that has exactly one crack. Thus, if the composition has exactly m parts with one crack, then the generating function has the structure

$$C(x) \left(\left(\frac{x}{1-x} \right)^2 C(x) \right)^m .$$

Accounting for all possibilities, the generating function for the number of cracked compositions that have at most one crack in each part is given by

$$A(x) = C(x) \sum_{m=0}^{\infty} \left(\left(\frac{x}{1-x} \right)^2 C(x) \right)^m = \frac{C(x)}{1 - \left(\frac{x}{1-x} \right)^2 C(x)}$$

$$= \frac{\dfrac{1-x}{1-2x}}{1 - \left(\dfrac{x}{1-x} \right)^2 \dfrac{1-x}{1-2x}} = \frac{(1-x)^2}{1 - 3x + x^2} = 1 + \frac{x}{1 - 3x + x^2}.$$

Again, we obtain a generating function of a known sequence, in this case the sequence $\{a_n\}_{n=1}^{\infty}$, where $a_0 = 1$ and $a_n = F_{2n}$ for $n \geq 1$, where F_n is the n-th Fibonacci number (see Exercise 2.12). Since F_{2n} counts the number of

tilings of a $1 \times (2n-1)$ board with squares and dominoes, we are looking for
a bijection between the cracked compositions of n with at most one crack per
part and the tilings of the $1 \times (2n-1)$ board.

We will now describe the bijection, starting with a tiling of a $1 \times (2n-1)$
board with squares and dominoes. Given such a tiling, make cuts at odd
positions, unless there is a domino that spans across the odd position. Note
that the leftmost piece resulting from those cuts covers an odd number of
cells, whereas all the other pieces have an even number of cells. In addition,

- the first piece is either a single square or consists of $k \geq 1$ dominoes
  followed by a single square; and

- the remaining pieces consist either of a pair of squares, a single domino,
  or a square followed by a sequence of $k \geq 1$ dominoes ending in a square.

For a given tiling, the bar graph is created as follows:

- If the tiling starts with a square, then create a column of height 1; if it
  starts with k dominoes followed by a square, create a column of height
  $k+1$ with a crack at level k.

- A pair of squares starts a new column of height 1.

- A single domino adds a cell on top of the rightmost column created so
  far.

- A sequence of $k \geq 1$ dominoes enclosed by a square on either side creates
  a new column of height $k+1$ with a crack at height k.

This process is reversible, and therefore we have found the desired bijection.
We illustrate the process in Example 3.25 using Figure 3.7.

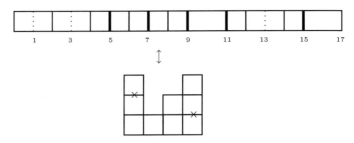

FIGURE 3.7: Tiling and associated cracked composition.

Example 3.25 *The first segment of the tiling consists of two dominoes followed by a square, which creates a bar of height three with a crack at height two. The next segment, consisting of two squares, starts a new column of height one, as does the next segment. The domino of the fourth segment adds a cell in the third bar. The fifth segment, consisting of one domino enclosed by squares starts a new bar of height two, with a crack at height one. Finally, the domino at the end of the tiling adds a cell on the rightmost bar.*

The third question about cracked compositions is a generalization of the second one, so let's see what we need to adjust when we set up the generating function. Let $C_m(x)$ be the generating function for cracked compositions that have parts with at most m cracks. Again we split up the composition, now according to the parts that have exactly m cracks, as indicated in Figure 3.8 for $m = 2$.

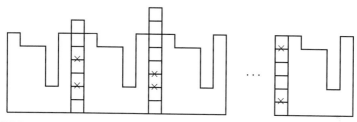

FIGURE 3.8: Cracked compositions with at most two cracks per part.

Then the compositions preceding the first such part, and the compositions between the parts with m cracks each have at most $m - 1$ cracks, and so their generating function is given by $C_{m-1}(x)$. The generating function for the part with exactly m cracks is given by $(x/(1 - x))^{m+1}$ (why?), and we get that

$$
C_m(x) = C_{m-1}(x) \sum_{i=0}^{\infty} \left(\left(\frac{x}{1-x} \right)^{m+1} C_{m-1}(x) \right)^i
$$
$$
= \frac{C_{m-1}(x)}{1 - \left(\dfrac{x}{1-x} \right)^{m+1} C_{m-1}(x)}.
$$

Dividing the numerator and denominator of the right-hand side by $C_{m-1}(x)$ and then cross-multiplying results in a recurrence relation for $1/C_m(x)$ instead of the one for $C_m(x)$:

$$
\frac{1}{C_m(x)} = \frac{1}{C_{m-1}(x)} - \left(\frac{x}{1-x} \right)^{m+1}.
$$

This is a useful trick to have in one's toolbox for solving recurrence relations, and we will see it again in a later section. Now we use iteration to obtain that for $m \geq 1$,

$$
\begin{aligned}
\frac{1}{C_m(x)} &= \frac{1}{C_{m-1}(x)} - \left(\frac{x}{1-x}\right)^{m+1} \\
&= \frac{1}{C_{m-2}(x)} - \left(\frac{x}{1-x}\right)^m - \left(\frac{x}{1-x}\right)^{m+1} \\
&= \cdots \\
&= \frac{1}{C_0(x)} - \sum_{i=2}^{m+1} \left(\frac{x}{1-x}\right)^i,
\end{aligned}
$$

which implies that

$$
\frac{1}{C_m(x)} = \frac{1-2x}{1-x} - \left(\frac{x}{1-x}\right)^2 \cdot \sum_{i=0}^{m-1} \left(\frac{x}{1-x}\right)^i \qquad (3.6)
$$

since $C_0(x) = C(x)$. The desired generating function for the number of cracked compositions with at most m cracks is given by the reciprocal. If we now let $m \to \infty$, we obtain that

$$
\frac{1}{C_m(x)} \to \frac{1-2x}{1-x} - \left(\frac{x}{1-x}\right)^2 \frac{1}{1-\frac{x}{1-x}} = \frac{(1-2x)^2 - x^2}{(1-2x)(1-x)} = \frac{1-3x}{1-2x}.
$$

Thus, the generating function for the number of cracked compositions that allow any number of cracks is given by $(1-2x)/(1-3x)$. This is, however, the generating function for words of length $n-1$ on the alphabet $\{1, 2, 3\}$ (see Exercise 2.4), so another bijection is lurking in the background waiting to be discovered – here it is! We start with a word of length $n-1$ and describe the associated cracked composition of n.

For $n = 2$, we associate the three types of cracked compositions with the letters $1, 2$, and 3 as follows:

$$
\boxed{} \leftrightarrow 1, \quad \boxed{\ast} \leftrightarrow 2, \quad \boxed{} \leftrightarrow 3.
$$

To create the words of length n we append either a 1, 2 or 3 to each of the words of length $n-1$. Here is how the associated cracked compositions of $n+1$ are created from the cracked compositions of n:

- If a 1 is appended to the word, then we add a cell on top of the rightmost column of the bar graph corresponding to the word of length $n-1$.

- If a 2 is appended to the word, then we add a crack and a cell on the top of the rightmost column of the bar graph corresponding to the word of length $n - 1$.

- If a 3 is appended to the word, then we add a new column of height one to the right end of the bar graph corresponding to the word of length $n - 1$.

Figure 3.9 shows the correspondence of cracked compositions of $n = 3$ and words of length two. The process described above is reversible (the last column of a bar graph is either of height 1, or has a cell with a crack or a cell without a crack at the top of a column of height at least 2). Therefore, we have found a bijection between the cracked compositions of n with any number of cracks and the words of length $n - 1$ on the alphabet $\{1, 2, 3\}$.

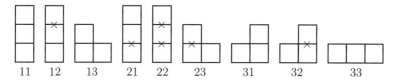

11   12   13   21   22   23   31   32   33

FIGURE 3.9: Bijection between cracked compositions and words.

We summarize the results on cracked compositions in the following theorem.

Theorem 3.26 *Let $C(x)$ be the generating function for the number of compositions of n, $\tilde{C}(x)$ be the generating function for the number of nonempty compositions of n, and $C_m(x)$ be the generating function for the cracked compositions of n with at most m cracks per part. Then*

(1) *The generating function for the number of cracked compositions with exactly one crack is given by $\tilde{C}^2(x) = \frac{x^2}{(1-2x)^2}$.*

(2) *$C_1(x) = 1 + x/(1 - 3x + x^2)$, and therefore, $[x^n]C_1(x) = F_{2n}$ for $n \geq 1$.*

(3) *More generally,*

$$C_m(x) = \frac{(1-x)^2}{1 - 3x + x^2 - x^2 \sum_{i=1}^{m-1} \dfrac{x^i}{(1-x)^i}} \rightarrow \frac{1 - 2x}{1 - 3x} \quad \text{as } m \to \infty.$$

Thus, the number of cracked compositions of n with any number of cracks is given by 3^{n-1} for $n \geq 1$.

3.6   Exercises

Exercise 3.1 *Let* $A(x, y, \vec{q}) = \sum_{n \geq 0} \sum_{\sigma \in C_{n,m}} x^n y^m \prod_{i=1}^{\infty} q_i^{r_i(\sigma)}$ *be the generating function for the number of compositions of* n *with* r_i *parts of size* i *and* $\sum_i r_i = m$. *Prove that*

$$A(x, y, \vec{q}) = \frac{1}{1 - y \sum_{i=1}^{\infty} q_i x^i}.$$

Exercise 3.2 *Let* $A(x, y, \vec{q}) = \sum_{n \geq 0} \sum_{\sigma \in P_{n,m}} x^n y^m \prod_{i=1}^{\infty} q_i^{r_i(\sigma)}$ *be the generating function for the number of palindromic compositions of* n *with* r_i *parts of size* i *and* $\sum_i r_i = m$. *Prove that*

$$A(x, y, \vec{q}) = \frac{1 + y \sum_{i=1}^{\infty} q_i x^i}{1 - y^2 \sum_{i=1}^{\infty} q_i^2 x^{2i}}.$$

Exercise 3.3* *Find an explicit formula for* $\dfrac{C_{\mathbb{N} \setminus \{1\}}(n)}{C_{\mathbb{N}}(n)}$ *and determine its limit when* $n \to \infty$.

Exercise 3.4 *Find the generating function for the number of compositions of* n *with* m *parts in* $\mathbb{N}$ *in which no part is unique (or equivalently, every part appears at least twice). (Hint: use the exponential generating function for* y, *that is,* $f(x, y) = \sum_{m \geq 0} f(m; x) \frac{y^m}{m!}$, *where* $f(m; x)$ *is the ordinary generating function for the number of compositions with* m *parts in which no part is unique.)*

Exercise 3.5 *A composition* σ *is said to be* odd-Carlitz *(respectively, even-Carlitz) if it is Carlitz and all its parts are odd (respectively, even) numbers. Prove that*

(1) *the generating function for the number of odd-Carlitz compositions of* n *with* m *parts in* $\mathbb{N}$ *is given by*

$$\frac{1}{1 - \sum_{i \geq 0} \dfrac{x^{2i+1} y}{1 + x^{2i+1} y}};$$

(2) *the generating function for the number of even-Carlitz compositions of* n *with* m *parts in* $\mathbb{N}$ *is given by*

$$\frac{1}{1 - \sum_{i \geq 0} \dfrac{x^{2i+2} y}{1 + x^{2i+2} y}}.$$

Exercise 3.6 *Prove that the generating function for the number of palindromic Carlitz compositions of n with m parts is given by*

$$CP(x,y) = \sum_{m \geq 0} CP(m;x)y^m = 1 + \frac{\sum_{i \geq 1} \frac{x^i y}{1 + x^{2i}y^2}}{1 - \sum_{i \geq 1} \frac{x^{2i}y^2}{1 + x^{2i}y^2}}.$$

Exercise 3.7

(1) *Find an explicit formula for the number of times the summand k occurs in all palindromic compositions of n with parts in* $\mathbb{N}$.

(2) *More generally, for any ordered subset A of* $\mathbb{N}$, *find the generating function for the number of times that the coefficient* a_i *occurs in all palindromic compositions with parts in A.*

Exercise 3.8 *Find an explicit formula for the number of Carlitz words (words in which no two adjacent letters are the same) of length n on the alphabet* $[k]$.

Exercise 3.9 *Derive the generating function for the number of rises in all compositions with parts in* $\{1,2\}$ *from the recurrence relation given in Theorem 3.11.*

Exercise 3.10 *Fill in the details of the proof of Part (2) of Theorem 3.12.*

Exercise 3.11 *Write a program that uses the recursive creation described in Theorem 3.12 to create the compositions of n with odd parts.*

Exercise 3.12 *A composition* $\sigma = \sigma_1 \cdots \sigma_m$ *of n with m parts is said to be limited if* $1 \leq \sigma_i \leq n_i$ *for all* $i = 1, 2, \ldots, n$.

(1) *Derive a formula for the generating function for the number of limited compositions of n.*

(2) *Using Part (1), obtain a simple formula for the case* $n_i = k$ *for all i.*

(3) *Prove that the number of limited compositions of n is given by* F_{n+1} *when* $n_i = 2$ *for all i.*

Exercise 3.13 *Prove that the generating function for the number of compositions of n with exactly k odd parts is given by*

$$\frac{x^k(1-x^2)}{(1-2x^2)^{k+1}}.$$

Exercise 3.14 *Let* $\overline{CP}(m,n)$ *be the number of n-color palindromic compositions of n with m parts. Show that*

(1) $\overline{CP}(2m-1, 2n-1) = \sum_{i=1}^{n-1}(2i-1)\binom{n+m-2-i}{2m-3}$, *for* $m > 1$.

(2) $\overline{CP}(2m-1, 2n) = \sum_{i=1}^{n-1} 2i\binom{n+m-2-i}{2m-3}$, *for* $m > 1$.

(3) $\overline{CP}(2m, 2n) = \binom{n+m-1}{2m-1}$.

Exercise 3.15 *An* odd-even *partition of* n *is a partition in which the parts alternate in parity, starting with the smallest odd part (see [9]). For instance, the odd-even partitions of* 7 *are* 16, 34, 52, *and* 7. *Use either generating function techniques or a combinatorial argument to show that the number of compositions of* n *with an odd number of parts is the same as the number of odd-even partitions with largest part* n.

Exercise 3.16* *A word* $w = w_1 \cdots w_n$ *is said to be* gap free *if* w *contains all the letters between the smallest and largest letter in* w. *For example,* 342523 *is gap free, but* 154212 *is not, as the letter* 3 *is missing. Find an explicit formula for the generating function for the number of gap free words of length* n *on the alphabet* $[k]$.

3.7   Research directions and open problems

We now suggest several research directions, which are motivated both by the results and exercises of this chapter. In later chapters we will revisit some of these research directions and pose additional questions.

Research Direction 3.1 *In Exercise 3.16, we defined the notion of a gap free word. Obviously, this notion can be applied to compositions as well, and can be generalized.*

Definition 3.27 *For a sequence (word, composition, partition)* $s = s_1 \cdots s_k$, *we define*

$$\mathrm{gap}(s) = \max_i s_i - \min_i s_i + 1 - |\{s_i | i = 1, 2, \ldots, k\}|,$$

which gives the number of parts missing in s *between the minimal and maximal parts of* s. *Note that if* $\mathrm{gap}(s) = 0$, *then the sequence is* gap free.

Example 3.28 *The word* $w = 13361755$ *has a gap of* 2, *which is easily determined by inspection as the letters* 2 *and* 4 *are missing. We can also obtain this result from the above formula as* $\mathrm{gap}(w) = 7 - 1 + 1 - |\{1, 3, 5, 6, 7\}| = 7 - 5 = 2$.

We can now formulate several research questions around this concept.

(1) *What is the explicit formula for the generating function for the number of k-ary words w of length n such that* $\text{gap}(w) = \ell$? *(Note that when* $\ell = 0$*, we obtain the generating function for the number of gap free words of length n on* $[k]$*.)*

(2) *More generally, can one find an explicit formula for the generating function for the number of compositions of n with m parts such that* $\text{gap}(\sigma) = \ell$?

(3) *A variation on the theme: Is there an explicit formula for the generating function for the number of k-ary circular words of length n (see Exercise 1.14) such that* $\text{gap}(w) = \ell$?

(4) *Generalize (3) to compositions, that is, find an explicit formula for the generating function for the number of circular compositions (see Exercise 1.15) of n with m parts such that* $\text{gap}(\sigma) = \ell$.

Research Direction 3.2 *In Exercise 3.4 we investigated compositions in which no part is unique, and derived an explicit formula for the mixed (exponential in y, ordinary in x) generating function*

$$f(x, y) = \sum_{m \geq 0} f_k(m; x) \frac{y^m}{m!}$$

for the number of compositions of n with m parts in $\mathbb{N}$ *in which no part is unique. Here are some questions related to this concept.*

(1) *Is there an explicit formula for the ordinary generating function*

$$f(x, y) = \sum_{m \geq 0} f_k(m; x) y^m$$

for the number of compositions of n with m parts in $\mathbb{N}$ *in which no part is unique?*

(2) *What is the explicit formula for the number of compositions of n with m parts in* $\mathbb{N}$ *such that if x is a part of* σ*, then* σ *contains x at least* ℓ *times?*

(3) *Answer Part (2) for palindromic or Carlitz compositions.*

Research Direction 3.3 *A variation on the notion of circular words and compositions is the notion of a* cyclic *word or composition, which we will investigate in the context of Carlitz compositions and Carlitz words.*

Definition 3.29 *A cyclic Carlitz composition of n with m parts in A is a Carlitz composition* $\sigma_1 \cdots \sigma_m$ *of n with m parts in A such that* $\sigma_1 \neq \sigma_m$. *A cyclic Carlitz word of length n on the alphabet* $[k]$ *is a Carlitz word (see Exercise 2.6)* $w_1 \cdots w_n$ *such that* $w_1 \neq w_n$.

Note that for a cyclic composition, we look at the connection between the first and the last part, but we do not put the composition on a circle.

Example 3.30 1231 *is a Carlitz composition of 7, but not a cyclic Carlitz composition. The cyclic Carlitz compositions of* $n = 1, \ldots, 5$ *are:*

$$1; \qquad 2; \qquad 3, 12, 21; \qquad 4, 13, 31; \qquad 5, 41, 14, 32, 23.$$

The cyclic Carlitz words of length 3 *on* [3] *are* 123, 132, 213, 231, 312, *and* 321.

The obvious questions to ask are:

(1) *Find an explicit generating function for the number of* k-*ary cyclic Carlitz words of length* n.

(2) *Find an explicit generating function for the number of cyclic Carlitz compositions of* n *with* m *parts in* $\mathbb{N}$ *(or more generally, in a set* A).

Research Direction 3.4 *In Section 3.4 we have seen that there is a strong connection between tilings and compositions. In particular, the number of* n-*color compositions of* n *is given by* F_{2n} *(see Theorem 3.23). Recall that* F_n *counts the number of tilings of a* $1 \times (n-1)$ *board with squares and dominoes, or equivalently, compositions of* $n - 1$ *with* 1*s and* 2*s (see Example 3.18). Give a bijection between the two types of compositions (or equivalently between tilings with squares and dominoes and* n-*color tilings).*

Chapter 4

Statistics on Compositions

4.1 History and connections

A statistic on a composition is a characteristic such as the numbers of parts, levels, rises, odd summands, etc. In this chapter we will present some results and techniques to obtain generating functions for statistics on compositions.

The earliest statistics to be considered, apart from the number of parts of the composition, were rises, levels and drops, considered by Carlitz, Hoggatt and their various co-authors [7, 36, 39, 42, 43]. Abramson considered longest runs and rises of a given size [1, 2]. Rawlings investigated weak rises (level or rise) and down-up alternating compositions [168].

Rises, levels and drops can be regarded as the simplest of patterns, namely 2-letter subword patterns (see Definition 1.24). A rise corresponds to the subword pattern 12, a level to the subword pattern 11, and a drop to the subword pattern 21. In this chapter we use the term pattern exclusively for subword patterns, enumerating compositions which include or avoid them. More general patterns will be discussed in Chapter 5.

In Section 4.2 we will present results on the enumeration of compositions in an ordered set A with regard to the statistics parts, rises, levels, and drops [90]. These results contain as special cases results on the sets considered in [7, 50, 51, 76, 77, 87, 98], as well as results on Carlitz compositions [36, 39, 119] and partitions. The approach we will take is to derive a generating function for several statistics of interest simultaneously, and then obtain results for individual statistics as special cases.

Following a line of research that arose from the study of permutation patterns, Heubach and Mansour [92] enumerated compositions which contained and/or avoided 2-letter and 3-letter patterns. Some of their results were further generalized to families of ℓ-letter patterns by Mansour and Sirhan [145]. These results will be given in Section 4.3, starting with results that only exist for 3-letter patterns or sets of 3-letter patterns, followed by the results for the more general ℓ-letter patterns.

4.2   Subword patterns of length 2: rises, levels and drops

Throughout the remainder of this chapter we let A be an ordered (finite or infinite) set of positive integers, say $A = \{a_1, \ldots, a_k\}$, where $a_1 < \cdots < a_k$, with the obvious modifications in the case $|A| = \infty$. In the theorems and proofs we will treat the two cases together if possible, and will note if the case $|A| = \infty$ requires additional steps. We will derive general results involving a number of statistics, which then can be applied to special cases.

Definition 4.1 *For any ordered set $A \subseteq \mathbb{N}$, we denote the set of all compositions (respectively palindromic compositions) of n with parts in A by $\mathcal{C}_n^A$ (respectively $\mathcal{P}_n^A$). For any composition σ, we denote the number of rises, levels, and drops by $\mathrm{ris}(\sigma)$, $\mathrm{lev}(\sigma)$, and $\mathrm{dro}(\sigma)$, respectively. Let $\overrightarrow{y} = (y, r, \ell, d)$ and $\overrightarrow{s}_\sigma = (\mathrm{par}(\sigma), \mathrm{ris}(\sigma), \mathrm{lev}(\sigma), \mathrm{dro}(\sigma))$. The generating functions for the number of compositions (respectively palindromic compositions) of n with parts in A according to the order of σ and the statistic $\overrightarrow{s}_\sigma$ are given by*

$$C_A(x, y, r, \ell, d) = \sum_{n \geq 0} \sum_{\sigma \in \mathcal{C}_n^A} x^n y^{\mathrm{par}(\sigma)} r^{\mathrm{ris}(\sigma)} \ell^{\mathrm{lev}(\sigma)} d^{\mathrm{dro}(\sigma)}$$

and

$$P_A(x, y, r, \ell, d) = \sum_{n \geq 0} \sum_{\sigma \in \mathcal{P}_n^A} x^n y^{\mathrm{par}(\sigma)} r^{\mathrm{ris}(\sigma)} \ell^{\mathrm{lev}(\sigma)} d^{\mathrm{dro}(\sigma)}.$$

In the proofs, we also need to consider generating functions for compositions that start with specified parts (as in the proof of Theorem 3.7).

Definition 4.2 *We will denote the generating functions for the number of compositions (respectively palindromic compositions) of n with parts in A according to the order of σ and the statistic $\overrightarrow{s}_\sigma$ that start with $\sigma_1 \cdots \sigma_\ell$ by*

$$C_A(\sigma_1 \cdots \sigma_\ell | x, y, r, \ell, d) = \sum_{n \geq 0} \sum_{\sigma_1 \cdots \sigma_\ell \sigma' \in \mathcal{C}_n^A} x^n y^{\mathrm{par}(\sigma)} r^{\mathrm{ris}(\sigma)} \ell^{\mathrm{lev}(\sigma)} d^{\mathrm{dro}(\sigma)}$$

and

$$P_A(\sigma_1 \cdots \sigma_\ell | x, y, r, \ell, d) = \sum_{n \geq 0} \sum_{\sigma_1 \cdots \sigma_\ell \sigma' \in \mathcal{P}_n^A} x^n y^{\mathrm{par}(\sigma)} r^{\mathrm{ris}(\sigma)} \ell^{\mathrm{lev}(\sigma)} d^{\mathrm{dro}(\sigma)}.$$

From these definitions, it follows immediately that

$$C_A(x, y, r, \ell, d) = 1 + \sum_{i=1}^{k} C_A(a_i | x, y, r, \ell, d) \tag{4.1}$$

and

$$P_A(x, y, r, \ell, d) = 1 + \sum_{i=1}^{k} P_A(a_i | x, y, r, \ell, d), \tag{4.2}$$

where the summand 1 corresponds to the case $n = 0$.

4.2.1 Results for compositions

We now state the results for subword patterns of length two for compositions.

Theorem 4.3 *The generating function for compositions according to order and the statistic $\overrightarrow{s}_\sigma = (\mathrm{par}(\sigma), \mathrm{ris}(\sigma), \mathrm{lev}(\sigma), \mathrm{dro}(\sigma))$ for any ordered subset A of $\mathbb{N}$ is given by*

$$C_A(x, y, r, \ell, d) = \frac{1 + (1-d)\sum_{j=1}^{k}\left(\dfrac{x^{a_j}y}{1 - x^{a_j}y(\ell - d)}\prod_{i=1}^{j-1}\dfrac{1 - x^{a_i}y(\ell - r)}{1 - x^{a_i}y(\ell - d)}\right)}{1 - d\sum_{j=1}^{k}\left(\dfrac{x^{a_j}y}{1 - x^{a_j}y(\ell - d)}\prod_{i=1}^{j-1}\dfrac{1 - x^{a_i}y(\ell - r)}{1 - x^{a_i}y(\ell - d)}\right)}.$$

Proof We start by deriving a recursion for $C_A(a_i | x, y, r, \ell, d)$. Note that the compositions of n starting with a_i with at least two parts can be created recursively by prepending a_i to a composition of $n - a_i$ that starts with a_j for some j. This either creates a rise (if $i < j$), a level (if $i = j$), or a drop (if $i > j$), and in each case, results in one more part. Thus,

$$C_A(a_i a_j | x, y, r, \ell, d) = \begin{cases} r\, x^{a_i}y\, C_A(a_j | x, y, r, \ell, d), \ i < j \\ \ell\, x^{a_i}y\, C_A(a_j | x, y, r, \ell, d), \ i = j \\ d\, x^{a_i}y\, C_A(a_j | x, y, r, \ell, d), \ i > j. \end{cases}$$

Summing over j and accounting for the single composition with exactly one part, namely a_i, gives

$$C_A(a_i | x, y, r, \ell, d) = x^{a_i}y + x^{a_i}y\, d \sum_{j=1}^{i-1} C_A(a_j | x, y, r, \ell, d)$$

$$+ x^{a_i}y\, \ell\, C_A(a_i | x, y, r, \ell, d)$$

$$+ x^{a_i}y\, r \sum_{j=i+1}^{k} C_A(a_j | x, y, r, \ell, d) \tag{4.3}$$

for $i = 1, 2, \ldots, k$. Let

$$t_0 = C_A(x, y, r, \ell, d), \quad t_i = C_A(a_i | x, y, r, \ell, d), \quad \text{and} \quad b_i = x^{a_i}y$$

for $i = 1, 2, \ldots, k$. Then (4.1) and (4.3) result in a system of $k + 1$ equations in $k + 1$ variables. By Cramer's rule (see Theorem B.19), $t_0 = \frac{\det(N_k)}{\det(M_k)}$, where M_k is the $(k + 1) \times (k + 1)$ matrix of the system of equations and N_k is the $(k+1) \times (k+1)$ matrix which results from replacing the first column in M_k by the vector of the right-hand side of the system. The next step is to derive a recursion for $\det(M_k)$ from which one can obtain the following explicit formula (for details see [90]):

$$\det(M_k) = \begin{vmatrix} 1 - b_1\ell & -b_1 r & -b_1 r & \cdots & -b_1 r & -b_1 r \\ -b_2 d & 1 - b_2\ell & -b_2 r & \cdots & -b_2 r & -b_2 r \\ -b_3 d & -b_3 d & 1 - b_3\ell & \cdots & -b_3 r & -b_3 r \\ \vdots & & \vdots & & & \vdots \\ -b_{k-1}d & -b_{k-1}d & -b_{k-1} & \cdots & 1 - b_{k-1}\ell & -b_{k-1}r \\ -b_k d & -b_k d & -b_k d & \cdots & -b_k d & 1 - b_k\ell \end{vmatrix}$$

$$= \prod_{j=1}^{k}(1 - b_j(\ell - d)) - d\sum_{j=1}^{k} b_j \prod_{i=1}^{j-1}(1 - b_i(\ell - r)) \prod_{i=j+1}^{k}(1 - b_j(\ell - d)).$$

A similar formula is obtained for $\det(N_k)$, which completes the proof when A is a finite set. Note that if $|A| = \infty$, then the result follows by taking limits as $k \to \infty$. $\qquad\square$

We now apply Theorem 4.3 to specific sets, and also focus on individual statistics by selectively setting some of the indeterminates y, r, ℓ, and d to 1.

Example 4.4 *Disregarding rises, levels and drops by setting $r = \ell = d = 1$ in Theorem 4.3, we recover the results of Theorem 3.13, namely that the generating function for the number of compositions of n with m parts in A is given by*

$$C_A(x, y, 1, 1, 1) = \frac{1}{1 - y\sum\limits_{j=1}^{k} x^{a_j}} = \sum_{m \geq 0}\left(\sum_{j=1}^{k} x^{a_j}\right)^m y^m. \tag{4.4}$$

Moreover, setting $y = 1$ in (4.4) gives the generating function for the number of compositions of n with parts in A as $1/(1 - \sum_{j=1}^{k} x^{a_j})$. Additional examples for specific choices of A can be found in [89].

Note that the number of rises equals the number of drops in all compositions of n. For each nonpalindromic composition there exists a composition in reverse order, matching up rises and drops, while for palindromic compositions, symmetry matches up rises and drops within the composition. Therefore, we will derive results only for rises, and in the remainder of the chapter not mention drops anymore. Any result that we derive for rises immediately gives the identical result for drops.

Corollary 4.5 *Let $A = \{a_1, \ldots, a_k\}$ and define*

$$r_A(x, y) = \sum_{n \geq 0} \sum_{\sigma \in C_n^A} \text{ris}(\sigma) x^n y^{\text{par}(\sigma)}$$

and

$$l_A(x, y) = \sum_{n \geq 0} \sum_{\sigma \in C_n^A} \text{lev}(\sigma) x^n y^{\text{par}(\sigma)}.$$

Then

$$r_A(x, y) = \left(\sum_{1 \leq i < j \leq k} x^{a_i + a_j} \right) \sum_{m \geq 0} (m+1) \left(\sum_{j=1}^{k} x^{a_j} \right)^m y^{m+2} \qquad (4.5)$$

and

$$l_A(x, y) = \left(\sum_{j=1}^{k} x^{2a_j} \right) \sum_{m \geq 0} (m+1) \left(\sum_{j=1}^{k} x^{a_j} \right)^m y^{m+2}.$$

Note that the generating functions given in Corollary 4.5 have the statistic of interest as the coefficient of $x^n y^{\text{par}(\sigma)}$. How can we obtain such a generating function from the general result stated in Theorem 4.3? The obvious choice is to take partial derivatives with respect to the relevant indeterminate, and then set that indeterminate to 1, a common technique in enumerative combinatorics. We will illustrate this general technique in detail in the proof of Corollary 4.5. The following rule for partial derivatives of products, which holds whenever $f_i(r) \neq 0$, will come in handy.

$$\frac{\partial}{\partial r} \prod_{i=1}^{m} f_i(r) = \sum_{i=i}^{m} \frac{\partial}{\partial r} f_i(r) \prod_{j \neq i} f_j(r) = \left(\prod_{i=1}^{m} f_i(r) \right) \sum_{i=1}^{m} \frac{\frac{\partial}{\partial r} f_i(r)}{f_i(r)}. \qquad (4.6)$$

Proof We derive the expression for $r_A(x, y)$. Since we are only interested in rises, we set $\ell = d = 1$ in Theorem 4.3 to obtain

$$C_A(x, y, r, 1, 1) = \frac{1}{1 - \sum_{j=1}^{k} \left(x^{a_j} y \prod_{i=1}^{j-1} (1 - x^{a_i} y (1 - r)) \right)}. \qquad (4.7)$$

Since

$$\sum_{n \geq 0} \sum_{\sigma \in C_n^A} \text{ris}(\sigma) x^n y^{\text{par}(\sigma)} = \frac{\partial}{\partial r} C_A(x, y, r, 1, 1) \Big|_{r=1},$$

we use (4.6) with $f_i(r) = (1 - x^{a_i}y(1 - r))$ and obtain that

$$
\frac{\partial}{\partial r} C_A(x, y, r, 1, 1) \bigg|_{r=1}
$$

$$
= \frac{\sum_{j=1}^{k} \left(x^{a_j} y \left[\prod_{i=1}^{j-1}(1 - x^{a_j}y(1 - r)) \right] \sum_{i=1}^{j-1} \frac{x^{a_j}y}{1 - x^{a_j}y(1 - r)} \right)}{\left(1 - \sum_{j=1}^{k} \left(x^{a_j} y \prod_{i=1}^{j-1}(1 - x^{a_i}y(1 - r)) \right) \right)^2} \Bigg|_{r=1}
$$

$$
= \frac{\sum_{j=1}^{k} x^{a_j} y \sum_{i=1}^{j-1} x^{a_i} y}{\left(1 - y \sum_{j=1}^{k} x^{a_j} \right)^2} = \frac{\sum_{1 \le i < j \le k} x^{a_i + a_j} y^2}{\left(1 - y \sum_{j=1}^{k} x^{a_j} \right)^2}. \tag{4.8}
$$

Using the power series expansion of $\frac{1}{(1-t)^2}$ (see (A.3)) and collecting the powers of y gives the desired result for $r_A(x, y)$. The proof for $l_A(x, y)$ is left as Exercise 4.1.   □

Applying Corollary 4.5 to specific sets A, we obtain the results given in [7, 50, 51, 77, 87] and some additional results for those sets given in [90]. We will illustrate the process with $A = \mathbb{N}$, and showcase a useful algebraic "trick" along the way that allows us to get a nice result. For ease of readability, we may use the shorthand notation $\frac{\partial}{\partial r}C_A(x, y, 1, 1, 1)$ in the remainder of the section to mean $\frac{\partial}{\partial r}C_A(x, y, r, 1, 1)\big|_{r=1}$. It is understood that the differentiation is done first, and then the value of the indeterminate is replaced.

Example 4.6 *Let $A = \mathbb{N}$. We first derive the generating function for rises in all compositions of n with a fixed number of parts $m \ge 2$, which is given by*

$$
[y^m] r_{\mathbb{N}}(x, y) = \sum_{j > i \ge 1} x^{i+j}(m - 1) \left(\sum_{j \ge 1} x^j \right)^{m-2}.
$$

The goal is to derive a nice expression, rather than just to stop at substituting values into the expression given in Corollary 4.5. We rewrite the first sum in such a way that the two indices can be separated, using a neat little trick. Namely we replace the condition $j \ge i + 1$ by $j = i + \ell$ and $\ell \ge 1$ to obtain

$$
[y^m] r_{\mathbb{N}}(x, y) = \sum_{i \ge 1} \sum_{j \ge i+1} x^{i+j}(m - 1) \left(\frac{x}{1 - x} \right)^{m-2}
$$

$$
= \sum_{i \ge 1} x^{2i} \sum_{\ell \ge 1} x^{\ell} \frac{(m - 1)x^{m-2}}{(1 - x)^{m-2}}
$$

$$
= \frac{x^3}{(1 - x)(1 - x^2)} \cdot \frac{(m - 1)x^{m-2}}{(1 - x)^{m-2}} = \frac{(m - 1)x^{m+1}}{(1 + x)(1 - x)^m}.
$$

Likewise, we can obtain the corresponding results for the number of levels in all compositions of n with m parts (see Exercise 4.3):

$$[y^m]l_{\mathbb{N}}(x,y) = \frac{(m-1)x^m}{(1+x)(1-x)^{m-1}}.$$

To derive the generating function for the number of rises and levels, respectively, in all compositions of n (see [87], Theorem 6), we substitute $y = 1$ in (4.5) and split up the inner sum as above which yields

$$r_{\mathbb{N}}(x,1) = \sum_{j>i\geq 1} x^{i+j} \sum_{m\geq 0} (m+1)\left(\frac{x}{1-x}\right)^m$$

$$= \frac{x^3}{(1-x)(1-x^2)} \cdot \frac{1}{(1-\frac{x}{1-x})^2} = \frac{x^3}{(1+x)(1-2x)^2}$$

and

$$l_{\mathbb{N}}(x,1) = \frac{x^2(1-x)}{(1+x)(1-2x)^2}.$$

We now turn our attention to Carlitz compositions. Recall that a Carlitz composition of n is a composition in which no adjacent parts are the same (see Definition 1.18). In other words, a Carlitz composition σ is a composition with $\text{lev}(\sigma) = 0$, and therefore, the generating function according to the statistic $\overrightarrow{s}_\sigma = (\text{par}(\sigma), \text{ris}(\sigma), \text{dro}(\sigma))$ is given by $C_A(x,y,r,0,d)$. Theorem 4.3 with $\ell = 0$ gives the following result.

Corollary 4.7 *Let $A = \{a_1, \ldots, a_k\}$ be any ordered subset of $\mathbb{N}$. Then the generating function for Carlitz compositions according to the statistic $\overrightarrow{s}_\sigma = (\text{par}(\sigma), \text{ris}(\sigma), \text{dro}(\sigma))$ is given by*

$$C_A(x,y,r,0,d) = 1 + \frac{\displaystyle\sum_{j=1}^{k}\left(\frac{x^{a_j}y}{1+x^{a_j}yd}\prod_{i=1}^{j-1}\frac{1+x^{a_i}yr}{1+x^{a_i}yd}\right)}{1 - \displaystyle\sum_{j=1}^{k}\left(\frac{x^{a_j}yd}{1+x^{a_j}yd}\prod_{i=1}^{j-1}\frac{1+x^{a_i}yr}{1+x^{a_i}yd}\right)}.$$

As before, we can obtain the enumeration of the total number of Carlitz compositions of n with m parts by setting the relevant indeterminates to 1.

Example 4.8 *[73, Proposition 2] Setting $r = d = 1$ in Corollary 4.7 we obtain the generating function for the number of Carlitz compositions with m parts in A (for the case $A = \mathbb{N}$, see [119]) as*

$$C_A(x,y,1,0,1) = \frac{1}{1 - \displaystyle\sum_{j=1}^{k}\frac{x^{a_j}y}{1+x^{a_j}y}}.$$

Since Carlitz compositions do not have any levels, the only statistic to be considered is the number of rises. Differentiating $C_A(x, y, r, 0, 1)$ with respect to r and then setting $r = 1$ gives the following result.

Corollary 4.9 *Let $A = \{a_1, \ldots, a_k\}$ be any ordered subset of $\mathbb{N}$. Then the generating function $\sum_{n \geq 0} \sum_\sigma \mathrm{ris}(\sigma) x^n y^{\mathrm{par}(\sigma)}$ (where the summation is over Carlitz compositions with parts in A) is given by*

$$\sum_{j=1}^{k} \left(\frac{x^{a_j} y}{1 + x^{a_j} y} \sum_{i=1}^{j-1} \frac{x^{a_i} y}{1 + x^{a_i} y} \right) \bigg/ \left(1 - \sum_{j=1}^{k} \frac{x^{a_j} y}{1 + x^{a_j} y} \right)^2.$$

Example 4.10 *Setting $A = \mathbb{N}$ and $y = 1$ in Corollary 4.9 yields that the generating function for the number of rises in the Carlitz compositions of n is given by*

$$\frac{\sum_{j \geq 1} \left(\frac{x^j}{1 + x^j} \sum_{i=1}^{j-1} \frac{x^i}{1 + x^i} \right)}{\left(1 - \sum_{j \geq 1} \frac{x^j}{1 + x^j} \right)^2}.$$

4.2.2 Results for palindromic compositions

Following the structure of Section 4.2.1, we will first prove the general result for palindromic compositions, and then derive the specific results for rises and levels. Note that even though in general the generating functions for palindromic compositions are not as simple as those for compositions, the derivation of the main result is less tedious in the case of palindromic compositions.

Theorem 4.11 *Let A be any ordered subset of $\mathbb{N}$. Then the generating function for palindromic compositions according to order and the statistic $\vec{s}_\sigma = (\mathrm{par}(\sigma), \mathrm{ris}(\sigma), \mathrm{lev}(\sigma), \mathrm{dro}(\sigma))$ is given by*

$$P_A(x, y, r, \ell, d) = \frac{1 + \sum_{i=1}^{k} \frac{x^{a_i} y + x^{2a_i} y^2 (\ell - d r)}{1 - x^{2a_i} y^2 (\ell^2 - d r)}}{1 - \sum_{i=1}^{k} \frac{x^{2a_i} y^2 d r}{1 - x^{2a_i} y^2 (\ell^2 - d r)}}.$$

Proof The proof is similar to that of Theorem 4.3. Palindromic compositions of n that start with a_i (for a fixed i) are of the form a_i (one part), $a_i a_i$ (two parts and one level), or $a_i \sigma' a_i$ where σ' is a nonempty palindromic composition. Palindromic compositions of n with at least three parts can be created by adding the part a_i both at the beginning and at the end of a palindromic composition of $n - 2a_i$ that starts with a_j. If $a_i = a_j$ (and therefore

$i = j$), then two additional levels are created; if $i \neq j$, then a rise and a drop are created, and in both cases we have two additional parts. Therefore,

$$P_A(a_i|x, y, r, \ell, d) = x^{a_i}y + x^{2a_i}y^2\ell + x^{2a_i}y^2\ell^2 P_A(a_i|x, y, r, \ell, d)$$
$$+ x^{2a_i}y^2 d r \sum_{j \neq i, j=1}^{k} P_A(a_j|x, y, r, \ell, d)$$
$$= x^{a_i}y + x^{2a_i}y^2\ell + x^{2a_i}y^2(\ell^2 - dr)P_A(a_i|x, y, r, \ell, d)$$
$$+ x^{2a_i}y^2 d r(P_A(x, y, r, l, d) - 1),$$

where the last equality follows from (4.2). We solve for $P_A(a_i|x, y, r, \ell, d)$ and substitute the result into (4.2) to obtain that

$$P_A(x, y, r, \ell, d) = \frac{1 + \sum_{i=1}^{k} \dfrac{x^{a_i}y + x^{2a_i}y^2(\ell - dr)}{1 - x^{2a_i}y^2(\ell^2 - dr)}}{1 - \sum_{i=1}^{k} \dfrac{x^{2a_i}y^2 dr}{1 - x^{2a_i}y^2(\ell^2 - dr)}}.$$

That the proof for palindromic compositions is less tedious stems from the fact that we only need to distinguish between the cases $i = j$ and $i \neq j$ when deriving $P_A(a_i|x, y, r, \ell, d)$. This allows us to solve for $P_A(a_i|x, y, r, \ell, d)$ in terms of $P_A(x, y, r, \ell, d)$. □

Example 4.12 *Applying Theorem 4.11 for $r = \ell = d = 1$ we get that the generating function for the number of palindromic compositions of n with m parts in A is given by $\frac{1+y\sum_{i=1}^{k} x^{a_i}}{1-y^2\sum_{i=1}^{k} x^{2a_i}}$. Setting $y = 1$ we obtain the generating function for the number of palindromic compositions of n with parts in A (see [98, Theorem 1.2]). More specifically, for $A = \mathbb{N}$ we recover the result of Theorem 3.4.*

We now turn our attention to the number of rises and levels. Using Theorem 4.11 to compute $\frac{\partial}{\partial r}P_A(x, y, 1, 1, 1)$ and $\frac{\partial}{\partial \ell}P_A(x, y, 1, 1, 1)$, we obtain the following result.

Corollary 4.13 *Let $A = \{a_1, \ldots, a_k\}$ be any ordered subset of $\mathbb{N}$, and let $f(x) = \sum_{i=1}^{k} x^{a_i}$, $f_j = f(x^j)$. Then the generating functions*

$$r_A(x, y) = \sum_{n \geq 0} \sum_{\sigma \in \mathcal{P}_n^A} \mathrm{ris}(\sigma)x^n y^{\mathrm{par}(\sigma)}$$

and

$$l_A(x, y) = \sum_{n \geq 0} \sum_{\sigma \in \mathcal{P}_n^A} \mathrm{lev}(\sigma)x^n y^{\mathrm{par}(\sigma)}$$

are given by

$$r_A(x, y) = \frac{y^3(f_1 f_2 - f_3) + y^4(f_2 f_2 - f_4) + y^5(f_2 f_3 - f_1 f_4)}{(1 - y^2 f_2)^2} \quad (4.9)$$

and

$$l_A(x, y) = \frac{y^2 f_2 + 2y^3 f_3 + y^4(2f_4 - f_2 f_2) + 2y^5(f_1 f_4 - f_2 f_3)}{(1 - y^2 f_2)^2}.$$

Example 4.14 *If we apply Corollary 4.13 for* $A = \mathbb{N}$ *and* $y = 1$, *we have that* $f(x) = \frac{x}{1-x}$ *and therefore, the generating function for the number of rises is given by*

$$r_{\mathbb{N}}(x, 1) = \frac{x^4(4x^4 + 4x^3 + 4x^2 + 3x + 1)}{(1 + x^2)(1 + x + x^2)(1 - 2x^2)^2}$$

(see [87], Theorem 6). However, if we compute the generating functions according to the number of rises and parts, then we encounter the usual phenomenon for palindromic compositions, namely that we have to distinguish between odd and even m. *To determine the coefficients of* y^m, *we first substitute* $f(x) = \frac{x}{1-x}$ *into (4.9), then expand the numerator of* $r_{\mathbb{N}}(x, y)$ *and collect terms according to the powers of* m. *The numerator of* $r_{\mathbb{N}}(x, y)$ *is given by*

$$\frac{x^4 y^3}{(1 - x)^2 (1 + x)} \left[\frac{2x + 1}{(x^2 + x + 1)} + \frac{2x^2 y}{(x + 1)(x^2 + 1)} - \frac{x^2 y^2}{(x^2 + x + 1)(x^2 + 1)} \right],$$

and the denominator of $r_{\mathbb{N}}(x, y)$ *is given by*

$$\frac{1}{\left(1 - \dfrac{y^2 x^2}{1 - x^2}\right)^2} = \sum_{m \geq 0} \frac{(m + 1)x^{2m}}{(1 - x^2)^m} y^{2m}.$$

All together, we obtain that

$$r_{\mathbb{N}}(x, y) = \sum_{m \geq 0} \frac{(m + 1)x^{2m+4}}{(1 - x)^2(1 + x)(1 - x^2)^m} y^{2m+3}$$

$$\cdot \left[\frac{2x + 1}{(x^2 + x + 1)} + \frac{2x^2}{(x + 1)(x^2 + 1)} y - \frac{x^2}{(x^2 + x + 1)(x^2 + 1)} y^2 \right].$$

If m *is even, only the terms with factor* $y^{2m+3}y$ *need to be taken into account, whereas for odd* m, *the terms with factors* y^{2m+3} *and* $y^{2m+3}y^2$ *need to be considered. Thus,* $[y^m]r_{\mathbb{N}}(x, y)$, *the number of rises in the palindromic compositions of* n *with* m *parts, is given by*

$$\begin{cases} \dfrac{(2j - 2)x^{2j+2}}{(1 + x^2)(1 - x^2)^j} & \text{for } m = 2j \\[4mm] \dfrac{x^{2j}(1 - x)(1 + (2j - 2)(x + x^3) + (2j - 3)x^2)}{(1 + x^2)(1 + x + x^2)(1 - x^2)^j} & \text{for } m = 2j - 1. \end{cases}$$

As before, we can apply Corollary 4.13 to specific sets, but the resulting generating functions generally do not factor nicely (see [90]). We conclude by giving results for palindromic Carlitz compositions.

Corollary 4.15 *Let $A = \{a_1, \ldots, a_k\}$ be any ordered subset of $\mathbb{N}$. Then the generating function for palindromic Carlitz compositions according to rises, order and parts is given by*

$$P_A(x, y, r, 0, 1) = 1 + \frac{\sum\limits_{i=1}^{k} \dfrac{x^{a_i} y}{1 + x^{2a_i} y^2 r}}{1 - \sum\limits_{i=1}^{k} \dfrac{x^{2a_i} y^2 r}{1 + x^{2a_i} y^2 r}}.$$

4.2.3 Partitions with parts in A

Last but not least, we apply Theorem 4.3 to obtain results for partitions. Since partitions are customarily written in decreasing order, they are exactly those compositions that have no rises, and therefore, their generating function is given by $C_A(x, y, 0, l, d)$.

Corollary 4.16 *Let $A = \{a_1, \ldots, a_k\}$ be any ordered subset of $\mathbb{N}$. Then the generating function for partitions according to the statistic*

$$\overrightarrow{s}_\sigma = (\mathrm{par}(\sigma), \mathrm{ris}(\sigma), \mathrm{lev}(\sigma), \mathrm{dro}(\sigma))$$

is given by

$$C_A(x, y, 0, \ell, d) = 1 + \frac{\sum_{j=1}^{k} \left(\dfrac{x^{a_j} y}{1 - x^{a_j} y(\ell - d)} \prod_{i=1}^{j-1} \dfrac{1 - x^{a_i} y \ell}{1 - x^{a_i} y(\ell - d)} \right)}{1 - d \sum_{j=1}^{k} \left(\dfrac{x^{a_j} y}{1 - x^{a_j} y(\ell - d)} \prod_{i=1}^{j-1} \dfrac{1 - x^{a_i} y \ell}{1 - x^{a_i} y(\ell - d)} \right)}.$$

Example 4.17 *Applying Corollary 4.16 for $A = \mathbb{N}$ and $\ell = d = 1$ and using the identity*

$$1 - \alpha \sum_{j=1}^{k} x^{a_j} \prod_{i=1}^{j-1} (1 - x^{a_i} \alpha) = \prod_{j=1}^{k} (1 - x^{a_j} \alpha), \tag{4.10}$$

we obtain that the generating function for the number of partitions of n with m parts in $\mathbb{N}$ is given by

$$C_A(x, y, 0, 1, 1) = \prod_{j \geq 1} (1 - x^j y)^{-1}.$$

In particular, setting $y = 1$, we recover the result from Example 2.65 for the generating function for the number of partitions of n.

Example 4.18 (*Partitions with distinct parts*) *Another interesting application of Corollary 4.16 concerns partitions with distinct parts, which just means that the number of levels is zero. Setting $\ell = 0$ and $d = 1$ in Corollary 4.16, we obtain that the generating function for the number of partitions of n with m distinct parts in A is given by*

$$C_A(x, y, 0, 0, 1) = \frac{1}{1 - \sum_{j=1}^{k} x^{a_j} y \prod_{i=1}^{j}(1 + x^{a_i}y)^{-1}} = \prod_{j=1}^{k}(1 + x^{a_j}y),$$

where the second equality is easily proved by induction. In particular, the generating function for the number of partitions of n with distinct parts is given by $\prod_{j \geq 1}(1 + x^j)$. This result can be proved directly by using the approach of Example 2.65, as each part j occurs at most once.

As before, we can compute the partial derivatives $\frac{\partial}{\partial \ell} C_A(x, y, 0, 1, 1)$ and $\frac{\partial}{\partial d} C_A(x, y, 0, 1, 1)$ to obtain results on the number of levels and drops in partitions (see Exercise 4.7 for a special case).

4.3   Longer subword patterns

As mentioned at the beginning of this chapter, rises, levels and drops can be considered as 2-letter subword patterns. Because there are only three such patterns, we were able to derive results for all patterns at once in Theorem 4.3. As the length of the pattern increases, the corresponding number of patterns increases rapidly, so we will derive generating functions for one pattern at a time. We will use the following notation to keep track of the number of occurrences of specific patterns.

Definition 4.19 *We denote the number of occurrences of the pattern τ in a composition σ by $\mathrm{occ}_\tau(\sigma)$.*

Before we dive into longer patterns, we will summarize the results obtained for rises, levels and drops in terms of 2-letter patterns. Setting $r = d = 1$ and $\ell = q$ in Theorem 4.3 for the pattern 11, and $\ell = d = 1$ and $r = q$ for the patterns 12 and 21, we obtain the results listed in Table 4.1.

We are now ready to look at longer subword patterns. The first paper on 3-letter subword patterns was a direct extension of the statistics rise, level and drop [92]. The authors enumerated compositions according to 3-letter subword patterns formed by the concatenation of any two of the patterns level, rise, and drop. For example, a level followed by a level corresponds to the single pattern 111 and we will refer to this statistic by the shorthand level+level. On the other hand, the statistic rise+drop is comprised of three different patterns, namely 121, 132, and 231.

TABLE 4.1: Generating functions for 2-letter patterns

τ	$\sum\limits_{\sigma \in \mathcal{C}^A} x^{\mathrm{ord}(\sigma)} y^{\mathrm{par}(\sigma)} q^{\mathrm{occ}_\tau(\sigma)}, \; A = \{a_1, \dots, a_k\}$
11	$1 \Big/ \left(1 - \sum\limits_{j=1}^{k} \dfrac{x^{a_j} y}{1 - x^{a_j} y (q-1)} \right)$
12, 21	$1 \Big/ \left(1 - \sum\limits_{j=1}^{k} x^{a_j} y \prod\limits_{i=1}^{j-1} \left(1 - x^{a_i} y (1-q) \right) \right)$

Just as in the case of the single patterns rise and drop, symmetry makes life easier, as certain statistics or patterns will occur equally often among all the compositions of n (with m parts) in A, making the respective generating functions the same. We make this notion precise with the following definitions.

Definition 4.20 *Two subword patterns τ and ν are* strongly tight Wilf-equivalent *if the number of compositions of n with m parts in A that contain τ exactly r times is the same as the number of compositions of n with m parts in A that contain ν exactly r times, for all A, n, m and r. We denote two patterns that are strongly tight Wilf-equivalent by $\tau \overset{st}{\sim} \nu$.*

The term tight refers to the fact that we deal with subword patterns, where the parts of the composition corresponding to the pattern are consecutive, that is, occur "tightly" together. A weaker equivalence considers only the case $r = 0$ and requires that an equal number of compositions avoid either pattern.

Definition 4.21 *Two subword patterns τ and ν are* tight Wilf-equivalent *if the number of compositions of n with m parts in A that avoid τ is the same as the number of compositions of n with m parts in A that avoid ν, for all A, n and m. We denote two patterns that are tight Wilf-equivalent by $\tau \overset{t}{\sim} \nu$.*

It is easy to see that whenever a pattern τ occurs in a composition σ of n with m parts in A, then $R(\tau)$, the reverse of τ, also occurs by the same argument we used to show that the number of rises equals the number of drops. In addition, there is no other map f such that $\tau \overset{t}{\sim} f(\tau)$ for **all** patterns τ. However, we will see that some patterns are (strongly) tight Wilf-equivalent, even though one is not the reverse of the other (see Research Direction 4.2). To distinguish between the patterns where the equivalence comes for free and those where work is to be done, we define the symmetry class of a pattern.

Definition 4.22 *For any pattern τ, $\{\tau, R(\tau)\}$ is called the* symmetry class *of τ with regard to strongly tight Wilf-equivalence in compositions.*

We now look at patterns of length three. There are a total of 13 patterns that fall into eight symmetry classes (Exercise 4.10). However, if we think of them in terms of rises, levels and drops, then there are only six essentially different patterns (up to symmetry) which are listed in Table 4.2.

TABLE 4.2: Statistics and their associated patterns

Statistic	Pattern	Statistic	Pattern
level+level	111	rise+drop=peak	121+132+231
level+rise	112	drop+rise=valley	212+213+312
level+drop	221	rise+rise	123

We first give results for the 3-letter patterns peak, valley and 123, as their generalization to longer patterns is difficult. The remaining 3-letter patterns, namely 111, 112, 221, 121, 132, 231, 212, 213, and 312, are special cases of the more general ℓ-letter patterns discussed in Sections 4.3.3 – 4.3.5. We modify our notation for the generating functions, using only one indeterminate, q, to keep track of the pattern, as opposed to the three indeterminates r, ℓ, and d used in Section 4.2.

Definition 4.23 *For any ordered set $A \subseteq \mathbb{N}$, we denote the set of compositions of n with parts in A (respectively with m parts in A) that contain the pattern τ exactly r times by $\mathcal{C}_n^A(\tau; r)$ and $\mathcal{C}_{n,m}^A(\tau; r)$, respectively. We denote the number of compositions in these two sets by $C_A^\tau(n, r)$ (respectively $C_A^\tau(m; n, r)$). The corresponding generating functions are given by*

$$C_A^\tau(x, q) = \sum_{\sigma \in \mathcal{C}^A} x^{\mathrm{ord}(\sigma)} q^{\mathrm{occ}_\tau(\sigma)} = \sum_{n,r \geq 0} C_A^\tau(n, r) x^n q^r$$

and

$$C_A^\tau(x, y, q) = \sum_{\sigma \in \mathcal{C}^A} x^{\mathrm{ord}(\sigma)} y^{\mathrm{par}(\sigma)} q^{\mathrm{occ}_\tau(\sigma)}$$

$$= \sum_{n,m,r \geq 0} C_A^\tau(m; n, r) x^n y^m q^r$$

$$= \sum_{m \geq 0} C_A^\tau(m; x, q) y^m.$$

Moreover, let $C_A^\tau(\sigma_1 \cdots \sigma_\ell | n, r)$ (respectively $C_A^\tau(\sigma_1 \cdots \sigma_\ell | m; n, r)$) be the number of compositions of n with parts in A (respectively with m parts in A) which contain τ exactly r times and whose first ℓ parts are $\sigma_1, \ldots, \sigma_\ell$. The corresponding generating functions are given by

$$C_A^\tau(\sigma_1 \cdots \sigma_\ell | x, q) = \sum_{n,r \geq 0} C_A^\tau(\sigma_1 \cdots \sigma_\ell | n, r) x^n q^r$$

and

$$C_A^\tau(\sigma_1 \cdots \sigma_\ell | x, y, q) = \sum_{n,m,r \geq 0} C_A^\tau(\sigma_1 \cdots \sigma_\ell | m; n, r) x^n y^m q^r$$

$$= \sum_{m \geq 0} C_A^\tau(\sigma_1 \cdots \sigma_\ell | m; x, q) y^m.$$

The initial conditions are $C_A^\tau(m; x, q) = 0$ for $m < 0$, $C_A^\tau(0; x, q) = 1$, $C_A^\tau(\sigma_1 | m; x, q) = 0$ for $m \leq 0$ and $C_A^\tau(\sigma_1 \sigma_2 | m; x, q) = 0$ for $m \leq 1$. In addition,

$$C_A^\tau(x, y, q) = 1 + \sum_{a \in A} C_A^\tau(a | x, y, q). \tag{4.11}$$

Before we look at specific patterns, we define two sets that will be used in many of the proofs.

Definition 4.24 *For any ordered set $A = \{a_1, a_2, \ldots, a_d\} \subseteq \mathbb{N}$, we define*

$$A_k = \{a_1, a_2, \ldots, a_k\}$$

(k denotes the index of the largest element included) and

$$\bar{A}_k = \{a_{k+1}, a_{k+2}, \ldots, a_d\}$$

(k denotes the index of the largest element excluded). Note that $A = A_d$ and $\bar{A}_k = A - A_k$. As before, we write $[d]$ to denote the set $A = \{1, \ldots, d\}$, and we will use $[i, j]$ for the set $\{i, i+1, \ldots, j\}$. Clearly, $\mathbb{N} = \lim_{d \to \infty}[d]$.

4.3.1 The patterns peak and valley

We start by discussing the patterns *peak* $= \{121, 132, 231\}$ and *valley* $= \{212, 213, 312\}$. To derive the generating function for the pattern peak, we split the composition into parts according to where the largest part occurs. We will show the derivation for the pattern peak only, since the derivation for the pattern valley is very similar. As a first step we obtain a continued fraction expression (see Definition 2.61) for $C_A^{peak}(x, y, q)$.

Lemma 4.25 *For $A = \{a_1, \ldots, a_d\}$, $b_i = x^{a_i} y$, and $C_A = C_A^{peak}(x, y, q)$,*

$$C_A = \cfrac{1}{1 - b_d - \cfrac{1}{[b_d(1-q), b_{d-1}, b_{d-1}(1-q), \ldots, b_2, b_2(1-q), b_1]}}.$$

Proof To derive an expression for $C_A^{peak}(x, y, q)$, we concentrate on occurrences of a_d, the largest part in the set $A = \{a_1, a_2, \ldots, a_d\}$. If a_d is surrounded by smaller parts on both sides, then a peak occurs. If σ does not contain a_d, then the generating function is given by $C_{A_{d-1}}^{peak}(x, y, q)$. If there is at least one occurrence, then we need to look at three cases for the first occurrence of a_d. If the first occurrence is at the beginning of the composition, that is, $\sigma = a_d \sigma'$, where σ' is a (possibly empty) composition with parts in A, then no peak occurs, and the generating function is given by $x^{a_d} y\, C_A^{peak}(x, y, q)$. If the first (and only) occurrence of a_d is at the end of the composition, that is, $\sigma = \bar{\sigma} a_d$, where $\bar{\sigma}$ is a nonempty composition with parts in A_{d-1}, then the generating function is given by $x^{a_d} y (C_{A_{d-1}}^{peak}(x, y, q) - 1)$. Finally, if the first occurrence of a_d is in the interior of the composition, then $\sigma = \sigma^{(1)} a_d \sigma^{(2)}$, where $\sigma^{(1)}$ and $\sigma^{(2)}$ are nonempty compositions with parts in A_{d-1} and A, respectively. Furthermore, if $\sigma^{(2)}$ starts with a_d, then $\sigma = \sigma^{(1)} a_d\, a_d\, \sigma^{(3)}$, and both parts a_d can be split off without decreasing the number of occurrences of peaks; the generating function is given by $(C_{A_{d-1}}^{peak}(x, y, q) - 1) x^{2a_d} y^2 C_A^{peak}(x, y, q)$. If $\sigma^{(2)}$ does not start with a_d, then a peak occurs and the generating function is given by

$$\underbrace{(C_{A_{d-1}}^{peak}(x, y, q) - 1)}_{\sigma^{(1)}\ \text{nonempty}} x^{a_d} y\, q \big(\underbrace{C_A^{peak}(x, y, q) - 1}_{\sigma^{(2)}\ \text{nonempty}} \underbrace{-x^{a_d} y\, C_A^{peak}(x, y, q)}_{\text{does not start with } a_d} \big).$$

Now, let $C_A = C_A^{peak}(x, y, q)$. Combining the five cases above, we obtain

$$C_A = C_{A_{d-1}} + x^{a_d} y\, C_A + x^{a_d} y\, (C_{A_{d-1}} - 1) + x^{2a_d} y^2 C_A (C_{A_{d-1}} - 1)$$
$$+ x^{a_d} y q (C_A - 1 - x^{a_d} y C_A)(C_{A_{d-1}} - 1),$$

or, equivalently,

$$C_A = \frac{(1 + x^{a_d} y(1 - q)) C_{A_{d-1}} - x^{a_d} y(1 - q)}{1 - x^{a_d} y(1 - x^{a_d} y)(1 - q) - x^{a_d} y(x^{a_d} y(1 - q) + q) C_{A_{d-1}}}$$

$$= \cfrac{1}{1 - x^{a_d} y - \cfrac{C_{A_{d-1}} - 1}{(1 + x^{a_d} y(1 - q)) C_{A_{d-1}} - x^{a_d} y(1 - q)}}$$

$$= \cfrac{1}{1 - x^{a_d} y - [x^{a_d} y(1 - q), 1 - 1/C_{A_{d-1}}]},$$

where $[x^{a_d} y(1 - q), 1 - 1/C_{A_{d-1}}]$ is a finite continued fraction. Hence, by induction on d and using the fact that $C_{A_1} = \frac{1}{1 - x^{a_1} y}$, we obtain the desired result for $C_A^{peak}(x, y, q)$. □

Note that the proof given above assumes that d is finite. A similar continued fraction expression can be obtained for $C_A^{valley}(x, y, q)$ (see Exercise 4.13).

Before we can state the explicit result for the respective generating functions for the patterns peak and valley, we need some notation. For any set $B \subseteq A$

and for $s \geq 1$, we define

$$P^s(B) = \{(i_1, \ldots, i_s) | \, a_{i_j} \in B, j = 1, \ldots, s, \text{ and}$$
$$i_{2\ell-1} < i_{2\ell} \leq i_{2\ell+1} \text{ for } 1 \leq \ell \leq \lfloor s/2 \rfloor\},$$
$$Q^s(B) = \{(i_1, \ldots, i_s) | \, a_{i_j} \in B, j = 1, \ldots, s, \text{ and}$$
$$i_{2\ell-1} \leq i_{2\ell} < i_{2\ell+1} \text{ for } 1 \leq \ell \leq \lfloor s/2 \rfloor\},$$
$$M^s(B) = \sum_{(i_1, \ldots, i_s) \in P^s(B)} \prod_{j=1}^s b_{i_j} \quad \text{and} \quad N^s(B) = \sum_{(i_1, \ldots, i_s) \in Q^s(B)} \prod_{j=1}^s b_{i_j},$$

where $b_{i_j} = x^{a_{i_j}} y$.

Theorem 4.26 *Let $A = \{a_1, \ldots, a_d\}$, and let $P^s(A)$, $Q^s(A)$, $M^s(A)$, and $N^s(A)$ be defined as above. Then*

$$C_A^{peak}(x, y, q) = \frac{1 + \sum_{j \geq 1} M^{2j}(A)(1 - q)^j}{1 + \sum_{j \geq 1} M^{2j}(A)(1 - q)^j - \sum_{j \geq 0} M^{2j+1}(A)(1 - q)^j}$$

and

$$C_A^{valley}(x, y, q) = \frac{1 + \sum_{j \geq 1} M^{2j}(A)(1 - q)^j}{1 + \sum_{j \geq 1} M^{2j}(A)(1 - q)^j - \sum_{j \geq 0} N^{2j+1}(A)(1 - q)^j}.$$

Proof We derive an explicit formula for $C_A^{peak}(x, y, q)$ based on recursions for $M^s(A_d) = M^s(A)$ for odd and even s; if s is odd, both the last and second-to-last elements can equal a_d, whereas for even s, the second-to-last element can be at most a_{d-1}. By separating the elements of $P^s(A)$ according to whether the last element equals a_d or is less than a_d, we get the following two recursions:

$$M^{2s+1}(A) = b_d \, M^{2s}(A) + M^{2s+1}(A_{d-1}), \tag{4.12}$$
$$M^{2s}(A) = b_d \, M^{2s-1}(A_{d-1}) + M^{2s}(A_{d-1}).$$

Let $G_d = 1/[b_d(1 - q), b_{d-1}, b_{d-1}(1 - q), \ldots, b_2, b_2(1 - q), b_1]$, that is, G_d consists of that portion of C_A in the continued fraction expansion which has a repeating pattern. Using induction, we first show that for $d \geq 2$,

$$G_d = \frac{\sum_{j \geq 0} M^{2j+1}(A_{d-1})(1 - q)^j}{1 + \sum_{j \geq 1} M^{2j}(A)(1 - q)^j} = \frac{G_{d-1}^1}{G_d^2}. \tag{4.13}$$

For $d = 2$, $G_1^1 = b_1$ (only the term $j = 0$ contributes), $G_2^2 = 1 + b_1 b_2 (1 - q)$ (only the term $j = 1$ contributes) and therefore, the induction hypothesis is true. Now let $d \geq 3$ and assume that the hypothesis holds for $d - 1$, that is, $G_{d-1} = \frac{G_{d-2}^1}{G_{d-1}^2}$. From the definition of G_d it is easy to see that

$$G_d = 1/[b_d(1 - q), b_{d-1} + G_{d-1}].$$

Substituting the induction hypothesis for $d-1$ into this recursion for G_d and simplifying yields

$$G_d = \frac{b_{d-1}G_{d-1}^2 + G_{d-2}^1}{G_{d-1}^2 + b_d(1-q)(b_{d-1}G_{d-1}^2 + G_{d-2}^1)}.$$

Using the definitions for G_{d-2}^1 and G_{d-1}^2 in (4.13) together with (4.12) yields

$$b_{d-1}G_{d-1}^2 + G_{d-2}^1 = G_{d-1}^1.$$

This result together with the definitions of G_{d-2}^1 and G_{d-1}^2 and another application of (4.12) gives $G_{d-1}^2 + b_d(1-q)(b_{d-1}G_{d-1}^2 + G_{d-2}^1) = G_d^2$, which completes the proof of (4.13). Now we use (4.13) and $C_A = \frac{1}{1-b_d-G_d}$ to get (after simplification) that

$$C_A = \frac{1}{1 - b_d - \frac{G_{d-1}^1}{G_d^2}} = \frac{G_d^2}{G_d^2 - (b_d G_d^2 + G_{d-1}^1)} = \frac{G_d^2}{G_d^2 - G_d^1}$$

$$= \frac{1 + \sum_{j\geq 1} M^{2j}(A)(1-q)^j}{1 + \sum_{j\geq 1} M^{2j}(A)(1-q)^j - \sum_{j\geq 0} M^{2j+1}(A)(1-q)^j}.$$

This completes the proof for the pattern peak since the formula also holds for the case $d = \infty$. The result for the pattern valley follows with minor modifications, focusing on the smallest rather than the largest part, replacing A_k with $\bar{A}_k$, and using the recursions

$$M^s(\bar{A}_k) = b_{k+1} N^{s-1}(\bar{A}_{k+1}) + M^s(\bar{A}_{k+1})$$

and

$$N^s(\bar{A}_k) = b_{k+1} M^{s-1}(\bar{A}_k) + N^s(\bar{A}_{k+1}),$$

which are obtained by separating the elements of $P^s(\bar{A}_k)$ according to whether the first element equals a_{k+1} or is greater than a_{k+1}. □

We now apply Theorem 4.26 to $A = \mathbb{N}$.

Example 4.27 *To determine the generating function for the compositions of n, we first compute the specific expressions for $M^s(\mathbb{N})$ and $N^s(\mathbb{N})$ for $s \geq 1$. In each case, we have terms of the form $\sum_{i_{s-1}<i_s} x^{i_s}$ or $\sum_{i_{s-1}\leq i_s} x^{i_s}$ which we rewrite as $\sum_{j\geq 1} x^{i_{s-1}+j}$ or $\sum_{j\geq 0} x^{i_{s-1}+j}$ to successively reduce the sums.*

We obtain

$$M^{2s}(\mathbb{N}) = \sum_{1 \leq i_1} \sum_{i_1 < i_2} \cdots \sum_{i_{2s-2} \leq i_{2s-1}} \sum_{i_{2s-1} < i_{2s}} x^{i_1 + \cdots + i_{2s}} y^{2s}$$

$$= \frac{x}{1-x} y^{2s} \sum_{1 \leq i_1} \sum_{i_1 < i_2} \cdots \sum_{i_{2s-3} < i_{2s-2}} \sum_{i_{2s-2} \leq i_{2s-1}} x^{i_1 + \cdots + i_{2s-2} + 2 i_{2s-1}}$$

$$= \frac{x}{(1-x)} \frac{1}{(1-x^2)} y^{2s} \sum_{1 \leq i_1} \sum_{i_1 < i_2} \cdots \sum_{i_{2s-3} < i_{2s-2}} x^{i_1 + \cdots + i_{2s-3} + 3 i_{2s-2}}$$

$$= \cdots = \frac{x^{1+3+\cdots+2s-3} y^{2s}}{(1-x)(1-x^2)\cdots(1-x^{2s-2})} \sum_{1 \leq i_1} \sum_{i_1 < i_2} x^{i_1 + (2s-1) i_2}$$

$$= \frac{x^{1+3+\cdots+(2s-3)+(2s-1)} y^{2s}}{(1-x)(1-x^2)\cdots(1-x^{2s-1})} \sum_{1 \leq i_1} x^{2s \, i_1}$$

$$= \frac{x^{1+3+\cdots+(2s-3)+(2s-1)} x^{2s} y^{2s}}{(1-x)(1-x^2)\cdots(1-x^{2s})} = \frac{x^{s(s+2)} y^{2s}}{\prod_{i=1}^{2s}(1-x^i)}.$$

Likewise, we obtain that

$$M^{2s+1}(\mathbb{N}) = \sum_{1 \leq i_1} \sum_{i_1 < i_2} \sum_{i_2 \leq i_3} \cdots \sum_{i_{2s-1} < i_{2s}} \sum_{i_{2s} \leq i_{2s+1}} x^{i_1 + \cdots + i_{2s+1}} y^{2s+1}$$

$$= \frac{x^{2+4+\cdots+(2s-2)+(2s)} x^{(2s+1)} y^{2s+1}}{(1-x)(1-x^2)\cdots(1-x^{2s+1})} = \frac{x^{s^2+3s+1} y^{2s+1}}{\prod_{i=1}^{2s+1}(1-x^i)}$$

and

$$N^{2s+1}(\mathbb{N}) = \sum_{1 \leq i_1} \sum_{i_1 \leq i_2} \sum_{i_2 < i_3} \cdots \sum_{i_{2s-1} \leq i_{2s}} \sum_{i_{2s} < i_{2s+1}} x^{i_1 + \cdots + i_{2s+1}} y^{2s+1}$$

$$= \frac{x^{(s+1)^2} y^{2s+1}}{\prod_{i=1}^{2s+1}(1-x^i)}.$$

Substituting these expressions into Theorem 4.26 and setting $y = 1$ and $q = 0$ gives the generating functions for the number of compositions of n without peaks and without valleys, respectively, as

$$C_{\mathbb{N}}^{peak}(x,1,0) = \frac{1 + \sum_{j \geq 1} \dfrac{x^{j(j+2)}}{\prod_{i=1}^{2j}(1-x^i)}}{1 + \sum_{j \geq 1} \dfrac{x^{j(j+2)}}{\prod_{i=1}^{2j}(1-x^i)} - \sum_{j \geq 0} \dfrac{x^{j^2+3j+1}}{\prod_{i=1}^{2j+1}(1-x^i)}}$$

and

$$C_{\mathbb{N}}^{valley}(x,1,0) = \cfrac{1 + \sum_{j\geq 1} \cfrac{x^{j(j+2)}}{\prod_{i=1}^{2j}(1-x^i)}}{1 + \sum_{j\geq 1} \cfrac{x^{j(j+2)}}{\prod_{i=1}^{2j}(1-x^i)} - \sum_{j\geq 0} \cfrac{x^{(j+1)^2}}{\prod_{i=1}^{2j+1}(1-x^i)}}.$$

We approximate the respective generating function by using finite sums (up to $j = 20$) in either the Mathematica *function* Series *or the Maple function* taylor. *The sequence for the number of peak-avoiding compositions with parts in $\mathbb{N}$ for $n = 0$ to $n = 20$ is given by 1, 1, 2, 4, 7, 13, 22, 38, 64, 107, 177, 293, 481, 789, 1291, 2110, 3445, 5621, 9167, 14947 and 24366. The corresponding sequence for valley-avoiding compositions is given by 1, 1, 2, 4, 8, 15, 28, 52, 96, 177, 326, 600, 1104, 2032, 3740, 6884, 12672, 23327, 42942, 79052 and 145528. Note that the first time a peak occurs is for $n = 4$ (as the composition 121), while the first time a valley can occur is for $n = 5$ (as the composition 212).*

Even though the statistics peak and valley are in some sense symmetric, one cannot obtain the number of valley-avoiding compositions from the number of peak-avoiding compositions with parts in $\mathbb{N}$. However there is a connection. The number of valleys in compositions of n with m parts is equal to the number of peaks in compositions of $m(n + 1) - n$ with m parts. This can be easily seen as follows: in each composition of n with m parts, replace each part σ_i by $(n + 1) - \sigma_i$, which results in a composition of

$$m(n + 1) - \sum_{i=1}^{m} \sigma_i = m(n + 1) - n.$$

This connection will be important in Chapter 6 when we apply results derived for various patterns in compositions to words on k letters.

4.3.2 The pattern 123

We now derive the generating function for the number of compositions of n with m parts in A that contain the pattern 123 exactly r times. This is the first type of pattern in which we have to employ a different method. Unlike in the case of the pattern peak, where splitting up the composition according to the largest part and checking whether both neighbors were smaller was enough to produce a recursion, we will now have to keep a "memory" of the parts to be split off. This situation, common for patterns of length three or more, prompts us to define an auxiliary function that keeps track of the number of occurrences of τ in an augmented sequence.

Definition 4.28 *Let* $A = \{a_1, \ldots, a_d\}$ *be any ordered subset of* $\mathbb{N}$ *and let*

$$D_A^\tau(x, y, q) = \sum_{\sigma \in \mathcal{C}^A} x^{\mathrm{ord}(\sigma)} y^{\mathrm{par}(\sigma)} q^{\mathrm{occ}_\tau(\hat{\sigma})}$$

be the generating function for the number of compositions σ *according to the order and number of parts of* σ *and the number of occurrences of* τ *in the augmented sequence* $\hat{\sigma}$, *where* $\hat{\sigma}$ *is of the form* $a\sigma$ *or* σb *for values of* a *and* b *that depend on the specific pattern* τ *under consideration.*

We first derive an explicit expression for $D_A^{123}(x, y, q)$.

Lemma 4.29 *Let* $A = \{a_1, \ldots, a_d\}$ *be any ordered subset of* $\mathbb{N}$ *and let*

$$t^p(A) = \sum_{1 \le i_1 < i_2 < \cdots < i_p \le d} y^p \prod_{j=1}^p x^{a_{i_j}}$$

for $p = 1, \ldots, d$. *Then*

$$D_A^{123} = \frac{1 + \sum_{p=2}^d \sum_{j=0}^{p-2} \binom{p-2}{j} t^{p+j}(A)(q-1)^{p-1}}{1 - t^1(A) - \sum_{p=3}^d \sum_{j=0}^{p-3} \binom{p-3}{j} t^{p+j}(A)(q-1)^{p-2}}, \tag{4.14}$$

where $\hat{\sigma} = a\sigma$ *for any integer* $a < a_1$.

Proof Let $\tau = 123$. We derive an expression for $D_{\bar{A}_1}^\tau(x, y, q)$ by considering compositions σ with parts in $\bar{A}_1$ such that $a_1\sigma$ contains the pattern τ exactly r times. If σ does not contain the part a_2, then the generating function for σ is given by $D_{\bar{A}_2}^\tau(x, y, q)$. Otherwise, we write

$$\sigma = \sigma^{(1)} a_2 \sigma^{(2)} a_2 \sigma^{(3)} \cdots a_2 \sigma^{(\ell+2)}$$

with $\ell \ge 0$, where $\sigma^{(j)}$ is a (possibly empty) composition with parts in $\bar{A}_2$ for $j = 1, \ldots, \ell+2$. There are four subcases, depending on whether $\sigma^{(1)}$ and $\sigma^{(2)}$ are empty compositions or not. If $\sigma^{(1)}$ and $\sigma^{(2)}$ are both empty compositions, then $a_1\sigma = a_1 a_2$ or $a_1\sigma = a_1 a_2 a_2 \sigma^{(3)} \cdots a_2 \sigma^{(\ell+2)}$, $\ell \ge 1$. In either case we can split off the first a_2 of σ, which results in one less part, but no reduction in the number of occurrences of the pattern τ. Thus, the generating function for σ is given by

$$x^{a_2} y \sum_{\ell \ge 0} (x^{a_2} y D_{\bar{A}_2}^\tau(x, y, q))^\ell = \frac{x^{a_2} y}{1 - x^{a_2} y D_{\bar{A}_2}^\tau(x, y, q)}.$$

If $\sigma^{(1)}$ is the empty composition and $\sigma^{(2)}$ is nonempty, then $a_1\sigma = a_1 a_2 \sigma^{(2)}$ or $a_1\sigma = a_1 a_2 \sigma^{(2)} a_2 \sigma^{(3)} \cdots a_2 \sigma^{(\ell+2)}$, $\ell \ge 1$, and the generating function for σ is given by

$$x^{a_2} yq(D_{\bar{A}_2}^\tau(x, y, q) - 1) \sum_{\ell \ge 0} (x^{a_2} y D_{\bar{A}_2}^\tau(x, y, q))^\ell = \frac{x^{a_2} yq(D_{\bar{A}_2}^\tau(x, y, q) - 1)}{1 - x^{a_2} y D_{\bar{A}_2}^\tau(x, y, q)}.$$

If $\sigma^{(1)}$ is nonempty and $\sigma^{(2)}$ is the empty composition, then $\sigma = \sigma^{(1)}a_2$ or $\sigma = \sigma^{(1)}a_2a_2\sigma^{(3)}\cdots a_2\sigma^{(\ell+2)}$, $\ell \geq 1$, and the generating function for σ is given by

$$x^{a_2}y(D^{\tau}_{\bar{A}_2}(x,y,q)-1)\sum_{\ell\geq 0}(x^{a_2}yD^{\tau}_{\bar{A}_2}(x,y,q))^{\ell} = \frac{x^{a_2}y(D^{\tau}_{\bar{A}_2}(x,y,q)-1)}{1-x^{a_2}yD^{\tau}_{\bar{A}_2}(x,y,q)}.$$

Finally, if both $\sigma^{(1)}$ and $\sigma^{(2)}$ are nonempty, then the generating function for σ is given by

$$\frac{x^{a_2}y(D^{\tau}_{\bar{A}_2}(x,y,q)-1)^2}{1-x^{a_2}yD^{\tau}_{\bar{A}_2}(x,y,q)}.$$

Adding all four cases, using the shorthand D_A for $D^{\tau}_A(x,y,q)$, and solving for $D_{\bar{A}_1}$ we obtain that

$$D_{\bar{A}_1} = \frac{(1-x^{a_2}y(1-q))D_{\bar{A}_2}+x^{a_2}y(1-q)}{1-x^{a_2}yD_{\bar{A}_2}}. \qquad (4.15)$$

From this recurrence relation for the generating functions $D_{\bar{A}_i}$ we want to derive an explicit expression for D_A. To do this, we give an interpretation of $t^p(\bar{A}_k)$ as the generating function for the number of partitions with p distinct parts from the set $\bar{A}_k$. These partitions either contain the part a_{k+1} or not. In the first case, the generating function is given by $x^{a_{k+1}}y\,t^{p-1}(\bar{A}_{k+1})$, and in the second case, by $t^p(\bar{A}_{k+1})$. Thus,

$$t^p(\bar{A}_k) = t^p(\bar{A}_{k+1}) + x^{a_{k+1}}y\,t^{p-1}(\bar{A}_{k+1}). \qquad (4.16)$$

We now prove (4.14) by induction on d, the number of elements in A. For $d=0$ and $d=1$ we have $D_{\varnothing}=1$ and $D_{\{a_1\}}=\sum_{n\geq 0}x^{n\cdot a_1}y^n=1/(1-x^{a_1}y)$, respectively, and thus, (4.14) holds. Now assume that (4.14) holds for $d-1$ and define $s_A(i,d) = \sum_{p=i}^{d}\sum_{j=0}^{p-i}\binom{p-i}{j}t^{p+j}(A)(q-1)^{p-i+1}$. Note that the second parameter of $s_A(i,d)$ indicates the number of elements in the set A, and that the powers of $(q-1)$ range from 1 to $d-i+1$. With this notation, the induction hypothesis becomes $D_A = \frac{1+s_A(2,d)}{1-t^1(A)-s_A(3,d)}$. Using the recursion (4.15) together with the induction hypothesis for the set $\bar{A}_1$ gives

$$D_A = \frac{(1-x^{a_1}y(1-q))D_{\bar{A}_1}+x^{a_1}y(1-q)}{1-x^{a_1}yD_{\bar{A}_1}}$$

$$= \frac{(1-x^{a_1}y(1-q))(1+s_{\bar{A}_1}(2,d-1))}{1-t^1(\bar{A}_1)-s_{\bar{A}_1}(3,d-1)-x^{a_1}y(1+s_{\bar{A}_1}(2,d-1))}$$

$$+ \frac{x^{a_1}y(1-q)(1-t^1(\bar{A}_1)-s_{\bar{A}_1}(3,d-1))}{1-t^1(\bar{A}_1)-s_{\bar{A}_1}(3,d-1)-x^{a_1}y(1+s_{\bar{A}_1}(2,d-1))}$$

$$= \frac{f_1}{f_2}.$$

We first expand the denominator and obtain that

$$f_2 = 1 - t^1(\bar{A}_1) - x^{a_1}y - \sum_{p=3}^{d-1}\sum_{j=0}^{p-3} \binom{p-3}{j}t^{p+j}(\bar{A}_1)(q-1)^{p-2}$$

$$- \sum_{p=2}^{d-1}\sum_{j=0}^{p-2} \binom{p-2}{j}x^{a_1}y\,t^{p+j}(\bar{A}_1)(q-1)^{p-1}.$$

Combining the two double sums by reindexing the second one, adding the terms for $p = d$ to the first one (note that $t^{d+j}(\bar{A}_1) = 0$ for $j \geq 0$) and applying (4.16) twice gives $f_2 = 1 - t^1(A) - s_A(3, d)$. Next we expand the numerator and simplify to obtain that

$$f_1 = 1 + x^{a_1}y(q-1)t^1(\bar{A}_1) + s_{\bar{A}_1}(2, d-1)$$
$$+ x^{a_1}y(q-1)(s_{\bar{A}_1}(2, d-1) + s_{\bar{A}_1}(3, d-1)).$$

We now collect terms according to powers of $(q-1)$. For $m = 0$, the coefficient is 1. If $m = 1$, then $[q-1]f_1$ is given by $x^{a_1}y\,t^1(\bar{A}_1) + t^2(\bar{A}_1) = t^2(A)$ by (4.16). If $m = 2, 3, \ldots, d-1$, then the coefficient of $(q-1)^m$ is equal to

$$\sum_{j=0}^{m-1}\binom{m-1}{j}t^{m+1+j}(\bar{A}_1) + x^{a_1}y\sum_{j=0}^{m-2}\binom{m-2}{j}(t^{m+j}(\bar{A}_1) + t^{m+1+j}(\bar{A}_1))$$

$$= \sum_{j=0}^{m-1}\binom{m-1}{j}t^{m+1+j}(\bar{A}_1) + x^{a_1}y\sum_{j=0}^{m-1}\left(\binom{m-2}{j} + \binom{m-2}{j-1}\right)t^{m+j}(\bar{A}_1)$$

by reindexing the second term in the second sum and using that $\binom{m-2}{-1} = \binom{m-2}{m-1} = 0$. Since $\binom{a}{b-1} + \binom{a}{b} = \binom{a+1}{b}$, we get that

$$[(q-1)^m]f_1 = \sum_{j=0}^{m-1}\binom{m-1}{j}t^{m+1+j}(\bar{A}_1) + x^{a_1}y\sum_{j=0}^{m-1}\binom{m-1}{j}t^{m+j}(\bar{A}_1)$$

$$= \sum_{j=0}^{m-1}\binom{m-1}{j}\left(t^{m+1+j}(\bar{A}_1) + x^{a_1}y\,t^{m+j}(\bar{A}_1)\right)$$

$$= \sum_{j=0}^{m-1}\binom{m-1}{j}t^{m+j+1}(A),$$

where the last equality follows once more from using (4.16). Thus,

$$f_1 = 1 + t^2(A)(q-1) + \sum_{m=2}^{d-1}\sum_{j=0}^{m-1}\binom{m-1}{j}t^{m+j+1}(A)(q-1)^m$$

$$= 1 + s_A(2, d)$$

after reindexing. Combining this result with the result for f_2 completes the proof of the lemma. $\qquad\square$

We are now ready to derive an explicit expression for $C_A^{123}(x, y, q)$.

Theorem 4.30 *Let* $A = \{a_1, \ldots, a_d\}$ *be any ordered subset of* $\mathbb{N}$. *Then,*

$$C_A^{123}(x, y, q) = \frac{1}{1 - t^1(A) - \sum_{p=3}^{d} \sum_{j=0}^{p-3} \binom{p-3}{j} t^{p+j}(A)(q-1)^{p-2}},$$

where $t^p(A) = \sum_{1 \le i_1 < i_2 < \cdots < i_p \le d} y^p \prod_{j=1}^{p} x^{a_{i_j}}$ *for* $p = 1, \ldots, d$.

Proof Let σ be any composition of n with m parts in $A = \{a_1, \ldots, a_d\}$ that contains the pattern $\tau = 123$ exactly r times. For any composition σ with parts in A, there are two possibilities: either σ does not contain a_1, in which case the generating function is given by $C_{\bar{A}_1}^\tau(x, y, q)$, or the composition contains at least one occurrence of a_1, that is, $\sigma = \bar{\sigma} a_1 \sigma'$, where $\bar{\sigma}$ is a composition in $\bar{A}_1$ and σ' has parts in A. In this case, the generating function is given by

$$C_{\bar{A}_1}^\tau(x, y, q) C_A^\tau(a_1|x, y, q).$$

All together, we have

$$C_A^\tau(x, y, q) = C_{\bar{A}_1}^\tau(x, y, q) + C_{\bar{A}_1}^\tau(x, y, q) C_A^\tau(a_1|x, y, q). \qquad (4.17)$$

Now let us consider the compositions σ of n with m parts in A starting with a_1 that contain the pattern τ exactly r times. Again, there are two cases: either σ contains exactly one occurrence of a_1, or the part a_1 occurs at least twice in σ. In the first case, the generating function is given by $x^{a_1} y D_{\bar{A}_1}^\tau(x, y, q)$. If σ contains a_1 at least twice, then we decompose the composition according to the second occurrence of a_1, that is, $\sigma = a_1 \bar{\sigma} a_1 \sigma'$, where $\bar{\sigma}$ is a (possibly empty) composition with parts from $\bar{A}_1$. Splitting off the initial part a_1 results in the generating function $x^{a_1} y D_{\bar{A}_1}^\tau(x, y, q) C_A^\tau(a_1|x, y, q)$. Thus,

$$C_A^\tau(a_1|x, y, q) = x^{a_1} y D_{\bar{A}_1}^\tau(x, y, q) + x^{a_1} y D_{\bar{A}_1}^\tau(x, y, q) C_A^\tau(a_1|x, y, q).$$

Solving for $C_A^\tau(a_1|x, y, q)$ and substituting the result into (4.17), together with Lemma 4.29 gives

$$C_A^\tau(x, y, q) = \frac{C_{\bar{A}_1}^\tau(x, y, q)}{1 - x^{a_1} y D_{\bar{A}_1}^\tau(x, y, q)} = \frac{C_{\bar{A}_1}^\tau(x, y, q)}{1 - x^{a_1} y \dfrac{1 + s_{\bar{A}_1}(2, d-1)}{1 - t^1(\bar{A}_1) - s_{\bar{A}_1}(3, d-1)}}.$$

Using the same arguments as in the proof of Lemma 4.29 yields that

$$C_A^\tau(x, y, q) = C_{\bar{A}_1}^\tau(x, y, q) \frac{1 - t^1(\bar{A}_1) - s_{\bar{A}_1}(3, d-1)}{1 - t^1(A) - s_A(3, d)}.$$

Iterating this equation and using that $C^\tau_{\bar{A}_d}(x,y,q) = C^\tau_\varnothing(x,y,q) = 1$ results in

$$C^\tau_A(x,y,q) = \prod_{k=0}^{d-1} \frac{1 - t^1(\bar{A}_{k+1}) - s_{\bar{A}_{k+1}}(3, d-k-1)}{1 - t^1(\bar{A}_k) - s_{\bar{A}_k}(3, d-k)}.$$

Simplification together with $t^p(\bar{A}_d) = t^p(\varnothing) = 0$ completes the proof. $\square$

Example 4.31 *To apply Theorem 4.30 to $A = \mathbb{N}$, we first show that*

$$t^p(\mathbb{N}) = x^{\binom{p+1}{2}} y^p/(x;x)_p,$$

where we use the customary notation $(a;q)_n = \prod_{j=0}^{n-1}(1 - aq^j)$. The derivation of the expression for $t^p(\mathbb{N})$ is similar to the one in Example 4.27. However, we now have only strict inequalities for the indices in the summation, and therefore all powers, as opposed to only odd powers, occur in the numerator. Clearly, the formula for $t^p(\mathbb{N})$ holds for $p = 0$. Resolving the individual sums and using $a_i = i$ for $i \geq 1$ we get that

$$t^p(\mathbb{N}) = y^p \prod_{j=1}^p \frac{x^j}{(1-x^j)} = \frac{x^{\binom{p+1}{2}} y^p}{(x;x)_p}.$$

Setting $y = 1$ and $q = 0$ in Theorem 4.30 gives the generating function for the number of compositions with parts in $\mathbb{N}$ that avoid 123 as

$$C^{123}_{\mathbb{N}}(x,1,0) = \cfrac{1}{1 - \frac{x}{1-x} - \sum_{p \geq 3} \sum_{j=0}^{p-3} \binom{p-3}{j} \frac{x^{\binom{p+1+j}{2}}}{(x;x)_{p+j}}(-1)^{p-2}},$$

and the sequence for the number of 123-avoiding compositions with parts in $\mathbb{N}$ for $n = 0$ to $n = 20$ is given by 1, 1, 2, 4, 8, 16, 31, 61, 119, 232, 453, 883, 1721, 3354, 6536, 12735, 24813, 48344, 94189, 183506, and 357518. Note that the first time the pattern 123 can occur is for $n = 6$, as the composition 123.

We will now generalize to the enumeration of compositions that contain ℓ-letter subword patterns. For instance, we extend the question of counting the occurrences of the subwords 11 (level) and 111 (level+level) to the question of counting the occurrences of the pattern $1^\ell = 11 \cdots 1$ in the set of compositions of n with m parts in A. Likewise, we will derive generating functions for generalizations of the patterns 112 (level+rise) and 221 (level+drop).

4.3.3 The pattern $\tau = 1^\ell$

We start with the easiest pattern, namely $\tau = 1^\ell$, which consists of ℓ 1s. Note that for $\ell = 3$, we obtain the result for the statistic level+level.

Theorem 4.32 *Let $\tau = 1^\ell$. Then the generating function $C_A^\tau(x, y, q)$ is given by*

$$\frac{1}{1 - \displaystyle\sum_{a \in A} \frac{x^a y + x^{2a} y^2 + \cdots + x^{a(\ell-1)} y^{\ell-1} + x^{a\ell} y^\ell q/(1 - x^a y q)}{1 + x^a y + x^{2a} y^2 + \cdots + x^{a(\ell-1)} y^{\ell-1} + x^{a\ell} y^\ell q/(1 - x^a y q)}}.$$

Proof For all $s = 1, 2, \ldots, \ell - 1$,

$$C_A^\tau(a^s | x, y, q) = x^{as} y^s + \sum_{b \neq a \in A} C_A^\tau(a^s b | x, y, q) + C_A^\tau(a^{s+1} | x, y, q).$$

Observe that if $\sigma \in C_{n,m}^A(\tau; r)$ (that is, σ is a composition of n with m parts in A having r occurrences of τ) and $\sigma_1 \cdots \sigma_s \sigma_{s+1} = a^s b$, where $1 \leq s \leq \ell - 1$ and $b \neq a$, then no occurrence of the pattern τ in σ can involve any of the first s letters of σ. Thus,

$$\sigma \in C_{n,m}^A(\tau; r) \text{ if and only if } b\sigma_{s+2} \cdots \sigma_m \in C_{n-sa, m-s}^A(\tau; r),$$

and therefore, for $1 \leq s \leq \ell - 1$,

$$C_A^\tau(a^s | x, y, q) = x^{as} y^s + x^{sa} y^s \sum_{b \neq a \in A} C_A^\tau(b | x, y, q) + C_A^\tau(a^{s+1} | x, y, q).$$

Equivalently, using (4.11), we have that

$$C_A^\tau(a^s | x, y, q) = x^{sa} y^s (C_A^\tau(x, y, q) - C_A^\tau(a | x, y, q)) + C_A^\tau(a^{s+1} | x, y, q). \quad (4.18)$$

Moreover, for $s = \ell$,

$$C_A^\tau(a^\ell | x, y, q) = x^{a\ell} y^\ell q + \sum_{b \neq a \in A} C_A^\tau(a^\ell b | x, y, q) + C_A^\tau(a^{\ell+1} | x, y, q).$$

If $\sigma \in C_{n,m}^A(\tau; r)$ and $\sigma_1 \cdots \sigma_\ell \sigma_{\ell+1} = a^\ell b$ where $b \neq a$, then exactly one occurrence of the pattern τ can involve the first ℓ letters of σ. Therefore, $\sigma \in C_{n,m}^A(\tau; r)$ if and only if $b\sigma_{\ell+2} \cdots \sigma_m \in C_{n-\ell a, m-\ell}^A(\tau; r - 1)$. If $b = a$, we get that $\sigma \in C_{n,m}^A(\tau; r)$ if and only if $\sigma_2 \cdots \sigma_m \in C_{n-a, m-1}^A(\tau; r - 1)$. Hence,

$$C_A^\tau(a^\ell | x, y, q) = x^{a\ell} y^\ell q + x^{a\ell} y^\ell q \sum_{b \neq a \in A} C_A^\tau(b | x, y, q) + x^a y q C_A^\tau(a^\ell | x, y, q),$$

which, using (4.11), is equivalent to

$$C_A^\tau(a^\ell | x, y, q) = \frac{x^{a\ell} y^\ell q}{1 - x^a y q} (C_A^\tau(x, y, q) - C_A^\tau(a | x, y, q)). \quad (4.19)$$

Iterating (4.18) for $s, s+1, \ldots, \ell - 1$ and using (4.19) we obtain that

$$C_A^\tau(a^s | x, y, q)$$
$$= \left(x^{as} y^s + \cdots + x^{a(\ell-1)} y^{\ell-1} + \frac{x^{a\ell} y^\ell q}{1 - x^a yq} \right) \left(C_A^\tau(x, y, q) - C_A^\tau(a | x, y, q) \right)$$

for all $s = 1, 2, \ldots \ell$. Letting $s = 1$ and solving for $C_A^\tau(a | x, y, q)$ gives

$$C_A^\tau(a | x, y, q) = \frac{x^a y + x^{2a} y^2 + \cdots + x^{a(\ell-1)} y^{\ell-1} + \dfrac{x^{a\ell} y^\ell q}{1 - x^a yq}}{1 + x^a y + x^{2a} y^2 + \cdots + x^{a(\ell-1)} y^{\ell-1} + \dfrac{x^{a\ell} y^\ell q}{1 - x^a yq}} C_A^\tau(x, y, q),$$

which together with (4.11) completes the proof. □

We can now apply Theorem 4.32 to a generalization of the Carlitz compositions.

Definition 4.33 *A k-Carlitz composition is a composition in which no k consecutive parts are the same, for $k \geq 2$.*

Example 4.34 (*k-Carlitz compositions*) *Applying Theorem 4.32 with $q = 0$ and $A = \mathbb{N}$ gives that the generating function for the number of k-Carlitz compositions of n with m parts in $\mathbb{N}$ is given by*

$$C^{1^k}(x, y, 0) = \frac{1}{1 - \displaystyle\sum_{j \geq 1} \frac{x^j y + x^{2j} y^2 + \cdots + x^{j(k-1)} y^{k-1}}{1 + x^j y + x^{2j} y^2 + \cdots + x^{j(k-1)} y^{k-1}}}.$$

In particular, if we use $k = 3$ in Theorem 4.32, then we obtain the generating function for the statistic level+level [92, Theorem 2.1].

4.3.4 The patterns $1^{\ell-1}2$ and $2^{\ell-1}1$

The next simplest patterns are the patterns $1^{\ell-1}2$ and $2^{\ell-1}1$, which is the most obvious way to generalize the 3-letter patterns 112 and 221. Note that these two patterns are not strongly tight Wilf-equivalent, but that their generating functions have a very similar structure.

Theorem 4.35 *Let $\tau = 1^{\ell-1}2$ and $\nu = 2^{\ell-1}1$. Then*

$$C_A^\tau(x, y, q) = \frac{1}{1 - \displaystyle\sum_{j=1}^d x^{a_j} y \prod_{i=1}^{j-1} (1 - x^{(\ell-1)a_i} y^{\ell-1}(1 - q))}$$

and

$$C_A^\nu(x, y, q) = \frac{1}{1 - \displaystyle\sum_{j=1}^d x^{a_j} y \prod_{i=j+1}^d (1 - x^{(\ell-1)a_i} y^{\ell-1}(1 - q))}.$$

Proof Let $A = \{a_1, \ldots, a_d\}$. To derive a recurrence relation for $C_A^\tau(x, y, q)$, we once more decompose σ according to where the largest part occurs. If σ does not contain a_d, then the generating function is given by $C_{A_{d-1}}^\tau(x, y, q)$. Otherwise, $\sigma = \sigma^{(1)} a_d \sigma^{(2)}$, where $\sigma^{(1)}$ has parts in A_{d-1}. The generating function for these compositions is given by $x^{a_d} y D_{A_{d-1}}^\tau(x, y, q) C_A^\tau(x, y, q)$ for $\hat{\sigma} = \sigma a_d$. Adding the two cases and solving for $C_A^\tau(x, y, q)$ gives

$$C_A^\tau(x, y, q) = \frac{C_{A_{d-1}}^\tau(x, y, q)}{1 - x^{a_d} y D_{A_{d-1}}^\tau(x, y, q)}. \tag{4.20}$$

To derive a recurrence relation for $D_A^\tau(x, y, q)$, we again decompose σ according to where the part a_d occurs and let $\hat{\sigma} = \sigma b$ where b is bigger than the maximal element of A. Thus, $\sigma = \sigma^{(1)} a_d \sigma^{(2)} \ldots a_d \sigma^{(s)} a_d \sigma^{(s+1)}$, where $\sigma^{(i)}$ has parts in A_{d-1} for $1 \leq i \leq s+1$. Note that the extra part b in the sequence σb can only be part of an occurrence of τ if σ ends in $a_d^{\ell-1}$, which is equivalent to $\sigma^{(i)} = \varnothing$ for $i = s+3-\ell, \ldots, s+1$. This special case can only occur if $s \geq \ell - 1$, in which case we get an extra factor of q in the generating function. From the s parts a_d, we obtain a factor of $x^{a_d s} y^s$, and the segments $\sigma^{(i)} a_d$ and $\sigma^{(s+1)} b$ each create a factor of $D_{A_{d-1}}^\tau(x, y, q)$. If $s \geq \ell - 1$, then we have to take into account whether the special case occurs, so we distinguish between the compositions not ending in $a_d^{\ell-1}$ and those that do. Overall, we obtain that

$$D_A^\tau(x, y, q) = \sum_{s=0}^{\ell-2} x^{a_d s} y^s D_{A_{d-1}}^\tau(x, y, q)^{s+1}$$

$$+ \sum_{s=\ell-1}^{\infty} x^{a_d s} y^s \left(\left(D_{A_{d-1}}^\tau(x, y, q)^{s+1} - D_{A_{d-1}}^\tau(x, y, q)^{(s+1)-(\ell-1)} \right) \right.$$

$$\left. + q D_{A_{d-1}}^\tau(x, y, q)^{(s+1)-(l-1)} \right).$$

Simplifying these sums using geometric series yields

$$D_A^\tau(x, y, q) = \frac{(1 - x^{(\ell-1)a_d} y^{\ell-1}(1-q)) D_{A_{d-1}}^\tau(x, y, q)}{1 - x^{a_d} y D_{A_{d-1}}^\tau(x, y, q)}. \tag{4.21}$$

Deriving an explicit formula for $D_A^\tau(x, y, q)$ is rather formidable, so we derive a recurrence relation for $\frac{1}{D_A^\tau(x, y, q)}$ instead. (We employed the same "trick" for the generating function of cracked compositions.) Rewriting (4.21) in terms of $\frac{1}{D_A^\tau(x, y, q)}$ and defining $c_d = 1 - x^{(\ell-1)a_d} y^{\ell-1}(1-q)$ and $b_d = x^{a_d} y$, we obtain the linear recurrence relation

$$\frac{1}{D_A^\tau(x, y, q)} = \frac{1}{c_d} \cdot \frac{1}{D_{A_{d-1}}^\tau(x, y, q)} - \frac{b_d}{c_d},$$

which can be iterated easily. Using that $D_{\varnothing}^{\tau}(x, y, q) = 1$, we obtain

$$\frac{1}{D_A^{\tau}(x, y, q)} = \frac{1}{\prod_{i=1}^{d} c_i} - \sum_{j=1}^{d} \frac{b_j}{\prod_{i=j}^{d} c_i} = \frac{1 - \sum_{j=1}^{d} b_j \prod_{i=1}^{j-1} c_i}{\prod_{i=1}^{d} c_i}. \tag{4.22}$$

To complete the proof, we return to Equation (4.20). Combining (4.20) and (4.21) (both have the same denominator) gives

$$C_A^{\tau}(x, y, q) = \frac{D_A^{\tau}(x, y, q) C_{A_{d-1}}^{\tau}(x, y, q)}{c_d D_{A_{d-1}}^{\tau}(x, y, q)}. \tag{4.23}$$

Rewriting (4.23) in terms of $C_A^{\tau}(x, y, q)/D_A^{\tau}(x, y, q)$ produces yet another nice linear recurrence, which, when iterated d times with the initial conditions $D_{\varnothing}^{\tau}(x, y, q) = C_{\varnothing}^{\tau}(x, y, q) = 1$, results in

$$C_A^{\tau}(x, y, q) = \frac{D_A^{\tau}(x, y, q)}{c_d c_{d-1} \cdots c_1}.$$

Combining this result with (4.22) gives the explicit expression for $C_A^{\tau}(x, y, q)$. We obtain the general case for $|A| = \infty$ by taking limits as $n \to \infty$. The result for $\nu = 2^{\ell-1}1$ is obtained by replacing a_d with a_1, using $\hat{\sigma} = \sigma b$ with b less than or equal to the smallest element in σ, and adjusting the arguments accordingly. □

Example 4.36 *We can apply Theorem 4.35 for $q = 0$ and $A = \mathbb{N}$ to obtain the generating function for the number of compositions of n with m parts which avoid the ℓ-letter patterns $\tau = 1^{\ell-1}2$ and $\nu = 2^{\ell-1}1$, respectively:*

$$C_{\mathbb{N}}^{\tau}(x, y, 0) = \frac{1}{1 - \sum_{j \geq 1} x^j y \prod_{i=1}^{j-1} (1 - x^{(\ell-1)i} y^{\ell-1})}$$

and

$$C_{\mathbb{N}}^{\nu}(x, y, 0) = \frac{1}{1 - \sum_{j \geq 1} x^j y \prod_{i \geq j+1} (1 - x^{(\ell-1)i} y^{\ell-1})}.$$

Furthermore, if we use $\ell = 3$ in Theorem 4.35, then we obtain the generating function for the statistics level+rise and level+drop [92, Theorem 2.2]. Note that the proof given for the patterns 112 and 221 in [92] uses a different argument (based on solving a system of equations) which is hard to generalize for the longer patterns $1^{\ell-1}2$ and $2^{\ell-1}1$.

4.3.5   Families of patterns

In this section we consider several families of ℓ-letter patterns. For example, the easiest generalization of the pattern 212 is the pattern $21^{\ell-2}2$, but a more general version would be the pattern $p\tau'p$, where τ' is a pattern of length $\ell - 2$ with letters from the alphabet $[p - 1]$. Likewise, the pattern 121 can be generalized to the pattern $\tau = 1\tau'1$, where τ' is a pattern of length $\ell - 2$ with letters from $\{2, 3, \ldots\}$. The following theorem gives the generating functions for these two families of patterns.

Theorem 4.37 *Let* $\tau = p\tau'p$ *and* $\nu = 1\nu'1$, *where* τ' *is a pattern of length* $\ell - 2$ *in* $[p - 1]$ *and* ν' *is a pattern of length* $\ell - 2$ *in* $\{2, 3, \ldots\}$. *Then*

$$C_A^\tau(x, y, q) = \cfrac{1}{1 - \displaystyle\sum_{j=1}^d \cfrac{x^{a_j}y}{1 + x^{a_j}y^{\ell-1}(1 - q) \displaystyle\sum_{\alpha \in \mathcal{B}_j} x^{\mathrm{ord}(\alpha)}}},$$

where $\mathcal{B}_j$ *is the set of all compositions* $\alpha = a_{i_1} \cdots a_{i_{\ell-2}}$ *with parts in* A_j *such that* α *is order-isomorphic to* τ', *and*

$$C_A^\nu(x, y, q) = \cfrac{1}{1 - \displaystyle\sum_{j=1}^d \cfrac{x^{a_j}y}{1 + x^{a_j}y^{\ell-1}(1 - q) \displaystyle\sum_{\alpha \in \bar{\mathcal{B}}_j} x^{\mathrm{ord}(\alpha)}}},$$

where $\bar{\mathcal{B}}_j$ *is the set of all compositions* $\alpha = a_{i_1} \cdots a_{i_{\ell-2}}$ *with parts in* $\bar{A}_{j-1}$ *such that* α *is order-isomorphic to* ν'.

Proof We use arguments similar to those of the proof of Theorem 4.35, except we now use $D_A^\tau(x, y, q)$ with $\hat{\sigma} = a_d\sigma$. Therefore,

$$C_A^\tau(x, y, q) = C_{A_{d-1}}^\tau(x, y, q) + x^{a_d}yC_{A_{d-1}}^\tau(x, y, q)D_A^\tau(x, y, q). \qquad (4.24)$$

Now we derive a recurrence relation for $D_A^\tau(x, y, q)$. If σ does not contain the letter a_d, then the generating function for the number of such σ is given by $C_{A_{d-1}}^\tau(x, y, q)$. Otherwise, we write $\sigma = \sigma^{(1)}a_d\sigma^{(2)}$ where $\sigma^{(1)}$ has parts in A_{d-1}. If $a_d\sigma^{(1)}a_d$ is not order-isomorphic to τ, then we can split off $\sigma^{(1)}a_d$ and we express the generating function as the difference between the generating function for all such σ and those where $\sigma^{(1)}$ is order-isomorphic to τ':

$$x^{a_d}y\left(C_{A_{d-1}}^\tau(x, y, q) - y^{\ell-2}\sum_{\alpha \in \mathcal{B}_d} x^{\mathrm{ord}(\alpha)}\right)D_A^\tau(x, y, q).$$

For those compositions where $\sigma^{(1)}$ is order-isomorphic to τ' we have one occurrence of τ in the part that we are splitting off and therefore get an extra factor of q in the generating function given by

$$x^{a_d}qy^{\ell-1}\sum_{\alpha \in \mathcal{B}_d} x^{\mathrm{ord}(\alpha)}D_A^\tau(x, y, q).$$

Combining the three cases we obtain

$$D_A^\tau(x, y, q) = C_{A_{d-1}}^\tau(x, y, q)$$
$$+ x^{a_d} y \Big(C_{A_{d-1}}^\tau(x, y, q) - (1-q) y^{\ell-2} \sum_{\alpha \in \mathcal{B}_d} x^{\text{ord}(\alpha)} \Big) D_A^\tau(x, y, q).$$

Solving this equation for $D_A^\tau(x, y, q)$ and substituting into (4.24) yields

$$C_A^\tau(x, y, q) = \frac{\big(1 + (1-q) x^{a_d} y^{\ell-1} \sum_{\alpha \in \mathcal{B}_d} x^{\text{ord}(\alpha)}\big) C_{A_{d-1}}^\tau(x, y, q)}{1 + (1-q) x^{a_d} y^{\ell-1} \sum_{\alpha \in \mathcal{B}_d} x^{\text{ord}(\alpha)} - x^{a_d} y C_{A_{d-1}}^\tau(x, y, q)}.$$

Using induction on d together with the initial condition $C_{A_{p-1}}^\tau(x, y, q) = 1/\big(1 - y \sum_{j=1}^{p-1} x^{a_j}\big)$ gives the result for τ. Replacing a_d by a_1, using $\hat{\sigma} = a_1 \sigma$ and adjusting the arguments accordingly gives the result for ν. $\square$

Example 4.38 *We can apply Theorem 4.37 for $\tau = 21^{\ell-2}2$. Since the only compositions with $\ell-2$ parts that are order-isomorphic to $1^{\ell-2}$ are of the form $a_i^{\ell-2}$, the generating function is given by*

$$C_A^{21^{\ell-2}2}(x, y, q) = \frac{1}{1 - \sum_{j=1}^d \left(\dfrac{x^{a_j} y}{1 + x^{a_j} y^{\ell-1} (1-q) \sum_{i=1}^{j-1} x^{(\ell-2) a_i}} \right)}.$$

In particular, by setting $A = \mathbb{N}$, $y = 1$, $q = 0$, and $\ell = 3$ we obtain the generating function for the number of compositions of n that avoid 212 as

$$C_{\mathbb{N}}^{212}(x, 1, 0) = \left(1 - \sum_{j \geq 1} \frac{x^j (1-x)}{1 - x + x^j (x - x^j)} \right)^{-1}.$$

The sequence of values for $n = 0, \ldots, 20$ is given by 1, 1, 2, 4, 8, 15, 30, 58, 114, 222, 434, 846, 1655, 3230, 6310, 12322, 24067, 46997, 91791, 179262 and 350106. Note that the first time the pattern 212 can occur is for $n = 5$, as the composition 212. Similarly,

$$C_A^{12^{\ell-2}1}(x, y, q) = \frac{1}{1 - \sum_{j=1}^d \left(\dfrac{x^{a_j} y}{1 + x^{a_j} y^{\ell-1} (1-q) \sum_{i=j+1}^d x^{(\ell-2) a_i}} \right)},$$

and the generating function for the number of compositions of n that avoid 121 is given by

$$C_{\mathbb{N}}^{121}(x, 1, 0) = \left(1 - \sum_{j \geq 1} \frac{x^j (1-x)}{1 - x + x^{2j+1}} \right)^{-1}.$$

The corresponding values for $n = 0, \ldots, 20$ are 1, 1, 2, 4, 7, 13, 24, 44, 82, 153, 284, 528, 981, 1820, 3378, 6270, 11638, 21608, 40121, 74494 and 138317. Note that the first time the pattern 121 can occur is for $n = 4$, as the composition 121.

In [92], the two patterns 212 and 121 were not treated individually, but as part of the patterns peak and valley, as were the patterns $213 \stackrel{st}{\sim} 312$ and $132 \stackrel{st}{\sim} 231$. We will now look at families that generalize these patterns, namely the patterns $\tau = p\tau'(p+1)$, where τ' is a pattern with letters in $[p-1]$, and $\nu = 1\nu'2$, where ν' is a pattern with letters in $\{3, 4, \ldots\}$.

Theorem 4.39 *Let $\tau = p\tau'(p+1)$ and $\nu = 1\nu'2$, where τ' and ν' are patterns of length $\ell - 2$ in $[p-1]$ and $\{3, 4, \ldots\}$, respectively. Then*

$$C_A^\tau(x, y, q) = \cfrac{1}{1 - \sum_{i=1}^d x^{a_i} y \prod_{j=1}^{i-1} \left(1 + x^{a_j} y^{\ell-1}(q-1) \sum_{\alpha \in \mathcal{B}_{j-1}} x^{\mathrm{ord}(\alpha)}\right)},$$

where $\mathcal{B}_{j-1}$ is the set of all compositions $\alpha = a_{i_1} \cdots a_{i_{\ell-2}}$ with parts in A_{j-1} such that α is order-isomorphic to τ', and

$$C_A^\nu(x, y, q) = \cfrac{1}{1 - \sum_{i=1}^d x^{a_i} y \prod_{j=i+1}^d \left(1 + x^{a_j} y^{\ell-1}(q-1) \sum_{\alpha \in \bar{\mathcal{B}}_{j+1}} x^{\mathrm{ord}(\alpha)}\right)},$$

where $\bar{\mathcal{B}}_{j+1}$ is the set of all compositions $\alpha = a_{i_1} \cdots a_{i_{\ell-2}}$ with parts in $\bar{A}_j$ such that α is order-isomorphic to ν'.

The proof of Theorem 4.39 is similar to that of Theorem 4.37 and is left for the interested reader.

Example 4.40 *We can now obtain results on the individual 3-letter patterns 213 and 132. In both cases, the patterns τ' and ν' consist of a single part, and therefore, $\mathrm{ord}(\alpha) = \alpha$. For $\tau = 213$ and $\nu = 132$, this gives*

$$\sum_{\alpha \in \mathcal{B}_{j-1}} x^{\mathrm{ord}(\alpha)} = \sum_{1 \le \alpha \le j-1} x^\alpha \text{ and } \sum_{\alpha \in \bar{\mathcal{B}}_{j+1}} x^{\mathrm{ord}(\alpha)} = \sum_{\alpha \ge j+1} x^\alpha,$$

respectively. Applying Theorem 4.39 for $\tau = 213$ with $A = \mathbb{N}$, $y = 1$, and $q = 0$ gives

$$C_{\mathbb{N}}^{213}(x, 1, 0) = \frac{1}{1 - \sum_{i \ge 1} x^i \prod_{j=1}^{i-1} (1 - x^j(x - x^j)/(1-x))},$$

and the sequence for the number of compositions of n that avoid 213 for $n = 0, 1, \ldots, 20$ is given by 1, 1, 2, 4, 8, 16, 31, 61, 119, 232, 452, 881, 1716,

3342, 6508, 12674, 24681, 48062, 93591, 182251 *and* 354900. *For* $\nu = 132$, *we obtain*

$$C_{\mathbb{N}}^{132}(x, 1, 0) = \frac{1}{1 - \sum_{i \geq 1} x^i \prod_{j \geq i+1} (1 - x^{2j+1}/(1-x))},$$

and the corresponding sequence for $n = 0, 1, \ldots, 20$ *is given by* 1, 1, 2, 4, 8, 16, 31, 61, 119, 232, 452, 880, 1712, 3331, 6479, 12601, 24505, 47653, 92664, 180187 *and* 350372. *Note that the patterns* 213 *and* 132 *both can occur for the first time for* $n = 6$.

Note the curious fact that the two sequences agree up to $n = 10$, even though both patterns appear for $n \geq 6$. Is this a coincidence? What is going on? To explore this phenomenon and find answers to these questions, see Exercise 4.14.

A final word of caution: Theorems 4.37 and 4.39 are quite powerful in that they cover large families of patterns, but their usefulness for finding nice expressions for the respective generating functions depends heavily on one's success finding explicit expressions for the sums over compositions that are order-isomorphic to the inner patterns τ' and ν'. In Example 4.40, the inner pattern was of length one, so those sums were readily computed. If the inner pattern is longer, then the task becomes more difficult.

The results of this chapter also cover the case of avoidance of subword patterns by setting $q = 0$ in the respective formulas for the generating functions. We classify the 3-letter patterns according to tight Wilf-equivalence in Table 4.3 and list the Theorems that give the respective generating functions.

TABLE 4.3: Equivalence classes for 3-letter patterns

τ	Theorem
111	Theorem 4.32
112, 221	Theorem 4.35
212, 121	Theorem 4.37
123	Theorem 4.30
132, 213	Theorem 4.39

4.4 Exercises

Exercise 4.1 *Derive the results given for levels in Corollary 4.5.*

Exercise 4.2 *Derive the generating function for the number of rises and levels, respectively, in all compositions with parts in*

(1) $A = \{1, k\}$;

(2) $A = \{m \mid m = 2k + 1, k \geq 0\}$.

Exercise 4.3 *Derive the generating function for the number of levels in all compositions of n with m parts given in Example 4.6.*

Exercise 4.4 *For palindromic compositions of n,*

(1) *derive the generating function for the number of levels;*

(2) *derive that $[y^m]l_{\mathbb{N}}(x, y) =$*

$$
\begin{cases}
\dfrac{x^2}{1 - x^2} & \text{for } m = 2 \\[2ex]
\dfrac{(2j - 1 - (2j - 3)x^2)x^{2j}}{(1 + x^2)(1 - x^2)^j} & \text{for } m = 2j, \ j \geq 2 \\[2ex]
\dfrac{2(1 + x)(j + (j - 1)x + jx^2)x^{2j+1}}{(1 + x^2)(1 + x + x^2)(1 - x^2)^j} & \text{for } m = 2j + 1, \ j \geq 1.
\end{cases}
$$

Exercise 4.5 *Derive the generating function for the number of rises and levels, respectively, in all palindromic compositions with odd parts.*

Exercise 4.6 *Derive the generating function for the number of rises in all palindromic Carlitz compositions of n with parts in $\mathbb{N}$.*

Exercise 4.7 *Derive the generating function for the number of levels and drops in all partitions of n with parts in $\mathbb{N}$.*

Exercise 4.8 *Find the generating function for the number of compositions of n that have m parts according to the number of weak rises (occurrences of the subword patterns 11 and 12).*

Exercise 4.9 *A sequence (composition, word, partition) $s = s_1 s_2 \cdots s_m$ is called up-down alternating if $s_{2i-1} \leq s_{2i}$ and $s_{2i} \geq s_{2i+1}$, for all $i \geq 1$. Find the generating function for the number of up-down alternating compositions of n with m parts.*

Exercise 4.10 *List the 13 (subword) patterns of length three, and indicate which ones belong to the same symmetry class with regard to occurrence.*

Exercise 4.11 *Prove the identity given in Equation (4.10), namely*

$$1 - \alpha \sum_{j=1}^{k} x^{a_j} \prod_{i=1}^{j-1} (1 - x^{a_i}\alpha) = \prod_{j=1}^{k} (1 - x^{a_j}\alpha).$$

Exercise 4.12 *For G_d^1 and G_d^2 as defined in the proof of Theorem 4.26, show that*

$$G_{d-1}^1 = b_{d-1}G_{d-1}^2 + G_{d-2}^1$$

and

$$G_d^2 = G_{d-1}^2 + b_d(1-q)(b_{d-1}G_{d-1}^2 + G_{d-2}^1).$$

Exercise 4.13 *Derive a continued fraction expansion for $C_A^{valley}(x,y,q)$ that also works for the case $d = \infty$.*

Exercise 4.14 *This exercise relates to Example 4.40.*

(1) *Write a program in Mathematica or Maple (or any other language) to create the compositions of n that avoid 213 and 132, respectively. (Hint: in Mathematica, you can either create the respective compositions recursively, eliminating those that contain the pattern of interest when appending a 1 or increasing the rightmost part, or use built-in functions to create all compositions of n and then extract those that avoid the respective pattern).*

(2) *Use your program to create the compositions of $n = 6, 7, \ldots, 10$ that avoid 213 and 132, respectively. Find a bijection between these compositions for $n = 6, 7, \ldots, 10$.*

(3) *For $n = 11$, there is one more composition that avoids 213 than there are compositions that avoid 132. Explain why the bijection you have found in Part (2) no longer works and identify the 213-avoiding composition(s) that are not in one-to-one correspondence with a 132-avoiding composition.*

Exercise 4.15 *Find the generating function for the number of compositions of n with m parts in $\mathbb{N}$ that avoid the pattern 1332.*

Exercise 4.16* *A pattern $\tau = \tau_1 \cdots \tau_n$ of length n is a word that contains all the letters $1, 2, \ldots, \max_i \tau_i$. Prove that the number of patterns of length n is given by (see A000670 in [180])*

$$\sum_{k=1}^{n} \sum_{j=0}^{k} (-1)^j \binom{k}{j} (k-j)^n = \sum_{k=1}^{n} k! S(n,k),$$

where $S(n,k)$ denotes the Stirling numbers of the second kind (see Table A.1) which count the number of ways to partition a set of n elements into k nonempty subsets.

4.5   Research directions and open problems

We now suggest several research directions which are motivated both by the results and exercises of this chapter. In later chapters we will revisit some of these research directions and pose additional questions.

Research Direction 4.1 *Lemma 4.25 gives that for a finite ordered set $A = \{a_1, \ldots, a_d\}$, $b_i = x^{a_i}y$, and $C_A = C_A^{peak}(x, y, q)$,*

$$
C_A = \cfrac{1}{1 - b_d - \cfrac{1}{[b_d(1-q), b_{d-1}, b_{d-1}(1-q), \ldots, b_2, b_2(1-q), b_1]}}.
$$

By taking the limit $d \to \infty$ we can obtain a continued fraction also for the case when $|A| = \infty$, that is, A is a countably infinite set $A = \{a_1, a_2, \ldots\}$:

$$
C_A = \lim_{d \to \infty} \cfrac{1}{1 - b_d - \cfrac{1}{[b_d(1-q), b_{d-1}, b_{d-1}(1-q), \ldots, b_2, b_2(1-q), b_1]}}.
$$

Can this continued fraction expression (or any other approach) be used to obtain a closed formula for the generating function $C_A^{peak}(x, y, q)$ when A is countably infinite?

Research Direction 4.2 *We have completely classified the equivalence classes with respect to avoidance of 3-letter subword patterns (see Table 4.3). A natural extension is to find the equivalence classes for tight Wilf-equivalence for larger values of n.*

(1) *Classification of 4-letter patterns according to tight Wilf-equivalence: Table 4.4 contains the values of the sequences $\{AC_n(\tau)\}_{n=0}^{25}$ (obtained via an appropriate modification of the program given in Appendix G) for the 4-letter patterns τ up to the symmetry class of reversal which suggests that the nontrivial tight Wilf-equivalence classes are given by*

$$1322 \overset{t}{\sim} 1232,\ 2123 \overset{t}{\sim} 2213,\ 1342 \overset{t}{\sim} 1432,\ 3124 \overset{t}{\sim} 3214.$$

Prove that these are indeed the nontrivial Wilf-equivalence classes. (Hint: To prove that $\tau \overset{t}{\sim} \nu$, replace each occurrence of τ by an occurrence of ν).

(2) *Use the steps of Part (1) (namely creating a table of sequence values to obtain a conjecture about nontrivial classes and then to prove the equivalence) to classify the subword patterns of lengths five and six according to tight Wilf-equivalence.*

(3) *Let t_k be the number of equivalence classes for k-letter patterns according to tight Wilf-equivalence. Clearly, $t_1 = 1$ and $t_2 = 2$. Table 4.3 shows that $t_3 = 8$, and Part (1) gives that $t_4 = 35$. Part (2) gives two more values, t_5 and t_6. A very hard question is to find properties of t_n in general.*

Research Direction 4.3 *Table 4.3 gives a complete list of (the theorems for) the explicit formulas of the generating functions for the number of compositions of n with m parts that avoid a given 3-letter subword pattern. A natural extension in light of Research Direction 4.2 is to ask for a similar result for 4-letter subword patterns, namely a complete list of the respective generating functions.*

Research Direction 4.4 *A third research direction is to look at a different type of pattern avoidance, namely a cyclic version. We say that a sequence (composition, word, partition) $s_1 \cdots s_m$ cyclicly avoids a subword $\tau = \tau_1 \cdots \tau_\ell$ if $s_1 \cdots s_m s_1 \cdots s_{\ell-1}$ avoids τ. For example, the composition 33412 avoids the subword 123, but does not cyclicly avoid 123 (since 3341233 contains 123). A natural research question is to find the generating function for the number of compositions of n that cyclicly avoid a subword pattern of length k.*

TABLE 4.4: $\{AC_n(\tau)\}_{n=0}^{25}$ for 4-letter subword patterns τ

τ	$\{AC_n(\tau)\}_{n=0}^{25}$ for 4-letter subword patterns τ					
1112	1	1	2	4	8	15
	29	56	108	207	398	765
	1471	2826	5431	10437	20058	38544
	74070	142341	273538	525655	1010149	1941201
	3730401					
1121	1	1	2	4	8	15
	29	56	108	208	402	777
	1502	2902	5608	10836	20937	40451
	78154	150998	291738	563658	1089034	2104109
	4065327					
1111	1	1	2	4	7	15
	29	57	111	218	429	841
	1651	3239	6355	12473	24475	48029
	94249	184946	362932	712194	1397569	2742507
	5381729					
1122	1	1	2	4	8	16
	31	62	122	242	477	944
	1866	3689	7293	14417	28504	56349
	111402	220236	435401	860773	1701717	3364239
	6650985					

τ	$\{AC_n(\tau)\}_{n=0}^{25}$ for 4-letter subword patterns τ					
1221	1	1	2	4	8	16
	31	62	122	242	477	945
	1867	3694	7302	14442	28556	56470
	111664	220810	436639	863426	1707375	3376228
	6676294					
2112	1	1	2	4	8	16
	31	62	122	242	478	946
	1870	3700	7315	14470	28617	56598
	111938	221385	437850	865960	1712665	3387243
	6699164					
1123	1	1	2	4	8	16
	32	63	125	247	489	966
	1910	3774	7459	14738	29123	57543
	113700	224654	443885	877044	1732898	3423912
	6765071					
1132	1	1	2	4	8	16
	32	63	125	247	489	966
	1910	3774	7459	14738	29124	57547
	113713	224689	443975	877264	1733420	3425118
	6767805					
1212	1	1	2	4	8	16
	31	62	122	243	479	949
	1877	3715	7351	14548	28788	56965
	112730	223072	441433	873532	1728602	3420658
	6769015					
2113	1	1	2	4	8	16
	32	63	125	247	489	966
	1910	3775	7462	14748	29150	57612
	113869	225054	444808	879135	1737559	3434179
	6787449					
1231	1	1	2	4	8	16
	32	63	125	247	489	966
	1910	3775	7463	14752	29163	57650
	113968	225300	445396	880500	1740663	3441114
	6802743					
1213	1	1	2	4	8	16
	32	63	125	247	489	966
	1911	3777	7468	14762	29183	57687
	114037	225423	445612	880868	1741271	3442075
	6804169					
1312	1	1	2	4	8	16
	32	63	125	247	489	966
	1911	3777	7468	14762	29184	57691
	114050	225458	445702	881087	1741789	3443266
	6806859					

τ	$\{AC_n(\tau)\}_{n=0}^{25}$ for 4-letter subword patterns τ					
1222	1	1	2	4	8	16
	32	63	126	251	499	993
	1977	3934	7830	15583	31014	61725
	122847	244492	486595	968434	1927403	3835966
	7634439					
2122	1	1	2	4	8	16
	32	63	126	251	499	993
	1978	3936	7835	15595	31041	61787
	122988	244802	487275	969912	1930591	3842805
	7649034					
1322	1	1	2	4	8	16
1232	32	64	127	253	504	1003
	1996	3973	7907	15736	31317	62325
	124035	246846	491255	977660	1945667	3872122
	7706010					
2123	1	1	2	4	8	16
2213	32	64	127	253	504	1003
	1996	3973	7907	15736	31318	62329
	124045	246871	491315	977800	1945988	3872846
	7707618					
1223	1	1	2	4	8	16
	32	64	127	253	504	1003
	1996	3973	7907	15737	31320	62333
	124055	246893	491364	977909	1946224	3873353
	7708703					
2132	1	1	2	4	8	16
	32	64	127	253	504	1003
	1996	3973	7908	15741	31332	62365
	124135	247086	491817	978947	1948562	3878547
	7720116					
1233	1	1	2	4	8	16
	32	64	128	255	510	1018
	2033	4059	8106	16185	32319	64533
	128860	257304	513782	1025909	2048520	4090447
	8167736					
1332	1	1	2	4	8	16
	32	64	128	255	510	1018
	2033	4059	8106	16185	32319	64533
	128860	257304	513783	1025911	2048527	4090463
	8167777					
1323	1	1	2	4	8	16
	32	64	128	255	510	1018
	2033	4059	8106	16185	32319	64533
	128861	257306	513787	1025922	2048551	4090519
	8167901					

τ	$\{AC_n(\tau)\}_{n=0}^{25}$ for 4-letter subword patterns τ					
2133	1	1	2	4	8	16
	32	64	128	255	510	1018
	2033	4059	8106	16185	32319	64533
	128861	257307	513790	1025929	2048569	4090562
	8168003					
2313	1	1	2	4	8	16
	32	64	128	255	510	1018
	2033	4059	8106	16185	32319	64533
	128862	257309	513795	1025942	2048600	4090634
	8168167					
3123	1	1	2	4	8	16
	32	64	128	255	510	1018
	2033	4059	8106	16186	32321	64538
	128873	257336	513858	1026083	2048914	4091327
	8169680					
1243	1	1	2	4	8	16
	32	64	128	256	511	1021
	2039	4072	8131	16237	32422	64741
	129274	258133	515435	1029210	2055101	4103574
	8193908					
2134	1	1	2	4	8	16
	32	64	128	256	511	1021
	2039	4072	8131	16237	32422	64741
	129274	258133	515435	1029210	2055102	4103578
	8193921					
1342 1432	1	1	2	4	8	16
	32	64	128	256	511	1021
	2039	4072	8131	16237	32422	64741
	129274	258133	515435	1029210	2055102	4103578
	8193922					
1423	1	1	2	4	8	16
	32	64	128	256	511	1021
	2039	4072	8131	16237	32422	64741
	129274	258133	515435	1029211	2055104	4103583
	8193932					
2314	1	1	2	4	8	16
	32	64	128	256	511	1021
	2039	4072	8131	16237	32422	64741
	129274	258133	515435	1029211	2055105	4103587
	8193945					
2143	1	1	2	4	8	16
	32	64	128	256	511	1021
	2039	4072	8131	16237	32422	64741
	129274	258133	515436	1029216	2055121	4103632
	8194062					

τ	$\{AC_n(\tau)\}_{n=0}^{25}$ for 4-letter subword patterns τ					
1324	1	1	2	4	8	16
	32	64	128	256	511	1021
	2039	4072	8131	16237	32422	64741
	129275	258136	515444	1029235	2055165	4103731
	8194282					
2413	1	1	2	4	8	16
	32	64	128	256	511	1021
	2039	4072	8131	16237	32422	64741
	129275	258136	515445	1029240	2055182	4103780
	8194412					
3124 3214	1	1	2	4	8	16
	32	64	128	256	511	1021
	2039	4072	8131	16237	32422	64741
	129275	258137	515448	1029246	2055196	4103812
	8194484					
1234	1	1	2	4	8	16
	32	64	128	256	511	1021
	2039	4072	8131	16238	32425	64750
	129298	258194	515583	1029559	2055905	4105396
	8197980					

Chapter 5

Avoidance of Nonsubword Patterns in Compositions

5.1 History and connections

In Chapter 4, we looked at enumeration of subword patterns. In this chapter the focus will be on pattern avoidance rather than enumeration. In addition we will consider different types of patterns, namely subsequence patterns, generalized patterns, and partially ordered patterns.

Research on avoiding (subsequence) permutation patterns has become an important area of study in enumerative combinatorics, as evidenced by the publication of a special volume of the *Electronic Journal of Combinatorics* on permutation patterns (see http://www.combinatorics.org, volume 9:2) and hundreds of articles elsewhere in journals (see [25] and references therein for an overview). This area of research has applications to computer science and algebraic geometry and has proved to be useful in a variety of seemingly unrelated topics, including the theory of Kazhdan-Lusztig polynomials, singularities of Schubert varieties [128], Chebyshev polynomials [53, 126, 148], rook polynomials for a rectangular board [146], various sorting algorithms [123, Ch. 2.2.1], and sortable permutations [191].

Pattern avoidance was first studied for $\mathcal{S}_n$, the set of permutations of $[n]$, avoiding a subsequence pattern $\tau \in \mathcal{S}_3$. Knuth [122] found that the number of permutations of $[n]$ avoiding any $\tau \in \mathcal{S}_3$ is given by the n-th Catalan number. Later, Simion and Schmidt [179] determined $|\mathcal{S}_n(T)|$, the number of permutations of $[n]$ simultaneously avoiding any given set of subsequence patterns $T \subseteq \mathcal{S}_3$. Burstein [31] extended this to words of length n on the alphabet $\{1, 2, \ldots, k\}$, determining the number of words that avoid a set of subsequence patterns $T \subseteq \mathcal{S}_3$. Burstein and Mansour [33, 34] considered forbidden subsequence patterns with repeated letters.

Recently, subsequence pattern avoidance has been studied for compositions. Savage and Wilf [176] considered subsequence pattern avoidance in compositions for a single pattern $\tau \in \mathcal{S}_3$, and showed that the number of compositions of n avoiding $\tau \in \mathcal{S}_3$ is independent of τ. Heubach and Mansour [91] filled in the missing pieces by investigating avoidance of multi-permutation subsequence patterns of length three. These results are the focus of Section 5.2.

In Section 5.3 we present new results on generalized 3-letter patterns, those that have some adjacency requirements. We derive results for permutation and multi-permutation patterns of type $(2, 1)$, which are the only generalized patterns not investigated in previous sections. Finally, in Section 5.4, we discuss results on partially ordered patterns. Kitaev [108] introduced these patterns in the context of permutations, extending the generalized permutation patterns defined by Babson and Steingrímsson [11]. Later, Kitaev and Mansour [109] investigated avoidance of partially ordered patterns in words, and Heubach, Kitaev and Mansour [88] generalized to the case of compositions. The majority of papers on partially ordered patterns are concerned with pattern avoidance, but there are also some results on the enumeration of these patterns in compositions [111].

Since we deal almost exclusively with pattern avoidance in this chapter, we introduce notation for the set of compositions avoiding a certain pattern, the number of such compositions and the corresponding generating function. To highlight the fact that we talk about avoidance, we use AC instead of C – think A *for avoidance* – for the relevant sets and generating functions.

Definition 5.1 *For any ordered set $A \subseteq \mathbb{N}$, we denote the set of compositions of n with parts in A (respectively with m parts in A) that avoid the pattern τ by $\mathcal{AC}_n^A(\tau)$ and $\mathcal{AC}_{n,m}^A(\tau)$, respectively. We denote the number of compositions in these two sets by $AC_A^\tau(n)$ (respectively $AC_A^\tau(m; n)$). The corresponding generating functions are given by $AC_A^\tau(x) = \sum_{n \geq 0} AC_A^\tau(n) x^n$ and*

$$AC_A^\tau(x, y) = \sum_{m,n \geq 0} AC_A^\tau(m; n) x^n y^m = \sum_{m \geq 0} AC_A^\tau(m; x) y^m.$$

More generally, let $AC_A^\tau(\sigma_1 \cdots \sigma_\ell | n)$ (respectively $AC_A^\tau(\sigma_1 \cdots \sigma_\ell | m; n)$) be the number of compositions of n with parts in A (respectively with m parts in A) that avoid τ and start with $\sigma_1, \ldots, \sigma_\ell$. The corresponding generating functions are given by

$$AC_A^\tau(\sigma_1 \cdots \sigma_\ell | x) = \sum_{n \geq 0} AC_A^\tau(\sigma_1 \cdots \sigma_\ell | n) x^n$$

and

$$AC_A^\tau(\sigma_1 \cdots \sigma_\ell | x, y) = \sum_{n,m \geq 0} AC_A^\tau(\sigma_1 \cdots \sigma_\ell | m; n) x^n y^m$$
$$= \sum_{m \geq 0} AC_A^\tau(\sigma_1 \cdots \sigma_\ell | m; x) y^m.$$

From the definitions, we immediately get that

$$AC_A^\tau(x, y) = 1 + \sum_{a \in A} AC_A^\tau(a | x, y). \tag{5.1}$$

As before, we are interested in determining which patterns are equivalent in the sense that they are avoided by the same number of compositions (and therefore have the same generating functions). We follow the terminology in the research literature and refer to this equivalence as Wilf-equivalence (as distinguished from tight Wilf-equivalence which relates to subword patterns).

Definition 5.2 *Two nonsubword patterns τ and ν are called* Wilf-equivalent *if the number of compositions of n with m parts in A that avoid τ is the same as the number of compositions of n with m parts in A that avoid ν, for all A, n, and m. We denote two patterns that are Wilf-equivalent by $\tau \sim \nu$.*

Finally, note that in this chapter most of the results are given for $A = \mathbb{N}$, not only because that is the most interesting case, but also because results for arbitrary sets A are much harder to derive for the more general patterns. Likewise, enumeration for these more general patterns is more difficult, and in general does not lead to nice or even explicit formulas for the generating functions. We will indicate when the results can be generalized to an arbitrary ordered set A.

5.2 Avoidance of subsequence patterns

We now look at subsequence patterns that, unlike the subword patterns, do not have any adjacency requirements. This type of pattern was originally studied in the context of permutations, and is referred to in the literature as *classical* pattern.

Definition 5.3 *A sequence (permutation, word, composition or partition) σ contains a subsequence pattern $\tau = \tau_1 \tau_2 \cdots \tau_k$ if the reduced form of any k-term subsequence of σ equals τ. Otherwise we say that σ avoids the subsequence pattern τ or is τ-avoiding. The reversal map for subsequence patterns generalizes in the obvious manner from that of subword patterns.*

Example 5.4 *The composition $\sigma = 112532$ contains the subsequence pattern 1123 twice (as the subsequences $\sigma_1 \sigma_2 \sigma_3 \sigma_4 = 1125$ and $\sigma_1 \sigma_2 \sigma_3 \sigma_5 = 1123$) and avoids the subsequence pattern 1234. The reversal of the subsequence pattern $\tau = 1243$ is $R(1243) = 3421$.*

In subsequent sections we introduce dashes into patterns to indicate the positions in the pattern where there is no adjacency requirement. With this notation, a subsequence pattern $\tau = \tau_1 \tau_2 \cdots \tau_k$ could be written as $\tau = \tau_1-\tau_2-\cdots-\tau_k$, as there are no adjacency requirements at all. Since this notation is somewhat cumbersome and conflicts with usage in the literature, and since we deal exclusively with subsequence patterns in this section, we will omit the dashes.

We start by deriving $AC_{\mathbb{N}}^{\tau}(x, y)$, where τ is a pattern of length two. In this case there are only two patterns to consider, namely 11 and 12.

Theorem 5.5 *For patterns of length two we have*

$$AC_{\mathbb{N}}^{12}(x, y) = \prod_{j \geq 1} \frac{1}{1 - x^j y}$$

and

$$AC_{\mathbb{N}}^{11}(x, y) = \sum_{n, m \geq 0} m! p_{n,m} x^n y^m,$$

where $p_{n,m}$ denotes the number of partitions of n with m distinct parts in $\mathbb{N}$ with generating function $\prod_{j \geq 1}(1 + x^j y)$.

Proof To derive $AC_{\mathbb{N}}^{12}(x, y)$ we just recognize that avoiding the subsequence pattern 12 is the same as avoiding the subword pattern 12 (why?), and therefore obtain the result from Example 2.65. For the pattern 11, we have to work a little. First observe that avoiding the subsequence pattern 11 means that the composition consists of distinct parts. Furthermore, the number of compositions of n with m distinct parts can be obtained by computing the number of partitions of n with m distinct parts (for which we have derived the generating function in Example 4.18), then multiplying by $m!$. □

Example 5.6 *If we want to compute the sequence of values for compositions of n with distinct parts, then we cannot just set $y = 1$ in $AC_{\mathbb{N}}^{11}(x, y)$ because the coefficients of x also depend on m. Rather, we have to compute the values for $p_{n,m}$ from its generating function, then multiply by $m!$, and then sum over all values of m for a given n. Obviously, this is a task for a computer algebra system such as* Mathematica *or* Maple. *Below is the* Maple *code to compute the values for $n = 0, 1, \ldots, 20$.*

```
N:=20:
s:=vector(N+1,[]):
F:=expand(product((1+x^j*y),j=1..N+1)):
for n from 0 to N do
  s[n+1]:=0:
  for m from 0 to n do
    s[n+1]:=s[n+1]+m!*coeff(coeff(F,x,n),y,m):
  od:
od:
print(s);
```

The corresponding Mathematica *code is given by*

```
f[x_,y_]:= Series[Product[(1 + x^j y), {j, 1, 20}],
                {x, 0, 21}, {y, 0, 21}]
co[n_,m_]:= If[n < m, 0, Coefficient[f[x, y], x^n y^m]];
Apply[Plus, Table[m! co[n, m], {n, 1, 20}, {m, 1, n}], 2]
```

where the Mathematica *function* `Apply` *in conjunction with the function* `Plus` *adds the coefficients for a given n. Note that the Maple function* `mtaylor` *gives the Taylor series for a function of several variables. The sequence produced by these commands is* 1, 1, 1, 3, 3, 5, 11, 13, 19, 27, 57, 65, 101, 133, 193, 351, 435, 617, 851, 1177, *and* 1555 *for* $n = 0, 1, \ldots, 20$.

5.2.1 Subsequence patterns of length three in $\mathcal{S}_3$

We will now consider avoidance of subsequence patterns of length three. The first result on subsequence patterns in compositions was given by Savage and Wilf [176]. For the six permutation patterns of length three, namely 123, 132, 213, 231, 312 and 321, the reversal map immediately gives the following Wilf-equivalences:

$$123 \sim 321, \ 132 \sim 231 \ \text{and} \ 213 \sim 312.$$

However, Savage and Wilf showed a much stronger result, which we give in a slightly modified form that is useful for computations of the coefficients. The result given in [176] does not include enumeration according to parts, and gives the generating function for $A = \mathbb{N}$ (as opposed to $A = [n]$).

Theorem 5.7 *[176] The number of compositions of n with parts in $\mathbb{N}$ that avoid a given 3-letter permutation pattern τ is independent of τ, that is, $123 \sim 132 \sim 213$. The generating function is given by*

$$AC_{[n]}^\tau(x, y) = \sum_{i=1}^{n} \frac{x^{i(n-1)}(1 - x^i y)^{n-2}}{\prod_{j=1, \, j \neq i}^{n}(x^i - x^j)(1 - x^i y - x^j y)}.$$

The proof of this result uses the complement map on compositions.

Definition 5.8 *Let $A = [d]$. For any composition $\sigma = \sigma_1 \sigma_2 \cdots \sigma_m$ of n with m parts in A, we define its* complement *$c(\sigma) = c_A(\sigma)$ with respect to A as the composition where σ_i is replaced by $d + 1 - \sigma_i$.*

Note that $c(\sigma)$ is a composition of

$$n' = \sum_{i=1}^{m}(d + 1 - \sigma_i) = m(d + 1) - n$$

with m parts in A.

Proof We paraphrase the proof given in [176] using our notation. At the core of the proof are results on the enumeration of permutations of a multiset that avoid a given pattern. This enumeration was accomplished for the pattern 132 by Albert et al. in [6, Lemma 1]. To utilize these results, we need to connect compositions with multisets. For a vector $\overrightarrow{m} = (m_1, m_2, \ldots, m_k)$ of positive integers, we define $M(\overrightarrow{m})$ to be the multiset that contains exactly m_i copies of the part i, for $i = 1, 2, \ldots, k$. Then the permutations of the multiset give exactly all the compositions that are made up from the parts of the multiset. Now let $f(\overrightarrow{m}, \tau)$ be the number of permutations of the multiset $M(\overrightarrow{m})$ that avoid the pattern τ. In [6] it was shown that for every $\overrightarrow{m}$, $f(\overrightarrow{m}, 123) = f(\overrightarrow{m}, 132)$, and that $f(\overrightarrow{m}, 132)$ is a symmetric function of $\overrightarrow{m}$. The first assertion gives that $123 \sim 132$, since we can define an equivalence relation on the set of compositions by declaring two compositions equivalent if they have the same multiset of parts, and the Wilf-equivalence holds on each of the equivalence classes, therefore overall. All that is missing is to show that $132 \sim 312$. First observe that $f(\overrightarrow{m}, \tau) = f(R(\overrightarrow{m}), c(\tau))$, since the reversal of $\overrightarrow{m}$ replaces the number of occurrences of the letter i by the number of occurrences of the letter $k + 1 - i$, which matches the substitution of letters in the complement operation. In particular, $f(\overrightarrow{m}, 132) = f(R(\overrightarrow{m}), 312)$. Finally, we use that the counting function $f(\overrightarrow{m}, 132)$ is symmetric in the m_i, therefore $f(\overrightarrow{m}, 132) = f(R(\overrightarrow{m}), 132) = f(R(R(\overrightarrow{m})), 312) = f(\overrightarrow{m}, 312)$, which shows that the number of compositions avoiding any 3-letter permutation pattern τ is independent of τ. The generating function follows with appropriate modifications for incorporating the number of parts from the results given in [6] and Equation 2 in [176]. □

An alternative proof of this result based on a generalization of the Simion-Schmidt construction [179] was given by Myers [157].

Example 5.9 *We compute the sequence for the number of compositions that avoid any one of the permutation patterns of length three using* Mathematica *or* Maple *for $n = 20$. Here are the respective codes:*

```
f[x_] := Sum[x^(19 i) (1-x^i)^18/
    (Product[(x^i-x^j) (1-x^i-x^j), {j,1, i-1}]
        Product[(x^i-x^j) (1-x^i-x^j), {j, i+1,20}]), {i,1,20}]
CoefficientList[Series[f[x], {x, 0, 20}],x]
```

and

```
f:=taylor(sum(x^(i*19)*(1-x^i)^(18)/
    product((x^i-x^j)*(1-x^i-x^j),'j'=1..i-1)/
    product((x^i-x^j)*(1-x^i-x^j),'j'=i+1..k),'i'=1..20)
    ,x=0,21);
```

The resulting sequence is given by 1, 1, 2, 4, 8, 16, 31, 60, 114, 214, 398, 732, 1334, 2410, 4321, 7688, 13590, 23869, 41686, 72405 *and* 125144 *for $n = 0, 1, \ldots, 20$.*

5.2.2 Three-letter subsequence patterns with repetitions

What remains to be studied are the seven patterns of length three with repetitions, namely 111, 112 $\sim$ 211, 121, 221 $\sim$ 122, and 212. Heubach and Mansour [91] suggested algorithms showing that

$$112 \sim 121 \text{ and } 221 \sim 212$$

based on the algorithm given by Burstein and Mansour [33] for proving that the number of words avoiding 112 and 121, respectively, is the same. However, the authors and Augustine found an error in this algorithm; the correction appears as additional pages for the original paper [33, Page 15]. Here we present the results using the corrected algorithm.

Theorem 5.10 *For any ordered set* $A = \{a_1, \ldots, a_d\} \subseteq \mathbb{N}$, 112 $\sim$ 121 *and* 221 $\sim$ 212.

Proof We first analyze the structure of the compositions avoiding 121 and 112, respectively. If $\sigma \in \mathcal{AC}_n^A(121)$, then it cannot contain parts other than a_1 between any two occurrences of a_1, which means that if σ contains the part a_1 more than once, then all such a_1 have to be consecutive. Deletion of all parts a_1 from σ results in a composition σ' which avoids 121, and all occurrences of a_2 in σ', if any, have to be consecutive. In general, step-wise deletion of all parts a_1 through a_j leaves a (possibly empty) composition $\tilde{\sigma}$ with parts in $\{a_{j+1}, \ldots, a_d\}$ in which all parts a_{j+1}, if any, occur consecutively.

On the other hand, if $\sigma \in \mathcal{AC}_n^A(112)$, then only the leftmost a_1 of σ can occur before any larger part. The remaining parts a_1 must occur at the end of σ. In fact, just as in the previous case, step-wise deletion of all parts a_1 through a_j leaves a (possibly empty) composition $\bar{\sigma}$ with parts in $\{a_{j+1}, \ldots, a_d\}$ in which all occurrences of a_{j+1}, except possibly the leftmost one, occur at the end of $\bar{\sigma}$.

We will call all occurrences of a part a_j occurring after the the leftmost a_j and before any a_k with $k > j$ *excess* a_j. We define an *excess* a_j-*block* as follows: If the part a_j occurs at least twice in σ, then an excess a_j-block is the longest sequence of consecutive parts $\sigma_\ell \cdots \sigma_{\ell+v}$, $v \geq 0$ starting at the second a_j from the left (at position ℓ) satisfying these conditions:

(1) $\sigma_{\ell+r} \leq a_j$ for all $0 \leq r \leq v$.

(2) if $\sigma_{\ell+r} = a_i < a_j$ for some $0 \leq r \leq v$, then $\sigma_{\ell+r}$ is not an excess a_i, that is, a_i occurs for the first time in σ.

We now describe the algorithm for the bijection $\rho : \mathcal{AC}_n^A(121) \to \mathcal{AC}_n^A(112)$. Given a composition $\sigma \in \mathcal{AC}_n^A(121)$, we define $\sigma^{(0)} = \sigma$ and apply the following transformation of d steps. Let $\sigma^{(j-1)}$ be the composition that results from applying the transformation step $j-1$ times, where $\sigma^{(j)}$ is obtained from

$\sigma^{(j-1)}$ by cutting out the excess a_j-block and inserting it immediately before any excess a_i-blocks, where $i < j$, or at the end of $\sigma^{(j-1)}$ if there are no excess a_i-blocks. Then $\rho(\sigma) = \sigma^{(d)}$. Clearly, $\sigma^{(d)} \in \mathcal{AC}_n^A(112)$ at the end of the algorithm, and the definition of the excess a_j-blocks insures uniqueness. Here is an example illustrating the algorithm, where $\sigma = 3311132244 \in \mathcal{AC}_{24}^{[4]}(121)$ and $a_j = j$ for $j = 1, 2, 3, 4$. For emphasis, the excess j-blocks have been underlined in the transformation of 3311132244:

$$3311\underline{1}32244 \mapsto 3313\underline{22}4411 \mapsto \underline{3313}2\underline{44}211 \mapsto 34\underline{43}132211 \mapsto 3443132211.$$

The inverse map consists of moving each excess a_j-block to the position immediately after the leftmost a_j, for $j = d, d-1, \ldots, 1$. We have therefore shown that the patterns 121 and 112 are Wilf-equivalent on compositions.

For $A = [d]$, the second Wilf-equivalence, $221 \sim 212$, is easily proved by using the complement operation on the equivalence $112 \sim 121$. Let $\sigma \in \mathcal{AC}_{n,m}^A(221)$ be such that the part i occurs in σ exactly r_i times. Then $c(\sigma) \in \mathcal{AC}_{n',m}^A(112)$, and the part i occurs exactly r_{d+1-i} times. Since c is one-to-one and $112 \sim 121$, this implies the Wilf-equivalence of the patterns 221 and 212. For general A, we cannot use the complement map; instead, one can modify the bijection ρ given for the patterns 112 and 121 to show that $221 \sim 212$. $\square$

Thus we only need to obtain the generating functions for the number of compositions that avoid the patterns 111, 112 and 221, respectively. For avoidance of the pattern 111, it is most useful to state the exponential generating function.

Theorem 5.11 *For any ordered set $A = \{a_1, \ldots, a_d\} \subseteq \mathbb{N}$, the exponential generating function for the number of compositions of n with m parts in A that avoid the pattern 111 is given by*

$$\sum_{m \geq 0} AC_A^{111}(m; x) \frac{y^m}{m!} = \prod_{a \in A} \left(1 + x^a y + \frac{1}{2} x^{2a} y^2\right).$$

Proof Let $\sigma \in \mathcal{AC}_{n,m}^A(111)$. Thus, the number of occurrences of the letter a_1 in σ is 0, 1 or 2, and the number of such compositions is given by $AC_{\bar{A}_1}^{111}(m; n)$, $m \, AC_{\bar{A}_1}^{111}(m-1; n-a_1)$ and $\binom{m}{2} AC_{\bar{A}_1}^{111}(m-2; n-2a_1)$, respectively. Adding the three cases, multiplying by $\frac{1}{m!} x^n y^m$ and summing over all $n, m \geq 0$ we get that

$$\sum_{m \geq 0} AC_A^{111}(m; x) \frac{y^m}{m!} = \left(1 + x^{a_1} y + \frac{1}{2} x^{2a_1} y^2\right) \sum_{m \geq 0} AC_{\bar{A}_1}^{111}(m; x) \frac{y^m}{m!}.$$

Since $\sum_{m \geq 0} AC_{\{a_1\}}^{111}(m; x) \frac{y^m}{m!} = 1 + x^{a_1} y + \frac{1}{2} x^{2a_1} y^2$, we get the desired result by induction on d. $\square$

Example 5.12 *We use Theorem 5.11 to obtain the sequence of values for the number of compositions of n avoiding the subsequence pattern 111 by setting $A = [n]$. Since we have an exponential generating function, we have to be careful when using* Mathematica *or* Maple. *First we create the series in terms of x and y, then extract the coefficients of $x^n y^m$, multiply those by $m!$, and then sum over all values of m for a given n. The* Mathematica *code to accomplish this for $n \geq 1$ is given below.*

```
Apply[Plus,Table[Coefficient[Series[
    Product[1+x^i*y +x^(2i)*y^2/2, {i, 1, 20}],
    {x, 0, 20}, {y, 0, 20}],
    x^n*y^m ]*m!, {n, 1, 20}, {m, 1, 20}], 1]
```

The corresponding Maple *code is given by*

```
taylor(sum(coeff(simplify(taylor(
    product(1+x^i*y+x^(2*i)*y^2/2,'i'=1..20),y,21)
        ),y,j)*j!,'j'=0..20),x,21)
```

The resulting sequence of values for $n = 0, 1, \ldots, 20$ is given by 1, 1, 2, 3, 7, 11, 21, 34, 59, 114, 178, 284, 522, 823, 1352, 2133, 3739, 5807, 9063, 14074, and 23639.

For the remaining two patterns we are able to obtain an explicit formula involving partial derivatives. This formula is rather complicated and not very useful for actual computations when n is larger than 11. Finding "nice" explicit expressions for $AC_A^{112}(x,y)$ and $AC_A^{221}(x,y)$ remains an open question (see Research Direction 5.1). Note the similarity in structure for the two patterns 112 and 221, which comes from the fact that they are complementary.

Theorem 5.13 *Let $b_i = \dfrac{x^{a_i} y^2}{1 - x^{a_i} y}$, $c_i = \dfrac{1}{x^{a_i} y}$ and $A = \{a_1, a_2, \ldots, a_d\}$. Then*

$$AC_A^{112}(x,y)$$
$$= b_1 e^{c_1} \frac{\partial}{\partial y}\left(b_2 e^{c_2 - c_1} \frac{\partial}{\partial y}\left(\cdots \frac{\partial}{\partial y}\left(b_{d-1} e^{c_{d-1} - c_{d-2}} \frac{\partial}{\partial y}\left(\frac{e^{-c_{d-1}}}{1 - x^{a_d} y}\right)\right) \cdots \right)\right),$$

and

$$AC_A^{221}(x,y)$$
$$= b_d e^{c_d} \frac{\partial}{\partial y}\left(b_{d-1} e^{c_{d-1} - c_d} \frac{\partial}{\partial y}\left(\cdots \frac{\partial}{\partial y}\left(b_2 e^{c_2 - c_3} \frac{\partial}{\partial y}\left(\frac{e^{-c_2}}{1 - x^{a_1} y}\right)\right) \cdots \right)\right).$$

Proof We start by deriving a recursion for $AC_A^{112}(x, y)$. Let $H_{112}^A(x, y)$ be the generating function for compositions $\sigma \in \mathcal{AC}_{n,m}^A(112)$ which contain at least one part a_1. Thus,

$$H_{112}^A(x, y) = AC_A^{112}(x, y) - AC_{\bar{A}_1}^{112}(x, y). \tag{5.2}$$

On the other hand, each such σ either ends in a_1 or not. If σ ends in a_1, then deletion of this a_1 results in a composition $\tilde{\sigma} \in \mathcal{AC}_{n-a_1, m-1}^A(112)$, since the a_1 at the right end of σ cannot be part of an occurrence of the pattern 112. If σ does not end in a_1, then σ contains exactly one a_1. Deletion of the single a_1 produces a composition $\tilde{\sigma} \in \mathcal{AC}_{n-a_1, m-1}^{\bar{A}_1}(112)$, and there are exactly $m-1$ such compositions. Thus, we have an alternative expression for $H_{112}^A(x, y)$:

$$
\begin{aligned}
H_{112}^A(x, y) \\
&= x^{a_1} y \, AC_A^{112}(x, y) + \sum_{n \geq a_1, \, m \geq 1} (m-1) AC_{\bar{A}_1}^{112}(m-1; n-a_1) x^n y^m \\
&= x^{a_1} y \, AC_A^{112}(x, y) + x^{a_1} y^2 \frac{\partial}{\partial y} AC_{\bar{A}_1}^{112}(x, y).
\end{aligned} \tag{5.3}
$$

Combining (5.2) and (5.3) results in

$$(1 - x^{a_1} y) AC_A^{112}(x, y) = AC_{\bar{A}_1}^{112}(x, y) + x^{a_1} y^2 \frac{\partial}{\partial y} AC_{\bar{A}_1}^{112}(x, y).$$

If we multiply this recurrence by $e^{-c_1}/(x^{a_1} y^2)$, then the right-hand side becomes a single partial derivative, and we obtain that

$$\frac{1}{b_1} e^{-c_1} AC_A^{112}(x, y) = \frac{\partial}{\partial y} \left(e^{-c_1} AC_{\bar{A}_1}^{112}(x, y) \right),$$

which is equivalent to

$$AC_A^{112}(x, y) = b_1 e^{c_1} \frac{\partial}{\partial y} \left(e^{-c_1} AC_{\bar{A}_1}^{112}(x, y) \right).$$

Iterating this recurrence results in the explicit formula for $AC_A^{112}(x, y)$. The result for $AC_A^{221}(x, y)$ follows by replacing a_1 with a_d and $\bar{A}_1$ with A_{d-1} in the derivation for $AC_A^{112}(x, y)$. $\qquad \square$

Example 5.14 *The explicit formulas for the generating functions for the patterns 112 and 221 are not practical for obtaining the sequence values as computational effort increases rapidly with n. Using the explicit formula in Maple or Mathematica we obtain results only up to $n = 11$ in a reasonable amount of time. An alternative approach consists of creating the list of compositions of n in Mathematica, then using pattern matching to determine which of those contain the respective pattern. Subtracting the number of compositions that* *contain a given pattern from the total number of compositions of n gives the desired sequence of values. This approach allows us to get the values for the number of compositions that avoid either 112 or 221 for $n \leq 15$ before the computational effort becomes too large. Below we give the* Mathematica *code for the pattern matching approach for 112.*

```
comps[n_]:=Flatten[Map[Permutations,IntegerPartitions[n]],1];
Table[2^(k - 1) - Length[Position[comps[k],
        {___, x_, ___, x_, ___, y_, ___} /; x<y]], {k,1,15}]
```

The sequence of values $[x^n]AC_A^{112}(x, y)$ is given by 1, 1, 2, 4, 7, 13, 23, 40, 67, 115, 190, 311, 505, 807, 1285, and 2031 for $n = 0, 1, \ldots, 15$; the corresponding sequence for the pattern 221 is given by 1, 1, 2, 4, 8, 15, 29, 55, 103, 190, 347, 630, 1134, 2028, 3585, and 6291.

Since both *Mathematica* and Maple reach their limits rather quickly, we provide C++ and Java codes for finding the sequence of values for the number of compositions of n with parts in $\mathbb{N}$ that avoid a subsequence pattern τ of length less than or equal to seven for any n. These programs are given in Appendix G. For results on avoiding more than one pattern with repetitions we refer the reader to [91] and Exercise 5.5.

5.2.3 Wilf-classes for longer subsequence patterns

In the previous two subsections we completely classified the patterns of length three according to Wilf-equivalence. We summarize these results in Table 5.1 for the nontrivial Wilf-equivalence classes, that is, those that do not just contain a pattern and its reverse. For patterns of length three, the only trivial class consists of the pattern 111.

TABLE 5.1: Wilf-equivalence classes for patterns of length three

123~132~213	112~121	122~212

For longer patterns, Mansour and Jelínek [105, 104] provided classification results for patterns according to strong equivalence which implies Wilf-equivalence (see Definition 6.10). Utilizing the C++ code given in Appendix G together with their structure results (Propositions 6.18 and 6.19 and Theorem 6.20), Mansour and Jelínek gave a complete classification for patterns of lengths four and five for compositions. Tables 5.2 and 5.3 list the lexicographically minimal patterns for the nontrivial Wilf-equivalence classes.

More detailed tables where the equivalence classes are listed together with values of $AC^\tau(n)$ for selected values of n appear in Appendix G. The values

TABLE 5.2: Wilf-equivalence classes for patterns of length four

1234~1243~1432~2134~2143~3214	1112~1121	1123~1132
1223~1232~1322~2123~2132~2213	1233~2133	1222~2122

TABLE 5.3: Wilf-equivalence classes for patterns of length five

12223~12232~12322~13222~21223~21232 ~21322~22123~22132~22213	12134~12143
	12443~21443
12345~12354~12543~15432~21345~21354 ~21543~32145~32154~43215	12434~21434
	12534~21534
12234~12243~21234~21243~22134~22143	11123~11132
12334~12343~12433~21334~21343~21433	12333~21333
11112~11121~11211	12435~21435
11223~11232~11322	12453~21453
12222~21222~22122	13245~13254
11234~11243~11432	23145~23154
12233~21233~22133	21134~21143
12344~21344~32144	31245~31254

shown are those that take relatively long to compute and which clearly show that the classes are indeed different. Sequence values for smaller n can be quickly computed using the C++ code given in Appendix G. Finding explicit or recursive expressions for the respective generating functions is an open problem (see Research Direction 5.2).

5.3 Generalized patterns and compositions

In Chapter 4 and in Section 5.2 we discussed enumeration and avoidance of subword patterns and subsequence patterns. These two types of patterns are special cases of the so-called generalized patterns, which were introduced by Babson and Steingrímsson [11] in the context of permutations to study Mahonian statistics.

Definition 5.15 *A generalized pattern of length k is a word consisting of k letters in which two adjacent letters may or may not be separated by a dash. The absence of a dash between two adjacent letters in the pattern indicates that the corresponding letters in the permutation, word, composition, or partition must be adjacent. If the pattern τ is of the form $\tau = \tau^{(1)} - \tau^{(2)} - \cdots - \tau^{(\ell)}$ where the $\tau^{(i)}$ are all subword patterns of lengths j_i, then τ is of type $(j_1, j_2, \ldots, j_\ell)$. The reversal map for a generalized pattern generalizes from Definition 4.22 in the obvious manner, namely reversing the parts of the pattern, and keeping intact any adjacency requirements.*

Note that a subword pattern is a generalized pattern that has no dashes, and a subsequence pattern is a generalized pattern that has dashes between all pairs of adjacent letters.

Definition 5.16 *A sequence (permutation, word, composition or partition) σ contains a generalized pattern τ of length k if τ equals the reduced form of any k-term subsequence of σ that follows the adjacency requirements given by τ. Otherwise we say that σ avoids the generalized pattern τ or is τ-avoiding.*

Example 5.17 *Let $\tau = 13$–2–1. Then τ is of type $(2,1,1)$, since $\tau^{(1)} = 13$, $\tau^{(2)} = 2$, and $\tau^{(3)} = 1$, and $R(\tau) = 1$–2–31. Furthermore, τ occurs in σ if there is a subsequence $\sigma_i \sigma_{i+1} \sigma_j \sigma_\ell$ with $i+1 < j < \ell$ and $\sigma_i = \sigma_\ell < \sigma_j < \sigma_{i+1}$. Thus, the composition 253452 contains two occurrences of 13–2–1, namely $\sigma_1 \sigma_2 \sigma_3 \sigma_6 = 2532$ and $\sigma_1 \sigma_2 \sigma_4 \sigma_6 = 2542$.*

In this section we study generalized patterns of length three for compositions. The only 3-letter generalized patterns that we have not yet discussed are those with one dash (or one adjacent pair of letters), that is, the patterns of type $(2,1)$. Note that if τ is of type $(2,1)$, then $R(\tau)$ is of type $(1,2)$; thus, for the purpose of Wilf-equivalence, we only need to look at type $(2,1)$ patterns. These patterns were studied by Claesson [55] for permutations and Burstein and Mansour [33] for words.

We will first consider type $(2,1)$ permutation patterns, and then type $(2,1)$ patterns that have repeated letters.

5.3.1 Permutation patterns of type (2,1)

There are six permutation patterns of type $(2,1)$, namely 12–3, 13–2, 21–3, 23–1, 31–2 and 32–1. These patterns fall into three separate equivalence classes. Not surprisingly, the classes split up according to the last part of the pattern, that is, the two patterns that have the same last part form a Wilf equivalence class. One might expect that a bijection that reverses the parts of the composition corresponding to the adjacent pair in the pattern would do the trick of showing Wilf-equivalence. This is indeed the case for two of the equivalence classes, but does not work for showing that 13–2 $\sim$ 31–2. We will start with the easy cases.

Theorem 5.18 *For any ordered set $A = \{a_1, a_2, \ldots\} \subseteq \mathbb{N}$, 12–3 $\sim$ 21–3 and 23–1 $\sim$ 32–1.*

Proof We show that 12–3 $\sim$ 21–3 by giving a bijection ϕ between the sets of compositions of n with m parts in A avoiding the respective patterns. Let $\sigma \in \mathcal{AC}^A_{n,m}(12$–$3)$ and assume that σ has maximal part a_j which occurs s times. Thus, σ can be decomposed as

$$\sigma = \sigma^{(1)} a_j \sigma^{(2)} a_j \cdots a_j \sigma^{(s)} a_j \sigma^{(s+1)},$$

where each $\sigma^{(i)}$ is a nonincreasing composition with parts in A_{j-1} for $i = 1, \ldots, s$, and $\sigma^{(s+1)}$ is a composition with parts in A_{j-1} that avoids 12–3. We define $\phi(\sigma)$ recursively as

$$\phi(\sigma) = R(\sigma^{(1)})a_j R(\sigma^{(2)})a_j \cdots a_j R(\sigma^{(s)})a_j \phi(\sigma^{(s+1)}),$$

where R is the reversal map. Clearly, σ avoids 12–3 if and only if $\phi(\sigma)$ avoids 21–3, and σ and $\phi(\sigma)$ are both in $\mathcal{C}_{n,m}^A$. Thus, 12–3 $\sim$ 21–3; the proof for 23–1 $\sim$ 32–1 follows with appropriate adjustments. $\qquad\square$

Now we deal with the harder equivalence.

Theorem 5.19 *For any ordered set* $A = \{a_1, \ldots, a_d\} \subseteq \mathbb{N}$, 13–2 $\sim$ 31–2.

Proof We define an algorithm that transforms $\sigma \in \mathcal{AC}_{n,m}^A(13\text{–}2)$ into $\sigma' \in \mathcal{AC}_{n,m}^A(31\text{–}2)$ and vice versa, thereby giving a bijection between $\mathcal{AC}_{n,m}^A(13\text{–}2)$ and $\mathcal{AC}_{n,m}^A(31\text{–}2)$. The basic idea is to move blocks of "1"s from one side of the (single) "3" to the other, leaving the corresponding (single) "2" in place. This process transforms a 13–2 pattern into a 31–2 pattern and vice versa. We make this idea precise with the following definitions: An *ascent* in σ is an integer σ_i such that $\sigma_i < \sigma_{i+1}$ for $i \geq 1$. The ascent σ_i is called *active* if there is an integer σ_j such that $i + 1 < j$ and $\sigma_i < \sigma_j < \sigma_{i+1}$, that is, an active ascent is the "1" in an occurrence of the pattern 13–2 of *width* $j - i + 1$. Note that an active ascent can be part of more than one occurrence of 13–2, and that $\sigma \in \mathcal{AC}_{n,m}^A(13\text{–}2)$ cannot have an active ascent. For each occurrence of a pattern 13–2 we define the associated *ascent block* to be the maximal substring $\sigma_k \sigma_{k+1} \cdots \sigma_i$ such that $\sigma_\ell < \sigma_j$ for $\ell = k, k+1, \ldots, i$. Figure 5.1 visualizes the schematic structure of an ascent block.

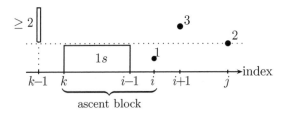

FIGURE 5.1: Structure of ascent block.

Similarly, a *descent* in a composition σ is an integer σ_i such that $\sigma_{i-1} > \sigma_i$ for $i > 1$. The descent σ_i is called *active* if there is an integer σ_j such that

$i < j$ and $\sigma_i < \sigma_j < \sigma_{i-1}$, that is, an active descent is the "1" in an occurrence of the pattern 31–2 of *width* $j - i$. An active descent can belong to more than one occurrence of 31–2, and $\sigma \in \mathcal{AC}_{n,m}^A(31\text{–}2)$ cannot have an active descent. For each occurrence of a pattern 31–2 we define the associated *descent block* to be the maximal substring $\sigma_i\sigma_{i+1}\cdots\sigma_k$ such that $\sigma_\ell < \sigma_j$ for $\ell = i,\ldots,k$. Figure 5.2 shows the structure of a descent block.

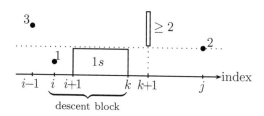

FIGURE 5.2:   Structure of descent block.

Both the ascent and the descent block consist of parts that can play the role of the "1" in the pattern. For a fixed ascent/descent block, the *innermost pattern* is the associated pattern of smallest width, while the *outermost pattern* is the associated one of largest width.

We can now describe the map $\rho:\mathcal{AC}_{n,m}^A(13\text{–}2) \mapsto \mathcal{AC}_{n,m}^A(31\text{–}2)$. Let $\sigma \in \mathcal{AC}_{n,m}^A(13\text{–}2)$ have r occurrences of the pattern 31–2. Let $\sigma^{(0)} = \sigma$ and $\sigma^{(j)}$ be the composition that results after j steps of the algorithm. Basically, each step transforms one of the active descents and removes at least one of its associated occurrences of the pattern 31–2, so that after at most r steps we obtain a composition $\rho(\sigma) \in \mathcal{AC}_{n,m}^A(31\text{–}2)$. Note that if $r = 0$, then $\sigma \in \mathcal{AC}_{n,m}^A(31\text{–}2)$ and $\rho(\sigma) = \sigma^{(0)} = \sigma$. Now assume that $r > 0$. Then $\sigma^{(j)}$ is obtained from $\sigma^{(j-1)}$ as follows: Let σ_{d_j} be the leftmost active descent in $\sigma^{(j-1)}$ and for $i = 1,\ldots,m$, let σ_i denote the i-th part in $\sigma^{(j-1)}$. For the active descent σ_{d_j} identify the associated innermost 31–2 pattern. Assume that it occurs at $\sigma_{d_j-1}\sigma_{d_j}\sigma_{j*}$. Since it has the smallest width, the descent block consists of $\sigma_{d_j}\cdots\sigma_{j*-1}$. Furthermore, since σ avoids 13–2 we have that $\sigma_\ell \leq \sigma_{j*-1} \leq \sigma_{d_j}$ or $\sigma_\ell \geq \sigma_{j*}$ for $\ell > j*$.

Next we cut out the descent block and move it to the left of σ_{d_j}, inserting it immediately after the rightmost part σ_{i*} with $\sigma_{i*} \leq \sigma_{j*}$, or at the beginning of $\sigma^{(j-1)}$ if such a σ_{i*} does not exist. This insertion may create a descent if $\sigma_{i*} > \sigma_{d_j}$, but the newly created descent cannot be active due to the definition of $i*$ and the consequences of the 13–2 avoidance. We have therefore reduced the number of 31–2 patterns by at least one. Let the resulting composition be $\sigma^{(j)}$.

Note that the movement of the descent block of the innermost pattern modifies other occurrences of 31–2 patterns associated with the active descent. Sometimes several patterns are removed at once. If not, then the part that previously played the role of the "2" for one of the associated patterns is now playing the role of the "1", and thus may become a new active descent. However, it occurs to the right of the previous active descent, and the set of values which can play the role of "2" for this potential active descent has shrunk. Therefore, after a finite number of applications of the algorithm, all occurrences of 31–2 have been removed from σ, and the resulting composition $\rho(\sigma)$ is in $\mathcal{AC}^A_{n,m}(31\text{–}2)$. In addition, $\rho(\sigma)$ has at least one active ascent (created from the active descent in the last step). The resulting composition $\rho(\sigma)$ is unique, and if $\sigma \neq \tilde{\sigma}$, then $\rho(\sigma) \neq \rho(\tilde{\sigma})$. This gives $|\mathcal{AC}^A_{n,m}(13\text{–}2)| \geq |\mathcal{AC}^A_{n,m}(31\text{–}2)|$.

To compute the image $\rho'(\sigma)$ of $\sigma \in \mathcal{AC}^A_{n,m}(31\text{–}2)$, modify the algorithm for ρ accordingly: in the j-th step identify the rightmost active ascent and its associated outermost 13–2 pattern. Assume that this 13–2 pattern occurs at $\sigma_{d_j}\sigma_{d_j+1}\sigma_{j*}$. Insert its ascent block immediately before σ_{j*}. Again, the resulting composition $\rho'(\sigma)$ is unique, and if $\sigma \neq \tilde{\sigma}$, then $\rho'(\sigma) \neq \rho'(\tilde{\sigma})$. This gives $|\mathcal{AC}^A_{n,m}(31\text{–}2)| \geq |\mathcal{AC}^A_{n,m}(13\text{–}2)|$ and therefore the two sets have the same number of compositions. □

We give a few examples to illustrate the two algorithms. Note that in each case, $\rho'(\rho(\sigma)) = \sigma$, even though the intermediate compositions are not necessarily the same. In addition, the number of patterns associated with active descents/ascents do not have to be the same in the composition and its image, and not even the number of active ascents and descents have to be the same. We will underline the active descent and ascent blocks and will show in bold font the σ_j that corresponds to the "2" of the pattern.

Example 5.20 *Let* $\sigma = 59424511241 \in \mathcal{AC}^{[9]}_{38,11}(13\text{–}2)$. *Note that* σ *has two active descents with associated 31–2 patterns 945, 512, and 514. It is transformed as follows:*

$$59\underline{424}\mathbf{5}11241 \rightarrow 5424\underline{95}\mathbf{11}241 \rightarrow 5421149\underline{5}\mathbf{2}41 \rightarrow 54211429541$$

corresponding to the movements of descent blocks (424) inserted after 5, (11) inserted after 2, and (2) inserted after 4. On the other hand, starting with $\sigma' = 54211429541 \in \mathcal{AC}^{[9]}_{38,11}(31\text{–}2)$ *(having two active ascents with associated 13–2 patterns 142, 295, and 294) we obtain*

$$5421142\underline{9}\mathbf{5}41 \rightarrow 5\underline{4211}49\mathbf{5}241 \rightarrow 5942\underline{11}4\mathbf{5}241 \rightarrow 59424511241$$

corresponding to the movements of ascent blocks (2) inserted before 4, (42114) inserted before 5, and (11) inserted before 2.

As a second example, we consider $\sigma = 9445421126718 \in \mathcal{AC}^{[9]}_{54,13}(13\text{--}2)$ with one active descent and associated 31–2 patterns 945, 946, 947, 948. *This composition is transformed as follows:*

$$9\underline{44}5421126718 \rightarrow 449\underline{542112}6718 \rightarrow 4454211296\underline{7}18$$
$$\rightarrow 4454211269\underline{7}18 \rightarrow 4454211267198$$

corresponding to movement of the blocks (44) *inserted before* 9, (542112) *inserted after* 4, (6) *inserted after* 2, *and* (71) *inserted after* 6. *The resulting composition* $4454211267198 \in \mathcal{AC}^{[9]}_{54,13}(31\text{--}2)$ *has one active ascent, but only a single associated* 13–2 *pattern, namely* 198. *The reverse map therefore has only one intermediate step, where the block* (44542112671) *is inserted before the* 8*:*

$$4454211267198 \rightarrow 9445421126718.$$

Finally, we give an example where the image has fewer active ascents. Let $\sigma = 6244582418191 \in \mathcal{AC}^{[9]}_{55,13}(13\text{--}2)$, *which has two active descents with associated* 31–2 *patterns* 624, 624, 625, *and* 824. *The image is created as follows:*

$$6\underline{24}4582418191 \rightarrow 26\underline{44}582418191 \rightarrow 244658\underline{2}418191 \rightarrow 2442658418191$$

corresponding to the movements of descent blocks (2) *inserted before* 6, (44) *inserted after* 2, *and* (2) *inserted after* 4. *The resulting composition has only one active ascent with two associated* 13–2 *patterns* 265 *and* 264. *The reverse map is given by*

$$244\underline{2}658418191 \rightarrow \underline{2446}582418191 \rightarrow 6244582418191$$

corresponding to the movements of ascent blocks (2) *inserted before* 4 *and* (244) *inserted before* 5.

So all together we have that 12–3 $\sim$ 21–3, 23–1 $\sim$ 32–1 and 13–2 $\sim$ 31–2. In fact, these are all the Wilf classes for patterns of type $(2,1)$ because the sequences for the number of compositions of n that avoid the respective patterns are different for $n = 8$ (see Examples 5.22, 5.24 and 5.26). Note that this situation is quite different from that for subsequence patterns, where all permutation patterns of length 3 are Wilf-equivalent. We now derive the corresponding generating functions for $A = [d]$ for the three Wilf classes.

Theorem 5.21 *The generating function for the number of compositions of n with m parts in $[d]$ that avoid* 12–3 *is given by*

$$AC^{12\text{--}3}_{[d]}(x,y) = \prod_{i=1}^{d}\left(1 - \frac{x^i y}{\prod_{j=1}^{i-1}(1 - x^j y)}\right)^{-1}.$$

Proof   Separating how the composition begins we obtain

$$AC_{[d]}^{12-3}(i|x,y)$$

$$= x^i y + \sum_{j=1}^{i} AC_{[d]}^{12-3}(ij|x,y) + \sum_{j=i+1}^{d} AC_{[d]}^{12-3}(ij|x,y)$$

$$= x^i y + x^i y \left(\sum_{j=1}^{i} AC_{[d]}^{12-3}(j|x,y) + \sum_{j=i+1}^{d} x^j y AC_{[j]}^{12-3}(x,y) \right).$$

Note that in the last sum the set of parts for the composition is restricted from $[d]$ to $[j]$ to guarantee avoidance of 12–3. From this recursion we get that the generating function $G_d(i) = AC_{[d]}^{12-3}(i|x,y) - AC_{[d-1]}^{12-3}(i|x,y)$ satisfies

$$G_d(i) = x^i y \sum_{j=1}^{i} G_d(j) + x^i y \cdot x^d y AC_{[d]}^{12-3}(x,y).$$

Solving for $G_d(i)$ leads to

$$G_d(i) = \frac{x^i y}{1 - x^i y} \left(\sum_{j=1}^{i-1} G_d(j) + x^d y AC_{[d]}^{12-3}(x,y) \right). \qquad (5.4)$$

We can prove by induction on i that

$$G_d(i) = \frac{x^{i+d} y^2 AC_{[d]}^{12-3}(x,y)}{\prod_{j=1}^{i} (1 - x^j y)},$$

for all $i = 1, 2, \ldots, d - 1$. The induction step uses that

$$1 + \sum_{i=1}^{d-1} \frac{x^i y}{\prod_{j=1}^{i} (1 - x^j y)} = \frac{1}{\prod_{j=1}^{d-1} (1 - x^j y)}, \qquad (5.5)$$

which is also readily proved by induction. Also, for $i = d$, we obtain from the definition that

$$G_d(d) = AC_{[d]}^{12-3}(d|x,y) - AC_{[d-1]}^{12-3}(d|x,y)$$

$$= x^d y AC_{[d]}^{12-3}(x,y) - 0 = x^d y AC_{[d]}^{12-3}(x,y).$$

Therefore, summing over $i = 1, 2, \ldots, d$ we obtain

$$AC_{[d]}^{12-3}(x,y) - AC_{[d-1]}^{12-3}(x,y) = x^d y \left(1 + \sum_{i=1}^{d-1} \frac{x^i y}{\prod_{j=1}^{i} (1 - x^j y)} \right) AC_{[d]}^{12-3}(x,y),$$

which by (5.5) is equivalent to

$$AC_{[d]}^{12-3}(x,y) - AC_{[d-1]}^{12-3}(x,y) = \frac{x^d y}{\prod_{j=1}^{d-1}(1-x^j y)} AC_{[d]}^{12-3}(x,y).$$

Hence, for all $d \geq 1$ we have

$$AC_{[d]}^{12-3}(x,y) = \left(1 - \frac{x^d y}{\prod_{j=1}^{d-1}(1-x^j y)}\right)^{-1} AC_{[d-1]}^{12-3}(x,y).$$

Iterating the above recurrence relation d times together with the initial condition $AC_{\varnothing}^{12-3}(x,y) = 1$ we get the desired result. □

Example 5.22 *By taking the limit as $d \to \infty$ in Theorem 5.21 we obtain the generating function for the number of compositions of n that avoid the generalized pattern 12–3 as*

$$AC_{\mathbb{N}}^{12-3}(x,1) = \prod_{i \geq 1}\left(1 - \frac{x^i}{\prod_{j=1}^{i-1}(1-x^j)}\right)^{-1}.$$

We evaluate this expression for $d = 20$ in Mathematica *or* Maple *and obtain the sequence of values for $n = 0$ to $n = 20$ as* 1, 1, 2, 4, 8, 16, 31, 60, 114, 215, 402, 746, 1375, 2520, 4593, 8329, 15036, 27027, 48389, 86314 *and* 153432.

Theorem 5.23 *The generating function for the number of compositions of n with m parts in $[d]$ that avoid 23–1 is given by*

$$AC_{[d]}^{23-1}(x,y) = \prod_{i=1}^{d}\left(1 - \frac{x^i y}{\prod_{j=i+1}^{d}(1-x^j y)}\right)^{-1}.$$

Proof Let σ be any composition of n with m parts in $A = [d]$ that avoids 23–1. Then σ either does not contain the part 1, or σ can be decomposed as

$$\sigma^{(1)}1\sigma^{(2)}1\cdots\sigma^{(s)}1\sigma',$$

where $\sigma^{(i)}$ and σ' have parts in $\bar{A}_1$, each $\sigma^{(i)}$ avoids 12 (why?) and σ' avoids 23–1. The generating function is given by $\left(xyAC_{\bar{A}_1}^{12}(x,y)\right)^s AC_{\bar{A}_1}^{23-1}(x,y)$ for $s \geq 1$. All together,

$$AC_{[d]}^{23-1}(x,y) = AC_{\bar{A}_1}^{23-1}(x,y) + \frac{xyAC_{\bar{A}_1}^{12}(x,y)}{1 - xyAC_{\bar{A}_1}^{12}(x,y)}AC_{\bar{A}_1}^{23-1}(x,y),$$

which is equivalent to

$$AC_{[d]}^{23-1}(x,y) = \frac{AC_{\bar{A}_1}^{23-1}(x,y)}{1 - xy AC_{\bar{A}_1}^{12}(x,y)}.$$

Using the above recurrence d times we obtain that

$$AC_{[d]}^{23-1}(x,y) = \prod_{i=1}^{d} \frac{1}{1 - x^i y AC_{\bar{A}_i}^{12}(x,y)}.$$

Arguing as in Example 4.17 we obtain that $AC_{\bar{A}_i}^{12}(x,y) = \prod_{j=i+1}^{d}(1 - x^j y)^{-1}$, which completes the proof. □

Notice the similarity in structure of $AC_{[d]}^{23-1}(x,y)$ and $AC_{[d]}^{12-3}(x,y)$, even though their derivations are different. The explanation for this phenomenon is that 12–3 $\sim$ 21–3, and 21–3 and 23–1 are complementary patterns. Having chosen to derive the generating function for the pattern 23–1 instead of the pattern 21–3, we needed to use a somewhat different method instead of doing a "proof by replacement." We intentionally gave the two different proofs to make the point that when deriving the generating function for a Wilf-equivalence class with more than one pattern, it is very useful to think about which pattern to use. Exercise 5.3 asks you to provide the simpler proof based on complementary patterns.

Example 5.24 *By taking the limit as $d \to \infty$ and substituting $y = 1$ in Theorem 5.23 we obtain that the generating function for the number of compositions of n that avoid the generalized pattern 23–1 is given by*

$$AC_{\mathbb{N}}^{23-1}(x,1) = \prod_{i \geq 1}\left(1 - \frac{x^i}{\prod_{j \geq i+1}(1 - x^j)}\right).$$

The sequence for the number of compositions of n that avoid 23–1 for $n = 0$ to $n = 20$ is given by 1, 1, 2, 4, 8, 16, 31, 61, 118, 228, 440, 846, 1623, 3111, 5955, 11385, 21752, 41530, 79250, 151161 and 288224.

Finally, we find the generating function for the compositions avoiding 13–2. Using arguments similar to those in the proofs of Theorems 5.21 and 5.23 we obtain a recursive result for $AC_{[d]}^{13-2}(x,y)$. Finding an explicit expression remains an open question (see Research Direction 5.1).

Lemma 5.25 *The generating function $AC_{[d]}^{13-2}(x,y)$ for the number of compositions of n with m parts in $[d]$ that avoid 13–2 satisfies*

$$AC_{[d]}^{13-2}(x,y) = 1 + \sum_{i=1}^{d} AC_{[d]}^{13-2}(i|x,y),$$

where

$$AC_{[d]}^{13-2}(i|x,y)$$

$$= x^i y \left(1 + \sum_{j=1}^{i+1} AC_{[d]}^{13-2}(j|x,y) + \sum_{j=i+2}^{d} AC_{\{1,\ldots,i,j,\ldots,d\}}^{13-2}(j|x,y) \right).$$

Example 5.26 *The expression in Lemma 5.25 is difficult to evaluate. Modifying the* Mathematica *code of Example 5.14, we may enumerate the number of compositions of* n *that avoid* 13–2 *directly and obtain the sequence for* $n = 0$ *to* $n = 15$ *as* 1, 1, 2, 4, 8, 16, 31, 60, 115, 218, 411, 770, 1434, 2656, 4897 *and* 8991 *(see Exercise 5.1).*

5.3.2 Patterns of type (2,1) with repeated letters

In this section we consider the three-letter generalized patterns of type $(2, 1)$ with repetitions, namely the patterns 11–1, 11–2, 12–1, 12–2, 21–1, 21–2 and 22–1. We start by determining the Wilf-equivalence classes. Since all patterns in the same equivalence class must have the same letters, 11–1 forms its own class. We will show that 12–2 $\sim$ 21–2 by giving an explicit bijection. Using the complement map (see Definition 5.8) on this equivalence, we immediately obtain that 21–1 $\sim$ 12–1. The remaining two patterns, 11–2 and 22–1, belong to distinct classes because the sequences for the number of compositions of n avoiding these two patterns are different from each other and from those for the patterns 12–2 and 21–1 (see Examples 5.42, 5.44 and 5.47). So all together we have five different Wilf classes.

To show that 12–2 $\sim$ 21–2, we start by determining the structure of the compositions that avoid 12–2 and 21–2, respectively. We give a bijection based on the structures of these compositions. To do so, we need some new terminology.

Definition 5.27 *For any composition* $\sigma \in \mathcal{C}_{n,m}^A$, *we define*

$$B(\sigma) = \cup_{i=1}^{m}\{\sigma_i\} = \{b_1, \ldots, b_s\},$$

that is, $B(\sigma)$ *is comprised of the parts of* σ. *The rightmost occurrence of each* $b_j \in B(\sigma)$ *in* σ *is called* b_j-critical, *and the subsequence of critical parts*

$$\sigma_{i_1} \cdots \sigma_{i_s} = \kappa_1 \cdots \kappa_s = \kappa(\sigma)$$

is called the core *of* σ. *Let* $\mathcal{T}_\kappa^B$ *be the set of all compositions that have core* κ *and whose values equal* B, *that is,* $\mathcal{T}_\kappa^B = \{\sigma | \kappa(\sigma) = \kappa, B(\sigma) = B\}$, *and let* $\mathcal{AT}_\kappa^B(\tau)$ *be the set of all compositions in* $\mathcal{T}_\kappa^B$ *that avoid* τ.

Note that $B(\sigma) \subseteq A$, and that $\kappa(\sigma)$ is a permutation of $B(\sigma)$.

Example 5.28 *For* $\sigma = 556632111233652$, $B(\sigma) = \{1,2,3,5,6\}$, *the critical parts of* σ *are* $\sigma_9 = 1$, $\sigma_{12} = 3$, $\sigma_{13} = 6$, $\sigma_{14} = 5$, $\sigma_{15} = 2$ *and* $\kappa(\sigma) = 13652$. *Note that* σ *contains the pattern* 12–2 *three times* ($\sigma_2\sigma_3\sigma_4 = 566$, $\sigma_9\sigma_{10}\sigma_{15} = 122$, *and* $\sigma_{10}\sigma_{11}\sigma_{12} = 233$).

Definition 5.29 *A* b_j-*block of* σ *is a subword* $\sigma_i \cdots \sigma_{i+k-1}$ *such that* $\sigma_i = \sigma_{i+1} = \cdots = \sigma_{i+k-1} = b_j$, $\sigma_{i-1} \neq b_j$, *and* σ_{i+k} *is either not equal to* b_j *or is* b_j-*critical. A block* $K = \sigma_i\sigma_{i+1} \cdots \sigma_{i+k-1}$ *is to the left of a block* $K' = \sigma_{i'}\sigma_{i'+1} \cdots \sigma_{i'+k'-1}$ *if* $i + k - 1 < i'$ *and we write* $K \lesssim K'$. *Likewise,* K *is to the right of* K' *if* $i' + k' - 1 < i$ *and we write* $K \gtrsim K'$.

Example 5.30 *The composition* $\sigma = 556632111233652$ *has the following structure: blocks* 55, 66, 3, 2 *and* 11, *critical part* 1, *blocks* 2 *and* 3, *and critical parts* 3, 6, 5, *and* 2. *Also,* $55 \lesssim 11$ *and* $33 \gtrsim 66$.

We can visualize any composition in terms of its critical parts and its blocks using a diagram like the one in Figure 5.3 for $\sigma = 556632111233652$, which we will refer to as a *block diagram*. Each horizontal line corresponds to a value of $B(\sigma)$, ordered from smallest to biggest from bottom to top. On each horizontal line we record the blocks and the critical parts for the respective b_j. Blocks are represented by rectangles and critical parts by dots, and they are placed from left to right in the order in which they occur in the composition. After the critical part b_j has occurred, there can be no further occurrences of b_j and therefore the horizontal line is terminated. We emphasize the critical parts by vertical (dotted) lines. We will use this type of diagram specifically to visualize the underlying structure of 12–2 and 21–1-avoiding compositions and to show the bijection between them. We are now ready to describe the structure of any 12–2-avoiding compositions.

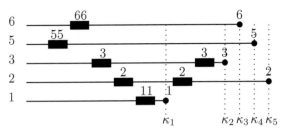

FIGURE 5.3: Block diagram for $\sigma = 556632111233652$.

Lemma 5.31 *Let* $B = \{b_1, \ldots, b_s\}$ *be an ordered set and let* κ *be a permutation of* B. *If* $\sigma \in \mathcal{AT}_\kappa^B(12\text{–}2)$, *then for* $j = 1, \ldots, s$, *there is at most one* b_j-*block to the left of* κ_1, *and also between any two consecutive critical parts.*

Furthermore, the blocks occurring to the left of κ_1 or between consecutive critical parts, respectively, are in decreasing order, that is, if K is a b-block and K' is a b'-block with $b < b'$, then $K \gtrsim K'$.

Proof We prove the statement by contradiction. Assume that there exist two b_j-blocks, $K = \sigma_i\sigma_{i+1}\cdots\sigma_{i+k-1}$ and $K' = \sigma_{i'}\sigma_{i'+1}\cdots\sigma_{i'+k'-1}$ with $K \lesssim K'$ between two consecutive critical parts. If $\sigma_{i+k-1} < \sigma_{i+k}$, then there is an occurrence of 12–2 at $\sigma_{i+k-1}\sigma_{i+k}\sigma_{i*}$, where σ_{i*} is σ_{i+k}-critical. (σ_{i*} exists since σ_{i+k} is not critical by assumption). If, on the other hand, $\sigma_{i+k-1} > \sigma_{i+k}$, then there exists a minimal p with $i + p \leq i'$ (since $\sigma_{i'} = b_j$) such that $\sigma_{i+k-1} > \sigma_{i+k} \geq \cdots \geq \sigma_{i+p-1} < \sigma_{i+p}$. Then $\sigma_{i+p-1}\sigma_{i+p}\sigma_{i*}$ is an occurrence of the pattern 12–2, where σ_{i*} is σ_{i+p}-critical. This shows that there cannot be two b_j-blocks between consecutive critical parts.

What remains to be shown is that the blocks are in decreasing order. Let K be a b-block and K' be a b'-block that occur between two consecutive critical parts, with $b < b'$. If $K \lesssim K'$, and $\sigma_{i+k} > b$, then $\sigma_{i+k-1}\sigma_{i+k}\sigma_{i*}$ is an occurrence of 12–2, where σ_{i*} is σ_{i+k}-critical. If $\sigma_{i+k} < b$, then there exists a minimal p with $i + p \leq i'$ (because $\sigma_{i'} = b' > b$) such that $\sigma_{i+k-1} > \sigma_{i+k} \geq \cdots \geq \sigma_{i+p-1} < \sigma_{i+p}$ and we get once more an occurrence of 12–2, where σ_{i*} is σ_{i+p}-critical, a contradiction.

This proof uses only the fact that there are no critical parts between the two blocks K and K', therefore the argument also applies when the two blocks are to the left of κ_1. □

Next we determine which blocks can actually occur for a given core κ.

Lemma 5.32 *Let $B = \{b_1, \ldots, b_s\}$ be an ordered set and κ be a permutation of B. Let $\sigma \in AT_\kappa^B(12\text{–}2)$ and let $\sigma_u = \kappa_i, \sigma_v = \kappa_{i+1}$, $i = 1, \ldots, s - 1$ be two consecutive critical parts in σ. If K is a b-block in $\sigma_{u+1}\sigma_{u+2}\cdots\sigma_{v-1}$, then $b \neq \kappa_j$ for all $j = 1, 2, \ldots, i$ and $b < \kappa_i$. If K is a b-block in $\sigma_1 \cdots \sigma_{u-1}$, where $\sigma_u = \kappa_1$, then there are no restrictions on b.*

Proof The definition of a critical part immediately gives that $b \neq \kappa_j$, for $j = 1, 2, \ldots, i$. The proof that $b < \kappa_i$ follows as in the proof of the second part of Lemma 5.31, where the assumption that $\sigma_{i+k-1} < \sigma_{i+k}$ leads to an occurrence of 12–2, and therefore to a contradiction. If $K \lesssim \kappa_1$, then no values are excluded for reason of critical values occurring to the left of K (there are none). By Lemma 5.31, all blocks occur in decreasing order, so they do not create an occurrence of 12, and hence 12–2 does not occur. □

Example 5.33 *(Continuation of Example 5.28)* *We saw in Example 5.28 that $\sigma = 556632111233652$ has three occurrences of 12–2. The first and third occurrences stem from the fact that the blocks to the left of κ_1 and those* *between κ_1 and κ_2 are not in decreasing order. The second pattern occurs since there is a b-block with $b \geq \kappa_1$ between κ_1 and κ_2.*

From Lemmas 5.31 and 5.32, we can derive the general structure for any 12–2 avoiding composition. Figure 5.4 shows the *structure diagram* for the core 4512. In this diagram, an oval indicates the positions where a block is allowed (but it may be empty), while an X indicates a position where such a block cannot occur. We refer to the oval positions as *allowable blocks* and the X positions as *forbidden blocks*.

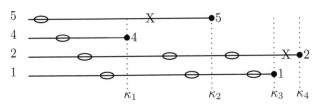

FIGURE 5.4:   Structure diagram for $\sigma \in \mathcal{AT}^B_{4512}(12\text{–}2)$.

Now we can determine the number of b_j-blocks that occur in a composition $\sigma \in \mathcal{AT}^B_\kappa(12\text{–}2)$.

Lemma 5.34 *Let $B = \{b_1, \ldots, b_s\}$ be an ordered set and let κ be a permutation of B. For $\sigma \in \mathcal{AT}^B_\kappa(12\text{–}2)$, the number of κ_j-blocks in σ is at most $|\{i | \kappa_j < \kappa_i \text{ and } i < j\}| + 1$.*

Proof This follows immediately from Lemmas 5.31 and 5.32 and can be visualized using Figure 5.4: a κ_j-block can occur at most once between two consecutive critical parts κ_i and κ_{i+1} only if $i < j$ and $\kappa_j < \kappa_i$. In addition, any b_j-block can occur at most once to the left of κ_1.   □

Lemma 5.34 will be the key to determining the generating function for the number of 12–2-avoiding compositions. We now look at the structure of the 21–2-avoiding compositions. Since the behavior of blocks on the left of κ_1 is the same as that between two consecutive critical parts, we refer to the section on the left of κ_1 as the section between κ_0 and κ_1. This allows us to give unified statements and simplify proofs. Using arguments similar to those in the proofs of Lemmas 5.31, 5.32 and 5.34, we obtain the following results.

Lemma 5.35 *Let $B = \{b_1, \ldots, b_s\}$ be an ordered set and let κ be a permutation of B. If $\sigma \in \mathcal{AT}^B_\kappa(21\text{–}2)$, then for $j = 1, \ldots, s$, there is at most one b_j-block between any two consecutive critical parts. Furthermore, the blocks*

occurring between consecutive critical parts are in increasing order, that is, if
K *is a b-block and* K' *is a b'-block with* $b < b'$, *then* $K \lesssim K'$.

As before, we can be more specific.

Lemma 5.36 *Let* $B = \{b_1, \ldots, b_s\}$ *be an ordered set and* κ *be a permutation*
of B. *Let* $\sigma \in \mathcal{AT}^B_\kappa(21\text{--}2)$ *and let* $\sigma_u = \kappa_{i-1}, \sigma_v = \kappa_i$, $i = 1, \ldots, s-1$ *be two*
consecutive critical parts in σ. *If* K *is a b-block in* $\sigma_{u+1}\sigma_{u+2}\cdots\sigma_{v-1}$, *then*
$b \neq \kappa_j$ *for all* $j = 1, 2, \ldots, i-1$ *and* $b \leq \kappa_i$.

We give an example of the structure diagram for 21–1-avoiding compositions
in Figure 5.5.

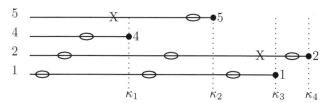

FIGURE 5.5: Structure diagram for $\sigma \in \mathcal{AT}^B_{4512}(21\text{--}2)$.

Using Lemmas 5.35 and 5.36 we obtain the number of b_j-blocks that occur
in $\sigma \in \mathcal{AT}^B_\kappa(21\text{--}2)$.

Lemma 5.37 *Let* $B = \{b_1, \ldots, b_s\}$ *be an ordered set and let* κ *be a permu-*
tation of B. *For* $\sigma \in \mathcal{AT}^B_\kappa(21\text{--}2)$, *the number of* κ_j-*blocks in* σ *is at most*
$|\{i | \kappa_j < \kappa_i \text{ and } i < j\}| + 1$.

This result is at first surprising, as it gives the same number of κ_j-blocks as
for the 12–2-avoiding compositions, even though the conditions for the blocks
are different.

Proof From Lemmas 5.35 and 5.36 we obtain that a κ_j-block can occur at
most once between two consecutive critical parts κ_{i-1} and κ_i only if $i-1 < j$
and $\kappa_j \leq \kappa_i$. Therefore the number of κ_j-blocks is

$$|\{i | \kappa_j \leq \kappa_i \text{ and } j > i-1\}| = |\{i | \kappa_j \leq \kappa_i \text{ and } j \geq i\}|$$
$$= |\{i | \kappa_j < \kappa_i \text{ and } j > i\}| + 1,$$

as claimed. $\square$

Lemmas 5.34 and 5.37 suggest that there is a bijection between the set of
12–2-avoiding compositions of n with m parts in A and the set of 21–2-avoiding
compositions of n with m parts in A. This is indeed the case.

Theorem 5.38 *For any ordered set* $A = \{a_1, a_2, \ldots\} \subseteq \mathbb{N}$, *12–2* $\sim$ *21–2 and* *21–1* $\sim$ *12–1.*

Proof   Let $\kappa = \kappa_1 \cdots \kappa_s$ be a permutation of B where B is a fixed subset of A. It is enough to establish a bijection

$$\psi_\kappa : \mathcal{C}_{n,m}^A \cap \mathcal{AT}_\kappa^B(12\text{–}2) \to \mathcal{C}_{n,m}^A \cap \mathcal{AT}_\kappa^B(21\text{–}2)$$

since each σ has a unique core $\kappa(\sigma)$ and set of values $B(\sigma)$. We will illustrate the algorithm with the 12–2-avoiding composition $\sigma = 55441111422251111112$. First we determine the block diagram for the given composition, as shown in Figure 5.6.

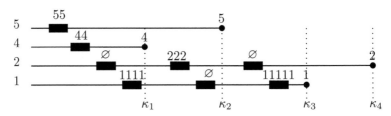

FIGURE 5.6:   Block diagram for $\sigma = 55441111422251111112$.

The critical parts separate the composition into s sections (the pieces between the vertical lines in the diagram). In each section, a b_j-block may occur (empty or nonempty) as indicated by a rectangle, or is forbidden. Note that if $\kappa_j = b_{i_j}$, then there are exactly j potential locations for a b_{i_j}-block. This is the case for both the 12–2-avoiding and the 21–2-avoiding compositions if they have the same core. For a given j, not only is the number of potential occurrences of b_j-blocks the same, but the number of forbidden b_j-blocks is also strictly determined by the core, and thus the same for the 12–2-avoiding and the 21–2-avoiding compositions. This can be seen as follows. If $\sigma \in \mathcal{AT}_\kappa^B(12\text{–}2)$, then there is no forbidden b_j-block to the left of κ_1 according to Lemma 5.32. In the segment created by κ_i and κ_{i+1}, a b_j-block is forbidden if $b_j \geq \kappa_i$ for $i = 1, 2, \ldots, s - 1$. By Lemma 5.36, this forbidden block corresponds to a forbidden block in the segment created by κ_{i-1} and κ_i in a composition $\sigma' \in \mathcal{AT}_\kappa^B(21\text{–}2)$ that has the same core, where the section created by κ_0 and κ_1 is to be interpreted as the section to the left of κ_1. Finally, there are no forbidden blocks in the section created by κ_{s-1} and κ_s for any $\sigma' \in \mathcal{AT}_\kappa^B(21\text{–}2)$, because there is only one possibility, namely κ_s-blocks, which are allowed. Therefore for each b_j the number of allowable blocks is identical in $\sigma \in \mathcal{AT}_\kappa^B(12\text{–}2)$ and $\sigma' \in \mathcal{AT}_\kappa^B(21\text{–}2)$ for a given core κ.

Now we can describe the algorithm. We start with the block diagram for the composition $\sigma \in \mathcal{AT}_{\kappa}^{B}(12\text{--}2)$ (see Figure 5.6) and mark the forbidden block locations with the letter X as shown in Figure 5.7.

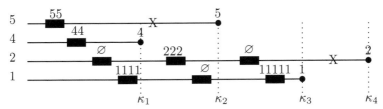

FIGURE 5.7:   Modified block diagram for $\sigma = 55441111422251111112$.

Next we use the structure diagram for the 21–2-avoiding compositions (see Figure 5.5). For each b_j we place the allowed (nonempty or empty) blocks of σ in the allowed positions of the 21–2 structure diagram. For example, the allowed 1-blocks are 1111, $\varnothing$, and 11111, and we place them in the three allowable 1-block positions of the structure diagram from left to right. For the 5-blocks there is one allowable block, 55, which is placed in the single allowable position between κ_1 and κ_2. The resulting 21–1-avoiding composition is $\sigma' = 11114442225551111112$, as shown in Figure 5.8.

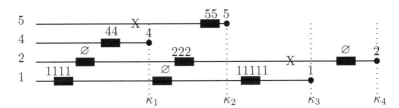

FIGURE 5.8:   Block diagram for $\sigma' = 11114442225551111112$.

Likewise we can create a 12–1-avoiding composition by starting with a 21–1-avoiding composition and following steps analogous to the ones described above. Since the structure diagrams uniquely determine the arrangements of the blocks, $\psi_{\kappa} : \mathcal{C}_{n,m}^{A} \cap \mathcal{AT}_{\kappa}^{B}(12\text{--}2) \to \mathcal{C}_{n,m}^{A} \cap \mathcal{AT}_{\kappa}^{B}(21\text{--}2)$ is a bijection. The Wilf-equivalence 21–1 $\sim$ 12–1 follows using the complement map.   □

We now derive the generating functions for the five Wilf-equivalence classes. First we give the results for the three patterns that are in their own equivalence classes (up to symmetry), namely 11–1, 11–2, and 22–1. For the first pattern we only obtain a recurrence relation, while for the other two patterns we give explicit formulas for the generating functions which are easy to compute. Finding an explicit expression for the pattern 11–1 remains an open question (see Research Direction 5.1).

Theorem 5.39 *Let $A = [d]$. Then*

$$AC_A^{11-1}(x,y) = \frac{1 + \sum_{i=1}^{d} \frac{x^{2i}y^2}{1+x^i y} AC_{A-\{i\}}^{11-1}(x,y)}{1 - \sum_{i=1}^{d} \frac{x^i y}{1 + x^i y}}.$$

Proof Considering the compositions that start with i, we obtain

$$AC_A^{11-1}(i|x,y)$$
$$= x^i y + \sum_{j \neq i} AC_A^{11-1}(ij|x,y) + AC_A^{11-1}(ii|x,y)$$
$$= x^i y(AC_A^{11-1}(x,y) - AC_A^{11-1}(i|x,y)) + x^{2i}y^2 AC_{A-\{i\}}^{11-1}(x,y),$$

which implies that

$$AC_A^{11-1}(i|x,y) = \frac{x^i y}{1+x^i y} AC_A^{11-1}(x,y) + \frac{x^{2i}y^2}{1+x^i y} AC_{A-\{i\}}^{11-1}(x,y).$$

Summing over $i = 1, 2, \ldots, d$, we get that

$$AC_A^{11-1}(x,y) = 1 + \sum_{i=1}^{d} \frac{x^i y}{1+x^i y} AC_A^{11-1}(x,y) + \sum_{i=1}^{d} \frac{x^{2i}y^2}{1+x^i y} AC_{A-\{i\}}^{11-1}(x,y),$$

which is equivalent to

$$AC_A^{11-1}(x,y) = \frac{1 + \sum_{i=1}^{d} \frac{x^{2i}y^2}{1+x^i y} AC_{A-\{i\}}^{11-1}(x,y)}{1 - \sum_{i=1}^{d} \frac{x^i y}{1 + x^i y}}, \tag{5.6}$$

as claimed. □

Example 5.40 *The generating function for the pattern 11–1 is lengthy to compute. For example, if $n = 15$, we would need all the generating functions $AC_B^{11-1}(x,y)$, where B is a proper subset of $[15]$, of which there are $2^{15} - 1 = 32767$. This is not very efficient, as each of these generating functions takes several steps to compute. Instead, we may obtain the number of 11–1-avoiding compositions of n by direct enumeration using a modification of the Mathematica code given in Example 5.14. The values for $n = 0, 1, \ldots, 15$ are 1, 1, 2, 3, 7, 12, 22, 40, 73, 136, 243, 433, 790, 1413, 2527 and 4516.*

We now give results for the patterns 11–2 and 22–1. Note the similarity of the two generating functions, which stems from the fact that the two patterns are complementary.

Theorem 5.41 *Let* $A = [d]$. *Then the generating functions for the number of compositions of* n *with* m *parts in* $[d]$ *that avoid* 11–2 *and* 22–1, *respectively, are given by*

$$AC_{[d]}^{11-2}(x,y) = \prod_{j=1}^{d} \frac{1 - \sum_{i=1}^{j-1} \frac{x^i y}{1 + x^i y}}{1 - x^j y - \sum_{i=1}^{j-1} \frac{x^i y}{1 + x^i y}}$$

and

$$AC_{[d]}^{22-1}(x,y) = \prod_{j=1}^{d} \frac{1 - \sum_{i=j+1}^{d} \frac{x^i y}{1 + x^i y}}{1 - x^j y - \sum_{i=j+1}^{d} \frac{x^i y}{1+x^i y}}.$$

Proof Looking at the compositions that start with i, we obtain that

$$AC_{[d]}^{11-2}(i|x,y) = x^i y + \sum_{j \neq i} AC_{[d]}^{11-2}(ij|x,y) + AC_{[d]}^{11-2}(ii|x,y)$$

$$= x^i y (AC_{[d]}^{11-2}(x,y) - AC_{[d]}^{11-2}(i|x,y)) + x^{2i} y^2 AC_{[i]}^{11-2}(x,y),$$

which implies that

$$AC_{[d]}^{11-2}(i|x,y) = \frac{x^i y}{1 + x^i y} AC_{[d]}^{11-2}(x,y) + \frac{x^{2i} y^2}{1 + x^i y} AC_{[i]}^{11-2}(x,y).$$

Summing over $i = 1, 2, \ldots, d$ and taking into consideration the empty composition, we obtain that

$$AC_{[d]}^{11-2}(x,y) = 1 + \sum_{i=1}^{d} \frac{x^i y}{1 + x^i y} AC_{[d]}^{11-2}(x,y) + \sum_{i=1}^{d} \left(\frac{x^{2i} y^2}{1 + x^i y} AC_{[i]}^{11-2}(x,y) \right).$$

To create a recursion, we look at the difference $AC_{[d]}^{11-2}(x,y) - AC_{[d-1]}^{11-2}(x,y)$. Using $AC_{[d]}$ as a shorthand for $AC_{[d]}^{11-2}(x,y)$, we obtain

$$AC_{[d]} - AC_{[d-1]} = \sum_{i=1}^{d} \frac{x^i y}{1 + x^i y} AC_{[d]} - \sum_{i=1}^{d-1} \frac{x^i y}{1 + x^i y} AC_{[d-1]} + \frac{x^{2d} y^2}{1 + x^d y} AC_{[d]},$$

which (after some algebraic manipulation) leads to

$$AC_{[d]}^{11-2}(x,y) = \frac{1 - \sum_{i=1}^{d-1} \frac{x^i y}{1 + x^i y}}{1 - x^d y - \sum_{i=1}^{d-1} \frac{x^i y}{1 + x^i y}} AC_{[d-1]}^{11-2}(x,y)$$

for all $d \geq 1$. Using the above recurrence relation d times together with the initial condition $AC_{\varnothing}^{11-2}(x,y) = 1$, we get the result for $AC_{[d]}^{11-2}(x,y)$.

We now derive the result for the generating function for the 22–1-avoiding compositions. The proof starts in the same way as the one for the 11–2-avoiding compositions and we obtain that

$$AC_{[d]}^{22-1}(i|x,y) = \frac{x^i y}{1 + x^i y} AC_{[d]}^{22-1}(x,y) + \frac{x^{2i} y^2}{1 + x^i y} AC_{[i,d]}^{22-1}(x,y). \quad (5.7)$$

Unfortunately, the second summand in (5.7) now depends on both i and d and will not cancel out when taking the difference $AC_{[d]}^{22-1}(x,y) - AC_{[d-1]}^{22-1}(x,y)$. Instead, we must first "normalize" this term so that it will once more cancel out. Note that we can map any composition of n with m parts in $\{i, \ldots, d\}$ to a composition of $n' = n - m(i-1)$ with m parts in $[d+1-i]$ by replacing σ_j by $\sigma_j - (i-1)$. In terms of generating functions, this means that $AC_{[i,d]}^{\tau}(x,y) = AC_{[d+1-i]}^{\tau}(x, x^{i-1}y)$ for any pattern τ (using the definition of the generating function). Replacing y by y/x^{i-1} in (5.7) yields

$$AC_{[d]}^{22-1}\left(i\Big|x, \frac{y}{x^{i-1}}\right)$$
$$= \frac{xy}{1+xy} AC_{[d]}^{22-1}\left(x, \frac{y}{x^{i-1}}\right) + \frac{x^2 y^2}{1+xy} AC_{[d+1-i]}^{22-1}(x,y).$$

Therefore,

$$AC_{[d]}^{22-1}\left(i\Big|x, \frac{y}{x^{i-1}}\right) - AC_{[d-1]}^{22-1}\left(i-1\Big|x, \frac{y}{x^{i-2}}\right)$$
$$= \frac{xy}{1+xy} AC_{[d]}^{22-1}\left(x, \frac{y}{x^{i-1}}\right) - \frac{xy}{1+xy} AC_{[d-1]}^{22-1}\left(x, \frac{y}{x^{i-2}}\right).$$

Replacing y by $x^{i-1}y$ we obtain

$$AC_{[d]}^{22-1}(i|x,y) - AC_{[d-1]}^{22-1}(i-1|x, xy)$$
$$= \frac{x^i y}{1+x^i y} AC_{[d]}^{22-1}(x,y) - \frac{x^i y}{1+x^i y} AC_{[d-1]}^{22-1}(x, xy).$$

Summing over $i = 2, 3, \ldots, d$ and using that $AC_{[d]}^{22-1}(1|x,y) = xy AC_{[d]}^{22-1}(x,y)$ we get (after doing some algebra) that

$$\left(1 - xy - \sum_{i=2}^{d} \frac{x^i y}{1 + x^i y}\right) AC_{[d]}^{22-1}(x,y)$$
$$= \left(1 - \sum_{i=2}^{d} \frac{x^i y}{1 + x^i y}\right) AC_{[d-1]}^{22-1}(x, xy).$$

Using the above recurrence relation d times together with the initial condition $AC_{\varnothing}^{22-1}(x,y) = 1$ yields the desired result. $\qquad \square$

Example 5.42 *By taking the limit as $d \to \infty$ and setting $y = 1$ in Theorem 5.41 we get that the generating functions for the number of compositions of n that avoid the generalized patterns 11–2 and 22–1, respectively, are given by*

$$AC_{\mathbb{N}}^{11-2}(x,1) = \prod_{j=1}^{\infty} \frac{1 - \sum_{i=1}^{j-1} \frac{x^i}{1+x^i}}{1 - x^j - \sum_{i=1}^{j-1} \frac{x^i}{1+x^i}}$$

and

$$AC_{\mathbb{N}}^{22-1}(x,1) = \prod_{j=1}^{\infty} \frac{1 - \sum_{i=j+1}^{d} \frac{x^i}{1+x^i}}{1 - x^j - \sum_{i=j+1}^{\infty} \frac{x^i}{1+x^i}}.$$

We use Mathematica *or* Maple *to obtain the sequence $[x^n]AC_{\mathbb{N}}^{11-2}(x,1)$ for $n = 0$ to $n = 20$ as 1, 1, 2, 4, 7, 13, 24, 43, 77, 139, 248, 441, 786, 1394, 2469, 4374, 7730, 13649, 24093, 42478, and 74847. The corresponding sequence for 22–1 is given by 1, 1, 2, 4, 8, 15, 30, 58, 112, 217, 420, 811, 1565, 3021, 5823, 11227, 21636, 41684, 80297, 154650, and 297816.*

Finally, we look at the generating functions for the remaining two classes. For the patterns 12–2 $\sim$ 21–2 we obtain a very nice result that connects pattern avoidance in compositions to permutations.

Theorem 5.43 *The generating function for the number of 12–2-avoiding compositions of n with m parts in $[d]$ is given by*

$$AC_{[d]}^{12-2}(x,y) = \sum_{B \subseteq [d]} \sum_{\kappa \in \mathcal{S}(B)} \prod_{j=1}^{|B|} \frac{x^{\kappa_j} y}{(1 - x^{\kappa_j} y)^{|\{i|\kappa_j < \kappa_i \text{ and } i < j\}| + 1}},$$

where $\mathcal{S}(B)$ is the set of permutations of B.

Proof Let $B \subseteq [d]$ and κ be fixed, where $\kappa \in \mathcal{S}(B)$. Using Lemmas 5.31, 5.32, and 5.34 we obtain that the generating function for the number of compositions of n with m parts in $\mathcal{AT}_{\kappa}^{B}(12\text{--}2)$ is given by

$$\prod_{j=1}^{|B|} \frac{x^{\kappa_j} y}{(1 - x^{\kappa_j} y)^{|\{i|\kappa_j < \kappa_i \text{ and } i < j\}| + 1}}.$$

The numerator of each term in the product accounts for the critical part, while the denominator results from the product of the generating functions of the $|\{i|\kappa_j < \kappa_i \text{ and } i < j\}| + 1$ (potentially empty) b_j-blocks. Now summing over all the permutations κ of B, and over all subsets B of $[d]$ we obtain that

$$AC_{[d]}^{12-2}(x,y) = \sum_{B \subseteq [d]} \sum_{\kappa \in \mathcal{S}(B)} \prod_{j=1}^{|B|} \frac{x^{\kappa_j} y}{(1 - x^{\kappa_j} y)^{|\{i|\kappa_j < \kappa_i \text{ and } i<j\}|+1}},$$

as claimed.   □

Note that this proof, and therefore the result given in Theorem 5.43, can be easily generalized to any ordered set $A = \{a_1, a_2, \ldots, a_d\} \subset \mathbb{N}$.

Example 5.44 *Even though Theorem 5.43 provides an explicit formula for the generating function, it is not a formula that can be efficiently implemented as a program. We have obtained the values for the number of 12–2-avoiding compositions via direct enumeration using* Mathematica *or* Maple *for $n = 0, 1, \ldots, 15$ as 1, 1, 2, 4, 8, 15, 29, 55, 104, 194, 359, 660, 1208, 2200, 3982, and 7166.*

The last Wilf-equivalence class consists of the patterns 12–1 and 21–1. We obtained this equivalence by applying the complement operation to the equivalence 21–2 $\sim$ 12–2. Past experience with complementary patterns (see for example Theorem 5.41) suggests that the generating functions of the complementary patterns are of very similar structure, but this is not the case. So far, we have only been able to obtain the recursion for the generating function for the pattern 12–1 given in Theorem 5.46, not an explicit formula as in Theorem 5.43. To find an explicit expression, especially one that involves permutations, remains an open question (see Research Direction 5.1). We make use of Lemma 5.45 to derive the generating function for the number of 12–1-avoiding compositions.

Lemma 5.45 *The solution of the system of equations $x_i = \beta_i + \alpha_i \sum_{j=1}^{i} x_j$, $i = 1, 2, \ldots, d$ is given by*

$$x_i = \frac{\beta_i}{1 - \alpha_i} + \alpha_i \sum_{j=1}^{i-1} \frac{\beta_j}{\prod_{k=j}^{i}(1 - \alpha_k)}, \quad i = 1, 2, \ldots, d.$$

Proof   The linear system of equations can be presented as

$$(\mathbf{I} - \mathbf{V}_d)(x_1, \ldots, x_d)^T = (\beta_1, \ldots, \beta_d)^T,$$

where $\mathbf{V}_d = (v_{ij})_{1 \le i,j \le d}$ with $v_{ij} = \alpha_i$ if $j \le i$, and 0 otherwise. By Cramer's Rule, $x_i = \frac{\det(\mathbf{U}_d^{(i)})}{\det(\mathbf{I} - \mathbf{V}_d)}$, where $\mathbf{U}_d^{(i)}$ is the matrix obtained from $\mathbf{I} - \mathbf{V}_d$ by replacing its i-th column by the column $(\beta_1, \ldots, \beta_d)^T$. The matrix $\mathbf{I} - \mathbf{V}_d$ is lower triangular, therefore $\det(\mathbf{I} - \mathbf{V}_d) = \prod_{j=1}^{d}(1 - \alpha_j)$. Since $\mathbf{U}_d^{(i)}$ is a special type of block matrix, we obtain that $\det(\mathbf{U}_d^{(i)}) = \prod_{j=i+1}^{d}(1-\alpha_j)\det(\mathbf{U}_i^{(i)})$ (see Theorem B.15). Using routine techniques from linear algebra (see Exercise 5.4), one can show that

$$\det(\mathbf{U}_i^{(i)}) = \beta_i \prod_{j=1}^{i-1}(1-\alpha_j) + \alpha_i \sum_{j=1}^{i-1}\beta_j \prod_{k=1}^{j-1}(1-\alpha_k).$$

Hence,

$$x_i = \frac{\det(\mathbf{U}_i^{(i)})}{\prod_{j=1}^{i}(1-\alpha_j)} = \frac{\beta_i}{1-\alpha_i} + \frac{\alpha_i}{\prod_{j=1}^{i}(1-\alpha_j)}\sum_{j=1}^{i-1}\beta_j\prod_{k=1}^{j-1}(1-\alpha_k),$$

which after simplification completes the proof. $\qquad\square$

Using this lemma, we can prove a result for the generating function of the patterns in the remaining Wilf-equivalence class.

Theorem 5.46 *For $A = [d] \subseteq \mathbb{N}$ and $\tau = 12\text{-}1$,*

$$AC_A^\tau(x,y) = 1 + \sum_{i=1}^{d}\frac{x^i y}{1-x^i y}\left(1 + \sum_{j=i+1}^{d}AC_{A-\{i\}}^\tau(j|x,y)\right)$$

$$+ \sum_{i=2}^{d}\sum_{j=1}^{i-1}\frac{x^{i+j}y^2}{\prod_{k=j}^{i}(1-x^k y)}\left(1 + \sum_{k=j+1}^{d}AC_{A-\{j\}}^\tau(k|x,y)\right),$$

where

$$AC_A^\tau(i|x,y) = x^i y + x^i y \sum_{j=1}^{i}AC_A^\tau(j|x,y) + x^i y \sum_{j=i+1}^{d}AC_{A-\{i\}}^\tau(j|x,y).$$

Proof Using arguments similar to those in the proofs of the other generating functions, we get that $AC_A^\tau(i|x,y)$ satisfies

$$AC_A^\tau(i|x,y) = x^i y + x^i y \sum_{j=1}^{i}AC_A^\tau(j|x,y) + x^i y \sum_{j=i+1}^{d}AC_{A-\{i\}}^\tau(j|x,y).$$

Applying Lemma 5.45 with $\alpha_i = x^i y$ and

$$\beta_i = x^i y\left(1 + \sum_{j=i+1}^{d}AC_{A-\{i\}}^\tau(j|x,y)\right)$$

gives

$$
AC_A^\tau(i|x,y) = \frac{x^i y}{1 - x^i y} \left(1 + \sum_{j=i+1}^d AC_{A-\{i\}}^\tau (j|x,y) \right)
$$
$$
+ \sum_{j=1}^{i-1} \frac{x^{i+j} y^2}{\prod_{k=j}^i (1 - x^k y)} \left(1 + \sum_{k=j+1}^d AC_{A-\{j\}}^\tau (k|x,y) \right)
$$

for all $i = 1, 2, \ldots, d$. The result for $AC_A^\tau(x,y)$ then follows from (5.1). □

Note that Theorem 5.46 can be easily generalized to $A = \{a_i, a_2, \ldots, a_d\}$ by replacing i by a_i.

Example 5.47 *The recursive formula for the generating function is not useful for computing the number of compositions avoiding 12–1. We obtained the sequence of values by direct enumeration (using* Mathematica *or* Maple*) as 1, 1, 2, 4, 7, 13, 23, 40, 68, 117, 195, 323, 531, 863, 1394, and 2234 for $n = 0, 1, \ldots, 15$.*

This completes the classification of all generalized patterns of length three. We now turn our attention to the last type of pattern that we will consider.

5.4 Partially ordered patterns in compositions

Most papers on partially ordered patterns consider avoidance of these patterns, but there are also some enumeration results. We start by presenting results on the number of occurrences of a given pattern in all compositions of n [111], and then give results on avoidance of partially ordered patterns in compositions of n [88]. Note that in [108] and [111], the patterns under study are called POGPs (*partially ordered generalized patterns*) and SPOPs (*segmented partially ordered patterns*), respectively. Segmented patterns are the same as subword patterns, so we refer to them as such, consistent with the rest of the book. However, what is common to POGPs and SPOPs is that they are defined on a partially ordered alphabet, and so we refer to them as partially ordered patterns or POPs.

Definition 5.48 *A partially ordered generalized pattern ξ is a word (in reduced form) consisting of letters from a partially ordered alphabet $\mathcal{T}$ (see Definition 7.44). If letters a and b are incomparable in a POP ξ, then the relative size of the letters corresponding to a and b in a composition σ is unimportant in an occurrence of ξ in the composition σ. Letters shown with the same number of primes are comparable to each other (for example, $1'$ and $3'$ are comparable), while letters shown without primes are comparable to all letters* *of the alphabet. Since such letters are either smaller or larger than all other letters (why?), we use either* 1 *or* ℓ *for such a letter (ℓ depends on the size of the primed letters in $\mathcal{T}$).*

Note that we always use ξ to refer to a pattern from a partially ordered set, and τ for a pattern from an (ordered) alphabet. Furthermore, in this section any pattern that does not contain dashes is a subword pattern. Before giving a more formal definition of the occurrence of a POP in a composition, we give an example.

Example 5.49 *Let $\mathcal{T} = \{1', 1'', 2''\}$, where the only relation among the elements is $1'' < 2''$. Then an occurrence of the (subword) pattern $\xi = 1'1''2''$ in σ means that there is a subword $\sigma_i\sigma_{i+1}\sigma_{i+2}$ with $\sigma_{i+1} < \sigma_{i+2}$ and no restrictions placed on σ_i. Thus, the sequence 31254 has two occurrences of $\xi = 1'1''2''$, namely 312 and 125, while the sequence 113425 contains seven occurrences of the (generalized) pattern $\xi = 1'-1''2''$, namely 113, 134 twice, 125 twice, 325, and 425.*

What is the reason that these partially ordered patterns were introduced? Just to see how far one can generalize and still get results? That in itself is always a reason for mathematicians, but a second reason is that POPs allow us to refer to a whole set of subword patterns via a single pattern.

Example 5.50 *In Section 4.3, we studied occurrences of the pattern peak, which is comprised of the patterns 121, 132 and 231. The pattern peak can be described by a single POP as follows: let $\mathcal{T} = \{1', 1'', 2\}$ and $\xi = 1'21''$. Likewise, the pattern valley $= \{212, 213, 312\}$ can be described as a POP by choosing $\mathcal{T} = \{1, 2', 2''\}$ and $\xi = 2'12''$. Thus, POPs are useful to describe "shapes" of patterns; however, not every POP results in a particular shape (see Example 5.52).*

This connection between a POP and its associated subword patterns can be used to give a formal definition of the occurrence of a POP.

Definition 5.51 *For a given POP $\xi = \xi_1\xi_2\cdots\xi_k$, we say that a subword pattern $\tau = \tau_1\tau_2\cdots\tau_k$ is a linear extension of ξ if $\xi_i < \xi_j$ implies that $\tau_i < \tau_j$. A generalized pattern is a linear extension of a generalized POP if it has the same adjacency requirements as the POP and obeys the rules for linear extension of subwords. A POP ξ occurs in a sequence (permutation, word, composition or partition) if any of its linear extensions occurs in the sequence. Otherwise, the sequence avoids the POP.*

Example 5.52 *For $\mathcal{T} = \{1', 1'', 2\}$ and $\xi = 1'21''$, the linear extensions are the patterns 121, 132 and 231. For $\mathcal{T} = \{1', 1'', 2''\}$ and $\xi = 1'1''2''$, the linear extensions are 112, 123, 212, 213, and 312. The occurrences of $\xi = 1'1''2''$ in 31254 indicated in Example 5.49 result from the occurrences of the linear extensions 312 and 125.*

5.4.1 Enumeration of POPs

We will present the main result given by Kitaev et al. in [111]. Note that in [111] the focus is on enumerating the number of occurrences of a given POP ξ among all compositions of n, rather than enumeration of compositions according to the number of occurrences of ξ within the composition. This is the same focus as in [50, 51, 76, 77, 87] for the subword patterns 11, 12, and 21. Kitaev et al. give additional results which we leave for the interested reader to investigate. To present the results of Kitaev et al. we need the following notation.

Definition 5.53 *For a subword pattern τ from the alphabet $[j]$, let μ_k be the number of parts of size k in τ. Then $\mu_\tau = (\mu_1, \mu_2, \ldots, \mu_j)$ is called the* content vector *of τ. In addition, let $\lambda_\tau(n)$ denote the number of compositions that are order-isomorphic to τ, that is, the reduced form of these compositions equals τ. The associated generating function is given by $\Lambda_\tau(x) = \sum_{n \geq 0} \lambda_\tau(n) x^n$.*

We will also make use of the following result.

Lemma 5.54 *For $\tau \in [j]$ with content vector μ_τ,*

$$\Lambda_\tau(x) = \prod_{k=1}^{j} \frac{x^{m_k}}{1 - x^{m_k}},$$

where $m_k = \mu_k + \cdots + \mu_j$ for $1 \leq k \leq j$.

Proof See Exercise 5.6. □

In order to enumerate the number of occurrences of a given POP in all compositions of n, we first compute a related generating function.

Definition 5.55 *For a given POP $\xi = \xi_1 \xi_2 \cdots \xi_k$, let $w_\xi(n, \ell, s)$ denote the number of occurrences of ξ among the compositions of n with $\ell + k$ parts such that the sum of the parts preceding the occurrence is s. Let $\Omega_\xi(x, y, u)$ denote the generating function for $w_\xi(n, \ell, s)$, that is,*

$$\Omega_\xi(x, y, u) = \sum_{n, l, s \geq 0} w_\xi(n, \ell, s) x^n y^\ell u^s.$$

Theorem 5.56 *[111, Theorem 2.1] Let ξ be a POP. Then*

$$\Omega_\xi(x, y, u) = \frac{(1 - x)(1 - xu)}{(1 - x - xy)(1 - xu - xyu)} \sum_\tau \Lambda_\tau(x),$$

where the sum is over all linear extensions τ of ξ.

Proof We first compute $\Omega_\tau(x, y, u)$ for a subword pattern τ. Each occurrence of τ determines a triple $\left(\sigma^{(1)}, \sigma^{(2)}, \sigma^{(3)}\right)$ such that $\sigma^{(1)}$ comprises the parts to the left of the occurrence of τ, the reduced form of $\sigma^{(2)}$ equals τ, and $\sigma^{(3)}$ comprises the parts to the right of the occurrence. Hence, for given $n, \ell, s \in \mathbb{N}$, $w_\xi(n, \ell, s)$ is the number of triples $\left(\sigma^{(1)}, \sigma^{(2)}, \sigma^{(3)}\right)$ of compositions such that $\operatorname{ord}(\sigma^{(1)}) + \operatorname{ord}(\sigma^{(2)}) + \operatorname{ord}(\sigma^{(3)}) = n$, $\operatorname{ord}(\sigma^{(1)}) = s$, $\sigma^{(2)}$ is order-isomorphic to τ, and $\sigma^{(1)}$ and $\sigma^{(3)}$ together have ℓ parts. Recall that $C(m; n)$ denotes the number of compositions of n with m parts in $\mathbb{N}$. Thus,

$$w_\tau(n, \ell, s) = \sum_{\substack{0 \le j \le \ell \\ 0 \le n' \le n - s}} C(j; s)\lambda_\tau(n')C(l - j; n - s - n').$$

From Exercise 2.13, we have that $\sum_{n \ge 0} C(m; n)x^n = x^m/(1 - x)^m$. Thus we can factor $\Omega_\tau(x, y, u)$ into a product of $\Lambda_\tau(x)$ and two geometric series:

$$\Omega_\tau(x, y, u) = \sum_{n, \ell, s \ge 0} w_\tau(n, \ell, s)x^n y^\ell u^s$$

$$= \Lambda_\tau(x)\left(\sum_{s, j \ge 0} C(j; s)(xu)^s y^j\right)\left(\sum_{\tilde{n}, \tilde{\ell} \ge 0} C(\tilde{\ell}; \tilde{n})x^{\tilde{n}} y^{\tilde{\ell}}\right)$$

$$= \Lambda_\tau(x)\left(\sum_{j \ge 0} \frac{(xu)^j}{(1 - xu)^j} y^j\right)\left(\sum_{\tilde{\ell} \ge 0} \frac{x^{\tilde{\ell}}}{(1 - x)^{\tilde{\ell}}} y^{\tilde{\ell}}\right)$$

$$= \Lambda_\tau(x)\frac{(1 - xu)(1 - x)}{(1 - xu - xyu)(1 - x - xy)}.$$

Note that for any POP ξ, $\Omega_\xi(x, y, u) = \sum_\tau \Omega_\tau(x, y, u)$, where the sum is over all linear extensions τ of ξ, which completes the proof. $\square$

Setting $y = u = 1$ in Theorem 5.56, we obtain a result for the number of occurrences of a POP ξ in all compositions of n.

Corollary 5.57 *[111, Corollary 2.2] Given a POP ξ, the number of occurrences of ξ among all compositions of n is given by*

$$[x^n]\Omega_\tau(x, 1, 1) = [x^n]\frac{(1 - x)^2}{(1 - 2x)^2} \sum_\tau \Lambda_\tau(x),$$

where the sum is over all linear extensions τ of ξ.

Example 5.58 *We illustrate the use of Corollary 5.57 with two patterns that we have considered before, namely $\tau = 112$ and $\nu = 221$. The content vector* *for τ is given by $\mu_\tau = (2,1)$, and the content vector for ν is $\mu_\nu = (1,2)$. Therefore, by Lemma 5.54,*

$$\Lambda_\tau(x) = \frac{x^4}{(1-x^3)(1-x)} \quad and \quad \Lambda_\nu(x) = \frac{x^5}{(1-x^3)(1-x^2)}.$$

The number of levels followed by a rise or drop, respectively, can then be computed as

$$[x^n]\frac{(1-x)^2x^4}{(1-2x)^2(1-x^3)(1-x)} \quad and \quad [x^n]\frac{(1-x)^2x^5}{(1-2x)^2(1-x^3)(1-x^2)}.$$

Using Mathematica or Maple, the sequences of values for the number of occurrences of either 112 or 221, for $n = 0, 1, \ldots, 20$ are found to be 0, 0, 0, 0, 1, 3, 8, 21, 51, 120, 277, 627, 1400, 3093, 6771, 14712, 31765, 68211, 145784, 310293, 658035, and 0, 0, 0, 0, 0, 1, 2, 6, 15, 36, 84, 193, 434, 966, 2127, 4644, 10068, 21697, 46514, 99270, 211023, respectively.

Note that we can recover many of the results in Chapter 4, for example, the generating function for the number of rises and levels given in Example 4.6, by using Corollary 5.57. However those results just concern individual subword patterns. To see the full power of POPs, we will now derive the generating function for the number of peaks.

Example 5.59 *A peak is represented by the POP $\xi = 1'21''$, with linear extensions 121, 132 and 231. These patterns have content vectors $(2,1)$, $(1,1,1)$ and $(1,1,1)$, respectively. We obtain*

$$\sum_\tau \Lambda_\tau(x) = \frac{x^4}{(1-x^3)(1-x)} + 2\frac{x^6}{(1-x^3)(1-x^2)(1-x)},$$

where the sum is over the linear extensions of ξ. The number of peaks may then be computed using Maple or Mathematica as 0, 0, 0, 0, 1, 3, 10, 27, 69, 168, 397, 915, 2074, 4635, 10245, 22440, 48781, 105363, 226330, 483867, and 1030149 for $n = 0, 1, \ldots, 20$.

In Section 4.3 we derived generating functions for patterns of length $\ell \geq 3$ according to order, number of parts and occurrences of τ. We can exploit these results to derive generating functions for the number of occurrences of a given pattern τ by using the technique in the proof of Corollary 4.5 (see Exercise 5.7).

5.4.2   Avoidance of POPs

We will present results on avoidance of two particular classes of POPs in compositions. These types of patterns were considered for permutations in [108] and for words in [109].

Definition 5.60 *Let* $\{\tau_0, \tau_1, \ldots, \tau_s\}$ *be a set of subword patterns. A multi-pattern is of the form* $\tau = \tau_0 - \tau_1 - \cdots - \tau_s$ *and a* shuffle *pattern is of the form* $\tau = \tau_0 - a_1 - \tau_1 - a_2 - \cdots - \tau_{s-1} - a_s - \tau_s$, *where each letter of* τ_i *is incomparable with any letter of* τ_j *whenever* $i \neq j$. *In addition, the letters* a_i *are either all greater or all smaller than any letter of* τ_j *for any* i *and* j.

Example 5.61 *Let* $\mathcal{T} = \{1', 1'', 2\}$, *that is* $1' < 2$, $1'' < 2$, *and* $1'$ *and* $1''$ *are incomparable. Then* $1'-2-1''$ *is a shuffle pattern, and* $1'-1''$ *is a multi-pattern. Clearly, we can get a multi-pattern from a shuffle pattern by removing the letters* a_i.

We begin by deriving the generating function for the simplest shuffle pattern, and then give a more general result.

Example 5.62 *Let* $A = \{a_1, a_2, \ldots, a_d\}$ *be any ordered set and let* $\tau = 1'-2-1''$. *Then we have*

$$AC_A^{1'-2-1''}(x, y) = \frac{1}{\prod_{a \in A}(1 - x^a y)^2} - \sum_{a \in A} \frac{x^a y}{\prod_{a \leq b \in A}(1 - x^b y)^2}. \quad (5.8)$$

This result follows from the specific structure of the compositions σ *that avoid* $\tau = 1'-2-1''$. *If* σ *avoids* τ *and contains* $s > 0$ *copies of the letter* a_d, *then the letters* a_d *can only appear as blocks on the left and right end of* σ. *If* σ *contains no* a_d, *then* $\sigma \in \mathcal{AC}_{n,m}^{A_{d-1}}(\tau)$ *where* $A_{d-1} = A - \{a_d\}$. *Therefore,*

$$AC_A^\tau(m; x) = \sum_{i=0}^{m-1} (i+1)x^{i\,a_d} AC_{A_{d-1}}^\tau(m - i; x) + x^{m\,a_d}, \quad (5.9)$$

for $n \geq 0$, *since there are* $(i + 1)$ *possibilities to separate* i *copies of* a_d *into two blocks when* $i < m$, *and there is exactly one composition when* $i = m$. *To obtain a recurrence in the sets* A_d, *we take second differences and substitute the expression given in* (5.9) *for each term. Simplifying the result gives that*

$$AC_A^\tau(m; x) - 2x^{a_d} AC_A^\tau(m - 1; x) + x^{2a_d} AC_A^\tau(m - 2; x) = AC_{A_{d-1}}^\tau(m; x).$$

Multiplying both sides of this recurrence by y^m *and summing over all* $m \geq 2$, *together with* $AC_A^\tau(0; x) = 1$ *and* $AC_A^\tau(1; x) = \sum_{a \in A} x^a y$, *we obtain a recurrence in the generating functions. The stated result then follows by induction on the number of elements of* A, *together with the fact that*

$$AC_{\{a_1\}}^\tau(x, y) = \frac{1}{1 - x^{a_1} y} = \frac{1}{(1 - x^{a_1} y)^2} - \frac{x^{a_1} y}{(1 - x^{a_1} y)^2}.$$

Now we look at the shuffle patterns $\tau-\ell-\nu$ and $\tau-1-\nu$, where ℓ and 1 are the greatest and smallest elements of the pattern, respectively.

Theorem 5.63 *Let $A = \{a_1, a_2, \ldots, a_d\}$ be any ordered set of positive integers.*

1. *Let ϕ be the shuffle pattern τ–ℓ–ν. Then for all $d \geq \ell$,*

$$AC_A^\phi(x, y) = \frac{AC_{A_{d-1}}^\phi(x, y) - x^{a_d} y AC_{A_{d-1}}^\tau(x, y) AC_{A_{d-1}}^\nu(x, y)}{(1 - x^{a_d} y AC_{A_{d-1}}^\tau(x, y))(1 - x^{a_d} y AC_{A_{d-1}}^\nu(x, y))}.$$

2. *Let ψ be the shuffle pattern τ–1–ν. Then for all $d \geq \ell$,*

$$AC_A^\psi(x, y) = \frac{AC_{\bar{A}_1}^\psi(x, y) - x^{a_1} y AC_{\bar{A}_1}^\tau(x, y) AC_{\bar{A}_1}^\nu(x, y)}{(1 - x^{a_1} y AC_{\bar{A}_1}^\tau(x, y))(1 - x^{a_1} y AC_{\bar{A}_1}^\nu(x, y))}.$$

Proof We derive a recurrence relation for $AC_A^\phi(x, y)$ where $\phi = \tau$–ℓ–ν. Let $\sigma \in \mathcal{AC}_{n,m}^A(\phi)$ be such that it contains exactly s copies of the letter a_d. If $s = 0$, then the generating function for the number of such compositions is $AC_{A_{d-1}}^\phi(x, y)$. If $s \geq 1$, then $\sigma = \sigma^{(0)} a_d \sigma^{(1)} a_d \cdots a_d \sigma^{(s)}$, where $\sigma^{(j)}$ is a ϕ-avoiding composition with parts in A_{d-1} for $j = 0, 1, \ldots, s$. Furthermore, either $\sigma^{(j)}$ avoids τ for all j, or there exists a j_0 such that $\sigma^{(j_0)}$ contains τ, $\sigma^{(j)}$ avoids τ for all $j = 0, 1, \ldots, j_0 - 1$, and $\sigma^{(j)}$ avoids ν for any $j = j_0 + 1, j_0 + 2, \ldots, s$. In the first case, the generating function for the number of such compositions is $x^{s a_d} y^s \left(AC_{A_{d-1}}^\tau(x, y)\right)^{s+1}$. In the second case, the generating function is given by

$$x^{s a_d} y^s \sum_{j=0}^{s} \left(AC_{A_{d-1}}^\tau(x, y)\right)^j \left(AC_{A_{d-1}}^\nu(x, y)\right)^{s-j} (AC_{A_{d-1}}^\phi(x, y) - AC_{A_{d-1}}^\tau(x, y)).$$

Defining

$$H(x, y) = \sum_{s \geq 0} x^{s a_d} y^s \sum_{j=0}^{s} \left(AC_{A_{d-1}}^\tau(x, y)\right)^j \left(AC_{A_{d-1}}^\nu(x, y)\right)^{s-j},$$

we obtain that

$$\begin{aligned}
AC_A^\phi(x, y) &= AC_{A_{d-1}}^\phi(x, y) + \sum_{s \geq 1} x^{s a_d} y^s \left(AC_{A_{d-1}}^\tau(x, y)\right)^{s+1} \\
&\quad + (H(x, y) - 1)(AC_{A_{d-1}}^\phi(x, y) - AC_{A_{d-1}}^\tau(x, y)) \\
&= \frac{AC_{A_{d-1}}^\tau(x, y)}{1 - x^{a_d} y AC_{A_{d-1}}^\tau(x, y)} \\
&\quad + H(x, y)(AC_{A_{d-1}}^\phi(x, y) - AC_{A_{d-1}}^\tau(x, y)).
\end{aligned}$$

Using the identity $\sum_{n\geq 0} x^n \sum_{j=0}^{n} p^j q^{n-j} = \left((1-xp)(1-xq)\right)^{-1}$ gives that

$$H(x,y) = \frac{1}{(1 - x^{a_d} y AC^\tau_{A_{d-1}}(x,y))(1 - x^{a_d} y AC^\nu_{A_{d-1}}(x,y))},$$

and therefore,

$$AC^\phi_A(x,y) = \frac{AC^\phi_{A_{d-1}}(x,y) - x^{a_d} y AC^\tau_{A_{d-1}}(x,y) AC^\nu_{A_{d-1}}(x,y)}{(1 - x^{a_d} y AC^\tau_{A_{d-1}}(x,y))(1 - x^{a_d} y AC^\nu_{A_{d-1}}(x,y))},$$

which gives the first result. Replacing a_d by a_1 and using similar arguments, we obtain the second result. □

Example 5.64 *We can use Theorem 5.63 to recover the result of Example 5.62 for $\phi = 1'-2-1''$. Since $AC^1_A(x,y) = 1$, the recurrence reduces to*

$$AC^\phi_A(x,y) = \frac{AC^\phi_{A_{d-1}}(x,y) - x^{a_d} y}{(1 - x^{a_d} y)^2},$$

which can be easily iterated and leads to the explicit formula given in (5.8).

We now give two corollaries to Theorem 5.63. The first one tells us that we can reverse the letters within the two outside patterns, whereas the second one allows for a swap of patterns. We therefore obtain Wilf-equivalence for a whole family of patterns. (Note that the definition of Wilf-equivalence given in Definition 5.2 also includes POPs.) The proofs are left as Exercise 5.8.

Corollary 5.65 *Let $\phi = \tau-\ell-\nu$ (respectively $\phi = \tau-1-\nu$) be a shuffle pattern, and let $f(\phi) = f_1(\tau)-\ell-f_2(\nu)$ (respectively $f(\phi) = f_1(\tau)-1-f_2(\nu)$), where $f_1, f_2 \in \{R, I\}$ are any trivial bijections (the reversal map or the identity). Then $\phi \sim f(\phi)$.*

Corollary 5.66 *For any shuffle pattern $\tau-\ell-\nu$ (respectively $\tau-1-\nu$) we have that*

$$\tau-\ell-\nu \sim \nu-\ell-\tau \quad (\text{respectively } \tau-1-\nu \sim \nu-1-\tau).$$

Example 5.67 *Using these two corollaries, we obtain for example that*

$$1'2'3'-4-2''1''1'' \sim 1'2'3'-4-1''1''2'' \sim 1''1''2''-4-1'2'3'.$$

We now look at the second class of patterns, multi-patterns. As before, the generating function can be easily computed for simple multi-patterns, and we give an example for the simplest such pattern.

Example 5.68 *The simplest nontrivial example of a multi-pattern is the pattern $\phi = 1'\text{-}1''2''$. Avoiding ϕ is the same as avoiding the patterns $1\text{-}12$, $1\text{-}23$, $2\text{-}12$, $2\text{-}13$, and $3\text{-}12$ simultaneously. To count the number of compositions in $\mathcal{AC}_{n,m}^A(1'\text{-}1''2'')$, we split up the composition: the first letter can be anything, and the rest of the composition has to avoid 12. Using Example 2.65, we obtain that*

$$AC_A^{1'\text{-}1''2''}(x,y) = 1 + \frac{y\sum_{a\in A} x^a}{\prod_{a\in A}(1-x^a y)}.$$

In order to prove general results for multi-patterns (and other POPs), it is convenient to introduce the notion of quasi-avoidance, which embodies what happens when creating compositions in a recursive manner. Specifically, adding a part at the end of a composition that avoids the pattern τ can create an single occurrence of τ in the rightmost parts of the composition.

Definition 5.69 *Let τ be a subword pattern. A sequence (composition, word, or partition) σ quasi-avoids τ if σ has exactly one occurrence of τ and this occurrence consists of the $|\tau|$ rightmost parts of σ, where $|\tau|$ denotes the number of letters in τ.*

Example 5.70 *The composition 4112234 quasi-avoids the pattern 1123, while the compositions 5223411 and 1123346 do not.*

Lemma 5.71 *Let τ be a nonempty subword pattern. Let $QC_A^\tau(x,y)$ denote the generating function for the number of compositions in $\mathcal{C}_{n,m}^A$ that quasi-avoid τ. Then*

$$QC_A^\tau(x,y) = 1 + AC_A^\tau(x,y)\left(y\sum_{a\in A} x^a - 1\right). \qquad (5.10)$$

Proof Adding the part a to the right end of a composition with $m-1$ parts that avoids τ creates either a composition with m parts that still avoids τ or one that quasi-avoids τ. Thus, for $m \geq 1$,

$$QC_A^\tau(m;x) = \left(\sum_{a\in A} x^a\right) AC_A^\tau(m-1;x) - AC_A^\tau(m;x).$$

Multiplying both sides of this equality by y^m and summing over all natural numbers m we get the desired result. $\qquad \square$

We now obtain a general theorem that is a good auxiliary tool for calculating the generating function for the number of compositions that avoid a given POP, which we will apply specifically to shuffle patterns.

Theorem 5.72 *Suppose $\tau = \tau'\text{-}\phi$, where ϕ is an arbitrary POP, and the letters of τ' are incomparable with the letters of ϕ. Then*

$$AC_A^\tau(x,y) = AC_A^{\tau'}(x,y) + QC_A^{\tau'}(x,y)AC_A^\phi(x,y).$$

Proof In order to find $AC_A^\tau(x, y)$, we observe that there are two possibilities: either σ avoids τ' or σ does not avoid τ'. In the first case the generating function is given by $AC_A^{\tau'}(x, y)$. If σ does not avoid τ' then we can write σ in the form $\sigma = \sigma^{(1)}\sigma^{(2)}\sigma^{(3)}$, where $\sigma^{(1)}\sigma^{(2)}$ quasi-avoids the pattern τ' and $\sigma^{(2)}$ is order-isomorphic to τ'. Clearly, $\sigma^{(3)}$ must avoid ϕ. So the generating function is equal to $QC_A^{\tau'}(x, y)AC_A^\phi(x, y)$ and we obtain the stated result. $\square$

Applying Lemma 5.71 and Theorem 5.72 to multi-patterns, we obtain the following theorem. The proof is left as Exercise 5.9.

Theorem 5.73 *Let* $\tau = \tau_1\text{-}\tau_2\text{-}\cdots\text{-}\tau_s$ *be a multi-pattern. Then*

$$AC_A^\tau(x, y) = \sum_{j=1}^{s} AC_A^{\tau_j}(x, y) \prod_{i=1}^{j-1}\left[\left(y\sum_{a \in A} x^a - 1\right) AC_A^{\tau_i}(x, y) + 1\right]. \quad (5.11)$$

Example 5.74 *We can easily obtain the result of Example 5.68 from Theorem 5.73. Since $AC_A^{12}(x, y) = \prod_{a \in A}(1 - x^a y)^{-1}$ and $AC_A^1(x, y) = 1$, the result follows. For an example that is more difficult to obtain directly, let $\tau = \tau_1\text{-}\tau_2\text{-}\cdots\text{-}\tau_s$ be a multi-pattern such that τ_j is equal to either 12 or 21, for $j = 1, 2, \ldots, s$. Then Theorem 5.73 gives*

$$AC_A^\tau(x, y) = \frac{1 - \left(1 + \dfrac{y\sum_{a \in A} x^a - 1}{\prod_{a \in A}(1 - x^a y)}\right)^s}{1 - y\sum_{a \in A} x^a}.$$

We now give results for a whole family of patterns. The first result concerns reversal of the individual patterns, while the second one gives Wilf-equivalence for the permutations of the subword patterns that make up the multi-pattern.

Theorem 5.75 *Let τ_0 and τ_1 be multi-patterns, $\tau = \tau_0\text{-}\tau_1$, and f_1 and f_2 be any trivial bijections. Then $\tau \sim f_1(\tau_0)\text{-}f_2(\tau_1)$.*

Proof We first show that $\tau_0\text{-}\tau_1 \sim \tau_0\text{-}f(\tau_1)$, where f is one of the trivial bijections. If σ avoids $\tau_0\text{-}\tau_1$ then either σ has no occurrence of τ_0 and therefore σ also avoids $\tau_0\text{-}f(\tau_1)$, or σ can be written as $\sigma = \sigma^{(1)}\sigma^{(2)}\sigma^{(3)}$, where $\sigma^{(1)}\sigma^{(2)}$ has exactly one occurrence of τ_0, namely $\sigma^{(2)}$. Then $\sigma^{(3)}$ must avoid τ_1, so $f(\sigma^{(3)})$ avoids $f(\tau_1)$, which implies that $\sigma_f = \sigma^{(1)}\sigma^{(2)}f(\sigma^{(3)})$ avoids $\tau_0\text{-}f(\tau_1)$. The converse is also true, giving a bijection between the set of compositions avoiding τ and the set of those avoiding $\tau_0\text{-}f(\tau_1)$. This result and properties of trivial bijections give the desired equivalence. $\square$

Repeated use of Theorem 5.75 allows us to show equivalence of multi-patterns that are permutations of each other. Before stating the general result, we will give an example. To avoid drowning in a sea of primes we use $\hat{}$ and $\tilde{}$ to mark letters that can be compared instead of using $''$ and $'''$.

Example 5.76 *Consider the multi-pattern* $\tau = 1'2'\text{--}\tilde{1}\tilde{2}\tilde{2}\text{--}\hat{1}\hat{2}$. *Suppose we want to switch the first and last parts of the multi-pattern. This can be achieved by using the reversal operation on appropriate parts (shown in bold font) of the multi-pattern:*

$$\tau = 1'2'\text{--}\tilde{1}\tilde{2}\tilde{2}\text{--}\hat{1}\hat{2} \sim 1'2'\text{--}\hat{2}\hat{1}\text{--}\tilde{2}\tilde{2}\tilde{1} \sim 1'2'\text{--}\hat{2}\hat{1}\text{--}\tilde{1}\tilde{2}\tilde{2}$$
$$\sim \tilde{2}\tilde{2}\tilde{1}\text{--}\hat{1}\hat{2}\text{--}2'1' \sim \hat{2}\hat{1}\text{--}\tilde{1}\tilde{2}\tilde{2}\text{--}2'1' \sim \hat{2}\hat{1}\text{--}\tilde{1}\tilde{2}\tilde{2}\text{--}1'2' \sim \hat{1}\hat{2}\text{--}\tilde{1}\tilde{2}\tilde{2}\text{--}1'2'.$$

Corollary 5.77 *Suppose we have multi-patterns* $\tau = \tau_1\text{--}\tau_2\text{--}\cdots\text{--}\tau_s$ *and* $\phi = \phi_1\text{--}\phi_2\text{--}\cdots\text{--}\phi_s$, *where* $\tau_1\tau_2\cdots\tau_s$ *is a permutation of* $\phi_1\phi_2\cdots\phi_s$. *Then* $\tau \sim \phi$.

Proof See Exercise 5.10. □

We conclude the section on POPs with an application of Theorem 5.73.

Definition 5.78 *Let* τ *be an arbitrary subword pattern and let* σ *be a composition. We say that two patterns overlap in* σ *if they contain any of the same letters of* σ. *We denote by* $\tau\text{--}nlap(\sigma)$ *the maximum number of nonoverlapping occurrences of* τ *in* σ.

Example 5.79 *The simplest subword pattern is a drop or descent in a composition, which occurs at position* i *if* $\sigma_i > \sigma_{i+1}$. *Clearly, two descents at positions* i *and* j *overlap if* $j = i+1$. *In particular, we can define the statistic maximum number of nonoverlapping descents, or MND, in a composition. For example,* $MND(333211) = 1$ *whereas* $MND(13321111432111) = 3$ *(namely either 32 or 21, together with 43 and 21).*

Obviously, this statistic, maximum number of nonoverlapping patterns, can be defined for any subword pattern τ. Using Theorem 5.73 for the multi-pattern $\tau\text{--}\tau\text{--}\cdots\text{--}\tau$ allows us to obtain the generating function for the entire distribution of the maximum number of nonoverlapping occurrences of a pattern τ in compositions.

Theorem 5.80 *For* $A = \{a_1, \ldots, a_d\}$ *and a subword pattern* τ, *we have*

$$\sum_{n,m\geq 0} \sum_{\sigma\in\mathcal{C}_{n,m}^A} q^{\tau\text{--}nlap(\sigma)} x^n y^m = \frac{AC_A^\tau(x,y)}{1 - q\left[\left(y\sum_{a\in A} x^a - 1\right) AC_A^\tau(x,y) + 1\right]}.$$

Proof Let Φ_s be the multi-pattern $\tau\text{--}\tau\text{--}\cdots\text{--}\tau$ consisting of s copies of τ. If $\sigma \in AC_{n,m}^A(\Phi_s)$, then it has at most $s-1$ nonoverlapping occurrences of τ. Theorem 5.73 yields

$$AC_A^{\Phi_s}(x,y) = \sum_{j=1}^{s} AC_A^\tau(x,y) \prod_{i=1}^{j-1}\left[\left(y\sum_{a\in A} x^a - 1\right) AC_A^\tau(x,y) + 1\right].$$

Therefore, the generating function for the number of compositions that have exactly s nonoverlapping occurrences of the pattern τ is given by

$$AC_A^{\Phi_{s+1}}(x,y) - AC_A^{\Phi_s}(x,y) = AC_A^\tau(x,y)\left[\left(y\sum_{a\in A}x^a - 1\right)AC_A^\tau(x,y) + 1\right]^s.$$

Hence,

$$\sum_{n,m\geq 0}\sum_{\sigma\in\mathcal{C}_{n,m}^A} q^{\tau-\mathrm{nlap}(\sigma)}x^n y^m$$

$$= \sum_{s\geq 0} q^s AC_A^\tau(x,y)\left[\left(y\sum_{a\in A}x^a - 1\right)AC_A^\tau(x,y) + 1\right]^s$$

$$= \frac{AC_A^\tau(x,y)}{1 - q\left[\left(y\sum_{a\in A}x^a - 1\right)AC_A^\tau(x,y) + 1\right]},$$

which completes the proof. □

We use Theorem 5.80 to obtain the distribution for *MND*, the maximum number of nonoverlapping descents.

Example 5.81 *For descents (the pattern 21) we modify the argument given in Example 2.65 for a general set A to obtain that*

$$AC_A^{21}(x,y) = \prod_{a\in A}(1 - x^a y)^{-1}.$$

This implies that the distribution of MND is given by

$$\frac{1}{\displaystyle\prod_{a\in A}(1 - x^a y) + q\left(1 - y\sum_{a\in A}x^a - \prod_{a\in A}(1 - x^a y)\right)}.$$

5.5 Exercises

Exercise 5.1 *Either modify the* Mathematica *code given in Example 5.14 or write your own code in* Maple *or any other programming language to directly enumerate the number of compositions that avoid 13–2. Give the sequence of values for $n = 0, 1, \ldots, 10$.*

Exercise 5.2 *Fill in the details of the proof of Theorem 5.21. That is, show that*

$$1 + \sum_{i=1}^{d-1}\frac{x^i y}{\prod_{j=1}^i(1 - x^j y)} = \frac{1}{\prod_{j=1}^{d-1}(1 - x^j y)} \tag{*}$$

and

$$G_d(i) = \frac{x^{i+d} y^2 AC_{[d]}^{12-3}(x,y)}{\prod_{j=1}^{i}(1 - x^j y)} \qquad (**)$$

for $i = 1, 2, \ldots, d-1$.

Exercise 5.3 *Modify the proof of Theorem 5.23 to obtain*

$$AC_{[d]}^{21-3}(x,y) = \prod_{i=1}^{d}\left(1 - \frac{x^i y}{\prod_{j=1}^{i-1}(1 - x^j y)}\right)^{-1}.$$

Exercise 5.4 *Show that*

$$\det(\mathbf{U}_i^{(i)}) = \beta_i \prod_{j=1}^{i-1}(1 - \alpha_j) + \alpha_i \sum_{j=1}^{i-1}\beta_j \prod_{k=1}^{j-1}(1 - \alpha_k)$$

for the matrix $\mathbf{U}_i^{(i)}$ *defined in the proof of Lemma 5.45.*

Exercise 5.5 *Find the generating function for the number of compositions of n with m parts in A that simultaneously avoid 12–3 and 21–3. Then use the generating function to compute the number of compositions of n that simultaneously avoid 12–3 and 21–3 for* $n = 0, \ldots, 20$.

Exercise 5.6 *Prove Lemma 5.54.*

Exercise 5.7 *Use the generating functions for* $\tau = 112$ *and* $\nu = 221$ *given in Theorem 4.35 to derive generating functions for the number of occurrences of* τ *and* ν, *respectively, among all compositions of n, thereby deriving the results of Example 5.58 in a different way.*

Exercise 5.8 *Prove Corollaries 5.65 and 5.66.*

Exercise 5.9 *Prove Theorem 5.73.*

Exercise 5.10 *Fill in the details of the proof of Corollary 5.77.*

5.6   Research directions and open problems

We now suggest several research directions which are motivated both by the results and exercises of this chapter. In later chapters we will revisit some of these research directions and pose additional questions.

Research Direction 5.1 *Throughout this chapter there were several enumeration problems for which we did not succeed to find an explicit formula for the generating function. These lead us to ask the following questions:*

(1) *Derive an explicit expression for the generating function for the number of compositions avoiding the subsequence pattern 112, for which a recursion was given in Theorem 5.13.*

(2) *Derive an explicit expression for the generating function for the number of compositions avoiding the subsequence pattern 221, for which a recursion was given in Theorem 5.13.*

(3) *Derive an explicit expression for the generating function for the number of compositions avoiding the pattern 13-2, for which a recursion was given in Lemma 5.25.*

(4) *Find an explicit expression for the generating function for the number of compositions avoiding the pattern 11-2, in terms of permutations if possible. (A recursion for this generating function was given in Theorem 5.41.)*

(5) *Derive an explicit expression for the generating function for the number of compositions avoiding the pattern 11-1, for which a recursion was given in Theorem 5.39.*

(6) *Find an explicit expression for the generating function for the number of compositions avoiding the pattern 12-1, in terms of permutations if possible. (A recursion for this generating function was given in Theorem 5.46.)*

Research Direction 5.2 *In this chapter we completely classified the subsequence patterns of length three according to Wilf-equivalence, and gave explicit expressions for their generating functions. Tables 5.2 and 5.3 show the Wilf-equivalence classes for patterns of length four and five, respectively, but finding expressions for the respective generating functions remains an open question.*

Research Direction 5.3 *In this chapter, we completely classified the 3-letter patterns of type* $(1, 2)$, *and gave explicit or recursive expressions for their generating functions. Extending to 4-letter generalized patterns, we need to consider four different types, namely*

$$(3, 1), (2, 2), (1, 2, 1) \text{ and } (1, 1, 2).$$

This leads to the following questions:

(1-a) *Classify the 4-letter patterns of type* $(3, 1)$ *according to Wilf-equivalence by using the values of the sequence* $\{AC_n(\tau)\}_{n=0}^{23}$ *given in Table 5.4. Either give a bijection for the patterns that have the same sequence of values or show that they are not Wilf-equivalent by computing additional sequence values and finding a value of n for which the sequences are different.*

(1-b) *Find expressions for the generating functions for each Wilf-equivalence class. As an example, we give the generating function $AC^\tau_{[d]}(x, y)$ for any type $(3, 1)$ pattern in which the single letter is the largest one. That is, $\tau = \tau' \text{-} k$ is a pattern such that τ' is a 3-letter subword pattern on $[k-1]$. Then each composition σ of n with parts in $[d]$ is either a composition with parts in $[d-1]$ or σ can be written as $\sigma^{(1)} d \sigma^{(2)} d \cdots \sigma^{(s-1)} d \sigma^{(s)}$ for $s \geq 2$, where $\sigma^{(i)}$ is a (potentially empty) composition with parts in $[d-1]$ that avoids τ'. Therefore,*

$$AC^\tau_{[d]}(x, y) = \frac{AC^\tau_{[d-1]}(x, y)}{1 - x^d y AC^{\tau'}_{[d-1]}(x, y)},$$

which, when iterated d times yields

$$AC^\tau_{[d]}(x, y) = \frac{AC^\tau_{[k-1]}(x, y)}{\prod_{j=k-1}^{d-1}(1 - x^{j+1} y AC^{\tau'}_{[j]}(x, y))}.$$

Since

$$AC^\tau_{[k-1]}(x, y) = C_{[k-1]}(x, y) = \frac{1}{\left(1 - y \sum_{i=1}^{k-1} x^i\right)}$$

(see Theorem 3.13), we obtain that

$$AC^\tau_{[d]}(x, y) = \frac{1}{\left(1 - y \sum_{i=1}^{k-1} x^i\right) \prod_{j=k-1}^{d-1}(1 - x^{j+1} y AC^{\tau'}_{[j]}(x, y))}.$$

Similar ideas lead to an expression for the generating function for avoidance of $\tau = \tau'\text{-}1$, where τ' is a three letter-subword on the letters $2, 3, \ldots, k$ for $k \leq 4$.

(2) *Classify the 4-letter patterns of type $(2, 2)$ according to Wilf-equivalence by using the values of the sequence $\{AC_n(\tau)\}_{n=0}^{23}$ given in Table 5.5. Give expressions for the respective generating functions.*

(3) *Classify the 4-letter patterns of type $(1, 1, 2)$ according to Wilf-equivalence by using the values of the sequence $\{AC_n(\tau)\}_{n=0}^{23}$ given in Table 5.6. Give expressions for the respective generating functions.*

(4) *Classify the 4-letter patterns of type $(1, 2, 1)$ according to Wilf-equivalence by using the values of the sequence $\{AC_n(\tau)\}_{n=0}^{23}$ given in Table 5.7. Give expressions for the respective generating functions.*

TABLE 5.4: $\{AC_n(\tau)\}_{n=0}^{23}$ for 4-letter patterns of type $(3,1)$

τ	$\{AC_n(\tau)\}_{n=0}^{23}$ for 4-letter patterns of type $(3,1)$					
112–1	1	1	2	4	8	15
211–1	29	55	104	195	364	677
	1257	2323	4282	7876	14459	26490
	48460	88522	161512	294353	535932	974947
121–1	1	1	2	4	8	15
	29	55	105	199	377	713
	1346	2533	4760	8931	16737	31328
	58591	109488	204463	381573	711718	1326846
111–2	1	1	2	4	8	15
	29	56	108	207	397	761
	1459	2794	5349	10237	19590	37477
	71685	137098	262176	501313	958492	1832476
111–1	1	1	2	4	7	15
	28	55	105	203	391	752
	1448	2778	5329	10225	19609	37575
	71987	137849	263962	505241	966973	1850204
211–2	1	1	2	4	8	16
	31	61	119	232	449	869
	1673	3215	6162	11784	22486	42816
	81365	154331	292215	552368	1042501	1964645
112–2	1	1	2	4	8	16
	31	61	119	232	449	869
	1674	3218	6170	11803	22531	42917
	81589	154814	293244	554525	1046964	1973768
112–3	1	1	2	4	8	16
211–3	32	63	124	242	471	912
	1760	3386	6495	12425	23715	45161
	85829	162816	308320	582919	1100431	2074464
121–2	1	1	2	4	8	16
	31	61	119	233	452	877
	1694	3270	6298	12116	23263	44595
	85358	163161	311514	594103	1131873	2154269
121–3	1	1	2	4	8	16
	32	63	124	242	472	916
	1775	3430	6615	12731	24461	46918
	89865	171887	328368	626586	1194381	2274451
113–2	1	1	2	4	8	16
311–2	32	63	124	243	475	925
	1799	3491	6764	13085	25285	48805
	94115	181339	349151	671828	1291996	2483410
131–2	1	1	2	4	8	16
	32	63	124	243	475	926
	1804	3508	6815	13225	25644	49685
	96202	186161	360063	696105	1345251	2598854

τ	$\{AC_n(\tau)\}_{n=0}^{23}$ for 4-letter patterns of type $(3,1)$					
122–1	1	1	2	4	8	16
	31	62	121	239	467	917
	1793	3510	6860	13406	26181	51115
	99761	194650	379711	740561	1444115	2815632
221–1	1	1	2	4	8	16
	31	62	121	239	467	917
	1794	3512	6865	13418	26210	51178
	99902	194950	380351	741894	1446886	2821316
212–1	1	1	2	4	8	16
	31	62	121	240	469	923
	1808	3547	6944	13607	26628	52113
	101952	199408	389942	762420	1490473	2913372
231–1	1	1	2	4	8	16
132–1	32	63	125	246	485	952
	1868	3659	7161	13999	27348	53387
	104163	203128	395962	771583	1503089	2927366
123–1	1	1	2	4	8	16
321–1	32	63	125	246	485	952
	1868	3661	7168	14022	27412	53558
	104593	204181	398460	777388	1516316	2957073
213–1	1	1	2	4	8	16
	32	63	125	246	485	952
	1869	3662	7171	14029	27427	53587
	104652	204288	398649	777691	1516756	2957547
312–1	1	1	2	4	8	16
	32	63	125	246	485	952
	1869	3663	7174	14037	27447	53636
	104765	204542	399204	778889	1519292	2962864
122–2	1	1	2	4	8	16
221–2	32	63	125	248	490	967
	1908	3760	7403	14563	28628	56241
	110422	216677	424962	833074	1632425	3197516
122–3	1	1	2	4	8	16
221–3	32	64	127	252	499	986
	1945	3833	7544	14834	29144	57216
	112253	220104	431352	844959	1654471	3238327
212–2	1	1	2	4	8	16
	32	63	125	248	491	970
	1916	3780	7454	14687	28920	56914
	111951	220099	432532	849648	1668392	3274977
212–3	1	1	2	4	8	16
	32	64	127	252	499	987
	1949	3846	7580	14928	29378	57780
	113574	223137	438194	860180	1687935	3311175

τ	$\{AC_n(\tau)\}_{n=0}^{23}$ for 4-letter patterns of type $(3,1)$					
132–2	1	1	2	4	8	16
	32	64	127	252	500	990
	1957	3864	7621	15016	29559	58136
	114250	224365	440320	863615	1692902	3316822
231–2	1	1	2	4	8	16
	32	64	127	252	500	990
	1957	3864	7621	15017	29563	58148
	114282	224445	440513	864069	1693941	3319151
312–2	1	1	2	4	8	16
	32	64	127	252	500	990
	1957	3864	7621	15016	29561	58148
	114293	224494	440676	864552	1695281	3322684
213–2	1	1	2	4	8	16
	32	64	127	252	500	990
	1957	3864	7621	15017	29565	58159
	114321	224562	440836	864919	1696102	3324483
123–2	1	1	2	4	8	16
321–2	32	64	127	252	500	990
	1957	3864	7621	15018	29569	58171
	114354	224650	441064	865492	1697498	3327805
223–1	1	1	2	4	8	16
322–1	32	64	127	253	503	999
	1983	3936	7808	15486	30707	60879
	120681	239199	474070	939499	1861764	3689208
232–1	1	1	2	4	8	16
	32	64	127	253	503	999
	1983	3937	7812	15498	30741	60968
	120903	239736	475331	942399	1868328	3703863
222–1	1	1	2	4	8	16
	32	63	126	251	498	990
	1968	3910	7768	15430	30651	60880
	120915	240142	476919	947132	1880906	3735216
231–3	1	1	2	4	8	16
	32	64	128	255	509	1014
	2020	4020	7997	15898	31589	62731
	124514	247020	489832	970872	1923487	3809194
132–3	1	1	2	4	8	16
	32	64	128	255	509	1014
	2020	4020	7997	15898	31589	62732
	124517	247030	489858	970940	1923653	3809592
312–3	1	1	2	4	8	16
213–3	32	64	128	255	509	1014
	2020	4020	7997	15898	31589	62731
	124515	247027	489859	970961	1923753	3809941

τ	$\{AC_n(\tau)\}_{n=0}^{23}$ for 4-letter patterns of type $(3,1)$					
321–3	1	1	2	4	8	16
123–3	32	64	128	255	509	1014
	2020	4020	7997	15898	31589	62733
	124521	247047	489914	971110	1924135	3810897
231–4	1	1	2	4	8	16
132–4	32	64	128	256	511	1020
	2034	4053	8070	16058	31931	63455
	126027	250162	496306	984150	1950595	3864361
213–4	1	1	2	4	8	16
312–4	32	64	128	256	511	1020
	2034	4053	8070	16058	31932	63460
	126044	250214	496454	984548	1951621	3866921
321–4	1	1	2	4	8	16
123–4	32	64	128	256	511	1020
	2034	4053	8070	16059	31935	63469
	126069	250280	496620	984951	1952577	3869147
412–3	1	1	2	4	8	16
	32	64	128	256	511	1020
	2034	4054	8074	16073	31978	63594
	126411	251179	498908	990629	1966371	3902068
214–3	1	1	2	4	8	16
	32	64	128	256	511	1020
	2034	4054	8074	16073	31978	63594
	126411	251179	498908	990630	1966375	3902082
124–3	1	1	2	4	8	16
421–3	32	64	128	256	511	1020
	2034	4054	8074	16073	31978	63594
	126411	251179	498908	990631	1966379	3902096
142–3	1	1	2	4	8	16
	32	64	128	256	511	1020
	2034	4054	8074	16073	31978	63594
	126411	251180	498913	990648	1966430	3902235
241–3	1	1	2	4	8	16
	32	64	128	256	511	1020
	2034	4054	8074	16073	31978	63594
	126412	251183	498923	990676	1966505	3902426
133–2	1	1	2	4	8	16
331–2	32	64	128	255	509	1015
	2023	4030	8028	15986	31828	63354
	126090	250912	499242	993233	1975835	3930185
313–2	1	1	2	4	8	16
	32	64	128	255	509	1015
	2023	4031	8031	15994	31850	63410
	126229	251246	500025	995040	1979944	3939409

τ	$\{AC_n(\tau)\}_{n=0}^{23}$ for 4-letter patterns of type $(3,1)$				
143–2	1	1	2	4	8	16
	32	64	128	256	511	1020
	2035	4058	8088	16116	32102	63929
	127280	253360	504244	1003410	1996449	3971801
341–2	1	1	2	4	8	16
	32	64	128	256	511	1020
	2035	4058	8088	16116	32102	63929
	127280	253360	504244	1003411	1996453	3971814
134–2	1	1	2	4	8	16
431–2	32	64	128	256	511	1020
	2035	4058	8088	16116	32102	63929
	127280	253360	504244	1003412	1996459	3971838
314–2	1	1	2	4	8	16
	32	64	128	256	511	1020
	2035	4058	8088	16116	32103	63933
	127294	253402	504362	1003723	1997246	3973765
413–2	1	1	2	4	8	16
	32	64	128	256	511	1020
	2035	4058	8088	16116	32103	63934
	127297	253411	504386	1003785	1997398	3974126
332–1	1	1	2	4	8	16
233–1	32	64	128	255	510	1017
	2030	4049	8079	16113	32139	64098
	127832	254928	508373	1013764	2021546	4031098
323–1	1	1	2	4	8	16
	32	64	128	255	510	1017
	2030	4049	8080	16115	32145	64113
	127871	255019	508590	1014260	2022672	4033616
243–1	1	1	2	4	8	16
342–1	32	64	128	256	511	1021
	2038	4068	8117	16196	32309	64448
	128544	256373	511289	1019634	2033324	4054693
423–1	1	1	2	4	8	16
324–1	32	64	128	256	511	1021
	2038	4068	8117	16196	32309	64449
	128548	256385	511323	1019724	2033551	4055247
234–1	1	1	2	4	8	16
432–1	32	64	128	256	511	1021
	2038	4068	8117	16197	32312	64458
	128572	256447	511475	1020085	2034390	4057162

TABLE 5.5: $\{AC_n(\tau)\}_{n=0}^{23}$ for 4-letter patterns of type $(2,2)$

τ	$\{AC_n(\tau)\}_{n=0}^{23}$ for 4-letter patterns of type $(2,2)$					
11–12	1	1	2	4	8	15
11–21	29	55	104	195	364	677
	1256	2318	4268	7839	14361	26248
	47889	87205	158545	287815	521754	944655
11–11	1	1	2	4	7	15
	28	54	104	198	379	721
	1371	2599	4908	9278	17489	32877
	61788	115892	216978	405878	758128	1414100
12–12	1	1	2	4	8	16
	31	61	118	230	443	853
	1634	3118	5933	11247	21269	40065
	75303	141041	263537	490969	912455	1691299
12–21	1	1	2	4	8	16
21–12	31	61	118	230	443	853
	1634	3120	5939	11273	21343	40277
	75845	142395	266759	498485	929551	1729501
13–12	1	1	2	4	8	16
12–13	32	63	124	242	471	911
	1757	3374	6459	12321	23435	44440
	84047	158533	298302	559967	1048824	1960242
21–13	1	1	2	4	8	16
	32	63	124	242	471	911
	1757	3374	6459	12323	23444	44475
	84161	158873	299242	562437	1055042	1975384
12–31	1	1	2	4	8	16
	32	63	124	242	471	911
	1757	3375	6463	12337	23485	44587
	84445	159565	300861	566123	1063231	1993229
11–32	1	1	2	4	8	16
11–32	32	63	125	246	484	949
	1857	3628	7076	13776	26785	52008
	100856	195368	378048	730824	1411522	2723929
11–22	1	1	2	4	8	16
	31	62	122	240	470	923
	1807	3534	6904	13471	26265	51165
	99586	193693	376469	731246	1419538	2754174
12–32	1	1	2	4	8	16
12–23	32	64	127	252	499	986
21–23	1944	3827	7520	14755	28909	56563
21–32	110527	215713	420517	818884	1593004	3095931
12–22	1	1	2	4	8	16
21–22	32	63	125	248	490	967
	1907	3755	7386	14512	28485	55860
	109448	214261	419112	819191	1600033	3123029

τ	$\{AC_n(\tau)\}_{n=0}^{23}$ for 4-letter patterns of type $(2,2)$					
13–22 31–22	1 32 1953 112739	1 64 3852 220762	2 127 7586 431884	4 252 14921 844170	8 500 29312 1648669	16 989 57516 3217364
13–32	1 32 2019 124017	1 64 4017 245785	2 128 7988 486825	4 255 15871 963681	8 509 31515 1906545	16 1014 62536 3769795
13–23 23–13	1 32 2019 124018	1 64 4017 245790	2 128 7988 486842	4 255 15871 963736	8 509 31515 1906706	16 1014 62536 3770239
31–23	1 32 2019 124020	1 64 4017 245795	2 128 7988 486853	4 255 15871 963757	8 509 31515 1906737	16 1014 62537 3770268
23–14 14–23	1 32 2034 125927	1 64 4053 249884	2 128 8069 495560	4 256 16055 982214	8 511 31920 1945695	16 1020 63421 3852230
32–14	1 32 2034 125927	1 64 4053 249885	2 128 8069 495565	4 256 16055 982233	8 511 31920 1945759	16 1020 63421 3852427
14–32	1 32 2034 125931	1 64 4053 249898	2 128 8069 495604	4 256 16055 982342	8 511 31920 1946046	16 1020 63422 3853154
31–24 13–42	1 32 2034 125981	1 64 4053 250025	2 128 8070 495921	4 256 16057 983116	8 511 31927 1947915	16 1020 63441 3857610
24–13 13–24	1 32 2034 125982	1 64 4053 250029	2 128 8070 495936	4 256 16057 983165	8 511 31927 1948064	16 1020 63441 3858036
21–33 12–33	1 32 2021 125198	1 64 4023 248771	2 128 8008 494194	4 255 15930 981512	8 509 31683 1948971	16 1014 62990 3869295
21–43 12–34 12–43 21–34	1 32 2034 126458	1 64 4054 251318	2 128 8074 499290	4 256 16074 991636	8 511 31982 1968926	16 1020 63609 3908376

TABLE 5.6: $\{AC_n(\tau)\}_{n=0}^{23}$ for 4-letter patterns of type $(1,1,2)$

τ	\multicolumn{6}{c}{$\{AC_n(\tau)\}_{n=0}^{23}$ for 4-letter patterns of type $(1,1,2)$}
1–1–12	1	1	2	4	8	15
1–1–21	29	54	101	185	336	603
	1080	1906	3340	5813	10050	17220
	29364	49698	83722	140091	233265	386112
1–2–11	1	1	2	4	8	15
	29	55	104	195	363	672
	1242	2282	4176	7619	13860	25132
	45465	82044	147749	265556	476419	853293
2–1–11	1	1	2	4	8	15
	29	55	104	195	363	672
	1242	2282	4176	7619	13860	25132
	45465	82044	147749	265556	476419	853293
1–1–11	1	1	2	4	7	15
	27	53	99	186	348	647
	1212	2232	4095	7549	13838	25211
	45962	83449	151345	273537	494147	890332
1–2–21	1	1	2	4	8	16
1–2–12	31	61	118	229	438	836
	1582	2981	5585	10414	19319	35654
	65497	119732	217951	395021	713137	1282315
2–1–12	1	1	2	4	8	16
2–1–21	31	61	118	229	438	836
	1582	2981	5585	10414	19319	35654
	65497	119733	217957	395043	713213	1282516
1–3–12	1	1	2	4	8	16
	32	63	124	241	467	896
	1710	3241	6106	11436	21302	39469
	72769	133534	243939	443746	803949	1450981
1–3–21	1	1	2	4	8	16
	32	63	124	241	467	896
	1710	3242	6109	11447	21332	39550
	72966	134005	245007	446117	809059	1461798
3–1–12	1	1	2	4	8	16
3–1–21	32	63	124	241	467	896
	1710	3242	6110	11450	21346	39592
	73093	134349	245899	448310	814278	1473803
2–2–11	1	1	2	4	8	16
	31	61	119	231	443	848
	1616	3066	5790	10878	20353	37942
	70485	130511	240892	443257	813292	1488266
2–1–31	1	1	2	4	8	16
2–1–13	32	63	124	242	470	907
	1742	3327	6324	11962	22524	42224
	78826	146574	271523	501195	922021	1690784

τ	$\{AC_n(\tau)\}_{n=0}^{23}$ for 4-letter patterns of type $(1,1,2)$					
2–3–11	1	1	2	4	8	16
3–2–11	32	63	124	242	470	907
	1741	3325	6320	11958	22533	42293
	79092	147411	273872	507311	937130	1726623
1–2–13	1	1	2	4	8	16
1–2–31	32	63	124	242	470	907
	1742	3328	6329	11981	22586	42407
	79324	147856	274676	508670	939209	1729308
2–2–12	1	1	2	4	8	16
2–2–21	32	63	124	244	478	932
	1810	3501	6745	12940	24725	47067
	89281	168762	317883	596671	1116127	2080849
2–3–21	1	1	2	4	8	16
2–3–12	32	64	127	251	494	968
3–2–21	1888	3666	7085	13632	26117	49831
3–2–12	94698	179267	338082	635270	1189487	2219620
2–2–13	1	1	2	4	8	16
2–2–31	32	64	127	251	495	972
	1900	3698	7166	13830	26586	50910
	97119	184587	349582	659812	1241290	2327918
1–1–32	1	1	2	4	8	16
1–1–23	32	63	125	245	481	938
	1825	3543	6859	13247	25539	49154
	94450	181276	347510	665526	1273449	2434942
1–1–22	1	1	2	4	8	16
	31	62	121	238	463	904
	1760	3419	6634	12852	24889	48130
	93039	179702	346925	669383	1291167	2489581
2–1–23	1	1	2	4	8	16
1–2–23	32	64	127	252	499	985
2–1–32	1939	3809	7465	14601	28502	55535
1–2–32	108023	209788	406825	787862	1523868	2944030
1–2–22	1	1	2	4	8	16
2–1–22	32	63	125	248	489	963
	1895	3722	7299	14293	27956	54615
	106581	207777	404680	787499	1531260	2975276
3–1–22	1	1	2	4	8	16
1–3–22	32	64	127	252	499	985
	1940	3814	7483	14657	28663	55973
	109163	212648	413798	804455	1562573	3032783
3–3–12	1	1	2	4	8	16
3–3–21	32	64	128	255	508	1009
	2002	3962	7827	15425	30335	59523
	116540	227673	443814	863278	1675600	3245434

τ	$\{AC_n(\tau)\}_{n=0}^{23}$ for 4-letter patterns of type $(1,1,2)$					
2–3–13	1	1	2	4	8	16
2–3–31	32	64	128	255	508	1010
	2005	3973	7860	15521	30596	60202
	118248	231847	453783	886622	1729357	3367413
3–2–13	1	1	2	4	8	16
3–2–31	32	64	128	255	508	1010
	2005	3973	7860	15521	30596	60203
	118253	231868	453856	886854	1730038	3369304
3–4–12	1	1	2	4	8	16
3–4–21	32	64	128	256	511	1019
4–3–12	2028	4029	7987	15802	31195	61454
4–3–21	120805	236983	463932	906402	1767403	3439723
2–4–31	1	1	2	4	8	16
	32	64	128	256	511	1019
	2029	4033	8002	15849	31333	61831
	121789	239454	469960	920748	1800855	3516374
2–4–13	1	1	2	4	8	16
	32	64	128	256	511	1019
	2029	4033	8002	15849	31333	61831
	121790	239460	469984	920828	1801096	3517049
4–2–13	1	1	2	4	8	16
4–2–31	32	64	128	256	511	1019
	2029	4033	8002	15849	31333	61832
	121796	239486	470078	921134	1802016	3519658
3–2–14	1	1	2	4	8	16
3–2–41	32	64	128	256	511	1019
	2029	4034	8006	15864	31382	61979
	122210	240598	472955	928353	1819675	3561937
2–3–14	1	1	2	4	8	16
2–3–41	32	64	128	256	511	1019
	2029	4034	8006	15864	31382	61980
	122216	240623	473043	928633	1820504	3564257
3–1–32	1	1	2	4	8	16
	32	64	128	255	509	1013
	2016	4005	7952	15767	31234	61808
	122191	241330	476193	938771	1849100	3639108
3–1–23	1	1	2	4	8	16
	32	64	128	255	509	1013
	2016	4005	7952	15767	31234	61809
	122194	241341	476225	938860	1849333	3639699
1–3–23	1	1	2	4	8	16
1–3–32	32	64	128	255	509	1013
	2016	4005	7952	15767	31234	61809
	122194	241344	476233	938895	1849433	3639995

τ	$\{AC_n(\tau)\}_{n=0}^{23}$ for 4-letter patterns of type $(1,1,2)$					
4–1–32	1	1	2	4	8	16
	32	64	128	256	511	1020
	2033	4048	8050	15993	31734	62902
	124547	246355	486806	961036	1895514	3735398
1–4–23	1	1	2	4	8	16
	32	64	128	256	511	1020
	2033	4048	8050	15993	31734	62902
	124547	246355	486806	961036	1895514	3735399
4–1–23	1	1	2	4	8	16
	32	64	128	256	511	1020
	2033	4048	8050	15993	31734	62902
	124547	246356	486810	961051	1895563	3735545
1–4–32	1	1	2	4	8	16
	32	64	128	256	511	1020
	2033	4048	8050	15993	31734	62903
	124550	246366	486839	961129	1895759	3736020
3–1–24	1	1	2	4	8	16
	32	64	128	256	511	1020
	2033	4049	8054	16007	31779	63035
	124917	247341	489348	967413	1911157	3773045
3–1–42	1	1	2	4	8	16
	32	64	128	256	511	1020
	2033	4049	8054	16007	31779	63035
	124918	247345	489363	967460	1911293	3773416
1–3–24	1	1	2	4	8	16
1–3–42	32	64	128	256	511	1020
	2033	4049	8054	16007	31779	63035
	124918	247346	489367	967477	1911348	3773586
1–2–33	1	1	2	4	8	16
2–1–33	32	64	128	255	509	1014
	2020	4019	7995	15893	31583	62733
	124564	247255	490660	973434	1930810	3829021
2–1–34	1	1	2	4	8	16
2–1–43	32	64	128	256	511	1020
1–2–34	2034	4053	8069	16056	31926	63449
1–2–43	126029	250222	496592	985185	1953864	3873890

TABLE 5.7: $\{AC_n(\tau)\}_{n=0}^{23}$ for 4-letter patterns of type $(1, 2, 1)$

τ	$\{AC_n(\tau)\}_{n=0}^{23}$ for 4-letter patterns of type $(1, 2, 1)$					
1–12–1	1	1	2	4	8	15
	29	54	101	185	336	603
	1080	1906	3340	5813	10050	17220
	29364	49698	83722	140091	233265	386112
1–11–2	1	1	2	4	8	15
	29	55	104	195	363	672
	1242	2282	4176	7619	13860	25132
	45465	82044	147749	265556	476419	853293
1–11–1	1	1	2	4	7	15
	27	53	99	186	348	647
	1212	2232	4095	7549	13838	25211
	45962	83449	151345	273537	494147	890332
1–12–2	1	1	2	4	8	16
1–21–2	31	61	118	229	438	836
	1582	2981	5583	10408	19298	35601
	65358	119400	217167	393237	709158	1273657
2–11–2	1	1	2	4	8	16
	31	61	119	231	443	848
	1614	3056	5760	10800	20155	37444
	69265	127631	234310	428609	781378	1419850
1–21–3	1	1	2	4	8	16
1–12–3	32	63	124	241	467	896
	1710	3242	6110	11450	21344	39584
	73062	134251	245617	447556	812367	1469172
1–31–2	1	1	2	4	8	16
1–13–2	32	63	124	242	470	907
	1742	3327	6324	11964	22533	42258
	78931	146875	272318	503192	926820	1701939
2–11–3	1	1	2	4	8	16
	32	63	124	242	470	907
	1741	3325	6319	11952	22511	42223
	78887	146856	272451	503819	928843	1707517
2–12–2	1	1	2	4	8	16
	32	63	124	244	478	932
	1810	3501	6745	12940	24725	47067
	89281	168762	317883	596671	1116127	2080849
2–12–3	1	1	2	4	8	16
2–21–3	32	64	127	251	494	968
	1888	3666	7085	13632	26117	49831
	94698	179267	338082	635270	1189487	2219620
2–13–2	1	1	2	4	8	16
	32	64	127	251	495	972
	1900	3698	7166	13830	26586	50910
	97119	184587	349582	659812	1241290	2327918

τ	$\{AC_n(\tau)\}_{n=0}^{23}$ for 4-letter patterns of type $(1,2,1)$					
1–23–1	1	1	2	4	8	16
	32	63	125	245	481	938
	1825	3541	6853	13229	25489	49022
	94128	180502	345718	661496	1264569	2415704
1–22–1	1	1	2	4	8	16
	31	62	121	238	463	904
	1758	3415	6624	12824	24817	47962
	92645	178808	344949	665095	1281951	2470037
1–23–2	1	1	2	4	8	16
1–32–2	32	64	127	252	499	985
	1939	3809	7465	14601	28502	55535
	108023	209788	406825	787862	1523868	2944030
1–22–2	1	1	2	4	8	16
	32	63	125	248	489	963
	1895	3722	7299	14293	27956	54615
	106581	207777	404680	787499	1531260	2975276
1–22–3	1	1	2	4	8	16
	32	64	127	252	499	985
	1940	3814	7483	14657	28663	55973
	109163	212648	413798	804455	1562573	3032783
3–12–3	1	1	2	4	8	16
	32	64	128	255	508	1009
	2002	3962	7827	15425	30335	59521
	116530	227631	443674	862852	1674404	3242232
2–31–3	1	1	2	4	8	16
2–13–3	32	64	128	255	508	1010
	2005	3973	7860	15521	30596	60202
	118248	231847	453785	886634	1729407	3367586
3–12–4	1	1	2	4	8	16
3–21–4	32	64	128	256	511	1019
	2028	4029	7987	15802	31195	61454
	120805	236982	463926	906375	1767304	3439398
2–31–4	1	1	2	4	8	16
2–13–4	32	64	128	256	511	1019
	2029	4033	8002	15849	31333	61832
	121796	239486	470078	921133	1802010	3519632
2–41–3	1	1	2	4	8	16
2–14–3	32	64	128	256	511	1019
	2029	4034	8006	15864	31382	61979
	122210	240597	472950	928333	1819607	3561725
1–32–3	1	1	2	4	8	16
1–23–3	32	64	128	255	509	1013
	2016	4005	7952	15767	31234	61809
	122194	241342	476227	938871	1849362	3639789

τ	$\{AC_n(\tau)\}_{n=0}^{23}$ for 4-letter patterns of type $(1,2,1)$					
1–32–4	1	1	2	4	8	16
1–23–4	32	64	128	256	511	1020
	2033	4048	8050	15993	31734	62902
	124547	246355	486806	961036	1895514	3735399
1–42–3	1	1	2	4	8	16
1–24–3	32	64	128	256	511	1020
	2033	4049	8054	16007	31779	63035
	124918	247346	489368	967480	1911360	3773622
1–33–2	1	1	2	4	8	16
	32	64	128	255	509	1014
	2020	4019	7995	15893	31582	62730
	124553	247225	490576	973219	1930270	3827723
1–43–2	1	1	2	4	8	16
1–34–2	32	64	128	256	511	1020
	2034	4053	8069	16056	31926	63448
	126025	250207	496546	985052	1953506	3872968

Chapter 6

Words

6.1 History and connections

Chronologically, enumeration of words according to the occurrence of a pattern or a set of patterns grew out of the corresponding research questions in permutations, and was followed by enumeration of compositions according to the occurrence of a given (set of) pattern(s). In this book, we have presented the more general results first.

The main focus of this section will be pattern avoidance in words for different types of patterns: subword, subsequence, generalized, and partially ordered patterns. A systematic investigation of this research area started in the late 1990s. The first paper on subsequence patterns was by Regev [169], who found the asymptotic behavior of the number of k-ary words avoiding an increasing subsequence of length ℓ. In 1999, Burstein [31] presented an explicit formula for the generating functions for the number of k-ary words of length n that avoid a set of permutation (subsequence) patterns $T \subset \mathcal{S}_3$, together with a complete classification of the permutation patterns of length three. The main technique used in his thesis is an extension of the Noonan-Zeilberger algorithm [163] that was originally created for enumerating permutations satisfying certain conditions. Burstein and Mansour [33] extended this line of inquiry to the case of subsequence patterns with repetitions.

A different generalization was given by Mansour [141], who derived the generating function for the number of k-ary words of length n that avoid both 132 and another pattern τ for several interesting cases of τ. Recently, Firro and Mansour [64] recovered the results of Burstein for avoidance of the subsequence pattern 123 by using a new technique, the scanning-element algorithm, which allows for an easier proof. Moreover, they gave an explicit formula for the number of k-ary words containing a subsequence pattern τ exactly once. A third proof of Burstein's result was given with yet another approach by Mansour [141], who adapted the block decomposition method used for permutations to words. Using finite automata theory, Brändén and Mansour [27] gave a combinatorial explanation for the number of k-ary words of length n avoiding a subsequence pattern of length three (see Chapter 7). Most recently, Jelínek and Mansour [105] classified the subsequence patterns up to length six with regard to Wilf-equivalence.

Several authors have considered other types of patterns besides subsequence patterns. Burstein and Mansour [33, 34] enumerated words according to generalized patterns of length three as well as words containing a subword pattern of length ℓ exactly r times, while Kitaev and Mansour [109] studied partially ordered generalized patterns.

In this chapter we sample some of the results. After defining some notation in Section 6.2, we give examples of how results for words can be recovered from those for compositions in Section 6.3. We also summarize results for subword patterns of length three and give a result for the pattern peak.

The focus of Sections 6.4 and 6.5 is on subsequence patterns. We start by giving classifications of the patterns up to length five, based on the results by Jelínek and Mansour [105], who connected Wilf-equivalence with equivalence of fillings of Ferrers diagrams by expressing words as matrices. The connection between the two types of equivalence gives surprisingly general results using previously known results on fillings of Ferrers diagrams [127]. Using systematic computer enumeration of small patterns we verify that these results, together with the reversal and complement maps, are sufficient to describe all Wilf-equivalence classes for patterns of length six or less.

Next, we give results for generating functions for subsequence permutation and multi-permutation patterns of length three. One of the major results in pattern avoidance of subsequence patterns was the derivation of the generating function for the pattern 132. We give three different proofs for this result. The first proof will illustrate the Noonan-Zeilberger algorithm as it applies to words, the second one will utilize block decomposition, a very useful tool in enumeration of words, and the third proof will showcase the scanning-element algorithm. Each of these approaches has advantages and disadvantages, as will become clear in this parallel treatment. We conclude the discussion of permutation patterns of length three by giving results on avoidance of pairs of patterns and their generalizations, as well as results for the generating functions of words containing the patterns 123 and 132, respectively exactly once.

In the last two sections we present results on generalized patterns and on partially ordered patterns obtained from the results for compositions presented in Chapter 5.

6.2   Definitions and basic results

Before giving any results, we will state the basic notation, nearly identical to that for compositions, for the generating functions of interest. Recall that $[k]$ denotes an ordered alphabet, and that $[k]^n$ denotes the set of k-ary words of length n (see Definition 1.7).

Definition 6.1 *We denote the set of k-ary words of length n that contain the pattern τ exactly r times by $\mathcal{W}_n^{[k]}(\tau; r)$, and the number of words in $\mathcal{W}_n^{[k]}(\tau; r)$ by $W_{[k]}^\tau(n, r)$. The corresponding generating function is given by*

$$W_{[k]}^\tau(x, q) = \sum_{w \in \mathcal{W}_n^{[k]}} x^n q^{\mathrm{occ}_\tau(w)} = \sum_{n,r \geq 0} W_{[k]}^\tau(n, r) x^n q^r,$$

where $\mathrm{occ}_\tau(w)$ denotes the number of occurrences of τ in the word w. Furthermore, we denote the set of k-ary words of length n that avoid the pattern τ by $\mathcal{AW}_n^{[k]}(\tau)$, and the number of words in $\mathcal{AW}_n^{[k]}(\tau)$ by $AW_{[k]}^\tau(n)$. The associated generating functions are given by

$$AW_{[k]}^\tau(x) = \sum_{n \geq 0} AW_{[k]}^\tau(n) x^n$$

and (keeping track of the alphabet as well)

$$AW_\tau(x, y) = \sum_{k \geq 0} AW_{[k]}^\tau(x) y^k = \sum_{n,k \geq 0} AW_{[k]}^\tau(n) x^n y^k.$$

If we consider several patterns at once, we replace τ by T and r by $\vec{r}$ in the notation defined above.

As mentioned before, pattern enumeration and pattern avoidance have been studied first for permutations, then for words, and finally for compositions, generalizing the combinatorial object under consideration in each step. Therefore words can be considered to be a special case of compositions. Specifically, we can think of k-ary words of length n as compositions with n parts on the set $[k]$ where we disregard the order of the composition. Thus, we immediately can obtain results on pattern enumeration or avoidance in words from the respective results for compositions by setting $A = [k]$, $x = 1$, and $y = x$ (whenever that is an allowed operation) to obtain that

$$W_{[k]}^\tau(x, q) = C_{[k]}^\tau(1, x, q) \quad \text{and} \quad AW_{[k]}^\tau(x) = AC_{[k]}^\tau(1, x).$$

In some instances, we cannot obtain the results for k-ary words from those for compositions. One such example, where we cannot just substitute $x = 1$ but have to do some additional work, is finding the generating function for the number of words that avoid the pattern *peak* (see Example 4.27).

Many of the results presented in Chapters 4 and 5 have been obtained by generalizing proofs for words, but not all results can be extended from words to compositions. This is in part due to the fact that when we restrict the set $A = [k]$ under consideration, we can map the restricted set, no matter whether it is $A\backslash\{1\}$, $A\backslash\{i\}$ or $A\backslash\{k\}$, to the alphabet $[k-1]$, resulting in much easier recurrence relations, with a much better chance of leading to explicit results. A second factor that makes analysis of pattern avoidance in words simpler than that for compositions is that there are fewer classes of patterns to consider. To make this statement more precise, we define Wilf-equivalence for patterns in words.

Definition 6.2 *Two subword patterns τ and ν are* tight Wilf-equivalent *if the number of k-ary words of length n that avoid τ is the same as the number of k-ary words of length n that avoid ν, for all n and k. We denote two patterns that are tight Wilf-equivalent by $\tau \overset{t}{\sim} \nu$.*

Definition 6.3 *Two nonsubword patterns τ and ν are called* Wilf-equivalent *if the number of k-ary words of length n that avoid τ is the same as the number of k-ary words of length n that avoid ν, for all n and k. We denote two patterns that are Wilf-equivalent by $\tau \sim \nu$.*

Having defined Wilf-equivalence, we can now discuss the symmetry classes for patterns τ in words. Since we are no longer concerned with the sum of the parts, a word and its complement are Wilf-equivalent, resulting in fewer Wilf classes.

Definition 6.4 *For any pattern τ, the* symmetry class *of τ with respect to enumeration or avoidance in words is given by $\{\tau, R(\tau), c(\tau), R(c(\tau))\}$.*

It is easy to see that $R(c(\tau)) = c(R(\tau))$ for any type of pattern τ with respect to pattern avoidance or enumeration in words. The fact that the complement of a pattern is in the same symmetry class also explains the structural phenomenon that we have seen in the results for enumeration or avoidance of patterns in compositions. Oftentimes we had expressions for generating functions that were very similar in structure, but nevertheless different – now we see why! The two patterns in question are complementary patterns, and therefore the results have to agree once we set $x = 1$. This phenomenon can be seen for example in Theorems 5.21 and 5.23.

We will revisit some of the results for compositions from the previous two chapters and present the corresponding results for words in this section, together with the original references (if the results were already known for words).

6.3   Subword patterns

We start by looking at the simplest subword patterns, namely 11 and 12.

Example 6.5 *Since 11 and 12 can be thought of as level and rise, we can exploit Theorem 4.3 with $A = [k]$, $x = 1$, and $y = x$. Defining $W_{[k]}(x, t, \ell, d)$ as* *the generating function for k-ary words according to length n and the statistic* $\overrightarrow{s}_w = (\mathrm{ris}(w), \mathrm{lev}(w), \mathrm{dro}(w))$, *we see that*

$$W_{[k]}(x, r, \ell, d) = 1 + \frac{x \sum_{j=1}^{k} \frac{(1 - x(\ell - r))^{j-1}}{(1 - x(\ell - d))^j}}{1 - dx \sum_{j=1}^{k} \frac{(1 - x(\ell - r))^{j-1}}{(1 - x(\ell - d))^j}}$$

$$= 1 - \frac{1 - \left(\frac{1 - x(\ell - r)}{1 - x(\ell - d)}\right)^k}{r - d\left(\frac{1 - x(\ell - r)}{1 - x(\ell - d)}\right)^k}.$$

In particular, the generating functions for the number of k-ary words avoiding the subword patterns 11 and 12, respectively, are given by

$$AW_{[k]}^{11}(x) = W_{[k]}(x, 1, 0, 1) = \frac{1 + x}{1 - (k-1)x} = (1 + x) \sum_{n \geq 0} (k-1)^n x^n$$

and

$$AW_{[k]}^{12}(x) = W_{[k]}(x, 0, 1, 1) = 1/(1 - x)^k = \sum_{n \geq 0} \binom{n + k - 1}{n} x^n,$$

where we have used (A.3) *for* $AW_{[k]}^{12}(x)$. *Thus we obtain immediately that the number of k-ary words of length n that avoid 11 or 12, respectively, is given by*

$$AW_{[k]}^{11}(n) = k(k-1)^{n-1} \quad and \quad AW_{[k]}^{12}(n) = \binom{n + k - 1}{n}.$$

Note that these results can be derived easily by direct combinatorial arguments.

More generally, we can use the results of Chapter 5 by setting $A = [k]$, $x = 1$, and $y = x$ to obtain several known results on k-ary words containing or avoiding subword patterns of length three proved by Burstein and Mansour [33, 34].

Example 6.6 *Theorem 4.32 gives that the generating function for the number of k-ary words of length n that contain the subword pattern 111 exactly r times is given by*

$$W_{[k]}^{111}(x, q) = \frac{1 + q(1 + q)(1 - x)}{1 - (k - 1 + x)q - (k - 1)(1 - x)q^2}.$$

That is, we recover the results of [33, Example 2.2] and [34, Theorem 3.1]. Likewise, Theorem 4.35 gives that the generating function (after simplification) for the number of k-ary words of length n that contain the subword pattern 112 exactly r times is

$$W_{[k]}^{112}(x, q) = \frac{(1 - q)x}{(1 - q)x - 1 + (1 - (1 - q)x^2)^k},$$

giving the result of [34, Theorem 3.2]. In addition, for $q = 0$ we obtain the generating function for the number of words of length n that avoid the subword pattern 112 as

$$AW_{[k]}^{112}(x) = \frac{x}{x - 1 + (1 - x^2)^k},$$

which was originally proved in [33, Theorem 3.10].

We conclude the study of subword patterns of length three by considering the pattern peak, which was studied for compositions in [92]. The result on the pattern peak given in Theorem 4.26 does not allow for easy substitution to obtain a result for words, so we derive the generating function for the number of words that contain the pattern peak directly.

Theorem 6.7 *Let $W_{[k]}^{peak}(x, q)$ be the generating function for the number of k-ary words of length n according to the number of peaks (occurrence of the subwords 121, 132, or 231). Then*

$$W_{[k]}^{peak}(x, q) = \frac{1 + (1 + x - xq)(W_{[k-1]}^{peak}(x, q) - 1)}{1 - x - x(x + (1 - x)q)(W_{[k-1]}^{peak}(x, q) - 1)},$$

for all $k \geq 1$, with initial condition $W_{[0]}^{peak}(x, q) = 1$.

Proof Assume $k \geq 1$ and let $w \in [k]^n$. Then either w does not contain the letter k, $w = k$, $w = w^{(1)}k$, $w = kw^{(2)}$, or $w = w^{(1)}kw^{(2)}$, where $w^{(1)}$ is a nonempty word on the alphabet $[k - 1]$ and $w^{(2)}$ is a nonempty word on the alphabet $[k]$. Adding the contributions to the generating function from each case results in

$$W_{[k]}^{peak}(x, q) = W_{[k-1]}^{peak}(x, q) + x + x(W_{[k-1]}^{peak}(x, q) - 1)$$
$$+ x(W_{[k]}^{peak}(x, q) - 1) + G_k(x, q), \tag{6.1}$$

where $G_k(x, q)$ is the generating function for the number of k-ary words w of length n according to the number of peaks such that $w = w^{(1)}kw^{(2)}$ and $w^{(1)}$ and $w^{(2)}$ are nonempty words on the alphabets $[k - 1]$ and $[k]$, respectively.

Now we consider the words of the form $w^{(1)}kw^{(2)}$ and apply the same decomposition to the nonempty word $w^{(2)}$, with the same five cases. Therefore,

$$G_k(x, q) = xq(W_{[k-1]}^{peak}(x, q) - 1)^2 + x^2(W_{[k-1]}^{peak}(x, q) - 1)$$
$$+ x^2q(W_{[k-1]}^{peak}(x, q) - 1)^2 + x^2(W_{[k-1]}^{peak}(x, q) - 1)(W_{[k]}^{peak}(x, q) - 1)$$
$$+ xq(W_{[k-1]}^{peak}(x, q) - 1)G_k(x, q).$$

Solving for $G_k(x, q)$ and substituting the resulting expression into (6.1), we obtain that

$$W_{[k]}^{peak}(x, q) = \frac{1 + (1 + x - xq)(W_{[k-1]}^{peak}(x, q) - 1)}{1 - x - x(x + (1 - x)q)(W_{[k-1]}^{peak}(x, q) - 1)},$$

which completes the proof. $\square$

Example 6.8 *Using the recursion given in Theorem 6.7, we may obtain the generating functions for the first few values of k using* Mathematica *or* Maple *as*

$$W_{[1]}^{peak}(x,q) = \frac{1}{1-x}$$

$$W_{[2]}^{peak}(x,q) = \frac{1 - (q-1)x^2}{1 - 2x - (q-1)x^2 + (q-1)x^3}$$

$$W_{[3]}^{peak}(x,q) = \frac{-1 + 3q'x^2 - q'^2 x^4}{-1 + 3x + 3q'x^2 - 4q'x^3 - q'^2 x^4 + q'^2 x^5}$$

and

$$W_{[4]}^{peak}(x,q) = \frac{\left(1 - 6q'x^2 + 5q'^2 x^4 - q'^3 x^6\right)}{\left(1 - 4x - 6q'x^2 + 10q'x^3 + 5q'^2 x^4 - 6q'^2 x^5 - q'^3 x^6 + q'^3 x^7\right)},$$

where $q' = q - 1$.

We summarize the results for words containing a single subword pattern τ of length three in Table 6.1 by giving the complete list of the generating functions $W_{[k]}^\tau(x,q)$ together with the reference where those results first appeared. By setting $q = 0$, we obtain the corresponding results for $AW_{[k]}^\tau(x)$.

Similarly, results on longer subword patterns (see Theorems 4.35, 4.37 and 4.39) give the corresponding results for words [34]. We present the example of the complementary patterns $1^{\ell-1}2$ and $2^{\ell-1}1$ that are not Wilf-equivalent for compositions, but are Wilf-equivalent for words.

Example 6.9 *In Theorem 4.35 the generating functions* $C_{[k]}^{1^{\ell-1}2}(x,y,q)$ *and* $C_{[k]}^{2^{\ell-1}1}(x,y,q)$ *differ, but we have for* $W_{[k]}^\tau(x,q) = C_{[k]}^\tau(1,x,q)$ *that*

$$W_{[k]}^{1^{\ell-1}2}(x,q) = \frac{1}{1 - x\sum_{j=0}^{k-1}(1 - x^{\ell-1}(1-q))^j} = W_{[k]}^{2^{\ell-1}1}(x,q).$$

We next consider subsequence patterns, first giving the classification according to Wilf-equivalence, and then giving results on generating functions.

6.4 Subsequence patterns – Classification

For 3-letter permutation patterns, the complement and reversal maps induce exactly two symmetry classes, namely

$$\{123, 321\} \text{ and } \{132, 231, 312, 213\}.$$

TABLE 6.1: $W_{[k]}^{\tau}(x,q)$ for 3-letter subword patterns τ

τ	$W_{[k]}^{\tau}(x,q)$	Reference
111	$\dfrac{1+(1-q)x(1+kx)-(1-q)(k-1)x^2}{1-(k-1+q)x-(k-1)(1-q)x^2}$	[33, Thm 2.1]
112	$\dfrac{1-q}{1-\frac{1}{x}-q+\frac{1}{x}(1-x^2(1-q))^k}$	[33, Thm 3.10]
212	$\dfrac{1}{1-x-x\sum_{j=0}^{k}\frac{1}{1+jx^2(1-q)}}$	[33, Thm 3.12]
213	$\dfrac{1}{1-x-x\sum_{i=0}^{k-2}\prod_{j=0}^{i}(1-jx^2(1-q))}$	[34, Thm 3.4]
123	$\dfrac{1}{1-kx-\sum_{j=3}^{k}(-x)^j\binom{k}{j}(1-q)^{\lfloor j/2\rfloor}U_{j-3}(q)}$ where $U_0(q)=U_1(q)=1$, $U_{2j}(q)=(1-q)U_{2j-1}(q)-U_{2j-2}(q)$, $U_{2j+1}(q)=U_{2j}(q)-U_{2j-1}(q)$, and $\sum_{j\ge0}U_j(q)t^j=\dfrac{1+t+t^2}{1+(1+q)t^2+t^4}$.	[34, Thm 3.3]

Burstein [31] proved analytically that $123 \sim 132$ through equality of their respective generating functions, which implies that there is only one Wilf-equivalence class for permutation patterns of length three. This equivalence can also be recovered from the result by Savage and Wilf on compositions given in Theorem 5.7.

Secondly, for patterns of length three with repeated letters the complement and reversal maps induce three symmetry classes, namely $\{111\}$, $\{121, 212\}$, and $\{112, 211, 221, 122\}$. The equivalence $112 \sim 121$ for words was first shown by Burstein and Mansour [33] and can also be obtained as a special case of the same equivalence for compositions given in Theorem 5.10. An even more general result that has this equivalence as a special case has been proved by Jelínek and Mansour [104] (see Theorem 6.19). No further reductions are possible as all patterns belonging to a specific equivalence class have to consist of the same letters. We summarize this analysis in Table 6.2, where each class is given by its representatives (up to symmetry).

TABLE 6.2: Wilf-equivalence classes for patterns of length 3

111	$112 \sim 121$	$123 \sim 132$

We will now classify longer patterns according to their Wilf-equivalence, presenting the results of Jelínek and Mansour. To do so, we need the following definition.

Definition 6.10 *We define the* content *of a word w to be the unordered multiset of the letters appearing in w. We say that two patterns τ and τ' are* strongly equivalent, *denoted by $\tau \overset{s}{\sim} \tau'$, if for every k and n there is a bijection f between $\mathcal{AW}_n^{[k]}(\tau)$ and $\mathcal{AW}_n^{[k]}(\tau')$ with the property that for every $w \in \mathcal{AW}_n^{[k]}(\tau)$, the word $f(w)$ has the same content as w, that is, $f(w)$ can be obtained from w by a suitable rearrangement of the letters.*

Clearly, if two patterns are strongly equivalent then they are also Wilf-equivalent for either words or compositions. Each pattern is strongly equivalent to its reversal, and if two patterns τ and τ' are strongly equivalent, then their complements $c(\tau)$ and $c(\tau')$ are strongly equivalent as well. Note that τ and $c(\tau)$ are usually not strongly equivalent. In order to derive results on strong equivalence, we introduce a matrix representation of a word.

Definition 6.11 *Let w be a k-ary word of length n. We define $M(w, k)$ to be the $k \times n$ matrix with a 1 in row i and column j if and only if the j-th letter of w is equal to i, and 0 otherwise. We assume that the rows of a matrix are numbered bottom to top, and the columns are numbered left to right.*

Example 6.12 *Let $w = 124123$ be a word on the alphabet $[5]$. Then*

$$M(w,5) = \begin{bmatrix} 0 & 0 & 0 & 0 & 0 & 0 \\ 0 & 0 & 1 & 0 & 0 & 0 \\ 0 & 0 & 0 & 0 & 0 & 1 \\ 0 & 1 & 0 & 0 & 1 & 0 \\ 1 & 0 & 0 & 1 & 0 & 0 \end{bmatrix},$$

that is, the nonzero entries in the bottom (=first) row indicate the positions of the 1s, and the position of the nonzero entries in the top (= 5-th) row of the matrix indicate the positions of the 5s.

With this representation, we may use known bijections on fillings of Ferrers diagrams to obtain new equivalences for words.

Definition 6.13 *A* Ferrers diagram *is an array of cells whose columns have nonincreasing length, and the bottom cells of the columns appear in the same row. A* filling *of a Ferrers diagram is an assignment of zeros and ones into its cells such that every column has exactly one cell containing a 1, which we refer to as a* 1-cell.

Note that we usually only depict the 1-cells in a filling of a Ferrers diagram for easier readability. Figure 6.1 shows two Ferrers diagrams.

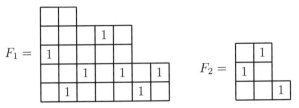

FIGURE 6.1:   Two Ferrers diagrams.

Definition 6.14 *A filling of a Ferrers diagram F contains a matrix M if F can be transformed via deletion of rows and columns to a rectangular diagram with a filling and size identical to M; otherwise, F avoids M. Two matrices M and M' are Ferrers-equivalent if for every Ferrers diagram F the number of M-avoiding fillings is equal to the number of M'-avoiding fillings; they are strongly Ferrers-equivalent if for every Ferrers diagram F there is a bijection between M-avoiding and M'-avoiding fillings of F that preserves the number of 1-cells in each row.*

Example 6.15 *Let*

$$M_1 = \begin{bmatrix} 1 & 0 & 0 & 0 \\ 0 & 0 & 1 & 1 \\ 0 & 1 & 0 & 0 \end{bmatrix} \quad and \quad M_2 = \begin{bmatrix} 1 & 0 \\ 0 & 1 \end{bmatrix}.$$

Then the Ferrers diagram F_1 of Figure 6.1 contains M_1, as F_1 can be transformed into M_1 via deletion of columns four, six and seven, and rows four and five. On the other hand, there are no choices of row and column deletions that will transform the Ferrers diagram F_2 into M_2.

In order to describe the connection between Ferrer's fillings and words, we need the following notation.

Definition 6.16 *For a word $w \in [k]^n$ and an integer ℓ, we let $w + \ell$ denote the word obtained by increasing each letter of w by ℓ. We also denote the concatenation of the words w and w' by ww'.*

Lemma 6.17 *Let τ and τ' be two patterns on k letters, and let ρ be a pattern on ℓ letters. If $M(\tau, k)$ and $M(\tau', k)$ are strongly Ferrers-equivalent, then the two patterns $\tau(\rho + k)$ and $\tau'(\rho + k)$ are strongly equivalent on the alphabet $[k + \ell]$.*

Proof   Let $\nu = \tau(\rho + k)$ and $\nu' = \tau'(\rho + k)$. For given m and n, choose a word $w \in \mathcal{AW}_n^{[m]}(\nu)$, and let $M = M(w, m)$ be its corresponding matrix. Clearly, since w avoids ν, M avoids the matrix $M(\nu, k + \ell)$. Now color the cells of M red and green, where a cell c is green if and only if the submatrix of M strictly to the right and strictly above c contains $M(\rho, \ell)$. All other cells are colored red. Figure 6.2 shows the coloring for the word 534212 with $\tau = 21$, $\rho = 1$ (and therefore $\nu = 213$), where the striped part corresponds to the color green.

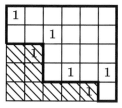

FIGURE 6.2: Coloring of the Ferrers diagram for the word 534212.

Note that the green cells form a Ferrers diagram (why?) and that the nonzero columns of this diagram induce an $M(\tau, k)$-avoiding filling. Using the strong Ferrers-equivalence of $M(\tau, k)$ and $M(\tau', k)$, we may transform this filling into a $M(\tau', k)$-avoiding filling. This operation transforms M into a matrix M' representing a ν'-avoiding word w' with the same content as w.

To see that this operation can be inverted, observe that the operation has only modified the filling of the green cells of M. Also, for every green cell c of M there is a copy of $M(\rho, \ell)$ strictly to the right and strictly above c which consists only of red cells. Thus the red cells of M coincide with the red cells of M' and we have found a bijection showing that $\nu \overset{s}{\sim} \nu'$. □

Lemma 6.17 allows us to translate results about fillings of Ferrers diagrams into results about words. Using known results about Ferrers-equivalence [58, 104, 127], we obtain the following results valid for any pattern ρ.

Proposition 6.18 *For $k \geq 1$, $M(12 \cdots k, k)$ is strongly Ferrers-equivalent to $M(k(k-1) \cdots 1, k)$ (see [127, Theorem 13]). This implies that*

$$12 \cdots k(\rho + k) \overset{s}{\sim} k(k-1) \cdots 1(\rho + k).$$

Proposition 6.19 *For $i, j \geq 0$, $M(2^i 12^j, 2)$ is strongly Ferrers-equivalent to $M(12^{i+j}, 2)$ (see [104, Lemma 39]). This implies that*

$$2^i 12^j(\rho + 2) \overset{s}{\sim} 12^{i+j}(\rho + 2).$$

These results do not account for all the equivalences among subword patterns of small length. For example, the pattern 132 does not have one of the above structures. To complete our classification we need another lemma whose proof uses an idea that has been previously applied in the context of pattern-avoiding set partitions [104, Theorem 48].

Theorem 6.20 *For any* $t \geq 3$, *all patterns that consist of a single* 1, *a single* 3 *and* $t - 2$ *letters* 2 *are strongly equivalent.*

Proof Let t be fixed. Let $\tau(i,j)$ denote the word of length t where the i-th letter is a 1, the j-th letter is a 3, and the remaining letters are equal to 2. Our aim is to show that all the patterns in the set $\{\tau(i,j), i \neq j, 1 \leq i,j \leq t\}$ are strongly equivalent. Since each word is strongly equivalent to its reversal, we only need to deal with the words $\tau(i,j)$ with $i < j$. Proposition 6.19 gives immediately that the words $\{\tau(i,t), i = 1, 2, \ldots, t - 1\}$ (those with a 3 at the end) are all strongly equivalent. Using the complement and reversal maps we can derive that the words $\{\tau(1,j), j = 2, 3, \ldots, t\}$ (those with a 1 at the beginning) are also strongly equivalent.

To prove the theorem, it suffices to show that for every $i < j < t$, the word $\tau(i,j)$ is strongly equivalent to the word $\tau(i+1, j+1)$. Let m be an integer. We will say that a word w *contains* $\tau(i,j)$ *at level* m if there is a pair of letters ℓ, h with $\ell < m < h$ such that the word w contains a subsequence s on the alphabet $\{\ell, m, h\}$ such that s is order-isomorphic to $\tau(i,j)$. For example, the word 132342 contains the pattern 1223 at level three (due to the subsequence 1334), while it avoids 1223 at level two.

Assume now that we are given a fixed pair of indices i and j, with $i < j < t$, and we want to provide a content-preserving bijection between $\tau(i,j)$-avoiding and $\tau(i+1, j+1)$-avoiding words of length n. We will say that a word w is an m-*hybrid* if for every $\overline{m} < m$, the word w avoids $\tau(i,j)$ at level $\overline{m}$, while for every $\widetilde{m} \geq m$, w avoids $\tau(i+1, j+1)$ at level $\widetilde{m}$. We will present, for any $m \geq 1$, a content-preserving bijection between m-hybrids and $(m+1)$-hybrids. By composing these bijections we obtain the required bijection between $\tau(i,j)$-avoiding and $\tau(i+1, j+1)$-avoiding words.

Let $m \geq 1$ be fixed and let w be an arbitrary word. A letter of w is called *low* if it is less than m, and a letter is called *high* if it is greater than m. A *low cluster* of w is a maximal block of consecutive low letters of w. A *high cluster* is defined analogously. Thus every letter of w different from m belongs to a unique high or low cluster. The *landscape* of w is a word over the alphabet $\{L, m, H\}$ obtained by replacing every low cluster of w by a single symbol L and every high cluster of w by a single symbol H. For example, the word $w = 534212$ has landscape $HmHL$ for $m = 3$, with two high clusters 5 and 4 and a single low cluster 212. Note that w contains $\tau(i,j)$ at level m if and only if the landscape of w contains the subsequence $m^{i-1}Lm^{j-i-1}Hm^{t-j}$.

We will now describe the bijection between m-hybrids and $(m+1)$-hybrids. Let w be an m-hybrid word and let X be its landscape. We split X into three parts, $X = PmS$, where the prefix P is formed by the letters of X that appear before the first occurrence of m in X, and where the suffix S is formed by the letters that appear after the first occurrence of m. Let $X' = SmP$. Then X' contains a subsequence $m^{i-1}Lm^{j-i-1}Hm^{t-j}$ if and only if X contains a subsequence $m^{i}Lm^{j-i-1}Hm^{t-j-1}$. Since X is the landscape of a word that avoids $\tau(i+1, j+1)$ at level m, we know that any word with landscape X' must avoid $\tau(i, j)$ at level m. Now define a word w' as follows:

1. The word w' has landscape X'.

2. For any p, the p-th low cluster of w' is identical to the p-th low cluster of w.

3. For any q, the q-th high cluster of w' is identical to the q-th high cluster of w.

Clearly there is a unique word w' satisfying these properties. For example, for the word $w = 534212$ considered above, w' has landscape $HLmH$, and is given by 521234. Note that the subsequence of all the low letters of w is the same as the subsequence of all the low letters of w', and these sequences are partitioned into low clusters in the same way. An analogous property holds for the high letters.

We claim that w' is an $(m+1)$-hybrid. We have already pointed out that w' avoids $\tau(i, j)$ at level m. Let us now argue that w' avoids $\tau(i, j)$ at level $\overline{m}$, for every $\overline{m} < m$, using proof by contradiction. Assume that w' contains a subsequence $s = \overline{m}^{i-1} \ell \, \overline{m}^{j-i-1} h \overline{m}^{t-j}$, for some $\ell < \overline{m} < h$. If $h < m$, then all the letters of s are low, and since w has the same subsequence of low letters as w', we know that w also contains s as a subsequence, contradicting the assumption that w is an m-hybrid.

Assume now that $h \geq m$. Let x and y be the two letters adjacent to h in the sequence s (note that h is not the last symbol of s, so x and y are well defined). Both x and y are low, and they belong to distinct low clusters of w', because the symbol h is not low. Since the low letters of w are the same as the low letters of w', and they are partitioned into clusters in the same way, we know that w contains a subsequence $\overline{m}^{i-1} \ell \, \overline{m}^{j-i-1} h' \overline{m}^{t-j}$, where h' is not low. This shows that w contains $\tau(i, j)$ at level $\overline{m}$, which is impossible, because w is an m-hybrid.

By an analogous argument, we may show that w' avoids $\tau(i+1, j+1)$ at any level $\widetilde{m} > m$. We conclude that the mapping described above transforms an m-hybrid w into an $(m+1)$-hybrid w'. It is easy to see that the mapping is reversible and therefore provides the required bijection between m-hybrids and $(m+1)$-hybrids for all $m \geq 1$. Therefore, combining the bijections for fixed m, we have a bijection between $\tau(i, j)$-avoiding and $\tau(i+1, j+1)$-avoiding words, and the claim follows by induction on i and j. $\square$

From Propositions 6.18 and 6.19 and Theorem 6.20 we can obtain classifications for patterns of lengths four, five, and six. Tables 6.3-6.5 show all the nontrivial Wilf-equivalence classes (those where the Wilf class is not equal to the symmetry class). In Tables G.3 and G.4 we list all of the Wilf-equivalence classes (including the trivial ones) together with a selection of values of $AW_{[k]}^{\tau}(n)$ which show that these Wilf-classes are indeed the only

TABLE 6.3:   Wilf-equivalence classes for patterns of length 4

1223~1232~1322~2132	1123~1132
1234~1243~1432~2143	1112~1121

ones. Note that for patterns of lengths four and five, Propositions 6.18 and 6.19 and Theorem 6.20, together with the symmetry operations, suffice to give a complete classification.

TABLE 6.4:   Wilf-equivalence classes for patterns of length 5

12223~12232~12322~13222~21232~21322	12435~13254
12345~12354~12543~15432~21354~21543	12443~21143
12234~12243~12343~12433~21243~21433	12134~12143
11234~11243~11432	11123~11132
11223~11232~11322	12534~21534
11112~11121~11211	12453~21453

In addition, we can apply Propositions 6.18 and 6.19 and Theorem 6.20 to compositions by defining an analogous matrix representation of a composition and creating bijections for compositions of n with a fixed number of parts. In the case of compositions the symmetry class consists only of τ and $R(\tau)$, so more patterns have to be checked. Tables G.1 and G.2 give the complete Wilf-classification for subsequence patterns of length four and five for compositions and list selected values of $AC_{\mathbb{N}}^{\tau}$ to show that the listed Wilf-classes are indeed different.

Now that we have given the complete classification according to Wilf-equivalence, we will derive the generating functions $AW_{[k]}^{\tau}(x)$ and $AW_{\tau}(x,y)$ for subsequence patterns of length three. Some of these results will be obtained by specializing generating functions derived in Chapter 5, but sometimes this will lead to complicated results. In those cases we will state the theorems that were originally derived for words.

6.5   Subsequence patterns – Generating functions

We give results for subsequence patterns of length three. Table 6.2 indicates that there are two Wilf-classes for patterns with repeated letters, namely 111

TABLE 6.5: Wilf-equivalence classes for patterns of length 6

112223~112232~112322~113222	124433~214433
112345~112354~112543~115432	124353~214353
122334~122343~122433~212343~212433~221433	124453~214453
122234~122243~123343~123433~124333	126435~216435
~212243~213433~214333	125634~215634
122223~122232~122322~123222~132222	126534~216534
~212232~212322~213222~221322	123443~211243
122345~122354~122543~123454~123544	125435~132154
~212354~212543~213544~221543	125344~215344
123456~123465~123654~126543~165432	123435~132354
~213465~213654~216543~321654	124335~133254
111223~111232~111322	123534~213534
112334~112343~112433	125354~213154
123554~211354~211543	125334~215334
122435~132454~132544	113245~113254
121345~121354~121543	125463~215463
111112~111121~111211	122134~122143
123645~213645~231654	126453~216453
123564~213564~312654	124343~212143
123546~132465~132654	121234~121243
124356~124365~214365	123543~213543
121334~121343~121433	123145~123154
122534~212534~221534	124443~211143
111234~111243~111432	124533~214533
123345~123354~213354	125534~215534
122453~212453~221453	124635~214635
122443~211343~211433	111123~111132
112234~112243	125364~215364
124535~131254	124553~214553
126354~216354	125436~143265
124563~214563	126345~216345
125346~134265	125543~215543
124354~213254	124543~214543
112134~112143	123245~123254
124435~132254	125434~215434
125433~215433	125643~215643
121134~121143	125453~215453
124653~214653	

and 112, and just one class for permutation patterns. Since the results for patterns with repeated letters are easier to derive, we present those results first, even though historically, permutation patterns were studied first. In addition, we give results for the generating functions for the number of words that avoid a pair of patterns $\{132, 12 \cdots \ell\}$, or contain 123 or 132 exactly once.

Theorem 6.21 *The exponential generating function for the number of k-ary words of length n that avoid the pattern 111 is given by*

$$\sum_{n \geq 0} AW_{[k]}^{111}(n) \frac{x^n}{n!} = (1 + x + \frac{1}{2}x^2)^k,$$

and therefore we obtain the number of k-ary words of length n that avoid 111 as

$$AW_{[k]}^{111}(n) = n! \sum_{j=0}^{k} \frac{\binom{k}{j}\binom{j}{n-j}}{2^{n-j}}.$$

Proof This follows from Theorem 5.11 with $x = 1$, $y = x$, and $A = [k]$, followed by two applications of the binomial theorem. □

Example 6.22 *Since we have an explicit formula, we can easily compute the sequence of values for $AW_{[k]}^{111}(n)$. The values for $k = 6$ and $n = 0, 1, \ldots, 12$ are 1, 6, 36, 210, 1170, 6120, 29520, 128520, 491400, 1587600, 4082400, 7484400, and 7484400. Note that for $n \geq 13$, a subsequence order-isomorphic to 111 has to occur because we have only six letters in the alphabet.*

For the pattern 112 we state the result and proof given in [33] which shows an interesting connection to the (unsigned) Stirling numbers of the first kind[1], which count the number of permutations of length n with k disjoint cycles (see for example [72]).

Theorem 6.23 *Let $F_{112}(x, y) = \sum_{k,n \geq 0} AW_{[k]}^{112}(n) \frac{x^n}{n!} y^k$. Then*

$$AW_{[k]}^{112}(n) = \sum_{j=0}^{k} \binom{n+k-j-1}{n} \begin{bmatrix} n \\ n-j \end{bmatrix}$$

and

$$F_{112}(x, y) = \frac{1}{1-y} \cdot \left(\frac{1-y}{1-y-xy} \right)^{1/y},$$

where $\begin{bmatrix} n \\ j \end{bmatrix}$ is the unsigned Stirling number of the first kind (see Table A.1).

---

[1]http://en.wikipedia.org/wiki/Stirling_numbers_of_the_first_kind

Proof The recurrence for $AW_{[k]}^{112}(n)$ is the same as the one used in the proof of Theorem 5.13 (except there the recurrence is in terms of the generating functions). Consider all words $w \in AW_n^{[k]}(112)$ which contain at least one 1. Their number is

$$g_{112}(n,k) = AW_{[k]}^{112}(n) - |\{w \in AW_n^{[k]}(112) : w \text{ has no 1s}\}|$$
$$= AW_{[k]}^{112}(n) - AW_{[k-1]}^{112}(n). \tag{6.2}$$

On the other hand, each such w either ends in 1 or not. If w ends in 1 then deletion of this 1 produces a word $\bar{w} \in AW_{n-1}^{[k]}(112)$, because addition of a 1 to the right end of any word $\bar{w} \in AW_{n-1}^{[k]}(112)$ does not produce extra occurrences of the pattern 112.

If w does not end in 1 then it can only contain a single 1 that does not occur at end of w. Deletion of this 1 produces a word $\bar{w} \in AW_{n-1}^{[2,k]}(112)$ and the single 1 can occur in $n-1$ positions (all except the rightmost one). Thus we have

$$g_{112}(n,k)$$
$$= AW_{[k]}^{112}(n-1) + (n-1)|\{w \in AW_{n-1}^{[k]}(112) : w \text{ has no 1s}\}|$$
$$= AW_{[k]}^{112}(n-1) + (n-1)AW_{[k-1]}^{112}(n-1). \tag{6.3}$$

Combining (6.2) and (6.3), we get for $n, k \geq 1$ that

$$AW_{[k]}^{112}(n) - AW_{[k-1]}^{112}(n) = AW_{[k]}^{112}(n-1) + (n-1)AW_{[k-1]}^{112}(n-1), \quad (6.4)$$

with initial conditions $AW_{[0]}^{112}(n) = \delta_{n0}$ for all $n \geq 0$, $AW_{[k]}^{112}(0) = 1$, and $AW_{[k]}^{112}(1) = k$ for all $k \geq 0$. Let $F_{112}(n;y) = \sum_{k \geq 0} AW_{[k]}^{112}(n)y^k$. Multiplying (6.4) by y^k and summing over $k \geq 0$, we get for $n \geq 1$ that

$$F_{112}(n;y) - \delta_{n0} - yF_{112}(n;y)$$
$$= F_{112}(n-1;y) - \delta_{n-1,0} + (n-1)yF_{112}(n-1;y).$$

Solving for $F_{112}(n;y)$ results in

$$F_{112}(n;y) = \frac{1 + (n-1)y}{1-y} F_{112}(n-1;y) \quad \text{for } n \geq 2. \tag{6.5}$$

Furthermore, $F_{112}(0;y) = \dfrac{1}{1-y}$ and $F_{112}(1;y) = \dfrac{y}{(1-y)^2}$. Iterating (6.5) yields

$$F_{112}(n;y) = \frac{y(1+y)(1+2y)\cdots(1+(n-1)y)}{(1-y)^{n+1}}$$
$$= (1+y)(1+2y)\cdots(1+(n-1)y)\left(\frac{y}{(1-y)^{n+1}}\right). \tag{6.6}$$

That is, $F_{112}(n; y)$ is a convolution of two functions. If we are able to determine the coefficients of each of these two functions, then we can compute $AW_{[k]}^{112}(n)$ as the convolution of the respective coefficients (see Definition 2.34). First,

$$(1 + y)(1 + 2y) \cdots (1 + (n-1)y) = y^n \prod_{j=0}^{n-1} \left(\frac{1}{y} + j \right)$$

$$= y^n \sum_{k=0}^{n} \begin{bmatrix} n \\ k \end{bmatrix} \left(\frac{1}{y} \right)^k = \sum_{k=0}^{n} \begin{bmatrix} n \\ k \end{bmatrix} y^{n-k} = \sum_{k=0}^{n} \begin{bmatrix} n \\ n-k \end{bmatrix} y^k.$$

Using (A.3) and reindexing, we obtain the series expansion of the second function as

$$\frac{y}{(1-y)^{n+1}} = \sum_{k=0}^{\infty} \binom{n+k-1}{n} y^k.$$

Therefore, $AW_{[k]}^{112}(n)$ can be computed as the convolution

$$AW_{[k]}^{112}(n) = \begin{bmatrix} n \\ n-k \end{bmatrix} * \binom{n+k-1}{n} = \sum_{j=0}^{k} \binom{n+k-j-1}{n} \begin{bmatrix} n \\ n-j \end{bmatrix}$$

which proves the first statement. The result for $F_{112}(x, y)$ can be obtained as the exponential generating function of $F_{112}(n; y)$ from the recursive relation (6.5). Alternatively, we can build the generating function $F_{112}(x, y)$ from the recurrence relation (6.4). Multiplying by $\frac{x^{n-1}}{(n-1)!}$, summing over $n \geq 1$, and recognizing the resulting series as a derivative yields

$$\frac{d}{dx} \sum_{n \geq 1} (AW_{[k]}^{112}(n) - AW_{[k-1]}^{112}(n)) \frac{x^n}{n!}$$

$$= \sum_{n \geq 1} (AW_{[k]}^{112}(n-1) - AW_{[k-1]}^{112}(n-1)) \frac{x^{n-1}}{(n-1)!}$$

$$+ \frac{d}{dx} \sum_{n \geq 1} AW_{[k-1]}^{112}(n-1) \frac{x^n}{(n-1)!}.$$

Defining $F(x; k) = \sum_{n \geq 0} AW_{[k]}^{112}(n) \frac{x^n}{n!}$ gives that

$$\frac{d}{dx} F_{112}(x; k) - \frac{d}{dx} F_{112}(x; k-1)$$

$$= F_{112}(x; k) - F_{112}(x; k-1) + \frac{d}{dx} (x F_{112}(x; k-1)),$$

or equivalently,

$$\frac{d}{dx} F_{112}(x; k) = F_{112}(x; k) + (1+x) \frac{d}{dx} F_{112}(x; k-1).$$

Multiplying this recurrence by y^k and summing over $k \geq 1$, we obtain

$$\frac{d}{dx} F_{112}(x, y) = \frac{1}{1 - y - yx} F_{112}(x, y).$$

Solving this equation and using the initial condition $F_{112}(0, y) = \dfrac{1}{1-y}$ yields the desired formula. $\qquad \square$

Example 6.24 *Since we have an explicit formula, we can compute the sequence of values for* $AW_{[3]}^{112}(n)$ *by using that* $\begin{bmatrix} n \\ n \end{bmatrix} = 1$, $\begin{bmatrix} n \\ n-1 \end{bmatrix} = \binom{n}{2}$, *and* $\begin{bmatrix} n \\ n-2 \end{bmatrix} = \frac{1}{4}(3n-1)\binom{n}{3}$ *(see Table A.1), as follows:*

$$\sum_{j=0}^{3} \binom{n+2-j}{n} \begin{bmatrix} n \\ n-j \end{bmatrix}$$

$$= \binom{n+2}{n} \begin{bmatrix} n \\ n \end{bmatrix} + \binom{n+1}{n} \begin{bmatrix} n \\ n-1 \end{bmatrix} + \binom{n}{n} \begin{bmatrix} n \\ n-2 \end{bmatrix}$$

$$= \frac{(n+2)(n+1)}{2} + (n+1) \binom{n}{2} + \frac{1}{4}(3n-1) \binom{n}{3}$$

$$= \frac{1}{24}(24 + 22n + 21n^2 + 2n^3 + 3n^4).$$

The values of $AW_{[3]}^{112}(n)$ *for* $n = 0, 1, \ldots, 20$ *are* 1, 3, 9, 24, 56, 116, 218, 379, 619, 961, 1431, 2058, 2874, 3914, 5216, 6821, 8773, 11119, 13909, 17196, *and* 21036.

Now we turn our attention to the case of the permutation patterns, which were historically the first ones to be studied. Three different methods, namely the Noonan-Zeilberger algorithm, the block decomposition method, and the scanning-element algorithm, have been used for deriving

$$AW_{123}(x, y) = AW_{132}(x, y).$$

We will present three different proofs to showcase the advantages and disadvantages of each of these approaches. Before doing so, we will state the result on the generating function for the number of words avoiding a permutation pattern.

Theorem 6.25 *[31, Theorem 3.2] The generating function for the number of k-ary words of length n that avoid 132 is given by*

$$AW_{132}(x, y) = 1 + \frac{y}{2x(1-x)} \cdot \left(1 - \sqrt{\frac{(1-2x)^2 - y}{1-y}}\right).$$

Note that in [31, Theorem 3.2], an explicit formula for $AW_{[k]}^{132}(n)$ is also given. We start with the original proof.

6.5.1 Noonan-Zeilberger algorithm

We discuss the Noonan-Zeilberger algorithm primarily for its historical importance, as its use in Burstein's thesis [31] started the research area of pattern avoidance in words. The Noonan-Zeilberger algorithm [163] was originally designed for enumerating permutations containing and avoiding permutation patterns. Burstein generalized this algorithm in two directions: by considering a set of patterns rather than a single pattern, and by adding the corrections needed to apply it to words instead of permutations. We will describe the generalized algorithm and illustrate it by deriving the generating function for the number of 123-avoiding words.

Noonan-Zeilberger algorithm for words
Let $\vec{\tau} = (\tau^{(1)}, \ldots, \tau^{(\ell)})$ be a vector of subsequence patterns and let $\vec{r} = (r_1, r_2, \ldots, r_\ell)$, where r_j indicates the number of times the pattern $\tau^{(j)}$ is to occur. If avoidance of a pattern is the goal, then $r_j = 0$.

Then the Noonan-Zeilberger algorithm for words can be described as follows:

1. Define a recurrence for the k-ary words of length n avoiding the patterns $\vec{\tau}$. This is a matter of art and may involve removing either the first or last letter of the word, removing specific smallest or largest letters of the word, and so on. Repeat if necessary.

2. Identify the auxiliary parameters of the recurrence which are associated with the auxiliary subpatterns that arise in Step 1.

3. Assign a main variable for each pattern $\tau^{(i)}$ of $\vec{\tau}$, and an auxiliary variable for each of the auxiliary parameters.

4. Let the weight of the word w be defined as $\mathrm{wt}(w) = \prod q_j^{\mathrm{occ}_j(w)}$, where $\mathrm{occ}_j(w)$ is the number of occurrences of the j-th pattern in w and the product is over all patterns (main and auxiliary) which need to be considered for the recurrence obtained after Step 1.

5. Let $\vec{s}(w) = (s_1, s_2, \ldots, s_k)$ be the content vector of w, that is, s_i is the number of occurrences of the letter i in w, and define

$$F_{n,\vec{s}} = \sum_{w \in [k]^n, \vec{s}(w) = \vec{s}} \mathrm{wt}(w) x^n.$$

Determine a functional equation for $F_{n,\vec{s}}$ by using the recurrence developed in Step 1.

6. Solve the equation obtained in Step 5. This usually involves taking r_j partial derivatives with respect to q_j for $j = 1, 2, \ldots, \ell$, that is, with respect to the variables corresponding to the main patterns.

7. Set $q_j = 0$ for $j = 1, 2, \ldots, \ell$ (main patterns) and $q_j = 1$ for $j > \ell$ (auxiliary patterns). This ensures avoidance of the main patterns and renders the auxiliary patterns irrelevant.

8. Sum the resulting equation over all content vectors $\vec{s}$ with

$$|\vec{s}| = s_1 + \cdots + s_k = n$$

to compute $F_n = \sum_{w \in [k]^n} \operatorname{wt}(w)$.

We are now ready give the proof of Theorem 6.25 using the Noonan-Zeilberger algorithm.

Proof Since there is only the single pattern 123 to be avoided, $\vec{r} = (0)$. Creating a recurrence relation by removing the last letter, we have to check for the occurrence of any pattern 12 which could have led to an occurrence of the pattern 123 with the letter that has been removed. Thus, for $j = 2, 3, \ldots, k$, let $\operatorname{occ}_j(w)$ be the number of occurrences of $w_{i_1} < w_{i_2} \le j$ such that $i_1 < i_2$. Introducing auxiliary variables $q_2, \ldots, q_k$ to keep track of the 2-letter patterns that end with $j = 2, \ldots, k$, we obtain the weight of w as

$$\operatorname{wt}(w) = q^{\operatorname{occ}_{123}(w)} \prod_{j=2}^{k} q_j^{\operatorname{occ}_j(w)}.$$

Define $F_n(q; q_2, q_3, \ldots, q_k) = \sum_{w \in [k]^n} \operatorname{wt}(w) = \sum_{\vec{s}} F_{n,\vec{s}}(q; q_2, q_3, \ldots, q_k)$ where

$$F_{n,\vec{s}}(q; q_2, q_3, \ldots, q_k) = \sum_{w \in [k]^n, \vec{s}(w) = \vec{s}} \operatorname{wt}(w)$$

and $\vec{s} = (s_1, \ldots, s_k)$ is the content vector of w. To create the recurrence, let w be any word of length n, w' be the word obtained by deleting the last letter from w, and assume that $w_n = i$. Then for $i = 1$, $\operatorname{wt}(w) = \operatorname{wt}(w')$. If $i > 1$ then the last letter i of w can form a pattern 12 with any letter $j < i$, which occurs s_j times in w. In addition, any occurrence of a 2-letter pattern 12 ending in $j < i$ can create an occurrence of 123. Therefore

$$\operatorname{wt}(w) = \operatorname{wt}(w') q_i^{\sum_{j=1}^{i-1} s_j} q^{\sum_{j=2}^{i-1} \operatorname{occ}_j(w')}$$

$$= q^{\operatorname{occ}_{123}(w')} q_i^{\sum_{j=1}^{i-1} s_j} \prod_{j=2}^{i-1} (q q_j)^{\operatorname{occ}_j(w')} \prod_{j=i}^{k} q_j^{\operatorname{occ}_j(w')}.$$

Summing over all w with a given content vector $\vec{s}$, we obtain that

$$F_{n,\vec{s}}(q; q_2, \ldots, q_k) = \delta(\vec{s} - \vec{e}_1 \ge \vec{0}) F_{n-1,\vec{s}-\vec{e}_1}(q; q_2, \ldots, q_k)$$

$$+ \sum_{i=2}^{k} q_i^{\sum_{j=1}^{i-1} s_j} \delta(\vec{s} - \vec{e}_i \ge \vec{0}) F_{n-1,\vec{s}-\vec{e}_i}(q; q q_2, \ldots, q q_{i-1}, q_i, \ldots, q_k), \quad (6.7)$$

where $\vec{e}_i$ is the i-th unit content vector, $\vec{0} = (0, \ldots, 0)$, δ denotes the truth value of the corresponding proposition, and inequalities hold componentwise. Note that the condition $\delta(\vec{s} - \vec{e}_i \geq \vec{0})$ ensures that the last letter can be an i. Let

$$F_{n,\vec{s}}(k, I) = F_{n,\vec{s}}(0; \underbrace{0, 0, \ldots, 0}_{I-1}, \underbrace{1, 1, \ldots, 1}_{k-I})$$

be the number of words with given content vector $\vec{s}$ that contain no occurrences of the pattern 123 and no instances of $w_{i_1} < w_{i_2} < I$ with $i_1 < i_2$. Then, substituting $q = q_2 = \cdots = q_I = 0$, $q_{I+1} = \cdots = q_k = 1$ for $I \in [k]$ into (6.7), we obtain that

$$F_{n,\vec{s}}(k, I) = \sum_{i=1}^{I} \delta(\vec{s} - \vec{e}_i \geq \vec{0}) F_{n-1,\vec{s}-\vec{e}_i}(k, I) \delta(s_1 = \cdots = s_{i-1} = 0)$$

$$+ \sum_{i=I+1}^{k} \delta(\vec{s} - \vec{e}_i \geq \vec{0}) F_{n-1,\vec{s}-\vec{e}_i}(k, i - 1). \qquad (6.8)$$

Note that we need the condition $\delta(s_1 = \cdots = s_{i-1} = 0)$ to ensure that we do not get zero factors when setting $q_2, q_3, \ldots, q_{i-1} = 0$. Now let J_w be the smallest letter that occurs in w. We see that if $J_w > I$ then the first sum on the right-hand side of (6.8) is zero, because $s_i = 0$ for all $i = 1, 2, \ldots, I$ implies that $\delta(\vec{s} - \vec{e}_i \geq \vec{0}) = 0$ for all $i \leq I$. What happens if $J_w \leq I$? If $i < J_w$ then $s_i = 0$, so $\delta(\vec{s} - \vec{e}_i \geq \vec{0}) = 0$. If $i > J_w$ then $\delta(s_1 = \cdots = s_{i-1} = 0) = 0$, because $s_{J_w} > 0$. Thus the only possible nonzero summand occurs when $i = J_w$, so our recursive relation takes the form

$$F_{n,\vec{s}}(k, I) = \delta(J_w \leq I) F_{n-1,\vec{s}-\vec{e}_{J_w}}(k, I) + \sum_{i=I+1}^{k} \delta(\vec{s} - \vec{e}_i \geq \vec{0}) F_{n-1,\vec{s}-\vec{e}_i}(k, i - 1).$$

Summing over all $\vec{s} \in \mathbb{N}^k$ such that $|\vec{s}| = n$, we obtain (after adjusting the index of the second sum) that

$$F_n(k, I) = \sum_{i=1}^{I} \sum_{|\vec{s}|=n, J_w=i} F_{n-1,\vec{s}-\vec{e}_i}(k, I) + \sum_{i=I}^{k-1} F_{n-1}(k, i). \qquad (6.9)$$

Note that in the first sum we have a restriction on the values of $\vec{s}$ by the definition of J_w, and therefore we do not obtain $F_{n-1}(k, I)$. Furthermore, since the smallest letter that occurs is i, the alphabet can be adjusted from $[k]$ to $[k - i + 1]$, with the resulting adjustment in the variables that are set to zero as well. With these modifications (6.9) becomes

$$F_n(k, I) = \sum_{i=1}^{I} F_{n-1}(k - i + 1, I - i + 1) + \sum_{i=I}^{k-1} F_{n-1}(k, i).$$

Therefore, for $I > 1$,

$$F_n(k+1, I) + F_n(k, I) - F_n(k, I - 1) - F_n(k+1, I+1)$$
$$= 2F_{n-1}(k+1, I) - F_{n-1}(k, I - 1) - F_{n-1}(k+1, I+1),$$

and for $I = 1$,

$$F_n(k+1, 1) - F_n(k+1, 2) = 2F_{n-1}(k+1, 1) - F_{n-1}(k, 1) - F_{n-1}(k+1, 2). \quad (6.10)$$

Next, Burstein used the clever trick of recasting the recurrence for F_n in a different form which results in a recurrence that can be solved. First, replace I by $k - I$ in the two recursions for F_n (which replaces $I > 1$ by $k > I + 1$), and then let $H_I(n, k) = F_n(k, k - I)$. Thus, for $k > I + 1$, $I \geq 1$,

$$H_{I+1}(n, k+1) - 2H_{I+1}(n-1, k+1) - H_{I+1}(n, k) + H_{I+1}(n-1, k)$$
$$= H_I(n, k+1) - H_I(n-1, k+1) - H_I(n, k),$$

and for $k = I + 1$, $I \geq 1$,

$$H_{I+1}(n, I + 2) - 2H_{I+1}(n-1, I+2)$$
$$= H_I(n, I+2) - H_I(n-1, I+2) - H_I(n-1, I+1).$$

Let $H_{I,k}(x) = \sum_{n \geq 0} H_I(n, k) x^n$ with initial conditions $H_I(0, k) = 1$ for all I, k such that $k \geq I + 1$. Then the above recurrence relations for $k \geq I + 1$ and $I \geq 1$ become

$$(1 - 2x)H_{I+1,k+1}(x) - (1 - x)H_{I+1,k}(x) = (1 - x)H_{I,k+1}(x) - H_{I,k}(x),$$

and

$$(1 - 2x)H_{I+1,I+2}(x) = (1 - x)H_{I,I+2}(x) - xH_{I,I+1}(x). \quad (6.11)$$

Now define $H_I(x, y) = \sum_{k \geq I+1} H_{I,k}(x) y^k$, a generating function that keeps track of the alphabet, and let $u_I(x) = H_{I,I+1}(x)$. Since

$$F_n(0; 1, 1, \ldots, 1) = F_n(k, 1) = H_{k-1}(n, k),$$

$u_I(x)$ is the function we really want, as it is the generating function for the number of words on the alphabet $[I + 1]$ that avoid the pattern 123 and that have no restrictions with respect to the auxiliary patterns. So our goal is to find an explicit expression for $u_I(x) = AW_{123}^{[I+1]}(x)$. In terms of $u_I(x)$, (6.11) becomes

$$H_{I,I+2}(x) = ((1 - 2x)u_{I+1} - xu_I)/(1 - x). \quad (6.12)$$

From the recurrence for $H_{I,k}(x)$ together with (6.12) we obtain a recurrence relation for $H_I(x, y)$ for $I \geq 1$ (see Exercise 6.4):

$$((1 - 2x) - (1 - x)y)H_{I+1}(x, y) - (1 - x - y)H_I(x, y)$$
$$= (1 - x)u_I(x)(y - 1)y^{I+1}. \quad (6.13)$$

Iterating this recurrence relation, we obtain that

$$H_I(x, y) = H_1(x, y) \left(\frac{1 - x - y}{1 - 2x - (1 - x)y} \right)^{I-1}$$

$$+ \sum_{j=1}^{I-1} \frac{(1 - x - y)^{I-1-j}}{(1 - 2x - (1 - x)y)^{I-j}} (1 - x) u_j(x)(y - 1) y^{j+1}.$$

To solve this recurrence relation, we need to find $H_1(x, y)$. Note that $H_1(n, k) = F_n(k, k-1)$ is the number of words of length n that avoid 123 and where the only occurrences of the pattern 12 are those containing the letter k. We can obtain all such words by taking a nonincreasing (possibly empty) sequence of letters from the alphabet $[k - 1]$ of length $m \le n$ and adding $n - m$ letters k with no restriction on position to obtain a word of length n. Therefore,

$$H_1(n, k) = \sum_{m=0}^{n} \binom{n}{m} \binom{m + k - 2}{m}, \tag{6.14}$$

and consequently (see Exercise 6.3) $H_1(x, y) = \frac{y^2}{(1 - 2x - (1 - x)y)}$. Hence,

$$H_I(x, y)$$
$$= \frac{(1 - x - y)^{I-1}}{(1 - 2x - (1 - x)y)^I} y^2 \tag{6.15}$$
$$+ \sum_{j=1}^{I-1} \frac{(1 - x - y)^{I-1-j}}{(1 - 2x - (1 - x)y)^{I-j}} (1 - x)(y - 1) y^{j+1} u_j(x, y).$$

Since $u_I(x) = [y^{I+1}] H_I(x, y)$, we can turn (6.15) into a recurrence for $u_I(x)$. To do so, we note that (see Exercise 2.23)

$$f_{\ell,m}(x) = [y^\ell] \left(\frac{(1 - x - y)^m}{(1 - 2x - (1 - x)y)^{m+1}} \right)$$
$$= \frac{(1 - x)^{\ell-m}}{(1 - 2x)^{\ell+1}} \sum_{i \ge 0} \binom{\ell + i}{i} \binom{m}{i} \frac{x^{2i}}{(1 - 2x)^i}. \tag{6.16}$$

Taking coefficients of y^{I+1} on both sides of (6.15) and using (6.16) we obtain the following recurrence relation for $u_I(x)$:

$$u_I(x) = \frac{1}{(1 - 2x)^I} \sum_{i \ge 0} \binom{I - 1 + i}{i} \binom{I - 1}{i} \frac{x^{2i}}{(1 - 2x)^i}$$

$$+ \sum_{j=1}^{I-1} \frac{(1 - x) u_j(x)}{(1 - 2x)^{I-j}} \sum_{i \ge 0} \binom{I - j - 1 + i}{i} \binom{I - j - 1}{i} \frac{x^{2i}}{(1 - 2x)^i}$$

$$- \sum_{j=1}^{I-1} \frac{(1 - x)^2 u_j(x)}{(1 - 2x)^{I+1-j}} \sum_{i \ge 0} \binom{I - j + i}{i} \binom{I - j}{i} \frac{x^{2i}}{(1 - 2x)^i}.$$

Since $u_1(x)$ is the generating function of words avoiding 123 on the alphabet [2] (all 2^n such words do), we obtain that $u_1(x) = \frac{1}{1-2x}$. Using this initial condition, one can find an explicit formula for $u_I(x) = AW^{132}_{[I+1]}(x)$, the generating function for the number of words on the alphabet $[I+1]$ of length n that avoid the pattern 123. □

6.5.2 Block decomposition for words

The block decomposition method was first used by Mansour and Vainshtein [147] to study the structure of 132-avoiding permutations and permutations containing a given number of occurrences of 132 [149]. Mansour [141] extended the use of the block decomposition approach for the set of permutations, described in [148], to the set of words. We will describe this approach and use it to provide a different and simpler proof of Theorem 6.25. In addition, the block decomposition method reveals more of the structure of the 132-avoiding words and can also be used to derive generating functions for words that avoid a set of patterns $\{132, \tau\}$ for families of patterns. To describe this approach, we need some definitions.

Definition 6.26 *Let $w = w_1 w_2 \cdots w_n \in AW^{[k]}_n(132)$ be a k-ary word of length n that avoids the subsequence pattern 132. Then we can decompose w as $w = w^{(1)} k w^{(2)} k w^{(3)} k \cdots k w^{(s+1)}$, where each of the $w^{(i)}$ is a possibly empty word on the alphabet $[k-1]$, and s denotes the number of occurrences of k in w. This representation of w is called the* block decomposition *of w.*

Example 6.27 *Let $w = 34345523525121 \in AW^{[5]}_{14}(132)$, then the block decomposition is given by $w^{(1)} = 3434$, $w^{(2)} = \varnothing$, $w^{(3)} = 23$, $w^{(4)} = 2$, and $w^{(5)} = 121$, and there are a total of $s = 4$ occurrences of 5 in w. Figure 6.3 shows the block decomposition for this particular word. The striped area indicates values that cannot be taken in a word that avoids 132. Note that the maximal value of $w^{(t_j+1)}$ determines the minimal value of $w^{(t_j)}$ and vice versa for $j = 1, 2, \ldots s$.*

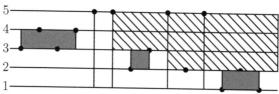

FIGURE 6.3: The block decomposition of $w = 34345523525121$.

Lemma 6.28, which follows immediately from Definition 6.26, is the basis for all the enumeration results in [141].

Lemma 6.28 Let $w = w^{(1)} k w^{(2)} k w^{(3)} k \cdots k w^{(s+1)} \in AW_n^{[k]}(132)$ be the block decomposition of w. Then one of the following assertions holds:

1. $s = 0$, that is, $w \in AW_n^{[k-1]}(132)$.

2. $s \geq 1$; then there exist t_j with $1 \leq t_1 < t_2 < \cdots < t_d \leq s+1$ such that

   (a) $w^{(t_j)} \neq \emptyset$ for $j = 1, 2, \ldots, d$;

   (b) $w^{(j)} = \emptyset$ for $j \in [s+1] \backslash \{t_1, t_2, \ldots, t_d\}$;

   (c) for $j = 1, 2, \ldots, d$, $w^{(t_j)}$ is a word on the letters

   $$q_j, q_j + 1, q_j + 2, \ldots, q_{j-1}$$

   that avoids 132 and in which the letter q_j occurs at least once, where $k - 1 = q_0 \geq q_1 \geq \cdots \geq q_d = 1$.

Note that the values q_j mentioned in Lemma 6.28 refer either to the maximal value of the word $w^{(t_j+1)}$ or the minimal value of $w^{(t_j)}$ for $j = 1, 2, \ldots, d-1$. Neither q_0 nor q_d have to be attained.

Example 6.29 (*Continuation of Example 6.27*) For $w = 34345523525121$ in $AW_{14}^{[5]}(132)$, the second case applies, with $t_1 = 1$, $t_2 = 3$, $t_3 = 4$, and $t_4 = 5$. The values of q_j are determined as follows: $q_0 = 4$, $q_1 = 3$ (the maximal value of $w^{(3)}$), $q_2 = 2$ (the maximal value of $w^{(4)}$), $q_3 = 2$ (the maximal value of $w^{(5)}$), and $q_d = 1$.

When proving Theorem 6.25 using the block decomposition method we will rely on the following lemma which gives results for two generating functions closely related to $AW_{132}(x, y)$.

Lemma 6.30 Let $H(x; k) = AW_{[k+1]}^{132}(x) - 1$ be the generating function for the number of $(k + 1)$-ary nonempty words of length n that avoid 132, and let $M(x; k) = AW_{[k+1]}^{132}(x) - AW_{[k]}^{132}(x)$ denote the generating function for the number of $(k + 1)$-ary words of length n that avoid 132 and in which the letter $k + 1$ occurs at least once. Then for $H(x, y) = \sum_{k \geq 0} H(x; k) y^k$ and $M(x, y) = \sum_{k \geq 0} M(x; k) y^k$ we have that

$$H(x, y) = \frac{1}{y(1 - y)} \left((1 - y) AW_{132}(x, y) - 1 \right)$$

and

$$M(x, y) = \frac{1}{y} \left((1 - y) AW_{132}(x, y) - 1 \right).$$

Proof See Exercise 6.5. □

We are now ready to present the alternative derivation for the generating function $AW_{132}(x, y)$ as given in [141].

Proof (of Theorem 6.25 using block decomposition method) By Lemma 6.28, there are exactly two possibilities for the block decomposition of an arbitrary $w \in \mathcal{AW}_n^{[k]}(132)$. The contribution to $AW_{[k]}^{132}(x)$ for the case $s = 0$ is $AW_{[k-1]}^{132}(x)$. Next we consider the case $s \geq 1$ and $0 \leq d \leq s + 1$. If $d = 0$, then the contribution to $AW_{[k]}^{132}(x)$ equals x^s since $w = k \cdots k$. For $1 \leq d \leq s + 1$, $w^{(t_j)}$ takes values in $q_j, \ldots, q_{j-1}$, where q_j is defined to be the maximal value of $w^{(t_{j+1})}$ for $j = 1, \ldots, d - 1$. Thus for counting purposes, we can consider $w^{(t_j)}$ to be a word from the alphabet $[q_{j-1} - q_j + 1]$. Note that for $w^{(t_1)}$ the maximal value does not have to be taken, but for the other subwords, the maximal value is achieved by definition of the q_j. Defining $a_j = q_{j-1} - q_j$, the contribution to $AW_{[k]}^{132}(x)$ for a given set of values t_j is given by

$$x^s \sum_{a_1 + a_2 + \cdots + a_d = k-2} H(x; a_1) \prod_{j=2}^{d} M(x; a_j).$$

Taking into consideration that the t_j can be selected in $\binom{s+1}{d}$ ways, we obtain that the contribution to $AW_{[k]}^{132}(x)$ for the case $s \geq 1$ is given by

$$\sum_{s \geq 1} \left(x^s + \sum_{d=1}^{s+1} \binom{s+1}{d} x^s \sum_{a_1 + a_2 + \cdots + a_d = k-2} H(x; a_1) \prod_{j=2}^{d} M(x; a_j) \right).$$

Adding the contributions from the two cases, multiplying by y^k, summing over all $k \geq 1$, and accounting for the empty word yields

$$AW_{132}(x, y) = 1 + yAW_{132}(x, y)$$
$$+ \sum_{s \geq 1} \left(\frac{x^s y}{1 - y} + \sum_{d=1}^{s+1} \binom{s+1}{d} x^s y^2 (M(x, y))^{d-1} H(x, y) \right)$$
$$= 1 + yAW_{132}(x, y) + \frac{xy}{(1 - y)(1 - x)}$$
$$+ \sum_{s \geq 1} \frac{x^s y^2 H(x, y)}{M(x, y)} \left((1 + M(x, y))^{s+1} - 1 \right)$$
$$= 1 + yAW_{132}(x, y) + \frac{xy}{1 - x} + \frac{x(1 - y)(AW_{132}(x, y) - 1)^2}{1 - \frac{x(1-y)}{y}(AW_{132}(x, y) - 1)},$$

where the last equality follows from Lemma 6.30. Letting $AW = AW_{132}(x, y)$, this equation can be reduced to

$$x(1 - x)(1 - y)AW^2 + (1 - y)(2x^2 - 2x - y)AW + x(1 - x)(1 - y) + y = 0,$$

from which we obtain the explicit formula, and if desired, a continued fraction expansion for $AW_{132}(x, y)$ (see Example 2.62). □

Note that the strength of the block decomposition method lies in the fact that all the nonempty words $w^{(j)}$ have the same structure. This is in part a result of the avoidance of the pattern 132. We already have encountered block decomposition (without calling it that) for compositions in the proof of Theorem 5.18, where it was used to derive a bijection to show Wilf-equivalence of the patterns 12–3 and 21–3. In general, the block decomposition method is more difficult to use for compositions than it is for words. The main reason is that for compositions we need to keep track of the order of the composition and therefore there is a dependence between the blocks.

The derivation of $AW_{132}(x, y)$ can be easily modified to yield results for $AW_{\{132,\tau\}}(x, y)$ as the core of the argument consists of the block decomposition in conjunction with Lemma 6.30, which carries over to a more general generating function $AW_{\{132,\tau\}}(x, y)$.

Theorem 6.31 *Let $\ell \geq 3$ and let $AW_\ell(x, y) = AW_{\{132,12\cdots\ell\}}(x, y)$. Then we have the recurrence*

$$AW_\ell(x, y) = 1 + \cfrac{y}{(1 - x)(1 - y) + \cfrac{xy(1 - x)(1 - y)}{x(1 - y) + \cfrac{y}{1 - AW_{\ell-1}(x, y)}}}$$

with $AW_2(x, y) = \frac{1-x}{1-x-y}$. Thus, $AW_\ell(x, y) - 1$ can be expressed as a continued fraction with ℓ levels

$$\cfrac{y}{(1 - x)(1 - y) - \cfrac{xy(1 - x)(1 - y)}{(1 - 2x)(1 - y) - \cfrac{xy(1 - x)(1 - y)}{(1 - 2x)(1 - y) - \cfrac{\ddots}{(1 - 2x)(1 - y) - xy}}}},$$

or in terms of Chebyshev polynomials of the second kind, as

$$AW_\ell(x, y) = 1 + \sqrt{xy^3(1 - x)(1 - y)} \frac{\alpha U_{\ell-2}(t) + \beta U_{\ell-3}(t)}{\alpha U_{\ell-3}(t) + \beta U_{\ell-4}(t)},$$

where $\alpha = (1 - 2x - y)$, $\beta = \frac{xy^2(1-x-y)}{\sqrt{xy^3(1-x)(1-y)}}$, and $t = -\frac{1-2x}{2}\sqrt{\frac{1-y}{xy(1-x)}}$.

Proof Let $H_\ell(x; k) = AW_{[k+1]}^{\{132,12\cdots\ell\}}(x) - 1$, $H_\ell(x, y) = \sum_{k \geq 0} H_\ell(x; k)y^k$, $M_\ell(x; k) = AW_{[k+1]}^{\{132,1\cdots\ell\}}(x) - AW_{[k]}^{\{132,1\cdots\ell\}}(x)$, and finally,

$$M_\ell(x, y) = \sum_{k \geq 0} M_\ell(x; k)y^k.$$

The cases $s = 0$ and $s \geq 1$ with $d = 0$ follow as in the block decomposition proof of Theorem 6.25. For $s \geq 1$ and $0 \leq d \leq s+1$, we now have to consider two cases, namely whether the block $w^{(s+1)}$ is empty or nonempty. Note that this time we define q_j to be the minimal value in $w^{(t_j)}$ for $j = 1, 2, \ldots, d-1$, which all have to occur. Note that if $w^{(s+1)}$ is nonempty then the pattern $12 \cdots \ell$ has to be avoided, while in all other blocks, the pattern $12 \cdots (\ell - 1)$ has to be avoided. Therefore,

- if $t_d \neq s+1$, then the contribution for $1 \leq d \leq s$ is given by

$$x^s \sum_{a_1 + \cdots + a_d = k-2} H_{\ell-1}(x; a_d) \prod_{j=1}^{d-1} M_{\ell-1}(x; a_j).$$

- if $t_d = s+1$, then the contribution for $1 \leq d \leq s$ is given by

$$x^s \sum_{a_1 + \cdots + a_d = k-2} H_\ell(x; a_d) \prod_{j=1}^{d-1} M_{\ell-1}(x; a_j).$$

Being careful about the number of ways the t_j can be chosen and then following the steps in the block decomposition proof of Theorem 6.25 gives the recurrence

$$AW_\ell(x, y) = 1 + \cfrac{y}{(1-x)(1-y) + \cfrac{xy(1-x)(1-y)}{x(1-y) + \cfrac{y}{1 - AW_{\ell-1}(x, y)}}} \tag{6.17}$$

(see Exercise 6.6). To obtain an explicit formula we derive $AW_2(x, y)$. This is the generating function for the number of words that simultaneously avoid the subsequence patterns 132 and 12. However, since any occurrence of 132 automatically results in an occurrence of 12, we have that $AW_2(x, y) = AW_{12}(x, y)$. From Theorem 5.5 we obtain that

$$AW_{12}(x, y) = \sum_{k \geq 0} AC_{[k]}^{12}(1, x) y^k = \sum_{k \geq 0} \prod_{j=1}^{k} \left(\frac{1}{1-x} \right) y^k = \sum_{k \geq 0} \left(\frac{y}{1-x} \right)^k.$$

This gives the result for $AW_2(x, y)$ and the finite continued fraction. For the expression involving the Chebyshev polynomials of the second kind, we write the recurrence relation as

$$f_\ell(x, y) = (1-x)(1-y) + \frac{xy(1-x)(1-y)}{x(1-y) - f_{\ell-1}},$$

where $f_\ell(x, y) = y/(AW_\ell(x, y) - 1)$. Using induction together with (C.2) as in the proof of Proposition C.8 we obtain that

$$f_\ell(x, y) = \frac{1}{\sqrt{xy^2(1-x)(1-y)}} \frac{\alpha U_{\ell-3}(t) + \beta U_{\ell-4}(t)}{\alpha U_{\ell-2}(t) + \beta U_{\ell-3}(t)}, \tag{6.18}$$

where

$$\alpha = 1 - 2x - y, \quad \beta = \frac{xy(1 - x - y)}{\sqrt{xy(1 - x)(1 - y)}}, \quad \text{and} \quad t = -\frac{1 - 2x}{2}\sqrt{\frac{1 - y}{xy(1 - x)}}.$$

Substituting $f_\ell = \frac{y}{AW_\ell(x,y)-1}$ in (6.18) and solving for $AW_\ell(x,y)$ gives the desired expression for $AW_\ell(x,y)$ in terms of Chebyshev polynomials of the second kind. $\qquad \square$

For additional examples of pattern avoidance in k-ary words of pairs of patterns of the form $\{132, \tau\}$ for several interesting cases of τ see [141]. In addition, the block decomposition approach as defined in Lemma 6.28 can be extended to describe the structure of k-ary words of length n that contain the pattern 132 exactly once. For more details see [65].

We now turn our attention to the third approach, the scanning-element algorithm. Again we will describe the approach and then use it to provide an alternative proof for Theorem 6.25, followed by examples that showcase the strength of the method. In fact we will present the proof for the pattern 123, which is Wilf-equivalent to 132 but which is hard to deal with via the block decomposition method.

6.5.3 Scanning-element algorithm

The scanning-element algorithm was first used for pattern avoidance and enumeration in permutations by Firro and Mansour [64], and then extended to study the number of k-ary words of length n that satisfy a certain set of conditions [65]. Not only does the approach work equally well for avoidance of the patterns 123 and 132, it can also be used more easily for the enumeration of patterns, especially for the pattern 123. To describe the algorithm we need the following definition.

Definition 6.32 *For $n, k \geq 0$, let $P(n, k) \subseteq [k]^n$, and let $p(n, k)$ denote the number of elements in $P(n, k)$. For $b_1, \ldots, b_m \in [k]$, let*

$$P(n, k; b_1, \ldots, b_m) = \{w_1 \cdots w_n \in P(n, k) \mid w_1 \cdots w_m = b_1 \cdots b_m\}$$

denote the set of elements in $P(n, k)$ that start with $b_1 \cdots b_m$, and let

$$p(n, k; b_1, \ldots, b_m)$$

denote the number of elements in $P(n, k; b_1, \ldots, b_m)$. We will use $P(n, k)$ and $P(n, k; \varnothing)$ interchangeably, and likewise, $p(n, k)$ and $p(n, k; \varnothing)$.

The main goal of this section is to describe how to derive a recursive structure for the family $\{P(n, k)\}_{n,k \geq 0}$. From Definition 6.32 we obtain immediately that

$$p(n, k; b_1, \ldots, b_m) = \sum_{j=1}^{k} p(n, k; b_1, \ldots, b_m, j), \tag{6.19}$$

and in particular that $p(n,k) = p(n,k;1) + \cdots + p(n,k;k)$. This suggests a way to create a recursion for the sequence $\{p(n,k)\}_{n,k}$.

Definition 6.33 *Let* $1 \leq s \leq m$, $1 \leq r \leq k$ *and* $c_1, \ldots, c_{m-s} \in [k]$. *If there exists a bijection between the sets*

$$P(n,k;b_1,\ldots,b_m) \text{ and } P(n-s,k-r;c_1,\ldots,c_{m-s}),$$

then the set $P(n-s,k-r;c_1,\ldots,c_{m-s})$ *is said to be a* reduction *of the set* $P(n,k;b_1,\ldots,b_m)$. *In this case we say that the set* $P(n,k;b_1,\ldots,b_m)$ *is* reducible (*otherwise it is* irreducible). *An element* b *is said to be an* $(m+1)$-inactive element ($(m+1)$-active element) *of the set* $P(n,k;b_1,\ldots,b_m)$ *if the set* $P(n,k;b_1,\ldots,b_m,b)$ *is reducible (irreducible).*

Example 6.34 *If* $P(n,k) = \mathcal{AW}_{[k]}^{1234}(n)$, *then for* $1 \leq j_1 < j_2 < j_3 \leq k$, *the set* $P(n-1,j_3;j_1,j_2)$ *is a reduction of the set* $P(n,k;j_1,j_2,j_3)$, *that is,* $j_3 > j_2$ *is an inactive 3-element of* $P(n,k;j_1,j_2)$ *in this case. Why? Because for any word of the form* $w = j_1 j_2 j_3 w_4 \cdots w_n$ *which avoids 1234, we know that* j_3 *has to be the maximal element in* w. *Thus, we can uniquely map* w *to the word* $w' = j_1 j_2 w_4 \cdots w_n \in [j_3]^{n-1}$ *which also avoids 1234. This process is reversible, therefore we have found the desired bijection.*

In order to set up the system of recurrence relations, we need to define some notation.

Definition 6.35 *Let* $I(n,k;b_1,\ldots,b_m)$ *denote the set of* $(m+1)$-active *elements of* $P(n,k;b_1,\ldots,b_m)$, *and* $\bar{I}(n,k;b_1,\ldots,b_m) = [k] - I(n,k;b_1,\ldots,b_m)$ *the set of* $(m+1)$-inactive *elements of* $P(n,k;b_1,\ldots,b_m)$. *In addition, let* $R(n,k;b_1,\ldots,b_m)$ *be the multi-set*

$$\left\{ (d,r,c_1,\ldots,c_{m-d}) \left| \begin{array}{l} P(n-d,k-r;c_1,\ldots,c_{m-d}) \text{ is a reduction} \\ \text{of the set } P(n,k;b_1,\ldots,b_m,j), 1 \leq d \leq m, \\ 1 \leq r \leq k, \; j \in \bar{I}(n,k;b_1,\ldots,b_m) \end{array} \right. \right\}.$$

Note the fact that $R(n,k;b_1,\ldots,b_m)$ is a multi-set because reductions of different sets may lead to the same reduced set, creating multiple entries in $R(n,k;b_1,\ldots,b_m)$.

We start with an overview of the algorithm before describing the step-by-step procedure. The first step is to determine the set of 1-inactive and 1-active elements of $P(n,k)$, that is, $\bar{I}(n,k)$ and $I(n,k)$, respectively. Using (6.19) and Definition 6.35 we have

$$p(n,k) = p(n,k;1) + \cdots + p(n,k;k)$$
$$= |\bar{I}(n,k)| \cdot p(n-1,k) + \sum_{j \in I(n,k)} p(n,k;j) \qquad (6.20)$$

for all $n \geq 1$. The first term accounts for the 1-inactive elements j, which are those that cannot be part of the pattern τ at the beginning of the word w. These values of j can be split off, resulting in a word of length $n-1$ without restriction on the alphabet for the remaining letters, and therefore we obtain a reduction to the set $P(n-1, k)$.

Equation 6.20 yields a recurrence relation for the sequence $\{p(n, k)\}_{n,k \geq 0}$ in terms of the sequences $\{p(n, k; j)\}_{n,k,j}$. We now apply the same idea to the sequence $p(n, k; b_1)$, and more generally, to $p(n, k; b_1, \ldots, b_m)$. To make the resulting set of equations more readable, we adopt the following notation.

Definition 6.36 *For any vector* $\vec{v}^s = (v_1, \ldots, v_s)$ *and scalars* $i_1, i_2, \ldots, i_\ell$,

$$(\vec{v}^s, i_1, \ldots, i_\ell) = (v_1, \ldots, v_s, i_1, \ldots, i_\ell),$$

and likewise

$$(i_1, \ldots, i_\ell, \vec{v}^s) = (i_1, \ldots, i_\ell, v_1, \ldots, v_s)$$

for $s, \ell \geq 1$.

Now we are ready to proceed. Given a set $P(n, k; \vec{b}^m)$, we first decide if it is reducible or irreducible. If it is irreducible, then we determine the $(m+1)$-inactive (active) elements of $P(n, k; \vec{b}^m)$. Using (6.19) again, we have for all $m \geq 0$ that

$$
\begin{aligned}
& p(n, k; \vec{b}^m) \\
&= \sum_{j \in \bar{I}(n,k;\vec{b}^m)} p(n, k; \vec{b}^m, j) + \sum_{j \in I(n,k;\vec{b}^m)} p(n, k; \vec{b}^m, j) \qquad (6.21) \\
&= \sum_{(d,r,\vec{c}^{(m-d)}) \in R(n,k;\vec{b}^m)} p(n-d, k-r; \vec{c}^{(m-d)}) + \sum_{j \in I(n,k;\vec{b}^m)} p(n, k; \vec{b}^m, j).
\end{aligned}
$$

Equation (6.21) suggests an algorithm for deriving finite or infinite systems of recurrence relations in terms of $p(n-d, k-r; \vec{b}^m)$ with $m, d, r \geq 0$. Let

$$q(n, k; \vec{b}^m) = \sum_{(d,r,\vec{c}^{(m-d)}) \in R(n,k;\vec{b}^m)} p(n-d, k-r; \vec{c}^{(m-d)}).$$

Then,

$$
\begin{cases}
0 : p(n, k) = |\bar{I}(n, k)| \cdot p(n-1, k) + \sum_{j_1 \in I(n,k)} p(n, k; j_1) \\[2mm]
1 : p(n, k; j_1) = q(n, k; j_1) + \sum_{j_2 \in I(n,k;j_1)} p(n, k; \vec{j}^2) \\[2mm]
2 : p(n, k; \vec{j}^2) = q(n, k; \vec{j}^2) + \sum_{j_3 \in I(n,k;\vec{j}^2)} p(n, k; \vec{j}^3) \\[2mm]
\vdots
\end{cases}
\qquad (6.22)
$$

Let us call the i-th row of (6.22) the i-th *level of* $p(n, k)$. If $I(n, k; \vec{j}^m) = \varnothing$ then there are no $(m + 1)$-active elements of the set $P(n, k; \vec{j}^m)$. Thus if $\cup_{\vec{j}^m} I(n, k; \vec{j}^m) = \varnothing$ then (6.22) contains only $m + 1$ levels and we can ask if there is an exact formula for the sequence $\{p(n, k)\}_{n \geq 0}$. To answer this question, we define for any sequence $\{p(n, k)\}_{n,k \geq 0}$ the generating functions

$$P_k(x; \vec{v}^s) = \sum_{n \geq 0} x^n p(n, k; \vec{v}^s)$$

$$:= \sum_{n \geq 0} x^n \left(\sum_{1 \leq j_1, \ldots, j_s \leq k} p(n, k; \vec{j}^s) \prod_{i=1}^{s} v_i^{j_i - 1} \right) \tag{6.23}$$

and

$$P(x, y; \vec{v}^s) = \sum_{k \geq 0} y^k P_k(x; \vec{v}^s).$$

Now we can describe the scanning-element algorithm whose input is a family of sets $\{P(n, k)\}_{n,k \geq 0}$ and whose output is an exact formula for the sequence $\{p(n, k)\}_{n,k \geq 0}$.

Scanning-element algorithm on words

1. Find a recurrence relation for the sequence $\{p(n, k)\}_{n,k \geq 0}$ as described in (6.22).

2. Decide whether the sequence $\{p(n, k)\}_{n,k \geq 0}$ results in a finite system of recurrence relations with $t + 1$ equations. If yes, continue; otherwise stop.

3. Rewrite the system of recurrence relations in the statement of (6.22) in terms of generating functions with t indeterminates $v_1, v_2, \ldots, v_t$. This is done by multiplying the recurrence relations by $\prod_{i=1}^{t} v_i^{j_i - 1}$ and summing over

$$j_1 \in I(n, k), j_2 \in I(n, k; j_1), \ldots, j_t \in I(n, k; j_1, \ldots, j_{t-1}).$$

4. Extract from Step 3 a system of functional equations in $t + 2$ variables $x, y, v_1, v_2, \ldots, v_t$.

5. Solve this system to get an expression for $P(x, y; 1, 1, \ldots, 1)$, which is a formula for the ordinary generating function $\sum_{n,k \geq 0} p(n, k) x^n y^k$, as desired.

The above algorithm suggests to refine the k-ary words according to the value(s) of the leftmost element(s), and then to apply algebraic techniques. Step 2 is the crucial one (as in the case of permutations, see [64]) because the system of equations often is not finite or has coefficients that do depend on n or $\vec{j}^s$.

We now give the third proof for Theorem 6.25, actually using $\tau = 123$ instead of $\tau = 132$. For the scanning-element algorithm, either pattern is equally easy. Note that we derive the generating function of the 123-avoiding words according to their first letter, that is, we are obtaining a refined result as a by-product of the proof.

Proof (of Theorem 6.25 using scanning-element algorithm) Define

$$P(n,k) = \mathcal{AW}_n^{[k]}(123)$$

and let $w = w_1 w_2 \cdots w_n \in \mathcal{AW}_n^{[k]}(123)$. If w starts with k or $k-1$, then w_1 cannot be part of any pattern 123, so the set $P(n-1,k)$ is a reduction of the sets $P(n,k;k)$ and $P(n,k;k-1)$. If $w_1 = j \le k-2$ then $j \in I(n,k)$. Now we look at the first two letters of w. If $w_1 w_2 = ij$ with $i \ge j$ then w avoids 123 if and only if $j w_3 \cdots w_n \in \mathcal{AW}_{n-1}^{[k]}(123)$, so the set $P(n-1,k;j)$ is a reduction of the set $P(n,k;i,j)$ for all $i \ge j$. If $i < j$ then the only letters that can follow are from the alphabet $[j]$, so we can map $ij w_3 \cdots w_n \in \mathcal{AW}_n^{[k]}(123)$ to $i w_3 \cdots w_n \in \mathcal{AW}_{n-1}^{[j]}(123)$. Thus the set $P(n-1,j;i)$ is a reduction of the set $P(n,k;i,j)$ for all $i < j$. Putting these all together, (6.19) leads to the following equations:

$$p(n,k;k) = p(n,k;k-1) = p(n-1,k) \tag{6.24}$$

$$p(n,k;i) = \sum_{j=1}^{i} p(n,k;i,j) + \sum_{j=i+1}^{k} p(n,k;i,j)$$

$$= \sum_{j=1}^{i} p(n-1,k;j) + \sum_{j=i+1}^{k} p(n-1,j;i), \tag{6.25}$$

for all $1 \le i \le k-2$. Multiplying (6.25) by v^{i-1} and summing over all $1 \le i \le k-1$, we obtain a recurrence in terms of $p(n,k;v)$:

$$p(n,k;v) - p(n,k;k)v^{k-1}$$
$$= \sum_{j=1}^{k-1} \frac{v^{j-1} - v^{k-1}}{1-v} p(n-1,k;j) + \sum_{j=2}^{k} (p(n-1,j;v) - p(n-1,j;j)v^{j-1}).$$

Therefore

$$p(n,k;v) - p(n-1,k;1)v^{k-1}$$
$$= \sum_{j=1}^{k-1} \frac{v^{j-1} - v^{k-1}}{1-v} p(n-1,k;j) + \sum_{j=2}^{k} (p(n-1,j;v) - p(n-1,j;1)v^{j-1})$$
$$= \frac{1}{1-v} \left(p(n-1,k;v) - v^{k-1} p(n-1,k;1) \right)$$
$$+ \sum_{j=2}^{k} \left(p(n-1,j;v) - p(n-2,j;1)v^{j-1} \right),$$

which implies that

$$p(n,k;v) = \frac{1}{1-v}\left(p(n-1,k;v) - v^k p(n-1,k;1)\right)$$

$$+ \sum_{j=2}^{k}\left(p(n-1,j;v) - p(n-2,j;1)v^{j-1}\right) \qquad (6.26)$$

for all $n \geq 2$. This leaves the cases $n = 0$ and $n = 1$. For $n = 0$, we obtain the generating function $P(0,k;v) = 1$ directly from the definition. For $n = 1$, we use that $p(1,k;j) = 1$ by definition and thus

$$p(1,k;v) = 1 + v + \cdots + v^{k-1} = \frac{1-v^k}{1-v}.$$

Multiplying (6.26) by x^n and summing over $n \geq 2$, we arrive at

$$P_k(x;v) - \frac{1-v^k}{1-v}x - 1 = \frac{x}{1-v}\left((P_k(x;v)-1) - v^k(P_k(x;1)-1)\right)$$

$$+ x\sum_{j=2}^{k}(P_j(x;v)-1) - x^2\sum_{j=2}^{k}v^{j-1}P_j(x;1),$$

which is equivalent to

$$P_k(x;v) = 1 - (k-1)x + \frac{x}{1-v}\left(P_k(x;v) - v^k P_k(x;1)\right)$$

$$+ x\sum_{j=2}^{k}P_j(x;v) - x^2\sum_{j=2}^{k}v^{j-1}P_j(x;1) \qquad (6.27)$$

for all $k \geq 2$. From the definitions we have that

$$P_0(x;v) = 1 \text{ and } P_1(x;v) = \frac{1}{1-x}.$$

Multiplying (6.27) by $\frac{y^k}{v^k}$ (to get rid of the factor v^k of $P_k(x;1)$, the function we want), summing over $k \geq 2$ and simplifying, we obtain that

$$P\left(x, \frac{y}{v};v\right) = \frac{(v+x^2)(v-y) - v^2 x}{(v-y)^2} + \frac{x}{1-v}\left(P\left(x,\frac{y}{v};v\right) - P(x,y;1)\right)$$

$$+ \frac{x}{v-y}\left(vP\left(x,\frac{y}{v};v\right) - xP(x,y;1)\right),$$

or equivalently

$$\left(1 - \frac{x}{1-v} - \frac{xv}{v-y}\right)P\left(x,\frac{y}{v};v\right) = \frac{(v+x^2)(v-y) - v^2 x}{(v-y)^2} \qquad (6.28)$$

$$- \left(\frac{x}{1-v} + \frac{x^2}{v-y}\right)P(x,y;1).$$

Equation (6.28) can be solved using the *kernel method* as described in [13]. This method consists of setting the coefficient of the function that we would like to "get rid of" to zero, so that the remainder of the equation can be solved for the function of interest. If substituting the solution so obtained into the remainder of the equation results in a solution that can be expanded into a Taylor series (as opposed to a Laurent series), then this is indeed a solution and it is unique. So let's get started. We want to eliminate $P(x, y/v; v)$ and then solve for $P(x, y; 1) = AW_{123}(x, y)$. Setting $1 - \frac{x}{1-v} - \frac{xv}{v-y} = 0$ results in

$$v_{+,-} = \frac{1 + y - 2x \pm \sqrt{(1-y)((1-2x)^2 - y)}}{2(1-x)}.$$

Now we substitute both solutions into the equation for $P(x, y; 1)$ and compute $\lim_{x \to 0} P(x, y; 1)$ for $v = v_+$ and $v = v_-$, respectively. The solution for which the limit equals 1 is the correct one. In our case this happens for $v = v_+$. Thus

$$AW_{123}(x, y) = P(x, y; 1) = \frac{(v_+ + x^2)(v_+ - y) - v_+^2 x}{x(v_+ - y)^2 \left(\frac{1}{1 - v_+} + \frac{x}{v_+ - y} \right)}.$$

Simplifying the resulting expression by rationalizing the denominator we obtain that

$$AW_{123}(x, y) = 1 + \frac{1}{1-x} y + \frac{y}{2x(1-x)} \left(1 - 2x - \sqrt{\frac{(1-2x)^2 - y}{1-y}} \right),$$

the result of Theorem 6.25. $\qquad \square$

Having computed an explicit expression for $P(x, y; 1)$, we can obtain an explicit expression for $P(x, y; v)$ from (6.28) by keeping v as a parameter instead of substituting the computed solution v_+. This result allows us to enumerate the words that avoid 123 according to the first letter, a result that is a by-product of the scanning-element algorithm and that is much harder to obtain from the other two methods, if at all.

6.5.4   Containing 123 or 132 **exactly once**

We conclude the discussion of permutation patterns of length three with results on the number of k-ary words of length n that contain the patterns 123 and 132, respectively, exactly once. Keep in mind that Wilf-equivalence applies to the avoidance of patterns. When counting according to the occurrence of patterns, the story is quite different. Not only are the generating functions different, but also the methods used in their derivation differ. Both results were proved by Firro and Mansour in [65]. They derived the result for the pattern 123 using the scanning-element algorithm and the result for the pattern 132 using the block decomposition approach. In each case, the other method is either difficult to use or cannot be applied. This once more illustrates the necessity of having several methods available in one's "toolbox." We will state the two results without proof.

Theorem 6.37 *The generating function for the number of k-ary words of length n that contain the pattern* 123 *exactly once is given by*

$$\frac{y\left[2x^3(x-1) + x^2(9 - 10x + 2x^2)(1 - y) - x(3 - 2x)(1 - y)^2 + (1 - y)^3\right]}{2x^3(1 - x)^2(1 - y)}$$

$$-\frac{y(1 - x - y)((1 - x)(1 - 2x) - y)}{2x^3(1 - x)^2}\sqrt{\frac{(1 - 2x)^2 - y}{1 - y}}.$$

Moreover, for $k \geq 3$ *and* $n \geq 0$*, the number of k-ary words of length n that contain the pattern* 123 *exactly once is given by*

$$h_{n+2,k} - 3h_{n+1,k} + 2h_{n,k} - 2(h_{n+2,k-1} - h_{n+1,k-1}) - 1 + \sum_{j=0}^{n+2} h_{n+2-j,k-2},$$

where

$$h_{n,k} = \sum_{j=0}^{n} \frac{(-1)^{n-j}}{1 + j}\binom{j}{n - j}\binom{k - 1 + j}{j}\binom{2j}{j}.$$

Theorem 6.38 *The generating function for the number of k-ary words of length n that contain the pattern* 132 *exactly once is given by*

$$\frac{y^2\left[x + y - 1 - ((1 - 2x)(1 - x) - y)\left(\frac{(1 - 2x)^2 - y}{1 - y}\right)^{-\frac{1}{2}}\right]}{2(1 - x)^2(1 - y)^2}.$$

6.6 Generalized patterns of type (2,1)

We start once more by classifying the patterns according to Wilf-equivalence. When considering avoidance of permutation patterns of type $(2, 1)$ in compositions, there were three equivalence classes, namely 12–3 $\sim$ 21–3, 23–1 $\sim$ 32–1, and 13–2 $\sim$ 31–2 (see Theorems 5.18 and 5.19). Using the complement map, the three classes reduce to two classes and we obtain the following results as a special case of compositions (see Theorem 5.21). For the pattern 13–2 we only obtain a recursion from Lemma 5.25 and there is no explicit formula known for this generating function.

Proposition 6.39 *The generating functions for the number of words avoiding the patterns 12–3 and 13–2, respectively, are given by*

$$AW_{[k]}^{12-3}(x) = \prod_{i=1}^{k} \frac{1}{1 - \frac{x}{(1-x)^{i-1}}}$$

and

$$AW_{[k]}^{13-2}(x) = 1 + \sum_{i=1}^{k} AW_{[k]}^{13-2}(i|x),$$

where

$$AW_{[k]}^{13-2}(i|x) = x \left(1 + \sum_{j=1}^{i+1} AW_{[k]}^{13-2}(j|x) + \sum_{j=i+2}^{k} AW_{[i+k-j+1]}^{13-2}(j|x) \right)$$

with $AW_{[k]}^{13-2}(k|x) = AW_{[k]}^{13-2}(k-1|x) = xAW_{[k]}^{13-2}(x).$

Note that $AW_{[1]}^{12-3}(x) = AW_{[1]}^{13-2}(x) = \frac{1}{1-x}$ and $AW_{[2]}^{12-3}(x) = AW_{[2]}^{13-2}(x) = \frac{1}{1-2x}$. Burstein and Mansour [33] derived the generating functions $AW_{[k]}^{13-2}(i|x)$, $i = 1, 2, 3, 4$, using the recursion given in Proposition 6.39.

Theorem 6.40 *The generating functions for the number of words that start with $i = 1, 2, 3, 4$ and avoid $\tau = 13$–2 are given by*

$$AW_{[k]}^{13-2}(1|x) = \frac{1}{1-x},$$

$$AW_{[k]}^{13-2}(2|x) = \frac{1}{1-2x},$$

$$AW_{[k]}^{13-2}(3|x) = \frac{(1-x)^2}{(1-2x)(1-3x+x^2)},$$

$$AW_{[k]}^{13-2}(4|x) = \frac{1-4x+6x^2-3x^3}{(1-2x)(1-3x)(1-3x+x^2)}.$$

We now consider patterns of type $(2,1)$ with repeated letters. In the case of compositions we had a total of five equivalence classes, namely 11–1, 11–2, 22–1, 12–2 ~ 21–2 and 21–1 ~ 12–1. In Theorem 5.38 we gave a combinatorial argument for the equivalence of the patterns 12–2 and 21–2. Burstein and Mansour [34] showed the same equivalence in the case of words using an analytical proof.

The complement map reduces the five equivalence classes to just three in the case of words, namely 11–1, 11–2 ~ 22–1, and the class consisting of the remaining patterns. We start by giving the generating function for 11–1, a result given originally by Burstein and Mansour in [34].

Theorem 6.41 *For all $k \geq 1$,*

$$AW_{[k]}^{11-1}(x) = \det \begin{vmatrix} B_k(x) & -A_k(x) & 0 & \cdots & 0 \\ B_{k-1}(x) & 1 & -A_{k-1}(x) & \cdots & 0 \\ B_{k-2}(x) & 0 & 1 & \cdots & 0 \\ \vdots & \vdots & \vdots & \ddots & \vdots \\ B_2(x) & 0 & 0 & \ddots & -A_2(x) \\ B_1(x) + A_1(x) & 0 & 0 & \cdots & 1 \end{vmatrix}$$

$$= \prod_{i=1}^{k} A_i(x) + \sum_{j=1}^{k} \left(B_j(x) \prod_{i=j+1}^{k} A_i(x) \right)$$

where $A_j(x) = \frac{jx^2}{1-(j-1)x}$ and $B_j(x) = \frac{1+x}{1-(j-1)x}$.

Note that the proof given in [34] proceeds from the recurrence relation that is obtained after appropriate simplification when setting $x = 1$ and $y = x$ in (5.6) of Theorem 5.39. The result follows by induction.

For the pattern 11–2, we obtain a nice explicit result from Theorem 5.41. This theorem is also an example of the phenomenon that complementary patterns have structurally similar generating functions for compositions, which coincide when specialized to the case of words.

Proposition 6.42 *The generating function for k-ary words of length n that avoid 11–2 is given by*

$$AW_{[k]}^{11-2}(x) = \prod_{j=0}^{k-1} \frac{1 - (j-1)x}{1 - (j+x)x}.$$

Finally, we present the result given in [33] for the pattern 21–1.

Theorem 6.43 *For $k \geq 0$,*

$$AW_{[k]}^{21-1}(x) = 1 + \sum_{d=0}^{k-1} \left(x^{d+1} AW_{[k-d]}^{21-1}(x) \sum_{i=d}^{k-1} (1-x)^{i-d} \binom{i}{d} \right).$$

Proof Let $AW_{[k]}^{21-1}(i_1, \ldots, i_d | n)$ be the number of words w in $AW_n^{[k]}(21\text{-}1)$ that start with $i_1 \cdots i_d$. Now consider $AW_{[k]}^{21-1}(i, j | n)$. If $j \geq i$ then the letter i cannot be part of any 21-1 in a word w beginning with ij and therefore places no further restrictions on the rest of w, so

$$AW_{[k]}^{21-1}(i, j | n) = AW_{[k]}^{21-1}(j | n - 1).$$

If $j < i$ then deleting the first i from a word w starting with ij results in a word $w' \in AW_{n-1}^{[k]}(21\text{-}1)$ which contains exactly one j, and j is the first letter.

Now let $w'' \in \mathcal{AW}_{n-1}^{[k-1]}(21\text{-}1)$ be the word obtained from w' by subtracting one from each letter that is bigger than j. Obviously, this mapping is a bijection onto the set of words in $\mathcal{AW}_{n-1}^{[k-1]}(21\text{-}1)$ starting with j (since neither it, nor the inverse mapping, which consists of adding one to each letter that is at least j and it is not the first letter j creates any new occurrences of 21-1). Thus

$$AW_{[k]}^{21-1}(i, j|n) = AW_{[k-1]}^{21-1}(j|n-1).$$

Using that

$$AW_{[k]}^{21-1}(i|n) = \sum_{j=1}^{k} AW_{[k]}^{21-1}(i, j|n)$$

we obtain that

$$
\begin{aligned}
AW_{[k]}^{21-1}&(i|n)\\
&= \sum_{j=1}^{i-1} AW_{[k-1]}^{21-1}(j|n-1) + \sum_{j=i}^{k} AW_{[k]}^{21-1}(j|n-1)\\
&= AW_{[k]}^{21-1}(n-1) + \sum_{j=1}^{i-1} (AW_{[k-1]}^{21-1}(j|n-1) - AW_{[k]}^{21-1}(j|n-1)).
\end{aligned}
\tag{6.29}
$$

Using (6.29) we can prove by induction on d that

$$
\begin{aligned}
AW_{[k]}^{21-1}&(d|n)\\
&= \sum_{j=0}^{d-1} \binom{d-1}{j} \sum_{i=0}^{d-1-j} \binom{d-1-j}{i} (-1)^i AW_{[k-j]}^{21-1}(n-1-i-j).
\end{aligned}
$$

Hence for all $n \geq 1$,

$$
\begin{aligned}
AW_{[k]}^{21-1}&(n)\\
&= \sum_{d=1}^{k} \sum_{j=0}^{d-1} \binom{d-1}{j} \sum_{i=0}^{d-1-j} \binom{d-1-j}{i} (-1)^i AW_{[k-j]}^{21-1}(n-1-i-j)
\end{aligned}
$$

with $AW_{[k]}^{21-1}(0) = 1$. Multiplying by x^n and summing over $n \geq 1$, yields

$$AW_{[k]}^{21-1}(x) - 1 = \sum_{d=1}^{k} \sum_{j=0}^{d-1} \binom{d-1}{j} x^{j+1} (1-x)^{d-1-j} AW_{[k-j]}^{21-1}(x),$$

and therefore

$$AW_{[k]}^{21-1}(x) - 1 = \sum_{d=0}^{k-1} \left(x^{d+1} AW_{[k-d]}^{21-1}(x) \sum_{i=d}^{k-1} (1-x)^{i-d} \binom{i}{d} \right),$$

as claimed. $\square$

We conclude the chapter on pattern avoidance in words with the most general type of pattern, namely partially ordered patterns.

6.7 Avoidance of partially ordered patterns

Like the other patterns, POPs were originally studied on the set of k-ary words [109], and then studied on the set of compositions [88]. We will present a sample of results that can be recovered by substituting $A = [k]$, $x = 1$ and $y = x$ into the results of Section 5.4.2 to recover the results originally given by Kitaev and Mansour [109].

Example 6.44 *From Example 5.62, we obtain a simple formula for the generating function for the number of k-ary words that avoid $1'$-2-$1''$, namely*

$$AW_{[k]}^{1'-2-1''}(x) = \frac{1}{(1-x)^{2k-1}} - \sum_{j=1}^{k-1} \frac{x}{(1-x)^{2j}}.$$

The main results for avoidance of shuffle patterns and multi-patterns in words follow from Theorems 5.63 and 5.73.

Example 6.45 *Let ϕ be the shuffle pattern τ-ℓ-ν. Then for all $k \geq \ell$,*

$$AW_{[k]}^{\phi}(x) = \frac{AW_{[k-1]}^{\phi}(x) - xAW_{[k-1]}^{\tau}(x)AW_{[k-1]}^{\nu}(x)}{(1 - xAW_{[k-1]}^{\tau}(x))(1 - xAW_{[k-1]}^{\nu}(x))}.$$

Example 6.46 *Let $\tau = \tau_1$-τ_2-$\cdots$-τ_s be a multi-pattern. Then*

$$AW_{[k]}^{\tau}(x) = \sum_{j=1}^{s} AW_{[k]}^{\tau_j}(x) \prod_{i=1}^{j-1} \left((kx-1)AW_{[k]}^{\tau_i}(x) + 1 \right).$$

Finally, we state the result on the maximum number of nonoverlapping patterns obtained from Theorem 5.80.

Example 6.47 *Let τ be any subword pattern. Then for all $k \geq 1$,*

$$\sum_{n \geq 0} \sum_{w \in [k]^n} q^{\tau-nlap(w)} x^n = \frac{AW_{[k]}^{\tau}(x)}{1 - q\left[(kx-1)AW_{[k]}^{\tau}(x) + 1 \right]}.$$

Using Table 6.1 and setting $q = 0$ in $W_{[k]}^{\tau}(x, q)$, we can easily obtain $AW_{[k]}^{\tau}(x)$ for any 3-letter subword pattern, and therefore a result on the maximum number of nonoverlapping 3-letter subword patterns. Since the (subsequence) pattern 123 has played so prominently in the previous section, we will use the (subword) pattern 123 as an example (given originally by Kitaev and Mansour [109]).

Example 6.48 *The distribution of the maximum number of nonoverlapping occurrences of the subword pattern* 123 *is given by*

$$\sum_{n\geq 0}\sum_{w\in[k]^n} q^{123-nlap(w)} x^n = \frac{1}{\sum_{j=0}^{k} a_j \binom{k}{j} x^j + q\left(1 - kx - \sum_{j=0}^{k} a_j \binom{k}{j} x^j\right)},$$

where $a_{3m} = 1$, $a_{3m+1} = -1$ *and* $a_{3m+2} = 0$, *for all* $m \geq 0$.

6.8   Exercises

Exercise 6.1 *Use the formula given in Table 6.1 to compute the sequence of values for the number of 5-ary words of length n that avoid the subword pattern* 123 *for* $n = 1, 2, \ldots, 10$.

Exercise 6.2 *Let* $w = w_1 \cdots w_n$ *be any word on the alphabet* $[k]$. *We say that* w_i *is a left-right-maximum in* w *if* $w_j < w_i$ *for all* $j = 1, 2, \ldots, i-1$.

(1) *Find an explicit formula for the generating function for the number of k-ary words of length n on the alphabet $[k]$ according to the number of left-right-maxima. (For more details see [43, 158].)*

(2) *Extend Part (1) to the case of compositions of n with m parts in $[d]$ according to the number of left-right-maxima.*

Exercise 6.3 *Show that the generating function for $H_1(n, k)$ is given by*

$$H_1(x, y) = \frac{y^2}{1 - 2x - (1-x)y},$$

where $H_1(n, k) = \sum_{m=0}^{n} \binom{n}{m}\binom{m+k-2}{m}$.

Exercise 6.4 *Show that the generating function $H_I(x, y)$ satisfies the recurrence relation given in (6.13), namely,*

$$(1 - 2x - (1-x)y)H_{I+1}(x, y) - (1 - x - y)H_I(x, y) = (1-x)u_I(x)(y-1)y^{I+1}$$

for all $I \geq 1$.

Exercise 6.5 *Give a proof of Lemma 6.30.*

Exercise 6.6 *Fill in the missing details of the proof of Theorem 6.31.*

Exercise 6.7 *Prove that the polynomials $U_j(q)$ in Table 6.1 given for the pattern 123 satisfy*

$$U_n(0) = \frac{1}{2}\left((-1)^{\lfloor n/3 \rfloor} + (-1)^{\lfloor (n+1)/3 \rfloor}\right).$$

Exercise 6.8 *Use the scanning-element algorithm to obtain that the generating function for words avoiding the pair of subsequence patterns $\{132, 122\}$ is given by*

$$AW_{\{132,122\}}(x,y) = \frac{(1-x)(1-y) - \sqrt{(1+y-x+xy)^2 - 4y}}{2x(1-y)}.$$

Exercise 6.9* *Prove that the generating function for the number of k-ary smooth words of length n (see Exercise 2.17) is given by*

$$1 + \frac{x(k - (3k+2)x)}{(1-3x)^2} + \frac{2x^2}{(1-3x)^2} \frac{1 + U_{k-1}\left(\frac{1-x}{2x}\right)}{U_k\left(\frac{1-x}{2x}\right)},$$

where U_s is the s-th Chebyshev polynomial of the second kind (see Definition C.1). Then show that the number of k-ary smooth words of length n is given by

$$\frac{1}{k+1} \sum_{j=1}^{k} (1 + (-1)^{j+1}) \cot^2\left(\frac{j\pi}{2(k+1)}\right)\left(1 + 2\cos\left(\frac{j\pi}{k+1}\right)\right)^{n-1}.$$

Exercise 6.10* *A k-ary word $w = w_1 w_2 \cdots w_n$ is said to be smooth cyclic if the word $w_1 w_2 \cdots w_n w_1$ is smooth. Prove that generating function for the number of k-ary smooth cyclic words of length n is given by*

$$1 + \frac{kx(1+3x)}{(1+x)(1-3x)} - \frac{2(k+1)x}{(1+x)(1-3x)} \frac{U_{k-1}\left(\frac{1-x}{2x}\right)}{U_k\left(\frac{1-x}{2x}\right)},$$

where U_s is the s-th Chebyshev polynomial of the second kind (see Definition C.1). Then show that the number of k-ary smooth words of length n is given by

$$\sum_{j=1}^{k} \left[1 + 2\cos\left(\frac{j\pi}{k+1}\right)\right]^n.$$

Exercise 6.11 *A (set) partition of $[n]$ is a collection $B_1, \ldots, B_d$ of nonempty disjoint sets, called blocks, whose union is the set $[n]$. We will assume that $B_1, B_2, \ldots, B_d$ are listed in increasing order of their minimal elements. To each partition we associate its canonical sequence, which is an integer sequence $w = w_1 w_2 \cdots w_n$ such that $w_i = k$ if and only if $i \in B_k$. For instance,* *the partition of* [7] *into the four blocks* $\{1, 4\}$, $\{2, 5, 7\}$, $\{3\}$, *and* $\{6\}$ *has associated canonical sequence* 1231242. *We will refer to the canonical sequence of a set partition of n as a* partition word *of length n.*

Note that a word w on $[d]$ represents a set partition with d blocks if and only if it has the following properties:

(1) *Each letter from the set* $[d]$ *appears at least once in* w.

(2) *For each* i, j *such that* $1 \leq i < j \leq d$, *the first occurrence of* i *precedes the first occurrence of* j.

We remark that sequences satisfying these properties are also known as restricted growth functions and they are often encountered in the study of set partitions and related topics (see [175, 190] and references therein). Let P_n *denote the set of partition words of length n. Then* $P_1 = \{1\}$, $P_2 = \{11, 12\}$ *and* $P_3 = \{111, 112, 121, 122, 123\}$. *Prove that the exponential generating function for the number of partition words of length n is given by* $e^{e^x - 1}$.

Exercise 6.12* *Using the definitions of Exercise 6.11, determine*

(1) *the generating functions for the number of partition words of size n that avoid a single pattern from the set* $P_3 = \{111, 112, 121, 122, 123\}$, *and*

(2) *the generating functions for the number of partition words of size n that avoid a single pattern from the set* $P_4 = \{1111, 1112, 1121, 1122, 1123,$ $1211, 1212, 1213, 1221, 1222, 1223, 1231, 1232, 1233, 1234\}$.

Note that a complete classification with regard to Wilf-equivalence for avoidance of partition word patterns in partition words has been given up to $n = 7$ *in [104].*

Exercise 6.13* *Combining the definitions in Exercises 6.9 and 6.11, prove that the generating function for the number of smooth partition words of* $[n]$ *is given by*

$$1 + \frac{x}{1 - 3x} \left(1 - \sum_{k \geq 1} \frac{1}{U_k \left(\frac{1-x}{2x} \right) U_{k-1} \left(\frac{1-x}{2x} \right)} \right),$$

where U_s *is the s-th Chebyshev polynomial of the second kind. Then, show that the number of smooth partition words of* $[n]$ *with k blocks is given by*

$$\frac{2}{k+1} \sum_{j=1}^{k} (1 + (-1)^{j+1}) \cos^2 \left(\frac{j\pi}{2(k+1)} \right) \left(1 + 2\cos \left(\frac{j\pi}{k+1} \right) \right)^{n-1}$$

$$- \frac{2}{k} \sum_{j=1}^{k-1} (1 + (-1)^{j+1}) \cos^2 \left(\frac{j\pi}{2k} \right) \left(1 + 2\cos \left(\frac{j\pi}{k} \right) \right)^{n-1}.$$

6.9   Research directions and open problems

We now suggest several research directions, which are motivated both by the results and exercises of this chapter. In later chapters we will revisit some of these research directions and pose additional questions.

Research Direction 6.1 *Let r_k be the number of Wilf-equivalence classes for pattern avoidance of k-letter subsequence patterns in words. Clearly, $t_1 = 1$ and $t_2 = 2$, and Tables 6.2 through 6.5 show that $r_3 = 3$, $r_4 = 4$, $r_5 = 12$ and $r_6 = 71$. Any nontrivial information on r_n is of interest!*

Research Direction 6.2 *We have obtained explicit formulas for the generating functions for the number of k-ary words of length n that avoid a subsequence pattern of length three (see Table 6.2 for the Wilf-equivalence classes and Theorems 6.21, 6.23 and 6.25 for the generating functions). A natural extension is to find explicit formulas for the generating functions for the number of k-ary words of length n that avoid a subsequence pattern of length four. Table 6.3 shows that there are four Wilf-equivalence classes, represented by the patterns 1223, 1234, 1123 and 1112. Burstein and Mansour [33] have already derived the generating function for the number of k-ary words of length n that avoid the subsequence pattern 1112, but finding the generating functions for the other three classes remains an open problem.*

Research Direction 6.3 *Table 6.3 gives the full classification of Wilf-equivalence classes for subsequence patterns (that is, patterns of type $(1,1,1,1)$) for k-ary words. This leads to the larger question of classifying the Wilf-equivalence classes for the remaining (generalized) patterns of length four, namely subword patterns of length four and generalized patterns of type $(3,1)$, $(2,2)$, $(1,2,1)$, and $(2,1,1)$. Using a computer program, we have obtained the sequences of the number of 5-ary words of length n for $n = 1, \ldots 9$ for the various patterns of length four. These sequences are listed in Tables 6.6 through 6.10 at the end of this section.*

(1) *Table 6.6 conjectures that for subword patterns*

- $1221 \sim 1121$,
- $1223 \sim 1213$,
- $1423 \sim 1243$,
- $1432 \sim 1342$,
- $2143 \sim 1324$,
- $1132 \sim 1232 \sim 1322 \sim 1332$.

To obtain the complete classification for the Wilf-equivalence classes for subword patterns of length four in k-ary words, prove or disprove each of the above Wilf-equivalences. That is, either compute additional values to show that the equivalence is not true, or give a proof that the patterns are indeed equivalent.

(2) *Table 6.7 conjectures that for type* $(3,1)$ *patterns*

- $132\text{--}2 \sim 132\text{--}1$,
- $213\text{--}2 \sim 213\text{--}1$,
- $133\text{--}2 \sim 113\text{--}2$,
- $123\text{--}4 \sim 234\text{--}1$,
- $112\text{--}1 \sim 122\text{--}1 \sim 122\text{--}2$,
- $123\text{--}1 \sim 123\text{--}2 \sim 123\text{--}3$,
- $121\text{--}3 \sim 212\text{--}3 \sim 131\text{--}2$,
- $122\text{--}3 \sim 112\text{--}3 \sim 211\text{--}3 \sim 221\text{--}3$,
- $124\text{--}3 \sim 134\text{--}2 \sim 214\text{--}3 \sim 143\text{--}2$,
- $213\text{--}4 \sim 132\text{--}4 \sim 142\text{--}3 \sim 231\text{--}4 \sim 241\text{--}3 \sim 243\text{--}1$.

To obtain the complete classification for the Wilf-equivalence classes for patterns of type $(3,1)$ *in k-ary words, prove or disprove each of the above Wilf-equivalences.*

(3) *Table 6.8 conjectures that for type* $(2,2)$ *patterns*

- $12\text{--}13 \sim 13\text{--}12$,
- $12\text{--}23 \sim 12\text{--}32 \sim 21\text{--}32$,
- $11\text{--}21 \sim 11\text{--}12$,
- $11\text{--}23 \sim 11\text{--}32$,
- $13\text{--}24 \sim 24\text{--}13$,
- $12\text{--}43 \sim 21\text{--}43 \sim 12\text{--}34$.

To obtain the complete classification for the Wilf-equivalence classes of patterns of type $(2,2)$ *in k-ary words, prove or disprove each of the above Wilf-equivalences.*

(4) *Table 6.9 conjectures that for type* $(1,2,1)$ *patterns*

- $1\text{--}12\text{--}2 \sim 1\text{--}21\text{--}2$,
- $1\text{--}31\text{--}2 \sim 1\text{--}13\text{--}2$,
- $1\text{--}33\text{--}2 \sim 1\text{--}22\text{--}3$,
- $2\text{--}14\text{--}3 \sim 2\text{--}41\text{--}3$,

    – 1–42–3 ∼ 1–24–3,

    – 1–23–2 ∼ 1–21–3 ∼ 1–32–2 ∼ 1–12–3,

    – 1–43–2 ∼ 1–23–4 ∼ 1–34–2 ∼ 1–32–4.

To obtain the complete classification for the Wilf-equivalence classes of patterns of type $(1, 2, 1)$ in k-ary words, prove or disprove each of the above Wilf-equivalences.

(5) *Table 6.10 conjectures that for type $(2, 1, 1)$ patterns*

    – 12–1–2 ∼ 12–2–1,

    – 11–1–2 ∼ 11–2–1,

    – 21–3–1 ∼ 21–1–3,

    – 11–3–2 ∼ 11–2–3,

    – 12–3–1 ∼ 12–1–3,

    – 13–3–2 ∼ 13–2–3,

    – 12–1–1 ∼ 12–2–2,

    – 21–3–3 ∼ 12–3–3,

    – 23–1–4 ∼ 23–4–1,

    – 13–2–4 ∼ 13–4–2,

    – 21–2–3 ∼ 12–3–2 ∼ 21–3–2 ∼ 12–2–3,

    – 12–4–3 ∼ 21–3–4 ∼ 21–4–3 ∼ 12–3–4.

To obtain the complete classification for the Wilf-equivalence classes of patterns of type $(2, 1, 1)$ in k-ary words, prove or disprove each of the above Wilf-equivalences.

Research Direction 6.4 *In Exercise 6.9 we obtained explicit formulas for the number of smooth k-ary words of length n. More generally, a word $w = w_1 \cdots w_n$ is said to be ℓ-smooth if $|w_i - w_{i+1}| \leq \ell$, for all $i = 1, 2, \ldots, n-1$. Clearly, a 1-smooth word is a smooth word. Find an explicit formula for the number of ℓ-smooth k-ary words of length n.*

Research Direction 6.5 *In Exercise 6.10 we obtained explicit formulas for the number of smooth cyclic k-ary words of length n. More generally, a word $w = w_1 \cdots w_n$ is said to be ℓ-smooth cyclic if $w_1 \cdots w_n w_1$ is an ℓ-smooth word. Clearly, a 1-smooth cyclic word is a smooth cyclic word. Find an explicit formula for the number of ℓ-smooth cyclic k-ary words of length n.*

TABLE 6.6: $\{AW_5^\tau(n)_{n=1}^9\}$ for 4-letter subword patterns τ

τ	$\{AW_5^\tau(n)\}_{n=1}^9$ for 4-letter subword patterns τ				
1123	5	25	125	615	3025
	14875	73126	359485	1767200	
1112	5	25	125	615	3025
	14875	73135	359575	1767875	
1132, 1232	5	25	125	615	3025
1322, 1332	14875	73138	359605	1768100	
1221, 1121	5	25	125	615	3025
	14875	73155	359775	1769375	
1223, 1213	5	25	125	615	3025
	14880	73176	359859	1769690	
1231	5	25	125	615	3025
	14875	73171	359935	1770575	
1312	5	25	125	615	3025
	14880	73188	359979	1770576	
2132	5	25	125	615	3025
	14880	73209	360189	1772131	
1212	5	25	125	615	3025
	14885	73235	360315	1772755	
1122	5	25	125	615	3025
	14885	73235	360320	1772815	
1423, 1243	5	25	125	620	3075
	15250	75625	375025	1859750	
1432, 1342	5	25	125	620	3075
	15250	75628	375055	1859975	
2143, 1324	5	25	125	620	3075
	15251	75635	375100	1860250	
2413	5	25	125	620	3075
	15251	75639	375140	1860546	
1234	5	25	125	620	3076
	15260	75700	375525	1862865	
1111	5	25	125	620	3080
	15300	76000	377520	1875280	

TABLE 6.7: $\{AW_5^\tau(n)_{n=1}^9\}$ for 4-letter patterns τ of type $(3, 1)$

τ	$\{AW_5^\tau(n)\}_{n=1}^9$ for 4-letter patterns τ of type $(3, 1)$				
112–1, 122–1	5	25	125	615	2985
122–2	14315	67945	319695	1493355	
213–3	5	25	125	615	2985
	14315	67971	320117	1497634	
132–2, 132–1	5	25	125	615	2985
	14315	67971	320117	1497638	

τ	$\{AW_5^\tau(n)\}_{n=1}^9$ for 4-letter patterns τ of type $(3,1)$				
112–2	5	25	125	615	2985
	14325	68065	320660	1499795	
213–2, 213–1	5	25	125	615	2985
	14320	68031	320601	1500861	
132–3	5	25	125	615	2985
	14325	68091	321081	1504044	
121–2	5	25	125	615	2985
	14325	68105	321335	1506885	
123–1, 123–2, 123–3	5	25	125	615	2985
	14325	68117	321508	1508438	
121–1	5	25	125	615	2985
	14345	68345	323185	1518765	
133–2, 113–2	5	25	125	615	2995
	14479	69639	333697	1594632	
122–3, 112–3, 211–3, 221–3	5	25	125	615	2995
	14484	69705	334283	1598959	
121–3, 212–3, 131–2	5	25	125	615	2995
	14499	69903	336005	1611495	
111–2	5	25	125	615	3015
	14745	71999	351216	1712064	
124–3, 134–2, 214–3, 143–2	5	25	125	620	3057
	15004	73386	357984	1742678	
213–4, 132–4, 142–3, 231–4, 241–3, 243–1	5	25	125	620	3057
	15006	73414	358246	1744710	
123–4, 234–1	5	25	125	620	3058
	15018	73517	359000	1749747	
111–1	5	25	125	620	3060
	15040	73660	359720	1752600	

TABLE 6.8: $\{AW_5^\tau(n)_{n=1}^9\}$ for 4-letter patterns τ of type $(2,2)$

τ	$\{AW_5^\tau(n)\}_{n=1}^9$ for 4-letter patterns τ of type $(2,2)$				
12–12	5	25	125	615	2975
	14155	66235	304690	1377400	
12–21	5	25	125	615	2975
	14155	66255	305130	1382900	
12–13, 13–12	5	25	125	615	2975
	14165	66467	307743	1407523	
13–32	5	25	125	615	2975
	14165	66467	307748	1407638	
12–31	5	25	125	615	2975
	14170	66542	308442	1412766	

τ	\multicolumn{5}{l}{$\{AW_5^\tau(n)\}_{n=1}^9$ for 4-letter patterns τ of type $(2,2)$}
12–23, 12–32 21–32	5 14175	25 66643	125 309636	615 1423624	2975
11–21, 11–12	5 14305	25 67765	125 317675	615 1475345	2985
13–22	5 14319	25 68005	125 320227	615 1496892	2985
11–23, 11–32	5 14400	25 68824	125 326871	615 1544221	2990
11–22	5 14455	25 69235	125 329430	615 1558430	2995
13–42	5 14893	25 72239	125 348300	620 1670218	3050
13–24, 24–13	5 14894	25 72258	125 348521	620 1672241	3050
14–23	5 14910	25 72420	125 349850	620 1681880	3051
14–32	5 14911	25 72437	125 350030	620 1683410	3051
12–43, 21–43 12–34	5 14927	25 72601	125 351401	620 1693579	3052
11–11	5 15020	25 73340	125 356280	620 1722200	3060

TABLE 6.9: $\{AW_5^\tau(n)_{n=1}^9\}$ for 4-letter patterns τ of type $(1,2,1)$

τ	\multicolumn{5}{l}{$\{AW_5^\tau(n)\}_{n=1}^9$ for 4-letter patterns τ of type $(1,2,1)$}
1–12–2, 1–21–2	5 13615	25 60385	125 255670	615 1029655	2945
1–12–1	5 13625	25 60555	125 257365	615 1042325	2945
2–13–2	5 13659	25 61301	125 266545	615 1125450	2945
1–23–1	5 13750	25 62270	125 274431	615 1179730	2950
1–31–2, 1–13–2	5 13819	25 62824	125 277717	615 1195051	2955
1–23–2, 1–21–3 1–32–2, 1–12–3	5 13834	25 63116	125 281025	615 1223497	2955
1–22–1	5 13835	25 63195	125 282545	615 1241645	2955
1–11–2	5 14145	25 66159	125 304797	615 1385315	2975
1–33–2, 1–22–3	5 14169	25 66593	125 309565	615 1426269	2975

τ	$\{AW_5^\tau(n)\}_{n=1}^9$ for 4-letter patterns τ of type $(1,2,1)$				
1–11–1	5	25	125	620	3040
	14700	69980	327620	1507580	
2–14–3, 2–41–3	5	25	125	620	3039
	14704	70240	331523	1547611	
1–42–3, 1–24–3	5	25	125	620	3039
	14705	70262	331803	1550313	
1–43–2, 1–23–4	5	25	125	620	3040
1–34–2, 1–32–4	14723	70459	333500	1562970	

TABLE 6.10: $\{AW_5^\tau(n)_{n=1}^9\}$ for 4-letter patterns τ of type $(2,1,1)$

τ	$\{AW_5^\tau(n)\}_{n=1}^9$ for 4-letter patterns τ of type $(2,1,1)$				
12–1–2, 12–2–1	5	25	125	615	2945
	13615	60405	256100	1034890	
11–1–2, 11–2–1	5	25	125	615	2945
	13625	60555	257365	1042325	
22–1–3	5	25	125	615	2945
	13659	61301	266545	1125450	
21–3–1, 21–1–3	5	25	125	615	2955
	13819	62814	277513	1192670	
11–3–2, 11–2–3	5	25	125	615	2950
	13760	62460	276491	1196542	
12–3–1, 12–1–3	5	25	125	615	2955
	13824	62920	278786	1204052	
13–1–2	5	25	125	615	2955
	13829	63025	280051	1215519	
13–2–1	5	25	125	615	2955
	13834	63110	280904	1222093	
21–2–3, 12–3–2	5	25	125	615	2955
21–3–2, 12–2–3	13834	63116	281025	1223497	
13–3–2, 13–2–3	5	25	125	615	2955
	13834	63128	281267	1226285	
11–2–2	5	25	125	615	2955
	13855	63595	286975	1277865	
12–1–1, 12–2–2	5	25	125	615	2975
	14145	66159	304797	1385315	
13–2–2	5	25	125	615	2975
	14169	66593	309565	1426269	
21–3–3, 12–3–3	5	25	125	615	2975
	14175	66697	310650	1435091	
11–1–1	5	25	125	620	3040
	14700	69980	327620	1507580	

τ	$\{AW_5^\tau(n)\}_{n=1}^9$ for 4-letter patterns τ of type $(2,1,1)$				
24–3–1	5	25	125	620	3039
	14703	70223	331354	1546331	
24–1–3	5	25	125	620	3039
	14704	70241	331546	1547914	
23–1–4, 23–4–1	5	25	125	620	3039
	14705	70261	331781	1550034	
13–2–4, 13–4–2	5	25	125	620	3039
	14705	70262	331803	1550313	
14–2–3	5	25	125	620	3040
	14723	70459	333499	1562946	
14–3–2	5	25	125	620	3040
	14724	70477	333692	1564552	
12–4–3, 21–3–4	5	25	125	620	3040
21–4–3, 12–3–4	14724	70479	333736	1565112	

Chapter 7

Automata and Generating Trees

7.1 History and connections

In this chapter we will discuss interrelated methods that are variations on the same theme, namely, ways to build sequences recursively and to then express the recursive structure in an associated graph. To apply this approach one has to construct a bijection between the objects in the question and walks in an appropriate edge-weighted directed multi-graph. To lay a foundation we provide the necessary graph theoretic background in Section 7.2.

The first approach we will discuss is an automaton (plural automata). This construct is widely used in computer science in connection with grammars and languages. Bränden and Mansour [27] were the first to apply automata to the enumeration of words avoiding subsequence patterns. They proved structural results that showed that the associated graphs have only finitely many vertices, and adapted the transfer matrix method to automata. The name *transfer matrix method* was coined by West in his thesis [191], and subsequently used by Stanley in his text [182] that has become the bible in enumerative combinatorics. The transfer matrix method uses the adjacency matrices of graphs to solve enumeration problems in combinatorics.

Bränden and Mansour [27] obtained asymptotic results for $AW_{[k]}^{\tau}(n)$ for any subsequence pattern τ using the transfer matrix method. In addition, Mansour implemented two programs, TOU_AUTO and TOU_FORMULA, which produce explicit results for the generating functions $AW_{[k]}^{\tau}(n)$, limited only by computational power. However, the most interesting result comes from the structure of the automaton of the increasing patterns $\tau_\ell = 12 \cdots (\ell + 1)$, which allowed Bränden and Mansour to give an alternative proof for Regev's result [169] on the asymptotic behavior of $AW_{[k]}^{\tau_\ell}(n)$. They provided a combinatorial proof which is based on a bijection between paths in the automaton and certain Young tableaux. We conclude Section 7.3 by describing how the automata for words avoiding a single pattern can be generalized to avoidance of patterns in compositions, as well as avoidance of sets of patterns in either words or compositions.

In Section 7.4 we discuss the second approach, generating trees, which consists of defining recursive rules that induce a tree. Generating trees were introduced by West [191] who used them to enumerate permutations avoiding certain subsequence patterns. Even though the two approaches look rather different at first, each generating tree (which usually has infinitely many vertices) can be mapped to a graph with a finite number of vertices in the form of an automaton. Using several examples, we describe how a recursive creation for the sequence translates into recursive rules and labels for the vertices of the tree.

We conclude this chapter by discussing the ECO method, which is a framework for describing processes to obtain recurrences for enumerating combinatorial objects. In essence, it is a road map (much like the Noonan-Zeilberger algorithm) in that it lays out the task to be done, but does not give a formulaic description of the steps. Essentially the same idea (but not with the name ECO method) was introduced in 1978 by Chung et al. [54] who enumerated Baxter permutations (permutations avoiding both 2-41-3 and 3-14-2) using a generating tree. In 1995, Barcucci and several co-authors presented the first enumerative application of the ECO method [14], enumerating Motzkin paths and various other combinatorial objects. Soon thereafter, these authors established a general methodology for plane tree enumeration [15], followed by a survey on the general theory [16]. We refer the interested reader to [16] and [63] for further details and results, concentrating here on the ECO method as applied to the enumeration of words and compositions. We will present the ECO method and results of a recent paper of Bernini at el. [23] who applied the method to pattern avoidance of pairs of generalized patterns in words, in Section 7.5.

7.2   Tools from graph theory

For all the topics in this chapter we need to be fluent in the basics of graph theory. We start with definitions and examples from basic graph theory and give results (without proof) that we need for our purpose. These definitions and theorems can be found in any standard text on graph theory, see for example [59].

Definition 7.1 *A (nondirected) graph $G = (V, E)$ consists of a set V of vertices and a set $E \subseteq V^2$ of edges, each of which connects two vertices. We denote an edge connecting vertices i and j by (i, j) or ij, where the order in which the vertices are written does not matter. Two vertices $i, j \in V$ are said to be* adjacent *if the edge (i, j) belongs to E. The* neighborhood *of a vertex $i \in V$, denoted by $N(i)$, is the set of all vertices adjacent to vertex i, that is, $N(i) = \{j \mid (i, j) \in E\}$. The* degree *of a vertex is the size of its neighborhood, denoted by $\deg(i) = |N(i)|$.*

The graphs we will encounter when discussing automata are a bit more general. They may have multiple edges between vertices, edges from a vertex to itself, or vertices with a direction and a weight.

Definition 7.2 *A directed graph or* digraph *is a graph in which the edges have a direction, that is, the edge (i, j) is different from the edge (j, i). An edge is called a* loop *if it connects a vertex with itself. A* multi-graph *is a graph that has multiple edges between a pair of vertices and also can have loops. A graph without multiple edges and loops is called a* simple graph. *The* neighborhood of *a vertex i in a digraph consists of* out-neighbors $N^+(i) = \{j \in V \mid (i, j) \in E\}$ *and* in-neighbors $N^-(i) = \{j \in V \mid (j, i) \in E\}$, *respectively. A vertex has both an* out-degree $\deg^+(i) = |N^+(i)|$ *and an* in-degree $\deg^-(i) = |N^-(i)|$.

To depict a graph we use dots for the vertices, and lines connecting the dots for the edges. If a graph is directed, then we indicate the direction by an arrow on the edge, where the directed edge (i, j) is drawn with the arrow pointing from i to j. When drawing a graph, it does not matter where the vertices are located and how the edges are connected.

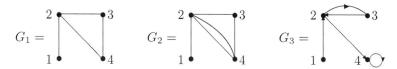

FIGURE 7.1: Simple graph, multi-graph, and digraph.

Example 7.3 *Figure 7.1 depicts a simple graph G_1 with $V = \{1, 2, 3, 4\}$ and $E = \{(1, 2), (2, 3), (2, 4), (3, 4)\}$. The degrees are given by $\deg(1) = 1$, $\deg(2) = 3$, and $\deg(3) = \deg(4) = 2$. The graph G_2 is a multi-graph, as there are two edges connecting vertices 2 and 4. G_3 is a digraph with vertex set $V = \{1, 2, 3, 4\}$ and edge set $E = \{(1, 2), (2, 3), (2, 4), (3, 2), (4, 4)\}$. The neighbors of vertex 3 in G_1 are the vertices 2 and 4, while vertex 3 in graph G_3 has in-neighbor 2 and out-neighbor 2, and thus, $\deg^+(3) = \deg^-(3) = 1$.*

The idea that the visualization of a graph is irrelevant to its structure is made precise with the notion of graph isomorphism.

Definition 7.4 *Two graphs $G = (V, E)$ and $G' = (V', E')$ are said to be* isomorphic *if there exists a bijection $f : V \to V'$ such that $(i, j) \in E$ if and only if $(f(i), f(j)) \in E'$. In other words, two graphs are isomorphic if they are really the same graph labeled or drawn in two different ways.*

Figure 7.2 presents the four nonisomorphic simple graphs on three vertices.

FIGURE 7.2:   The nonisomorphic simple graphs on three vertices.

Definition 7.5 *We say that a graph $G' = (V', E')$ is a subgraph of the graph $G = (V, E)$ if $V' \subseteq V$ and $E' \subseteq E$.*

Example 7.6 *The graph $G' = (\{1, 2, 4\}, \{(1, 2), (2, 4)\})$ is a subgraph of both the graphs G_1 and G_2 of Figure 7.1.*

Besides the graphical representation, we can use a matrix to encode the structure of the graph.

Definition 7.7 *For a graph G with n vertices, we define the adjacency matrix of G by $A = (a_{ij})_{n \times n}$, where a_{ij} is the number of edges from vertex i to vertex j. The labeling of rows and columns has to be in the same order, and is done in the obvious manner (row 1 corresponds to vertex v_1, etc.).*

Example 7.8 *The adjacency matrices of the graphs shown in Figure 7.1 are given by*

$$
A_1 = \begin{bmatrix} 0 & 1 & 0 & 0 \\ 1 & 0 & 1 & 1 \\ 0 & 1 & 0 & 1 \\ 0 & 1 & 1 & 0 \end{bmatrix}, \quad
A_2 = \begin{bmatrix} 0 & 1 & 0 & 0 \\ 1 & 0 & 1 & 2 \\ 0 & 1 & 0 & 1 \\ 0 & 2 & 1 & 0 \end{bmatrix}, \quad and \ A_3 = \begin{bmatrix} 0 & 1 & 0 & 0 \\ 0 & 0 & 1 & 1 \\ 0 & 1 & 0 & 0 \\ 0 & 0 & 0 & 1 \end{bmatrix}.
$$

Note that adjacency matrices of nondirected graphs are symmetric, and that for simple graphs, the adjacency matrix contains only zeros and ones, and the entries on the main diagonal are all zero.

Definition 7.9 *A* walk *from vertex i to vertex j in a graph $G = (V, E)$ is a sequence of vertices $v_0 v_1 \cdots v_k$ such that $v_0 = i$, $v_k = j$, and $(v_p, v_{p+1}) \in E$ for $p = 0, 1, \ldots, k - 1$. A walk where the only allowed repetitions are from loops, that is, $v_{\ell_1} = v_{\ell_2}$ only if $\ell_1 = \ell_2 \pm 1$, is called a* loop path, *and a walk without any repeated vertices is called a* path. *Instead of $v_0 v_1 \cdots v_k$, we may also use the notation $v_0 \to v_1 \to \cdots \to v_k$. The* length *of a walk is defined to be k, the number of edges in the walk. The* distance *between two vertices $i, j \in V$, denoted by $d(i, j)$, is the length of the shortest path between them. If there is no path between two vertices, then their distance is defined as $d(i, j) = \infty$. A* closed path, *or* cycle, *is a path where the first and last vertices are equal, and those are the only repeated vertices. A graph without any cycles is called* acyclic.

In the applications that follow, we will only encounter paths and loop paths, and will refer to them (in an abuse of notation) simply as paths, as opposed to writing (loop) paths all the time. If we want to emphasize that we are actually talking about a genuine path, we will use the term *simple path*.

Example 7.10 *In the graph G_1 of Figure 7.1, $1 \to 2 \to 4 \to 3$ is a path of length three. The distances between pairs of vertices are given by $d(1, 2) = 1$, $d(1, 3) = d(1, 4) = 2$ and $d(2, 3) = d(2, 4) = d(3, 4) = 1$.*

In fact, we can reach every vertex from every other vertex in the graph G_1. This is an important property for graphs and will feature prominently in the definition of a tree.

Definition 7.11 *The graph $G = (V, E)$ is said to be connected if for any two vertices $i, j \in V$ there exists a path from i to j. A graph that is not connected can be partitioned into components, each of which is a maximal connected subgraph.*

Example 7.12 *Graphs G_1 and G_2 in Figure 7.1 are connected, while graph G_3 has three components: $\{1\}$, $\{2, 3\}$ and $\{4\}$.*

A well-known fact concerning the adjacency matrix of a graph is that it provides an easy method for counting the number of walks in a graph.

Theorem 7.13 *Let A be the adjacency matrix of a graph G. Then the number of walks from vertex i to vertex j of length k in G is given by the entry $(A^k)_{ij}$ of the matrix A^k.*

Example 7.14 *For the adjacency matrices given in Example 7.8, the second powers are given by*

$$(A_1)^2 = \begin{bmatrix} 1 & 0 & 1 & 1 \\ 0 & 3 & 1 & 1 \\ 1 & 1 & 2 & 1 \\ 1 & 1 & 1 & 2 \end{bmatrix}, \quad (A_2)^2 = \begin{bmatrix} 1 & 0 & 1 & 2 \\ 0 & 6 & 2 & 1 \\ 1 & 2 & 2 & 2 \\ 2 & 1 & 2 & 5 \end{bmatrix} \quad and \quad (A_3)^2 = \begin{bmatrix} 0 & 0 & 1 & 1 \\ 0 & 1 & 0 & 1 \\ 0 & 0 & 1 & 1 \\ 0 & 0 & 0 & 1 \end{bmatrix}.$$

From these matrices, we can read off the number of walks of length two in the graphs in Figure 7.1. For example, for the graphs G_1 and G_2 there are three (closed) paths from 2 to 2, namely $2 \to 1 \to 2$, $2 \to 3 \to 2$, and $2 \to 4 \to 2$. Note that the multi-edge in graph G_2 increases the number of paths from 2 to 2 from three to six, since the multi-edges allow for different ways to create the path the $2 \to 3 \to 2$. If we label the edges as e_1 and e_2, then there are four paths (described via the respective edges): $e_1 e_1$, $e_2 e_2$, $e_1 e_2$, and $e_2 e_1$.

For the digraph G_3 there are overall fewer paths of length two, as direction plays a role. However, paths in G_3 may involve the loop at vertex 4. For *example, the path of length two from 2 to 4 involves going around the loop once,* $2 \to 4 \to 4$.

Note that the k-th power of a matrix can be computed and displayed in matrix format by using the following Mathematica *and* Maple *commands:*

```
MatrixPower[A,k]//MatrixForm
```

and

```
with(linalg):evalm(A^k)
```

respectively.

We need one more generalization of a graph in order to apply automata to compositions.

Definition 7.15 *An* (edge)-weighted directed graph *is a digraph in which each edge of the graph has been assigned a weight.*

By modifying the definition of the adjacency matrix, we can obtain a result similar to Theorem 7.13 for these more general graphs.

Definition 7.16 *For a weighted graph* G *with* n *vertices, we define the* weighted adjacency matrix *as* $A = (a_{ij})_{n \times n}$ *where* a_{ij} *is the sum of the weights of all edges from vertex* i *to* j. *For a walk* $P = i_0 i_1 \cdots i_k$ *with edges* $e_j = (i_j, i_{j+1})$ *and associated weights* c_j, *the weight of the walk* $c(P)$ *is defined as the product of the edge weights of all the edges of the walk, that is,* $c(P) = \prod_{j=0}^{k-1} c_j$.

Note that Definition 7.16 reduces to Definition 7.7 by assigning a weight of one to every edge of a nonweighted graph.

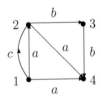

FIGURE 7.3:   An edge-weighted directed graph.

Example 7.17 *Figure 7.3 shows an example of a weighted digraph. Edges* $(1, 2)$, $(1, 4)$ *and* $(2, 4)$ *have weight* a, *the second* $(1, 2)$ *edge has weight* c, *and* *the two remaining edges have weight b. The weighted adjacency matrix for this graph is given by*

$$A_4 = \begin{bmatrix} 0 & a+c & 0 & a \\ 0 & 0 & b & a \\ 0 & 0 & 0 & b \\ 0 & 0 & 0 & 0 \end{bmatrix}.$$

The weight of the path $P = 1 \to 2 \to 4$ is either given by $c(P) = ca$ or $c(P) = a^2$, depending on which of the $(1,2)$ edges is chosen.

With this definition of the weight of a walk and the weighted adjacency matrix, we have the following generalization of Theorem 7.13.

Theorem 7.18 *Let A be the adjacency matrix of a weighted directed graph G. Then the total weight of all walks from vertex i to vertex j of length k in G is given by the entry $(A^k)_{ij}$ of the matrix A^k.*

Example 7.19 *For the adjacency matrix A_4 defined in Example 7.17, the second power is given by*

$$(A_4)^2 = \begin{bmatrix} 0 & 0 & b(a+c) & a(a+c) \\ 0 & 0 & 0 & b^2 \\ 0 & 0 & 0 & 0 \\ 0 & 0 & 0 & 0 \end{bmatrix}.$$

The rows consisting of zeros indicate that there are no walks from either vertex 3 or vertex 4 to anywhere in two steps. (In fact, there will be no walks to anywhere from vertices 3 or 4 in k steps with $k \geq 2$). The total weight of walks from vertex 1 to vertex 4 is given by $a(a+c)$, accounting for the two paths with weights a^2 and ca.

Now we look at a special type of graph that we will need in Section 7.4 when discussing generating trees.

Definition 7.20 *A tree T of order n is a simple, connected, acyclic graph on n vertices. A vertex of a tree with degree one is called a* leaf. *A tree is called a* rooted tree *if one vertex has been designated the* root, *in which case the edges have a natural orientation away from the root. The* tree order *is the partial ordering on the vertices of a tree with $u \leq v$ if and only if the unique path from the root to v passes through u. If there is a directed edge from vertex i to j, then j is called the* child *of i, and i is called the* parent *of j. The* height *or* level *of a vertex in a rooted tree is the length of the path from the root to the vertex. A* labeled tree *is a tree in which each vertex is labeled.*

Trees can be classified in many different ways.

Theorem 7.21 *The following statements are equivalent for a tree $\mathcal{T}$:*

1. *$\mathcal{T}$ is a tree.*

2. *There is a unique path in $\mathcal{T}$ between any two vertices in $\mathcal{T}$.*

3. *Removing any edge from $\mathcal{T}$ results in a disconnected graph.*

4. *Addition of any edge between two nonadjacent vertices in $\mathcal{T}$ results in a cycle.*

Example 7.22 *Figure 7.4 shows three trees. Tree $\mathcal{T}_1$ has order eight, while the labeled trees $\mathcal{T}_2$ and $\mathcal{T}_3$ both have order six. Looking at $\mathcal{T}_2$, we can identify three leafs, namely vertices 3, 5 and 6. If we designate vertex 1 as the root, then the height of the leaf 5 is 3 and the height of vertex 4 is 2.*

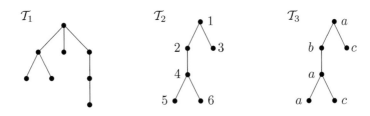

FIGURE 7.4:   A sampling of trees.

Note that the vertex labels of a tree of order n often consist of the numbers $1, 2, \ldots, n$ to uniquely identify each vertex of the tree, but can also consist of labels that identify what the vertex is representing (for example in a genealogical tree), or a property that the vertex has. Tree $\mathcal{T}_3$ is an example of the type of labeling we will be using for the generating trees in Section 7.4.

7.3   Automata

An automaton is a mathematical model for a finite state machine, which starts in an initial state and then is fed one input symbol at a time. It has an associated transition function (encoded for example in a look-up table) that tells the machine to which state to move based on the current state and the input read. Once the input stream is depleted, the automaton stops in either an accept or a reject state. Depending on the state in which the automaton stops, it either accepts or rejects the input stream (or word). The set of words accepted by an automaton is called the *language* accepted by the automaton.

We will use automata to enumerate words and compositions that avoid a given pattern. To do so, we will first define a general automaton, and then specify the components for the automaton that enumerates the number of words avoiding a specific pattern. Expressing the enumeration problem in terms of an automaton is useful when investigating a specific pattern (as each pattern has a tailored automaton), but does not work well for discovering results for families of patterns. One exception is the family of increasing patterns $\tau_\ell = 12 \cdots (\ell + 1)$, for which we will obtain structural results that allow us to compute the asymptotic behavior of $AW^{\tau_\ell}_{[k]}$.

Definition 7.23 *A* finite automaton *is given by*

$$\mathcal{A} = (\mathcal{E}, \Sigma, \Delta, S_0, F),$$

where

- $\mathcal{E} = \{S_0, S_1, \ldots, S_e\}$ *is a finite set of* states;

- Σ *is a finite set of symbols* $\{\Sigma_1, \Sigma_2, \ldots, \Sigma_\ell\}$ *called the* alphabet *of the automaton;*

- Δ *is the* transition function, *that is,* $\Delta : \mathcal{E} \times \Sigma^* \to \mathcal{E}$, *where* Σ^* *is the set of all finite words* $w = w_1 w_2 \cdots w_m$ *such that* $w_i \in \Sigma$ *for all* $i = 1, 2, \ldots, m;$

- $S_0 = \varepsilon$ *is the* initial state *of the automaton before any input has been processed; and*

- F *is a subset of* $\mathcal{E}$, *called the* accept states.

We identify the automaton $\mathcal{A}$ with a directed graph in which the vertex labels denote (the equivalence classes of) the states $S_0, S_1, \ldots, S_e$. If $\Delta(S_i, \Sigma_k) = S_j$, then the edge from S_i to S_j is labeled Σ_k. The adjacency matrix of this graph can then be used to count the number of walks of any length from state S_i to S_j. Note that the edge labels do not signify weights unless stated otherwise.

Example 7.24 *We will define an automaton to count the number of tilings of size $2 \times n$ with squares and dominos (see Example 3.18) that cannot be vertically split into two smaller tilings of sizes $2 \times n_1$ and $2 \times (n - n_1)$. These types of tilings have been referred to as* basic blocks *[84] or* indecomposable blocks *[30]. We denote the empty tiling by ε. In order to produce proper tilings, we have to put any new tile into the leftmost possible position without creating empty spots to the left of an occupied spot. The choices for placement* *of a tile are: placing either a square or a domino in the top row or in the bottom row, or placing a domino vertically. The latter choice is allowed only at the left end of the tiling (anywhere else, it would produce a place where the tiling can be split). These choices constitute our alphabet, and we depict them by*

Next we need to identify the states of the automaton. Placing a tile changes the shape of the right edge, and furthermore, the shape of the right edge allows us to detect when we have completed a tiling (when we have reached a straight right edge). Thus, the states constitute the equivalence classes of all tilings that have a particular shape of the right edge (as we do not care about the particular tiling, just the shape of the right edge). Figure 7.5 shows how placement of a tile changes the tiling, and how this is expressed in the associated graph in terms of the respective states.

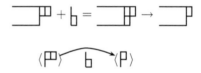

FIGURE 7.5: Example of transition function for tilings.

Now we are ready for drawing the graph of the tiling automaton. If the automaton is in the initial state, then we are allowed any of the five possible placements described above. The set of final states consists of only the vertical domino (this is the one that we are interested in – being in this state is equivalent to having completed a basic block). For any other state, the only choices we have are to place a square or a horizontal domino into the other row. Figure 7.6 shows the automaton graph G_A for the tiling problem.

After having warmed up with the visual example of tilings, we now focus on automata for pattern avoidance.

7.3.1 Automata and pattern avoidance in words

We will define an automaton to count the number of words avoiding a certain subsequence pattern (see Definition 5.2). As we have seen in Example 7.24, the states of the automaton correspond to equivalence classes that lump together combinatorial objects that are equivalent with regard to the characteristic that we want to enumerate. Here we are interested in pattern avoidance, so sequences that have the same structure with regard to avoidance of the given pattern should be equivalent. Specifically, two sequences (words

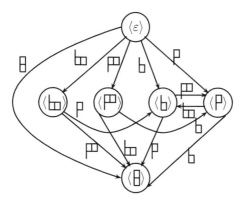

FIGURE 7.6: Automaton for tilings that are indecomposable.

or compositions) should be considered the "same" if the pattern of interest will occur or be avoided equally in both sequences when appending any letter to the sequences. We first consider words, and later modify the approach for compositions.

Definition 7.25 *For $k > 0$, let $[k]^* = \bigcup_n [k]^n$ be the set of all finite words on the alphabet $[k]$, and let τ be a subsequence pattern. For fixed k and τ, two words $v, w \in [k]^*$ are equivalent on $[k]^*$, denoted by $v \overset{\mathcal{A}}{\sim} w$, if for all words $r \in [k]^*$ we have*

$$vr \text{ avoids } \tau \quad \Leftrightarrow \quad wr \text{ avoids } \tau,$$

where vr denotes the concatenation of the words v and r. We denote the equivalence class of a word w by $\langle w \rangle$, and use $\mathcal{AW}^{[k]^*}(\tau)$ to denote the set of words in $[k]^*$ that avoid τ.

Such an equivalence is sometimes called the *Nerode equivalence* [101].

Example 7.26 *Let $\tau = 132$, $k \geq 4$, $v = 13$ and $w = 14$. Then $v \overset{\mathcal{A}}{\not\sim} w$, since 133 avoids τ, but 143 contains τ.*

At first sight it may seem difficult to determine if $v \overset{\mathcal{A}}{\sim} w$, since a priori there are an infinite number of words $r \in [k]^*$ to be checked. However, Lemma 7.27 (originally shown by Brändén and Mansour [27]) saves us, as it ensures that we have to check only finitely many words r.

Lemma 7.27 *Let τ be a pattern of length ℓ and let $v, w \in [k]^*$ be any two words. Then $v \overset{\mathcal{A}}{\sim} w$ if and only if for all words $r \in [k]^s$ with $0 \leq s \leq \ell$ we have that vr avoids $\tau \Leftrightarrow wr$ avoids τ.*

Proof Define an equivalence relation $\overset{A}{\sim}_{\ell}$ on $[k]^*$ by: $v \overset{A}{\sim}_{\ell} w$ if for all words $r \in [k]^s$, $0 \le s \le \ell$, we have that vr avoids $\tau \Leftrightarrow wr$ avoids τ. Clearly, $v \overset{A}{\sim} w$ implies $v \overset{A}{\sim}_{\ell} w$. Now assume that $v \overset{A}{\sim}_{\ell} w$, but $v \overset{A}{\not\sim} w$. Without loss of generality, we may assume that there is an $r \in [k]^*$ such that vr contains τ and wr avoids τ. Any occurrence of τ in vr can use at most ℓ letters of r. Thus, there is a subsequence r' of r of length at most ℓ such that vr' contains τ and wr' avoids τ, that is, $v \overset{A}{\not\sim}_{\ell} w$, a contradiction. $\qquad\square$

We can now define the automaton for pattern avoidance in words.

Definition 7.28 *Given a positive integer k and a subsequence pattern τ, we define the finite automaton $\mathcal{A}(\tau, k) = (\mathcal{E}(\tau, k), [k], \Delta, \langle \varepsilon \rangle, \mathcal{E}(\tau, k) \backslash \{\langle \tau \rangle\})$, where $\mathcal{E}(\tau, k)$ is the set of the equivalence-classes of $\overset{A}{\sim}$, $\Delta(\langle w \rangle, i) = \langle wi \rangle$, and ε is the empty word.*

Definition 7.29 *We identify $\mathcal{A}(\tau, k)$ with the directed graph G_A that has vertices $\mathcal{E}(\tau, k) \backslash \{\langle \tau \rangle\}$ and a (labeled) edge $\overset{i}{\longrightarrow}$ between $\langle v \rangle$ and $\langle w \rangle$ if $vi \overset{A}{\sim} w$. Multi-edges between two vertices are combined by placing their respective labels on a single edge connecting the two vertices. We call a path in G_A simple if it starts in $\langle \varepsilon \rangle$, does not use any loops, and does not end in $\langle \tau \rangle$. The associated adjacency matrix $A(\tau, k)$ of $\mathcal{A}(\tau, k)$ of size $(|\mathcal{E}(\tau, k)| - 1) \times (|\mathcal{E}(\tau, k)| - 1)$ is defined by*

$$A(\tau, k)_{ij} = |\{s \in [k] : \Delta(\langle x_i \rangle, s) = \langle x_j \rangle\}|,$$

that is, $A(\tau, k)_{ij}$ counts the number of edges between x_i and x_j.

Note that we omit the state $\langle \tau \rangle$ in the associated graph in the context of pattern avoidance, so the associated graph consists of only accept states. Is there any effect of this choice on the count of the number of words avoiding τ? The answer is no: the state $\langle \tau \rangle$ is irrelevant, as we are never interested in paths that pass through or end in the equivalence class $\langle \tau \rangle$. Therefore, we are permitted to disregard this state for the purpose of counting the number of words avoiding τ. This is quite different from the automaton we built in Example 7.24, where almost all paths passed through nonaccept states to the accept state. There, deletion of nonaccept states would not be possible.

In addition, in the setting of avoidance of subsequence patterns, we can list the states in *canonical order*, that is as $\langle x_1 \rangle, \langle x_2 \rangle, \ldots, \langle x_e \rangle$ in such a way that for $i < j$ there is no path from $\langle x_j \rangle$ to $\langle x_i \rangle$. Why is that so? When appending a letter from the alphabet, either the letter is relevant, in which case we create a new state, or the letter added is not relevant, in which case we have a loop. In fact, we claim that the automaton graph has only loop paths and paths. Assume there was a cycle, for example $\langle x_1 \rangle \overset{i}{\rightarrow} \langle x_2 \rangle \overset{j}{\rightarrow} \langle x_1 \rangle$. This would imply that $wij \overset{A}{\sim} w$, in which case the letter i cannot have been relevant and therefore $wi \overset{\mathcal{A}}{\sim} w$, and $\langle x_1 \rangle \overset{i}{\nrightarrow} \langle x_2 \rangle$. Thus, the graph has no cycles, which implies that we can find a canonical order for the states. However, the above argument for the absence of cycles works only for avoidance of subsequence patterns, not for generalized or subword patterns where adjacency requirements come into play and may result in cycles.

We start by deriving the automaton for avoidance of the (subsequence) pattern 123 in words on the alphabet [4].

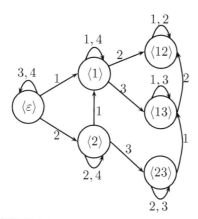

FIGURE 7.7: Final states in the automaton $\mathcal{A}(123, 4)$.

Example 7.30 *Let* $\tau = 123$. *Then in any word that starts with a 3 or a 4, the first letter cannot be part of an occurrence of* τ, *and therefore, such a word is equivalent to* ε. *However, a first letter 1 or 2 can be part of an occurrence of* τ, *leading to two new equivalence classes,* $\langle 1 \rangle$ *and* $\langle 2 \rangle$. *For both of these states, appending a 4 cannot create a pattern* τ. *Likewise, appending i to* $\langle i \rangle$ *(a repeated letter, but* τ *has no repeated letters) is irrelevant, and therefore,* $i4 \overset{\mathcal{A}}{\sim} i$ *and* $ii \overset{\mathcal{A}}{\sim} i$.

Now let's look at the other two possibilities for $\langle 1 \rangle$. *Appending either a 2 or a 3 leads to the new equivalence classes* $\langle 12 \rangle$ *and* $\langle 13 \rangle$ *(since* $123 \overset{\mathcal{A}}{\nsim} 133$). *Next we look at* $\langle 2 \rangle$ *and claim that* $21 \overset{\mathcal{A}}{\sim} 1$. *Why? Let's think in terms of occurrence of the pattern* τ. *If the occurrence of* τ *involves the 2 as the smallest part, then the 1 will also create a pattern* τ. *The other possibility is an occurrence of* τ *that does not involve the 2, but starts with the 1. Therefore,* $\Delta(\langle 2 \rangle, 1) = \langle 1 \rangle$. *The last case, appending the letter 3 to* $\langle 2 \rangle$ *leads to a new class,* $\langle 23 \rangle$. *We repeat the analysis for the states whose labels have two letters. Clearly,* $\Delta(\langle ij \rangle, i) = \Delta(\langle ij \rangle, j) = \langle ij \rangle$, *and for* $i = 3, 4$, $\Delta(\langle 12 \rangle, i) = \langle 123 \rangle$ *(which is not displayed in the graph). Likewise,* $\Delta(\langle i3 \rangle, 4) = \langle 123 \rangle$ *for* $i = 1, 2$.

The same argument that showed that $\Delta(\langle 2 \rangle, 1) = \langle 1 \rangle$ *gives that* $\Delta(\langle 13 \rangle, 2) =$
$\langle 12 \rangle$. *Finally,* $\Delta(\langle 23 \rangle, 1) = \langle 13 \rangle$ *follows from the definition of the equivalence.*
Figure 7.7 shows the graph $G_\mathcal{A}$.

Let's now look at the associated adjacency matrix. As a convention, we will
order the states as follows (if this results in a canonical order): according to
the length of their label in increasing order, and among the states with labels
of the same length in such a way that the edge (i, j) only exists for $i \leq j$.

Example 7.31 *(Continuation of Example 7.30) Using the canonical ordering*
$\langle \varepsilon \rangle$, $\langle 2 \rangle$, $\langle 1 \rangle$, $\langle 23 \rangle$, $\langle 13 \rangle$, *and* $\langle 12 \rangle$, *the first three powers of the associated*
adjacency matrix $A = A(123, 4)$ *are given by*

$$
A = \begin{bmatrix}
2 & 1 & 1 & 0 & 0 & 0 \\
0 & 2 & 1 & 1 & 0 & 0 \\
0 & 0 & 2 & 0 & 1 & 1 \\
0 & 0 & 0 & 2 & 1 & 0 \\
0 & 0 & 0 & 0 & 2 & 1 \\
0 & 0 & 0 & 0 & 0 & 2
\end{bmatrix},
\quad
A^2 = \begin{bmatrix}
4 & 4 & 5 & 1 & 1 & 1 \\
0 & 4 & 4 & 4 & 2 & 1 \\
0 & 0 & 4 & 0 & 4 & 5 \\
0 & 0 & 0 & 4 & 4 & 1 \\
0 & 0 & 0 & 0 & 4 & 4 \\
0 & 0 & 0 & 0 & 0 & 4
\end{bmatrix},
\quad \text{and } A^3 = \begin{bmatrix}
8 & 12 & 18 & 6 & 8 & 8 \\
0 & 8 & 12 & 12 & 12 & 8 \\
0 & 0 & 8 & 0 & 12 & 18 \\
0 & 0 & 0 & 8 & 12 & 6 \\
0 & 0 & 0 & 0 & 8 & 12 \\
0 & 0 & 0 & 0 & 0 & 8
\end{bmatrix}.
$$

The relevant entries are the elements in the first row of the respective matrix.
They correspond to the number of ways to go from the start state (empty word)
to any state in a given number of steps. Summing the entries of the first row
of A^n *therefore gives* $AW^{123}_{[4]}(n)$, *the total number of words of length* n *on the*
alphabet $[4]$ *avoiding the pattern* 123. *Thus,*

$$AW^{123}_{[4]}(1) = 2 + 1 + 1 = 4,$$
$$AW^{123}_{[4]}(2) = 4 + 4 + 5 + 1 + 1 + 1 = 16, \text{ and}$$
$$AW^{123}_{[4]}(3) = 8 + 12 + 18 + 6 + 8 + 8 = 60.$$

No surprise here – for $n = 1$ *or* $n = 2$ *the pattern* 123 *cannot occur, and for*
$n = 3$, *there are exactly four words that contain* 123, *namely the words* 123,
124, 134, *and* 234.

We give a second example of an automaton, this time for avoidance of the
pattern $\tau = 122$ in words on the alphabet $[3]$. Figure 7.8 shows the graph
corresponding to the automaton $\mathcal{A}(122, 3)$.

Comparing the graphs in Figures 7.7 and 7.8, we see similarities and differences between the two graphs. The most notable difference is that the
automaton $\mathcal{A}(122, 3)$ has a state that has a label with three letters, while the
graph of $\mathcal{A}(123, 4)$ has labels with at most two letters, even though in both
cases the pattern to be avoided is of length three. On the other hand, the two
graphs have in common that in each graph all states have the same number
of loops (two loops for $\mathcal{A}(123, 4)$ and one loop for $\mathcal{A}(122, 3)$). In Section 7.3.3
we will prove that this is always the case for automata of words avoiding an
increasing pattern $\tau_\ell = 12 \cdots \ell(\ell + 1)$.

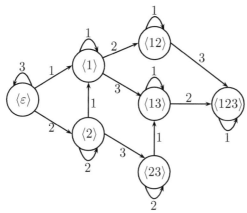

FIGURE 7.8: The graph of the automaton $\mathcal{A}(122, 3)$.

We know much less about the structure of automata for avoidance of patterns with repeated letters in words. Mansour developed a C++ program, TOU_AUTO, which computes the equivalence classes and the adjacency matrix for the automaton of a subsequence pattern τ. The program TOU_AUTO is listed in Appendix G.5 and is also available on Mansour's home page [139] for use by the readers. Using this program, we can obtain the adjacency matrices for the automata $\mathcal{A}(1132, 4)$ and $\mathcal{A}(1312, 4)$, respectively, as

$$
\begin{bmatrix}
2 & 1 & 1 & 0 & 0 & 0 & 0 & 0 & 0 & 0 & 0 & 0 \\
0 & 2 & 0 & 1 & 1 & 0 & 0 & 0 & 0 & 0 & 0 & 0 \\
0 & 0 & 2 & 0 & 1 & 1 & 0 & 0 & 0 & 0 & 0 & 0 \\
0 & 0 & 0 & 2 & 0 & 0 & 1 & 1 & 0 & 0 & 0 & 0 \\
0 & 0 & 0 & 1 & 2 & 0 & 0 & 0 & 1 & 0 & 0 & 0 \\
0 & 0 & 0 & 0 & 0 & 2 & 0 & 0 & 1 & 1 & 0 & 0 \\
0 & 0 & 0 & 0 & 0 & 0 & 2 & 1 & 0 & 0 & 0 & 0 \\
0 & 0 & 0 & 0 & 0 & 0 & 0 & 2 & 0 & 0 & 0 & 0 \\
0 & 0 & 0 & 1 & 0 & 0 & 0 & 0 & 2 & 0 & 1 & 0 \\
0 & 0 & 0 & 0 & 0 & 0 & 0 & 0 & 0 & 2 & 1 & 0 \\
0 & 0 & 0 & 0 & 0 & 0 & 0 & 0 & 0 & 0 & 2 & 1 \\
0 & 0 & 0 & 0 & 0 & 0 & 1 & 0 & 0 & 0 & 0 & 2
\end{bmatrix}
\quad \text{and} \quad
$$

$$
\begin{bmatrix}
2 & 1 & 1 & 0 & 0 & 0 & 0 & 0 & 0 & 0 & 0 & 0 & 0 & 0 & 0 & 0 & 0 & 0 & 0 & 0 \\
0 & 1 & 0 & 1 & 1 & 1 & 0 & 0 & 0 & 0 & 0 & 0 & 0 & 0 & 0 & 0 & 0 & 0 & 0 & 0 \\
0 & 0 & 2 & 1 & 0 & 0 & 1 & 0 & 0 & 0 & 0 & 0 & 0 & 0 & 0 & 0 & 0 & 0 & 0 & 0 \\
0 & 0 & 0 & 2 & 0 & 0 & 0 & 1 & 1 & 0 & 0 & 0 & 0 & 0 & 0 & 0 & 0 & 0 & 0 & 0 \\
0 & 0 & 0 & 0 & 1 & 1 & 0 & 1 & 0 & 1 & 0 & 0 & 0 & 0 & 0 & 0 & 0 & 0 & 0 & 0 \\
0 & 0 & 0 & 0 & 0 & 2 & 0 & 0 & 0 & 0 & 1 & 1 & 0 & 0 & 0 & 0 & 0 & 0 & 0 & 0 \\
0 & 0 & 0 & 0 & 0 & 0 & 2 & 0 & 0 & 0 & 0 & 0 & 1 & 1 & 0 & 0 & 0 & 0 & 0 & 0 \\
0 & 0 & 0 & 0 & 0 & 0 & 0 & 2 & 1 & 1 & 0 & 0 & 0 & 0 & 0 & 0 & 0 & 0 & 0 & 0 \\
0 & 0 & 0 & 0 & 0 & 0 & 0 & 0 & 2 & 0 & 1 & 0 & 0 & 0 & 1 & 0 & 0 & 0 & 0 & 0 \\
0 & 0 & 0 & 0 & 0 & 0 & 0 & 0 & 0 & 2 & 0 & 0 & 0 & 0 & 0 & 1 & 0 & 0 & 0 & 0 \\
0 & 0 & 0 & 0 & 0 & 0 & 0 & 0 & 0 & 0 & 2 & 0 & 0 & 0 & 0 & 0 & 0 & 0 & 0 & 0 \\
0 & 0 & 0 & 0 & 0 & 0 & 0 & 0 & 1 & 0 & 1 & 2 & 0 & 0 & 0 & 0 & 0 & 0 & 0 & 0 \\
0 & 0 & 0 & 0 & 0 & 0 & 0 & 0 & 1 & 0 & 0 & 0 & 1 & 0 & 0 & 0 & 0 & 0 & 1 & 1 \\
0 & 0 & 0 & 0 & 0 & 0 & 0 & 0 & 0 & 0 & 0 & 0 & 0 & 2 & 0 & 0 & 1 & 0 & 0 & 0 \\
0 & 0 & 0 & 0 & 0 & 0 & 0 & 0 & 0 & 1 & 0 & 0 & 0 & 2 & 0 & 0 & 0 & 0 & 0 & 0 \\
0 & 0 & 0 & 0 & 0 & 0 & 0 & 0 & 0 & 1 & 0 & 0 & 0 & 0 & 2 & 0 & 0 & 0 & 0 & 0 \\
0 & 0 & 0 & 0 & 0 & 0 & 0 & 0 & 0 & 0 & 0 & 0 & 0 & 1 & 0 & 0 & 2 & 0 & 0 & 0 \\
0 & 0 & 0 & 0 & 0 & 0 & 0 & 0 & 0 & 0 & 0 & 0 & 0 & 0 & 0 & 0 & 1 & 0 & 2 & 0 \\
0 & 0 & 0 & 0 & 0 & 0 & 0 & 0 & 0 & 0 & 0 & 0 & 0 & 0 & 0 & 0 & 0 & 0 & 1 & 0 \\
0 & 0 & 0 & 0 & 0 & 0 & 0 & 0 & 1 & 1 & 0 & 0 & 0 & 0 & 1 & 0 & 0 & 0 & 0 & 1
\end{bmatrix}.
$$

Note that the automaton for the pattern 1312 has two types of states, those with one and those with two loops.

As mentioned before, we can also use the program TOU_AUTO to obtain the states of the automata for each of the permutation patterns of length three for different values of k. Table 7.1 lists the states of the automaton $\mathcal{A}(\tau, k)$ when $\tau \in \mathcal{S}_3$ and $k = 3, 4$. For the corresponding equivalence classes for $k = 5$, see [27].

TABLE 7.1:   States of the automaton $\mathcal{A}(\tau, k)$ for $\tau \in \mathcal{S}_3$

k	τ	The equivalences classes in $\mathcal{E}(p, k)$
3	123	$\langle \varepsilon \rangle, \langle 1 \rangle, \langle 12 \rangle, \langle 123 \rangle$
	132	$\langle \varepsilon \rangle, \langle 1 \rangle, \langle 13 \rangle, \langle 132 \rangle$
	213	$\langle \varepsilon \rangle, \langle 2 \rangle, \langle 21 \rangle, \langle 213 \rangle$
	231	$\langle \varepsilon \rangle, \langle 2 \rangle, \langle 23 \rangle, \langle 231 \rangle$
	312	$\langle \varepsilon \rangle, \langle 3 \rangle, \langle 31 \rangle, \langle 312 \rangle$
	321	$\langle \varepsilon \rangle, \langle 3 \rangle, \langle 32 \rangle, \langle 321 \rangle$
4	123	$\langle \varepsilon \rangle, \langle 1 \rangle, \langle 2 \rangle, \langle 12 \rangle, \langle 13 \rangle, \langle 23 \rangle, \langle 123 \rangle$
	132	$\langle \varepsilon \rangle, \langle 1 \rangle, \langle 2 \rangle, \langle 13 \rangle, \langle 14 \rangle, \langle 24 \rangle, \langle 132 \rangle, \langle 241 \rangle$
	213	$\langle \varepsilon \rangle, \langle 2 \rangle, \langle 3 \rangle, \langle 21 \rangle, \langle 23 \rangle, \langle 31 \rangle, \langle 32 \rangle, \langle 213 \rangle$
	231	$\langle \varepsilon \rangle, \langle 2 \rangle, \langle 3 \rangle, \langle 23 \rangle, \langle 24 \rangle, \langle 32 \rangle, \langle 34 \rangle, \langle 231 \rangle$
	312	$\langle \varepsilon \rangle, \langle 3 \rangle, \langle 4 \rangle, \langle 31 \rangle, \langle 41 \rangle, \langle 42 \rangle, \langle 312 \rangle, \langle 314 \rangle$
	321	$\langle \varepsilon \rangle, \langle 3 \rangle, \langle 4 \rangle, \langle 32 \rangle, \langle 42 \rangle, \langle 43 \rangle, \langle 321 \rangle$

In light of the previous examples and Table 7.1, several questions about the structure of the automaton graphs arise:

1. From Table 7.1 we see that patterns in the same Wilf-equivalence class can have a different number of states, which implies that the respective graphs are not isomorphic. However, patterns that are complements of each other, and patterns that are the reverse of each other have the same number of states. So the questions are:

   - Since w avoids τ if and only if $R(w)$ avoids $R(\tau)$, are the automata $\mathcal{A}(\tau, k)$ and $\mathcal{A}(R(\tau), k)$ always isomorphic?

   - Since w avoids τ if and only if $c(w)$ avoids $c(\tau)$, are the automata $\mathcal{A}(\tau, k)$ and $\mathcal{A}(c(\tau), k)$ always isomorphic?

2. Since the number of paths of length n in a graph is given by the elements of the n-th power of its adjacency matrix, can we obtain exact or asymptotic results (as $n \to \infty$) from the adjacency matrix?

3. What kind of structural results are there for families of patterns, specifically the increasing patterns?

The first question is the easiest to answer. From the definition of $\overset{\mathcal{A}}{\sim}$ it is clear that $\mathcal{A}(\tau, k)$ and $\mathcal{A}(c(\tau), k)$ are isomorphic as automata (replace the state $\langle s \rangle$ by $\langle c(s) \rangle$ and the edge j by $c(j)$). This is generally not the case for $\mathcal{A}(\tau, k)$ and $\mathcal{A}(R(\tau), k)$. Indeed, using the program TOU_AUTO, Brändén and Mansour [27] found that for $\tau = 2314$ and $k = 5$ we have $|\mathcal{E}(2314, 5)| = 13$ and $|\mathcal{E}(4132, 5)| = 14$. For an example where the number of states is the same, but the automata are not isomorphic, see Exercise 7.2.

Before answering the remaining questions, we give another example, the automaton for avoiding $\tau = 3241$ in words on the alphabet $k = 5$. Note that the adjacency matrix given by the program TOU_AUTO may correspond to states not listed in canonical order (if $i < j$, then there is no path from $\langle x_j \rangle$ to $\langle x_i \rangle$), so the matrix is not necessarily upper triangular.

Example 7.32 *If $\tau = 3241$ and $k = 5$, then the accept states are $\langle \varepsilon \rangle$, $\langle 3 \rangle$, $\langle 4 \rangle$, $\langle 32 \rangle$, $\langle 34 \rangle$, $\langle 42 \rangle$, $\langle 43 \rangle$, $\langle 324 \rangle$, $\langle 325 \rangle$, $\langle 342 \rangle$, $\langle 432 \rangle$, $\langle 435 \rangle$, and $\langle 3243 \rangle$. The automaton $\mathcal{A}(3241, 5)$ has adjacency matrix*

$$A(3241, 5) = \begin{bmatrix}
3 & 1 & 1 & 0 & 0 & 0 & 0 & 0 & 0 & 0 & 0 & 0 & 0 \\
0 & 3 & 0 & 1 & 1 & 0 & 0 & 0 & 0 & 0 & 0 & 0 & 0 \\
0 & 0 & 3 & 0 & 0 & 1 & 1 & 0 & 0 & 0 & 0 & 0 & 0 \\
0 & 0 & 0 & 3 & 0 & 0 & 0 & 1 & 1 & 0 & 0 & 0 & 0 \\
0 & 0 & 0 & 0 & 3 & 0 & 1 & 0 & 0 & 1 & 0 & 0 & 0 \\
0 & 0 & 0 & 0 & 0 & 3 & 1 & 1 & 0 & 0 & 0 & 0 & 0 \\
0 & 0 & 0 & 0 & 0 & 0 & 3 & 0 & 0 & 0 & 1 & 1 & 0 \\
0 & 0 & 0 & 0 & 0 & 0 & 0 & 3 & 0 & 0 & 0 & 0 & 1 \\
0 & 0 & 0 & 0 & 0 & 0 & 0 & 1 & 3 & 0 & 0 & 0 & 0 \\
0 & 0 & 0 & 0 & 0 & 0 & 0 & 2 & 0 & 2 & 1 & 0 & 0 \\
0 & 0 & 0 & 0 & 0 & 0 & 0 & 0 & 0 & 0 & 3 & 1 & 1 \\
0 & 0 & 0 & 0 & 0 & 0 & 0 & 0 & 0 & 0 & 0 & 3 & 0 \\
0 & 0 & 0 & 0 & 0 & 0 & 0 & 0 & 0 & 0 & 0 & 1 & 3
\end{bmatrix}.$$

Note that almost all states in $\mathcal{A}(3241, 5)$ have three loops, with the exception of $\langle 342 \rangle$, which has two loops. In addition, there is a multi-edge between $\langle 342 \rangle$ and $\langle 324 \rangle$, namely $\langle 342 \rangle \xrightarrow{2,4} \langle 324 \rangle$.

We will now tackle the second question posed above, namely obtaining results on $AW_{[k]}^{\tau}(n)$ from the adjacency matrix. This is achieved by using the transfer matrix method, which can be used to obtain asymptotic results as well as explicit ones, with the asymptotic results usually easier to derive. The third question will be answered in Section 7.3.3.

7.3.2 Transfer matrix method

The transfer matrix method uses the adjacency matrices of graphs for enumeration in combinatorics. The main idea is to express the generating function for a sequence $\{a_n\}$ in terms of the adjacency matrix of an associated graph. To apply this approach, one has to construct a bijection between the objects in question and walks in an appropriate edge-weighted directed multi-graph. We state the general result, which will also be used in Section 7.4 for generating trees.

Theorem 7.33 *[182, Theorem 4.7.2] Consider an edge-weighted directed multi-graph G on p vertices $v_1, \ldots, v_p$, and let A denote its weighted adjacency matrix, that is, a_{ij} is the total weight of the edges from v_i to v_j. Then the generating function for the total weight of walks from v_r to v_s is given by*

$$\sum_{n \geq 0} (A^n)_{r,s} x^n = (I - xA)^{-1}_{r,s} = \frac{(-1)^{r+s} \det(I - xA : s, r)}{\det(I - xA)}, \qquad (7.1)$$

where I is the identity matrix and $\det(B : s, r)$ is the minor of B obtained when removing row s and column r. In particular, the generating function is a rational function of x whose degree is strictly less than the multiplicity n_0 of the value 0 as an eigenvalue of A.

Proof Let $F_{ij}(A, x) = \sum_{n \geq 0} (A^n)_{ij} x^n = (I - xA)^{-1}_{ij}$. From Theorem B.12, we have that

$$F_{ij}(A, x) = \frac{(-1)^{i+j} \mathrm{adj}(I - xA)_{ij}}{\det(I - xA)},$$

which gives the desired result for the generating function. To show the claim about the degree of the generating function, suppose that A is a $p \times p$ matrix and that n_0 is the multiplicity of the eigenvalue 0. Then λ^{n_0} is a factor of the characteristic polynomial (see Definition B.20) and

$$\det(\lambda I - A) = a_{p-n_0} \lambda^{n_0} + \cdots + a_1 \lambda^{p-1} + \lambda^p.$$

Thus by Theorem B.17, Part 1,

$$\det(I - xA) = x^p \det\left(\frac{1}{x} I - A\right) = a_{p-n_0} x^{p-n_0} + \cdots + a_1 x + 1,$$

which implies that as polynomials in x,

$$\deg(\det(I - xA)) = p - n_0 \quad \text{and} \quad \deg(\det(I - xA : j, i)) \leq p - 1.$$

Hence, $\deg F_{ij}(A, x) \leq p - 1 - (p - n_0) < n_0$, as claimed. $\qquad \square$

Since the generating function is a rational function whose maximal degree is bounded by a computable quantity (namely the degree of the matrix A), it can be used to derive asymptotic results. Theorem 7.34 is a version of Theorem 7.33 specific to automata.

Theorem 7.34 *Let* $k \in \mathbb{N}$, τ *be a pattern,* e_k *be the number of states in* $\mathcal{A}(\tau, k)$, *and let* $A(\tau, k)$ *be the adjacency matrix of the graph* $G_\mathcal{A}$ *with the states in canonical order. Then the generating function for* $AW^\tau_{[k]}(n)$ *is given by*

$$AW^\tau_{[k]}(x) = \frac{\sum_{j=1}^{e_k-1} (-1)^{1+j} \det(I - xA{:}j, 1)}{\prod_{i=1}^{e_k-1}(1 - \ell_i x)}, \qquad (7.2)$$

where ℓ_i *is the number of loops at state* $\langle v_i \rangle$, *and* $e_k = |\mathcal{E}(\tau, k)|$. *If the states are not in canonical order, then the denominator is the product of the eigenvalues of* $(I - xA)$.

Proof This follows immediately from Theorem 7.33 since we want to count the number of paths of length n from $\langle \varepsilon \rangle$ to any state other than $\langle \tau \rangle$ in $\mathcal{A}(\tau, k)$. Since the states are in canonical order, A is a triangular matrix and its determinant is given by the product of the entries on the diagonal. □

We will next find the exact asymptotics (up to a constant) for $AW^\tau_{[k]}(n)$ for all patterns τ. The following simple lemma given in [27], which tells us about the number of loops, will be helpful in finding the asymptotic growth of the sequence $AW^\tau_{[k]}(n)$ for fixed k and τ.

Lemma 7.35 *Let* $\mathcal{A}(\tau, k)$ *be an automaton,* d *be the number of distinct letters in* τ, *and suppose that* $k \geq d - 1$. *If* $\langle v \rangle$ *is any state different from* $\langle \tau \rangle$, *then the number of loops at* $\langle v \rangle$ *does not exceed* $d - 1$. *Moreover, there are exactly* $d - 1$ *loops at* $\langle \varepsilon \rangle$.

Proof Suppose that there are more than $d - 1$ loops at $\langle v \rangle$ for $v \overset{\mathcal{A}}{\not\sim} \tau$. Then the loops use at least d different edge labels. From these labels we can form a word w that is order-isomorphic to τ. But then $vw \overset{\mathcal{A}}{\sim} v$ which is a contradiction. To show the second statement, let τ_1 be the first letter of τ. Then, for $i < \tau_1$ or $i > k - d + \tau_1$, we have $i \overset{\mathcal{A}}{\not\sim} \varepsilon$. But there are $d - 1$ such i's, which proves the lemma. □

To express how a sequence behaves for large values of n, we use the following notation.

Definition 7.36 *For two sequences* $\{a_n\}$ *and* $\{b_n\}$ *of real numbers, we write* $a_n \approx b_n$ *if* $\lim_{n \to \infty} \frac{a_n}{b_n} = 1$. *If* $a_n \approx K^n$ *for fixed* K, *then we say that* $\{a_n\}$ *is of exponential order* K^n.

We can now state the asymptotics for any pattern τ with d letters.

Theorem 7.37 *Let τ be any pattern with d distinct (but possibly repeated) letters and let $k \geq d - 1$ be given. Then there is a constant $C > 0$ such that*

$$AW_{[k]}^{\tau}(n) \approx Cn^M(d-1)^n \quad (n \to \infty),$$

where $M + 1$ is the maximum number of states with $d - 1$ loops in any simple path.

Proof Let $P = v_1 v_2 \cdots v_j$ be a simple path in $\mathcal{A}(\tau, k)$ of length $j - 1$. Then $AW_{[k]}^{\tau}(n) = \sum_P N(P, n)$, where $N(P, n)$ is the number of paths of length n associated with the simple path P of length $j - 1$ with $1 \leq j \leq n + 1$. Each simple path of length $j - 1$ can be extended to a walk of length n by passing through a total of $n - j + 1$ of the loops. Let ℓ_i be the number of loops at vertex v_i, and α_i be the number of loops the walk passes through at vertex v_i. Then

$$N(P, n) = \sum_{\alpha_1 + \cdots + \alpha_j = n - j + 1} \ell_1^{\alpha_1} \ell_2^{\alpha_2} \cdots \ell_j^{\alpha_j}, \tag{7.3}$$

where the sum is over all weak ($\alpha_i \geq 0$) compositions of $n - j + 1$ into j parts. Since $N(P, n)$ is a convolution, the associated generating function is a product of geometric series, and therefore $N(P, n) = [t^{n-j+1}](1 - \ell_1 t)^{-1} \cdots (1 - \ell_j t)^{-1}$.

Let r be the number of indices i such that $\ell_i = d - 1$. Note that by Lemma 7.35, r is greater than or equal to one. Using partial fraction decomposition, we obtain that the dominant term of $(1 - \ell_1 t)^{-1} \cdots (1 - \ell_j t)^{-1}$ is equal to

$$\frac{f(t)}{(1 - (d-1)t)^r},$$

where $f(t)$ is a polynomial of degree less than r and $f((d-1)^{-1}) \neq 0$. By well known results from complex analysis (see Appendix E and Chapter 8) it follows that $N(P, n) \approx C(P)(d-1)^n n^{r-1}$, where $C(P) > 0$ is a constant that depends on P and k. Summing over all paths and taking limits, only the contribution of the paths with maximal r "survives." $\square$

If every state except $\langle \tau \rangle$ in $\mathcal{A}(\tau, k)$ has exactly $d - 1$ loops, then it follows from Equation (7.3) that $AW_{[k]}^{\tau}(n) = (d-1)^n Q(n)$, where Q is a polynomial in n. We have in fact the following result.

Corollary 7.38 *Let τ be a pattern with d distinct letters and let $\mathcal{A}(\tau, k)$ be an automaton where all states but $\langle \tau \rangle$ have exactly $d - 1$ loops. Then*

$$AW_{[k]}^{\tau}(n) = \sum_{j=0}^{M} a_j (d-1)^{n-j} \binom{n}{j},$$

where a_j counts the number of simple paths of length j in $\mathcal{A}(\tau, k)$. Moreover, if τ is a pattern of length $\ell + 1$, then $a_i = (k - d + 1)^i$ for all $i = 0, 1, \ldots, \ell$.

Proof The corollary follows from the proof of Theorem 7.37, since the longest simple path is of length M, and $N(P, n) = (d-1)^{n-j} \binom{n}{j}$ for a path of length j. If τ is a pattern of length $\ell + 1$, then all the words of length $j \leq \ell$ avoid τ, and therefore, $AW_{[k]}^{\tau}(j) = \sum_{i=0}^{j} a_i (d-1)^{j-i} \binom{j}{i} = k^j$. Comparing coefficients in the binomial expansions, we have that $a_i = (k-d+1)^i$ for $i = 0, 1, \ldots, \ell$. □

Thus we can obtain exact results very nicely for any pattern or family of patterns for which the assumptions of Corollary 7.38 apply. The increasing patterns to be discussed in Section 7.3.3 are one such family. Finding other patterns or families of patterns remains an open question (see Research Direction 7.2). On the other hand, specifying the alphabet as opposed to the pattern gives a very nice exact result.

Example 7.39 *If τ is any pattern of length $\ell + 1$ with exactly d different letters and $[k] = [d]$, then Corollary 7.38 gives that*

$$AW_{[d]}^{\tau}(n) = \sum_{j=0}^{\ell} (d-1)^{n-j} \binom{n}{j}.$$

We can also obtain explicit results using the program TOU_FORMULA, which utilizes Theorem 7.34 for computing an explicit formula for $AW_{[k]}^{\tau}(n)$. This Maple program is posted on Toufik Mansour's homepage [139] and also listed in Appendix G.5. It uses as input the adjacency matrix of the automaton $\mathcal{A}(\tau, k)$ from the C++ program TOU_AUTO and returns the exact formula for $AW_{[k]}^{\tau}(n)$. These two programs allow us to get an explicit formula for $AW_{[k]}^{\tau}(n)$ for any given τ and $k \geq 1$, limited only by computing power. Using $[b_0, b_1, \ldots, b_d]_x = \sum_{j=0}^{d} b_j \binom{n}{j} x^{n-j}$ as a shorthand notation, Table 7.2 presents the explicit formulas for $AW_{[k]}^{\tau}(n)$ for all patterns $\tau \in S_4$ for $k = 3, \ldots, 6$. The complete analysis for all the subsequence patterns of length four is posted on Toufik Mansour's homepage [140].

We now turn to a family of patterns, namely the increasing patterns of length $\ell + 1$.

7.3.3 The increasing patterns

In [169], Regev gave a complete answer for the asymptotic behavior of $AW_{[k]}^{\tau_\ell}(n)$ as $n \to \infty$ for the increasing pattern $\tau_\ell = 12 \cdots (\ell + 1)$. His proof made use of results from representation theory. We will present a different proof of Regev's result based on both the structure of the automaton for increasing patterns and on results on Young tableaux given by Brändén and Mansour [27].

We start by deriving the structural results for the automaton $\mathcal{A}(\tau_\ell, k)$ which allow us to apply Corollary 7.38 to obtain an exact result for the constant in the asymptotic expression.

TABLE 7.2:   Avoidance of permutation patterns $\tau \in \mathcal{S}_4$ in words

τ	k	$AW^{\tau}_{[k]}(n)$
1234, 1243	3	$[1]_3$
1432, 2143	4	$[1, 1, 1, 1]_3$
	5	$[1, 2, 4, 8, 11, 10, 5]_3$
	6	$[1, 3, 9, 27, 66, 126, 183, 189, 126, 42]_3$
1324	3	$[1]_3$
	4	$[1, 1, 1, 1]_3$
	5	$[1, 2, 4, 8, 11, 10, 5, 1]_3$
	6	$[1, 3, 9, 27, 66, 126, 183, 197, 152, 80, 26, 4]_3$
1342	3	$[1]_3$
	4	$[1, 1, 1, 1]_3$
	5	$[1, 2, 4, 8, 11, 10, 4]_3$
	6	$[1, 3, 9, 27, 66, 126, 176, 168, 96, 24]_3$
1423	3	$[1]_3$
	4	$[1, 1, 1, 1]_3$
	5	$[1]_2 + [0, 3, 3, 9, 10, 11, 3]_3$
	6	$[13, 1]_2 + [-12, 15, -2, 37, 57, 134, 169, 167, 76, 12]_3$
2413	3	$[1]_3$
	4	$[1, 1, 1, 1]_3$
	5	$[10, 4, 1]_2 + [-9, 8, 1, 9, 11, 10, 2]_3$
	6	$[96, 28, 5]_2 + [-95, 71, -36, 54, 52, 132, 167, 137, 44, 4]_3$

Lemma 7.40 *Let $k \geq \ell$ be given and let $\tau_\ell = 12\cdots(\ell+1)$. For any set $S = \{s_1 < \cdots < s_m\} \subseteq [k]$ and $j \in [k]$, let*

$$S^j = \{s_1 < \cdots < s_{i-1} < j < s_{i+1} < \cdots < s_m\},$$

where i is the index such that $s_{i-1} < j \leq s_i$ ($s_0 := 0, s_{m+1} := k+1$). In addition, let w_S be the word consisting of the elements of S listed in increasing order, where $\varepsilon = w_\varnothing$. Then the transition function Δ for the automaton $\mathcal{A}(\tau_\ell, k)$ is given by

$$\Delta(\langle w_S \rangle, j) = \begin{cases} \langle w_{S^j} \rangle & \text{if } |S^j| \leq \ell, \text{ and } j < k - \ell + 1 + m \\ \langle w_S \rangle & \text{if } j \geq k - \ell + 1 + m \\ \langle \tau_\ell \rangle & \text{otherwise} \end{cases}.$$

and the number of states is

$$|\mathcal{E}(\tau_\ell, k)| = \binom{k}{\ell} + 1.$$

In particular, the loops of w_S are the elements of $S \cup \{k, k-1, \ldots, k-\ell+m+1\}$, and therefore, $\langle w_S \rangle$ has exactly ℓ loops.

To follow the proof of this result it may be helpful to look at Figure 7.7 which shows the automaton $\mathcal{A}(\tau_2, k)$.

Proof It is clear that the words w_S are representatives of different classes. Since we are to avoid an increasing pattern of length $\ell + 1$, we have only $k - (\ell + 1) + 1 = k - \ell$ letters available in the alphabet from which to start the full pattern. At each step, one more letter is available as the partial pattern still to be avoided shrinks by one letter. In addition, there is the state $\langle \tau \rangle$. Thus,

$$|\mathcal{E}(\tau_\ell, k)| = 1 + \sum_{i=0}^{\ell} \binom{k - (\ell + 1) + i}{i} = 1 + \binom{k}{\ell},$$

where we have used (A.6) with $a = k - \ell - 1$.

Let $v \in \mathcal{AW}^{[k]^*}(\tau_\ell)$. We say that an increasing subword $x_1 x_2 \cdots x_j$ of v is *extendible* if $x_j \leq k + j - \ell - 1$, that is, $x_1 x_2 \cdots x_j$ can be extended into an occurrence of τ_ℓ using letters from $[k]$. Suppose that the longest extendible increasing subsequence in v consists of $s \leq \ell$ letters. For $1 \leq j \leq s$ let

$$r_j(v) := \min\{x_j \mid x_1 x_2 \cdots x_j \text{ is an extendible subword of } v\}.$$

Clearly $r_1(v) < r_2(v) < \cdots < r_s(v)$. Let

$$S = \{r_1(v), r_2(v), \ldots, r_s(v)\}.$$

Then we see that $w_S \overset{\mathcal{A}}{\sim} v$, that is, we have determined the equivalence class for any v. The statement about the transition function Δ follows from the construction of the sets S^j together with arguments analogous to those given in Example 7.30 for the automaton of the pattern $\tau_2 = 123$. The transition function immediately gives that the elements of S are loops for w_S. In addition, for $j \in \{k, k-1, \ldots, k-\ell+m+1\}$, $\langle w_S j \rangle$ is not extendible and therefore $\langle w_S j \rangle = \langle w_S \rangle$. □

In order to give the promised alternative proof of Regev's result we will need some standard notions and notation from the theory of partitions and symmetric functions. We will state the necessary definitions and refer the reader to an in-depth discussion of these topics in [183, Chapter 7].

Definition 7.41 *To each partition (see Definition 1.1) we can associate a Young diagram as follows: for a partition $\vec{\lambda} = (\lambda_1, \lambda_2, \ldots, \lambda_k) \in \mathbb{N}^k$ with $\sum_{i=1}^{k} = n$ and $\lambda_1 \geq \lambda_2 \geq \cdots \geq \lambda_k$, the Young diagram is an array of k left-justified boxes with λ_i boxes in the i-th row, where rows are labeled from top to bottom. The vector $\vec{\lambda}$ is referred to as the* shape *of the Young diagram. If we have a partition into k equal parts of size m, that is, $\vec{\lambda} = (m, m, \ldots, m)$, then we write $\lambda_{k,m}$ and refer to the diagram as having size $k \times m$. A* Young tableau *(plural tableaux) is a filling of a Young diagram with letters from some alphabet. If the entries in the Young tableau are all distinct and entries in all rows and columns are strictly increasing, then the filling is called a* standard Young tableau. *If the entries increase weakly in rows and increase strictly* *in columns, then the Young tableau is called a* semi-standard tableau. *The* weight *of a Young tableau is given by the sequence of values that record the number of times each integer appears in the Young tableau (similar to the definition of the content vector).*

Note that it is customary to refer to partitions as vectors $\vec{\lambda}$ in the context of Young tableaux, and to write them as a vector, while in other contexts, partitions are written as sequences.

Example 7.42 *Figure 7.9 shows the five semi-standard Young tableaux of shape* $(2,2)$. *The leftmost two tableaux are standard Young tableaux. The weights for the five tableaux are:* $(1,1,1,1)$, $(1,1,1,1)$, $(1,1,2)$, $(2,2)$, *and* $(2,1,1)$. *Clearly, standard Young tableaux always have weight vectors consisting of all ones.*

1	3
2	4

1	2
3	4

1	2
3	3

1	1
2	2

1	1
2	3

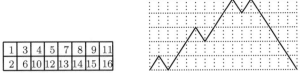

FIGURE 7.9: Young tableaux of shape $\lambda_{2,2}$ (or size 2×2).

Example 7.43 *Standard Young tableaux of size* $2 \times n$ *are enumerated by the Catalan numbers, which also enumerate Dyck paths (see Example 2.50). This can be seen using the following bijection: given a standard Young tableau of size* $2 \times n$, *read the integers from* 1 *to* $2n$ *in order, and create a path in the integer lattice starting at* $(0,0)$ *by placing an up-step if the integer occurs in the top row, and a down-step if the integer occurs in the bottom row. It is not hard to show that this mapping is a bijection (see [183, Exercise 6.19 (ww)]). Figure 7.10 shows a* 2×8 *standard Young tableau and its associated Dyck path.*

1	3	4	5	7	8	9	11
2	6	10	12	13	14	15	16

FIGURE 7.10: Standard Young tableau and associated Dyck path.

Young diagrams may remind you of Ferrers diagrams (see Definition 6.13), and indeed, they refer to the same type of combinatorial object. In drawing the Ferrers diagrams, we have followed the *French convention* (rows numbered from bottom to top, like the Cartesian coordinate system). By contrast, we will draw the Young diagrams and corresponding Young tableaux following the *English convention* (rows numbered from top to bottom, like matrix indices), thereby accommodating all tastes. Apparently, this nomenclature of English and French conventions started out as a joke. In his book on symmetric functions [135, p.2], MacDonald suggests to readers preferring the French convention to "read [his] book upside down in a mirror." (The French also prefer to have the column index listed before the row index – a convention we do NOT follow). The real difference between Ferrers diagrams and Young diagrams is in their content – fillings of Ferrers diagrams consist of zeros and ones, whereas Young tableaux are filled with positive integers.

Definition 7.44 *A partially ordered set or poset is a set P together with a binary order relation $\leq$ satisfying the following three properties:*

1. *For all $x \in P$, $x \leq x$ (reflexivity).*

2. *If $x \leq y$ and $y \leq x$, then $x = y$ (antisymmetry).*

3. *If $x \leq y$ and $y \leq z$, then $x \leq z$ (transitivity).*

We use $x \geq y$ to mean $y \leq x$, $x < y$ to mean that $x \leq y$ and $x \neq y$, and $x > y$ to mean $y < x$. Two elements x and y of P are comparable *if $x \leq y$ or $y \leq x$; otherwise, x and y are* incomparable. *The* dual P' *of a poset P is defined on the elements of P with order $x \leq y$ in $P' \Leftrightarrow y \leq x$ in P. If P and P' are isomorphic, then P is called* self-dual.

Example 7.45 *The following sets with their respective orders are familiar posets.*

(1) *Any collection of sets together with set-inclusion, that is, $S \leq T$ if $S \subseteq T$.*

(2) *X^k together with the component-wise order induced by the order of X: for $\vec{x}, \vec{y} \in X^k$, $\vec{x} \leq_{X^k} \vec{y}$ if and only if $x_i \leq_X y_i$ for $i = 1, 2, \ldots, k$.*

(3) *The set of Young diagrams together with the component-wise order. Visually, a Young diagram $\vec{\lambda} \leq \vec{\mu}$ if and only if the Young diagram for $\vec{\lambda}$ fits inside the Young diagram of $\vec{\mu}$. The poset of Young diagrams defined by $\{\vec{\mu} | \vec{\mu} \leq \vec{\lambda}\}$ is denoted by $\lambda_{\lambda_1, \lambda_2, \cdots, \lambda_k}$.*

Definition 7.46 *An (induced) subposet of P is a subset Q of P together with a partial ordering of Q such that for $x, y \in Q$ we have*

$$x \leq y \text{ in } Q \Leftrightarrow x \leq y \text{ in } P.$$

The order on Q is called the induced order. *In particular, the (closed) interval $[x, y] = \{z \in P \mid x \le z \le y\}$ is defined whenever $x \le y$.*

Definition 7.47 *A* chain (*or* totally ordered set *or* linearly ordered set) *is a poset in which any two elements are comparable. A subset C of a poset P is called a chain if C is a chain when regarded as a subposet of P. The* length *of a finite chain C is defined as $\ell(C) = |C| - 1$. A* maximal chain *is a chain of maximal length in P. An order-preserving bijection from P to the poset $[n]$ is called a* linear extension *of P. The number of linear extensions of P is denoted by $e(P)$. An* order ideal *of P is a subset I of P such that if $x \in I$ and $y \le x$, then $y \in I$. The set of all order ideals of P, ordered by inclusion, forms a poset denoted by $J(P)$.*

It can be shown (see [182, Section 3.5]) that $e(P)$ is equal to the number of maximal chains in $J(P)$.

Definition 7.48 *An* (order) lattice *is a poset P in which every pair of elements has both a unique* supremum (*or* least upper bound) *and an* infimum (*or* least lower bound). *A* bounded lattice *has a least element $\hat{0}$ and a largest element $\hat{1}$, where $\hat{0} \le x \le \hat{1}$ for all $x \in P$.*

Example 7.49 (*Young lattice*) *We will look at the poset $P = [\varnothing, \lambda_{2,2}]$, the set of Young diagrams whose shape is contained in $(2, 2)$. The order of the shapes is by component-wise comparison. This poset is a lattice, as can be clearly seen from Figure 7.11, and is referred to as the* Young lattice. *For example, the infimum of $(2, 1)$ and (1) is (1), and their supremum is $(2, 1)$. The largest element is $\hat{1} = (2, 2)$ and the least element is $\hat{0} = \varnothing$. The two maximal chains are $\{\varnothing, (1), (2), (2, 1), (2, 2)\}$ and $\{\varnothing, (1), (1, 1), (2, 1), (2, 2)\}$ of length four. The set $I = \{\varnothing, (1), (1, 1)\}$ constitutes one of the six order ideals of P. Note that by definition of the maximal chain, its elements consist of shapes that differ exactly by one box. Therefore, the maximal chain has length $2 \cdot 2 = 4$, and in general, the maximal chain in $[\varnothing, \lambda_{k,m}]$ has length $k \cdot m$.*

Definition 7.50 *Two posets P and Q are* isomorphic *if there exists an order-preserving bijection $\Phi : P \to Q$ whose inverse is order-preserving, that is*

$$x \le y \text{ in } P \Leftrightarrow \Phi(x) \le \Phi(y) \text{ in } Q.$$

We can now define a partial order on $\mathcal{A}(\tau_\ell, k)$ that is isomorphic to a well-known poset.

Theorem 7.51 *Let the partial order on the states in $\mathcal{A}(\tau_\ell, k)$ be defined as follows:*

$$\langle x \rangle \le \langle y \rangle \Leftrightarrow \text{ there exists a path from } \langle x \rangle \text{ to } \langle y \rangle \text{ in } \mathcal{A}(\tau_\ell, k).$$

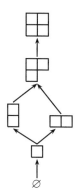

FIGURE 7.11:   Young lattice $[\varnothing, \lambda_{2,2}]$.

Then this partial order is isomorphic to $J([\ell] \times [k - \ell])$, the order ideal of the poset $[\ell] \times [k - \ell]$ (with the component-wise order).

Proof   Let $S, T \subseteq [k]$. Without loss of generality, we can assume that

$$S = \{s_1 < s_2 < \cdots < s_\ell\} \text{ and } T = \{t_1 < t_2 < \cdots < t_\ell\}.$$

(If $|S| < \ell$, then we can extend the set by appending the values $\{k + s - 1 - \ell, k + s + 2 - \ell, \ldots, k\}$, where $s = |S|$, and likewise for T.) We claim the following:

There exists a path from $\langle w_S \rangle$ to $\langle w_T \rangle \Leftrightarrow s_i \geq t_i$ for all $1 \leq i \leq \ell$.   (∗)

If there is an edge (that is, path of length one) from $\langle w_S \rangle$ to $\langle w_T \rangle$ then by Lemma 7.40 we are done. For paths of length $j \geq 2$, the claim follows by induction on the length of the path. To prove the other direction, assume that $s_i \geq t_i$ for all $1 \leq i \leq \ell$. Consider the path

$$\langle w_S \rangle \xrightarrow{t_1} \langle w_S t_1 \rangle \xrightarrow{t_2} \langle w_S t_1 t_2 \rangle \xrightarrow{t_3} \cdots \xrightarrow{t_\ell} \langle w_S t_1 t_2 \cdots t_\ell \rangle.$$

It is not hard to see that $\langle w_S t_1 t_2 \cdots t_\ell \rangle = \langle w_T \rangle$, as the path successively transforms S into T by definition of Δ, and therefore (∗) holds. We now establish a bijection between the partial order on the automaton and the Young lattice $[\varnothing, \lambda_{\ell, k-\ell}]$. Let $\vec{s} = (s_1, s_2, \ldots, s_\ell)$ consist of the elements of S ordered by size, and define

$$\Phi(\vec{s}) = (s_\ell - \ell, s_{\ell-1} - \ell + 1, \ldots, s_1 - 1).$$

Because the s_i are strictly increasing and $i \leq s_i \leq k$, the function Φ produces a Young diagram contained in $\ell \times (k - \ell)$. Likewise, we can obtain a vector

$\vec{s}$ with distinct elements ordered by size when starting from a Young diagram λ. Using the bijection

$$\Psi((\lambda_1, \lambda_2, \ldots, \lambda_\ell)) = (k - \ell - \lambda_\ell, \ldots, k - \ell - \lambda_1),$$

we see that $[\varnothing, \lambda_{\ell, k-\ell}]$ is its own dual. Therefore,

$$\langle w_S \rangle \leq \langle w_T \rangle \Leftrightarrow \Psi(\Phi(\vec{s})) \leq \Psi(\Phi(\vec{t})),$$

that is, the partial order on the automaton is order-isomorphic to the Young lattice $[\varnothing, \lambda_{\ell, k-\ell}]$. Now the statement follows from the simple fact that $[\varnothing, \lambda_{\ell, k-\ell}]$ is isomorphic to $J([\ell] \times [k - \ell])$. $\qquad\square$

Example 7.52 *Using the bijections given in the proof of Theorem 7.51, we can give the order lattice for the automaton $\mathcal{A}(123, 4)$. We illustrate how to associate states and Young shapes for the state $\langle 13 \rangle$, which has set $S = \{1, 3\}$ and vector $\vec{s} = (1, 3)$. Since $\ell = 2$ and $k = 4$, $\Phi((1, 3)) = (3 - 2, 1 - 1) = (1, 0)$. The dual is obtained by $\Psi((1, 0)) = (2 - 0, 2 - 1) = (2, 1)$. Thus the state $\langle 13 \rangle$ is associated with the Young shape $(2, 1)$. Figure 7.12 shows the Young lattice $[\varnothing, \lambda_{2,2}]$ together with the associated states of the automaton $\mathcal{A}(123, 4)$.*

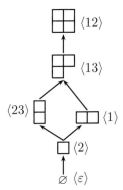

FIGURE 7.12: Young lattice $[\varnothing, \lambda_{2,2}]$ with associated states of $\mathcal{A}(123, 4)$.

So far we have established a correspondence between the states of the automaton and Young diagrams, but what are the counterparts of simple paths? Can they be "found" in this lattice? Not quite – we need to look at Young tableaux as opposed to just Young diagrams. We will illustrate that the simple paths of the automaton of an increasing pattern τ_ℓ correspond to chains in the set of all Young tableaux on $[\varnothing, \lambda_{\ell, k-\ell}]$ in Example 7.53. To assist with this task, Figure 7.13 shows the Young tableaux associated with each of the Young diagrams corresponding to the automaton $\mathcal{A}(123, 4)$. For the proof of this bijection, see Stanley [183, Chapter 7].

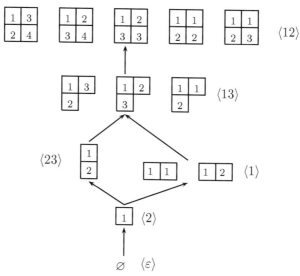

FIGURE 7.13: Young tableaux contained in the shape $\lambda_{2,2}$.

Example 7.53 *We will describe how the Young tableaux given in Figure 7.13 correspond to simple paths in the automaton. Each state is associated with a specific Young diagram, which in turn has several associated Young tableaux. For each such Young tableau corresponding to a given state, the largest element in the tableau indicates the length of the simple path from $\langle \varepsilon \rangle$ to that state. For example, the Young diagram $(2, 1)$ corresponds to the state $\langle 13 \rangle$ (see Figure 7.12). There are three associated Young tableaux, namely those filled with 132, 123 and 112 (where the fillings are always from top to bottom and from left to right in the given order).*

For the first one, looking at the filling we can see that the tableau was successively created by the sequence of tableaux with fillings 1 (corresponds to $\langle 2 \rangle$), 12 (corresponds to $\langle 23 \rangle$), and finally 132. Therefore the Young tableau with filling 132 corresponds to the simple path $\langle \varepsilon \rangle \rightarrow \langle 2 \rangle \rightarrow \langle 23 \rangle \rightarrow \langle 13 \rangle$ of length three, the maximal element in the Young tableau with filling 132. On the other hand, the Young tableau 11 corresponding to the state $\langle 1 \rangle$ has maximal entry one, so state $\langle 1 \rangle$ was reached directly from $\langle \varepsilon \rangle$, and therefore, the Young tableau 112 corresponds to the simple path of length 2 given by

$\langle\varepsilon\rangle \to \langle 1\rangle \to \langle 13\rangle$.

In addition, the longest simple paths correspond to those Young tableaux that fill the shape $\lambda_{2,2}$ completely and also contain all possible values. These tableaux correspond to a diagram that is the largest element in the Young lattice, and therefore, the length of the longest simple path equals the length of the maximal chain in the Young lattice.

We now give the alternative proof of Regev's result:

Theorem 7.54 *[169, Theorem 1] For all $k \geq \ell$ we have*

$$AW_{[k]}^{\tau_\ell}(n) \approx C_{\ell,k} n^{\ell(k-\ell)} \ell^n \quad (as\ n \to \infty),$$

where $C_{\ell,k}^{-1} = \ell^{\ell(k-\ell)} \prod_{i=1}^{\ell} \prod_{j=1}^{k-\ell} (i+j-1)$.

Proof By Lemma 7.40 each state except $\langle\tau_\ell\rangle$ has ℓ loops. Thus Corollary 7.38 applies and we have that

$$AW_{[k]}^{\tau_\ell}(n) = \sum_{j=0}^{M} a_j \ell^{n-j} \binom{n}{j},$$

where a_j counts the number of simple paths of length j in $\mathcal{A}(\tau_\ell, k)$ and M is the maximal number of states (since all have ℓ loops) in a simple path. From the proof of Theorem 7.51 we have that the maximal simple paths in the automaton correspond exactly to the maximal chains in the Young lattice $[\varnothing, \lambda_{\ell,k-\ell}]$. The length of the maximal chains in $[\varnothing, \lambda_{\ell,k-\ell}]$ is given by $M = \ell(k-\ell)$. Since the dominant term with regard to the limiting behavior is the one with the highest power, M, we obtain that

$$AW_{[k]}^{\tau_\ell}(n) \approx a_M \ell^{-M} \binom{n}{M} \ell^n \approx \frac{a_M}{M!} \ell^{-M} n^M \ell^n \quad (n \to \infty),$$

where a_M is equal to the number of maximal chains in $J([\ell] \times [k-\ell])$. Using results on maximal chains in $J([\ell] \times [k-\ell])$ [183, Proposition 7.10.3] and the hook-length formula [183, Corollary 7.21.6] we have that

$$a_{\ell(k-\ell)} = \frac{(\ell(k-\ell))!}{\prod_{i=1}^{\ell} \prod_{j=1}^{k-\ell} (i+j-1)},$$

which gives the desired result. $\qquad\square$

For the moment, we will leave asymptotics behind. However, we will draw on what we have derived so far to obtain an exact result for the special case of

$\ell = 2$, that is, avoidance of the pattern $\tau = 123$ via a combinatorial proof. It should be clear from the correspondence in Theorem 7.51 and Example 7.53 that the simple paths of length r in $\mathcal{A}(\tau_\ell, k + \ell)$ are in one-to-one correspondence with tableaux T of the following type:

(1) T is weakly increasing in rows and columns.

(2) No integer appears in more than one row.

(3) The entries of T are exactly $[r]$.

(4) The shape of T is contained in $\lambda_{\ell,k}$.

Before giving the result on $AW_{[k]}^{123}(n)$ we need some notation.

Definition 7.55 *A Young tableau that satisfies Properties (1) and (2) above is called a* segmented *tableau, and we denote the number of segmented tableaux satisfying properties (3) and (4) by* $a(\ell, k, r)$. *A segmented tableau of size* $[\ell] \times [k]$ *is called* primitive *if all columns are different (that is, every pair of columns differs in at least one entry). We denote the set of primitive segmented tableaux of size* $[\ell] \times [i]$ *with* r *different entries by* $\mathcal{PR}_{\ell,i,r}$, *and let* $\mathrm{pr}(\ell, i, r) = |\mathcal{PR}_{\ell,i,r}|$.

With these definitions, we can obtain an explicit formula for $AW_{[k+\ell]}^{\tau_\ell}(n)$ in terms of $\mathrm{pr}(\ell, i, r)$ which can be computed for special cases.

Proposition 7.56 *The number of words of length n on the alphabet $[k + \ell]$ that avoid the increasing pattern* $\tau_\ell = 12 \cdots (\ell + 1)$ *is given by*

$$AW_{[k+\ell]}^{\tau_\ell}(n) = \sum_{r=0}^{\ell k} \sum_{i=0}^{r} \ell^{n-r} \mathrm{pr}(\ell, i, r) \binom{k}{i} \binom{n}{r}. \tag{7.4}$$

Proof Since the number of words of length n on $[k + \ell]$ avoiding τ_ℓ is the same as the number of paths of length n in the automaton, we can count the number of words that avoid an increasing pattern τ_ℓ via the paths in the automaton. All such paths are loop paths that can be counted via the simple paths. Select the r vertices that define the simple path, which can be done in $\binom{n}{r}$ ways. Since the total length is n, there have to be $n - r$ loops. At each of the vertices, there are ℓ loops to choose from. Because of the structure of the loop paths, selecting m_i loops at the i-th vertex in the path results in a total of $\ell^{m_1+m_2+\cdots+m_r} = \ell^{n-r}$ choices. Finally, the simple paths are in one-to-one correspondence to primitive segmented tableaux, and therefore

$$AW_{[k+\ell]}^{\tau_\ell}(n) = \sum_{r=0}^{\ell k} \ell^{n-r} a(\ell, k, r) \binom{n}{r}. \tag{7.5}$$

Next, we can create the segmented tableau of size $[\ell] \times [k]$ satisfying properties (3) and (4) that have i different columns from the primitive segmented tableaux in $\mathcal{PR}_{\ell,i,r}$ by inserting α_1 copies of the first column to the left of the first column, α_2 copies of the second column between the first and the second column, and so on. After the i-th column we may insert α_{i+1} columns of all blanks, requiring that

$$\alpha_1 + \alpha_2 + \cdots + \alpha_{i+1} = k - i.$$

This process can be done in $\binom{k}{i}$ ways, and therefore, we can express the number of segmented tableaux in terms of the number of primitive segmented tableaux as

$$a(\ell, k, r) = \sum_{i=0}^{r} \mathrm{pr}(\ell, i, r) \binom{k}{i},$$

which gives (7.4). $\qquad\square$

Note that the function $a(\ell, k, r)$ is actually a polynomial in k of degree r due to the creation of the segmented tableaux from the primitive segmented tableaux. The numbers $\mathrm{pr}(\ell, i, r)$ are generally hard to compute, but there are two special cases in which we can obtain nice results, namely $\mathrm{pr}(\ell, n, n)$ and $\mathrm{pr}(2, i, r)$. The latter expression will help us give the promised combinatorial proof for a closed formula for $AW_{[k]}^{123}(n)$. We start by deriving $\mathrm{pr}(\ell, n, n)$.

Theorem 7.57 *There is a bijection between the permutations of n avoiding τ_ℓ and the primitive segmented tableaux of size $[\ell] \times [n]$ with n different entries, that is,*

$$\mathrm{pr}(\ell, n, n) = |\mathcal{S}_n(\tau_\ell)|.$$

Proof We will define a bijection between $\mathcal{S}_n(\tau_\ell)$ and $\cup_{m=0}^{\ell} \mathcal{PR}_{m,n,n}$ such that the height m of the tableau corresponds to the greatest increasing subsequence in a given permutation (thus, $m \leq \ell$). With $r_i(v)$ as defined in the proof of Lemma 7.40, let $r(v) = (r_1(v), r_2(v), \ldots, r_m(v))$, where m is the length of the longest increasing subsequence in v. Finally, assume that k is large enough so that all increasing subsequences in permutations in $\mathcal{S}_n(\tau_\ell)$ are considered extendible.

Now if $\pi = \pi_1 \pi_2 \cdots \pi_n$ is any permutation in $\mathcal{S}_n(\tau_\ell)$, define a tableau $T = T(\pi)$ as follows. Let the first column of T be $r(\pi)$, the second column be $r(\pi_1 \cdots \pi_{n-1})$, and so on. For example, for the permutation 351462, the first column has entries 1, 2, and 6, since the smallest end value for increasing subsequences of length one is 1, for those of length two it is 2 (from the subsequence $\pi_3\pi_6$), and for length three it is 6 (from either the subsequence $\pi_3\pi_4\pi_5$ or the subsequence $\pi_1\pi_2\pi_5$). In the same manner, we obtain the remaining five columns and

$$T(351462) = \begin{array}{|c|c|c|c|c|c|} \hline 1 & 1 & 1 & 1 & 3 & 3 \\ \hline 2 & 4 & 4 & 5 & 5 \\ \hline 6 & 6 \\ \hline \end{array}.$$

For each new column, one of two things happens: either the value that was removed from consideration was a smallest end value of the increasing subsequences of length s, and therefore the respective value will increase if there is still an increasing subsequence of that length; or, if the removed value came from the only subsequence of length s in the previous step, then the entry in row s will be empty for the current (and any subsequent) column. Thus, there is exactly one value that is different in adjacent columns. By construction, we have that $T(\pi) \in \cup_{m=0}^{\ell} \mathcal{PR}_{m,n,n}$. Moreover, from Lemma 7.40 we get that a tableau T is the image of some $\mathcal{S}_n(\tau_\ell)$ if and only if the following are all true:

(a) T has n columns and entries $1, 2, \ldots, n$.

(b) Let T^i denote the i-th column. If $i < j$ then T^i is smaller than T^j in component-wise order. (If T^i and T^j have different size fill the empty slots of T^j with $n + 1$).

(c) Exactly one new entry appears or is modified every time when moving from T^{i+1} to T^i, that is, when scanning the columns from right to left.

Furthermore, if $T \in \cup_{m=0}^{\ell} \mathcal{PR}_{m,n,n}$, then conditions (a) and (b) are trivially satisfied. At least one new entry appears every time we move from T^{i+1} to T^i, because otherwise $T^i = T^{i+1}$ and T fails to be primitive. On the other hand, if more than one new entry appears in a transition, then in some later transition there cannot appear a new entry, since T has n columns and n distinct entries. This verifies condition (c) and gives the statement. □

From Theorem 7.57 we obtain as a special case that

$$\mathrm{pr}(2, n, n) = |\mathcal{S}_n(123)| = C_n = \frac{1}{n+1} \binom{2n}{n},$$

the n-th Catalan number, which was originally shown by Knuth [122]. We now look at a more general case, the enumeration of $\mathrm{pr}(2, i, r)$, where we also obtain a connection to the ubiquitous Catalan numbers.

Theorem 7.58 *The number of primitive segmented tableaux of size $2 \times i$ with r different entries is given by*

$$\mathrm{pr}(2, i, r) = \frac{1}{i+1} \binom{2i}{i} \binom{i}{r-i} = C_i \binom{i}{r-i},$$

where C_i is the i-th Catalan number.

Before we give a proof of Theorem 7.58 we will need some definitions and a lemma. Let $\mathcal{PR}^+(2, s, r)$ be the set of tableaux in $\mathcal{PR}(2, s, r)$ that completely fill the shape $[2] \times [s]$, and let $\mathrm{pr}^+(2, s, r) = |\mathcal{PR}^+(2, s, r)|$. Then

$$\mathrm{pr}(2, s, r) = \mathrm{pr}^+(2, s, r) + \mathrm{pr}^+(2, s, r + 1),$$

since we can obtain the tableaux with r entries that are not completely filled from the completely filled ones with $r + 1$ entries by deleting all entries $r + 1$. To prove the theorem we will show that

$$\mathrm{pr}^+(2, s, r) = \binom{s - 1}{2s - r} C_s,$$

where C_s is the s-th Catalan number.

Definition 7.59 *Two arrays A and B are said to be* order-equivalent *if $A_{ij} \leq A_{i'j'}$ if and only if $B_{ij} \leq B_{i'j'}$ for all i, j, i', j'.*

We first define an operation $+$ that takes tableaux with r different entries to tableaux with $r + 1$ different entries. Let $T \in \mathcal{PR}^+(2, s, r)$. Suppose that j is an index such that $T_{ij} = T_{i(j+1)}$ for either $i = 1$ or $i = 2$. Write T as $T = LR$, where L consists of the first j columns and R consists of the remaining $s - j$ columns. Let R' be the array that is order-equivalent to R with entries from the set $\{$distinct entries in $R\} \cup \{r + 1\} - \{T_{i(j+1)}\}$. Order-equivalence will then dictate which element (if any) must occur more than once. We define $T + j$ to be the tableau LR'. For example, let

$$T = \begin{array}{|c|c|c|c|} \hline 1 & 2 & 4 & 4 \\ \hline 3 & 5 & 5 & 6 \\ \hline \end{array}.$$

Then there are two such indices, namely $j = 2$ (since $T_{22} = T_{23}$) and $j = 3$ (since $T_{13} = T_{14}$). For $j = 2$, R' has entries $\{4, 5, 6\} \cup \{7\} - \{5\} = \{4, 6, 7\}$. Since the smallest element in R occurs twice, so the smallest element in R' has to occur twice. Therefore,

$$T + 2 = \begin{array}{|c|c|c|c|} \hline 1 & 2 & 4 & 4 \\ \hline 3 & 5 & 6 & 7 \\ \hline \end{array}.$$

Note also that $T \in \mathcal{PR}^+(2, s, r)$ has exactly $2s$ cells. Once we have placed the r distinct values, then there are $2s - r$ cells where a newly placed value can lead to an equality of adjacent cells. Thus there are $2s - r$ pairs i and j with $j \in [s - 1]$ and $i = 1$ or $i = 2$ such that $T_{ij} = T_{i(j+1)}$. Let $S = \{s_1 < s_2 < \cdots < s_t\}$ be the set of these indices for j (since we can have equality in only one of the two rows) and define a function $\Phi_S : \mathcal{PR}^+(2, s, r) \to \mathcal{ST}_{2,s}$ by

$$\Phi_S(T) = (\cdots ((T + s_t) + s_{t-1}) + \cdots + s_1),$$

where $\mathcal{ST}_{2,s}$ is the set of standard tableaux of shape $[2] \times [s]$ and Φ_S consists of applying the operation $+$ a total of t times, starting with the tableau T and iteratively adding the elements in S from right to left. The fact that Φ is a bijection will prove the theorem, because $|\mathcal{ST}_{2,s}| = C_s$ (see [183, Exercise 6.19 (ww)]). To find the inverse of Φ we need to define an inverse operation to $+$.

Let $T \in \mathcal{PR}^+(2, s, r)$ and $1 \le b \le s - 1$ be such that $T_{1b} < T_{1(b+1)}$ and $T_{2b} < T_{2(b+1)}$ (because this is the situation produced by $+$). Define two arrays $T|_b$ and $T|^b$ as follows. For $T = LR$, where L consists of the first b columns and R of the $s - b$ last columns of T, define $T|^b = LR'$ to be the array where R' is the unique array order-equivalent to R with entries from the set $\{\text{distinct entries in } R\} \cup \{T_{1b}\} - \{r\}$. Similarly, let $T|_b = LR'$ be the array where R' is the unique array order-equivalent to R with entries from the set $\{\text{distinct entries in } R\} \cup \{T_{2b}\} - \{r\}$. For $b = 2$,

$$\left.\begin{array}{|c|c|c|c|}\hline 1 & 2 & 4 & 4 \\ \hline 3 & 5 & 6 & 7 \\ \hline \end{array}\right.^2 = \begin{array}{|c|c|c|c|}\hline 1 & 2 & 2 & 2 \\ \hline 3 & 5 & 4 & 6 \\ \hline \end{array}$$

$$\left.\begin{array}{|c|c|c|c|}\hline 1 & 2 & 4 & 4 \\ \hline 3 & 5 & 6 & 7 \\ \hline \end{array}\right|_2 = \begin{array}{|c|c|c|c|}\hline 1 & 2 & 4 & 4 \\ \hline 3 & 5 & 5 & 6 \\ \hline \end{array}$$

Note that exactly one of $T|^2$ and $T|_2$ is a primitive segmented tableau. This is no accident.

Lemma 7.60 *Let $T \in \mathcal{PR}^+(2, s, r)$ and $1 \le b \le s - 1$ be such that $T_{1b} < T_{1(b+1)}$ and $T_{2b} < T_{2(b+1)}$. Then*

$$T|_b \in \mathcal{PR}^+(2, s, r - 1) \iff T|^b \notin \mathcal{PR}^+(2, s, r - 1)$$
$$\iff T_{2(b+1)} = T_{2b} + 1.$$

Moreover, if $A = T|_b \in \mathcal{PR}^+(2, s, r - 1)$, then $A_{2b} = A_{2(b+1)}$, and if $B = T|^b \in \mathcal{PR}^+(2, s, r - 1)$, then $B_{1b} = B_{1(b+1)}$.

Proof Let $A = T|_b$. All entries in T that are smaller than T_{2b} will be mapped onto themselves, and $A_{ij} = T_{ij} - 1$ for $A_{ij} > T_{2b}$. This is the case because all values in the set $\{T_{2b} + 1, T_{2b} + 2, \ldots, r\}$ have to occur to the right of T_{2b} by construction of the map Φ. Therefore, $A \in \mathcal{PR}^+(2, s, r - 1)$ if and only if $T_{2(b+1)} = T_{2b} + 1$ (since otherwise the entry T_{2b} will appear in both the first and the second row). Now we look at $B = T|^b$. For $i \ge 1$, let y_i be the entries in T satisfying $T_{2b} < y_i \le T_{2(b+1)}$ ordered by size. Then the entry y_1 will be mapped to an element smaller than T_{2b} and y_i will be mapped to y_{i-1} for $i > 1$. Thus $B \in \mathcal{PR}^+(2, s, r - 1)$ if and only if $T_{2(b+1)} > T_{2b} + 1$ as claimed. The last statement of the lemma follows from the arguments just given. $\square$

We are now ready to give a proof of Theorem 7.58.

Proof of Theorem 7.58   If $T \in \mathcal{PR}^+(2, s, r)$ and $1 \leq b \leq s - 1$ are such that $T_{1b} < T_{1(b+1)}$ and $T_{2b} < T_{2(b+1)}$, we define $T - b$ to be the array $T|_b$ or $T|^b$ that is in $\mathcal{PR}^+(2, s, r - 1)$. By Lemma 7.60 we have that

$$
\begin{aligned}
(T + j) - j = T &\quad \text{if } T_{ij} = T_{i(j+1)} \quad \text{for either } i = 1 \text{ or } i = 2, \\
(T - j) + j = T &\quad \text{if } T_{ij} < T_{i(j+1)} \quad \text{for both } i = 1, 2.
\end{aligned} \tag{7.6}
$$

Now, for $S = \{s_1 < s_2 < \cdots < s_t\}$ with $t = 2s - r$ and $P \in \mathcal{ST}_{2,s}$, we define

$$
\Psi(S, P) = P - s_1 - s_2 - \cdots - s_t.
$$

It follows from (7.6) that Ψ is the inverse to Φ and therefore Theorem 7.58 follows. $\square$

Putting all the pieces together results in a combinatorial proof of the closed form for the number of words that avoid the subsequence pattern 123, which is different, but of course equivalent to the formula proved originally by Burstein [31] using a noncombinatorial argument (see Theorem 7.61). Note that the result of Theorem 7.61 applies to all permutation patterns $\tau \in \mathcal{S}_3$ (since we have shown in Section 6.3 that all subsequence permutation patterns of length three are equivalent).

Theorem 7.61   *For all $n, k \geq 0$ we have that*

$$
AW^{123}_{[k+2]}(n) = \sum_{r=0}^{2k} \sum_{i=0}^{r} 2^{n-r} C_i \binom{i}{r-i} \binom{n}{r} \binom{k}{i},
$$

where $C_i = \frac{1}{i+1} \binom{2i}{i}$ is the i-th Catalan number. Moreover, the generating function $F^{123}(x, y) = \sum_{n,k \geq 0} AW^{123}_{[k+2]}(n) x^n y^k$ is given by

$$
F^{123}(x, y) = \frac{1}{(1-y)(1-2x)} C\left(\frac{xy(1-x)}{(1-y)(1-2x)^2} \right),
$$

where $C(x)$ is the generating function for the Catalan numbers. Equivalently, $F^{123}(x, y)$ is algebraic and satisfies the equation:

$$
xy(1-y)(1-x)F^2 - (1-y)(1-2x)F + 1 = 0.
$$

Proof   The formula for $AW^{123}_{[k+\ell]}(n)$ follows immediately from Proposition 7.56 and Theorem 7.58. The result for the generating function is obtained by using the result of [31] for the generating function $AW^{123}_{[k]}(x, y)$, adjusted for the shift in the sequence $AW^{123}_{[k+2]}(n)$. $\square$

We close this section with a result on the sequence $\{AW^\tau_{[n]}(n)\}_{n \geq 0}$. For permutation patterns τ, the study of $\mathcal{S}_n(\tau)$ has been an area of very active research. Since $\mathcal{S}_n(\tau) \subset \mathcal{AW}_n^{[n]}(\tau)$, the sequence $\{AW_{[n]}^{\tau}(n)\}_{n \geq 0}$ is a natural extension and of particular interest. Typical questions posed involve finding an explicit formula for the number of words of length n on the alphabet $[n]$ avoiding a given pattern. If we are only able to derive an expression for the generating function, then typical questions concern the type of generating function for $\{AW_{[n]}^{\tau}(n)\}_{n \geq 0}$. We can give a partial answer for patterns from $\mathcal{S}_3$. To do so, we need the following definition.

Definition 7.62 *A sequence* $\{f(n)\}$ *is* polynomially recursive (*or* P-recursive) *if there are a finite number of polynomials* $p_i(n)$ *such that*

$$\sum_{i=0}^{N} p_i(n) f(n+i) = 0,$$

for all integers $n \geq 0$.

This property of a sequence is closely related to a property of its generating function.

Definition 7.63 *A generating function* $f(x)$ *of one variable is* D-finite *if there exist polynomials* $q_i(x)$, $i = 0, \ldots, m$ *with* $p_m(x) \neq 0$ *such that*

$$\sum_{i=0}^{m} q_i(x) u^{(i)} = 0,$$

where $u^{(i)} = d^i u / dx^i$ *is the* i-*th derivative of* u *with respect to* x.

Note that a sequence is P-recursive if and only if its generating function is D-finite (see [183, Proposition 6.4.1]). Furthermore, every algebraic generating function (see Definition 2.40) is also D-finite (see [183, Proposition 6.4.6]), so the sequence of coefficients of an algebraic generating function is always P-recursive. However, not every P-recursive sequence has an algebraic generating function. (For details, see [183, Chapter 6.4].) Thus the question is: Is the generating function for $\{AW_{[n]}^{\tau}(n)\}_{n \geq 0}$ algebraic? If not, is the sequence at least P-recursive?

Theorem 7.64 *Let* τ *be a permutation pattern of length three. Then the sequence* $f(n) = AW_{[n]}^{\tau}(n)$ *is P-recursive and satisfies the three term recurrence*

$$p(n)f(n-2) + q(n)f(n-1) + r(n)f(n) = 0,$$

where

$$p(n) = 3(n-3)(n-1)(3n-5)(3n-4)(5n-4),$$
$$q(n) = 288 - 1440n + 2780n^2 - 2435n^3 + 976n^4 - 145n^5, \; and$$
$$r(n) = 2(n-2)^2 n(n+1)(5n-9).$$

Proof The fact that $f(n)$ is P-recursive follows easily from the expansion of $f(n)$ as a double sum using Theorem 7.61 and the theory developed in [129]. The polynomials p, q and r were found using the package MULTISUM (see [172]) developed by Riese. □

Corollary 7.65 *The asymptotic behavior of* $f(n) = AW^\tau_{[n]}(n)$ *as* $n \to \infty$ *for* $\tau \in \mathcal{S}_3$ *is given by*

$$f(n) \approx Cn^{-2} \left(\frac{27}{2} \right)^n$$

for some constant $C > 0$.

Proof This is a direct consequence of Theorem 7.64 and the theory of asymptotics for P-recursive sequences (see [194]). □

Since the exponents of n in the asymptotic expansion of a sequence with an algebraic generating function are always nonnegative, Corollary 7.65 indicates that the generating function of $AW^\tau_{[n]}(n)$ is transcendental (that is, not algebraic).

7.3.4 Further results and generalizations

The results in the previous sections on avoidance of a pattern τ in words can be generalized in two different ways:

1. Avoidance of a set of patterns T in words.

2. Avoidance of a pattern or sets of patterns in compositions.

In both cases, the generalization is in the definition of the automaton. Once we have defined an appropriate automaton, we can then apply Theorem 7.34 to the respective adjacency matrix and replace τ by T and/or *words* by *compositions*. Note that the equivalence $\overset{A}{\sim}$ includes compositions, as each composition is also a word. Therefore, we only need to extend the equivalence for the case of pattern avoidance of a set of patterns.

Definition 7.66 *Given a set of patterns T, we define an equivalence relation $\sim_T$ on $[k]^*$ (where $[k]^*$ is the set of all finite words with letters from $[k]$) as $v \sim_T w$ if for all words $r \in [k]^*$ we have*

$$vr \text{ avoids } T \Leftrightarrow wr \text{ avoids } T,$$

where a word u avoids T if u avoids all patterns in T simultaneously.

Note that Lemma 7.27 generalizes to the equivalence relation $\sim_T$.

Definition 7.67 *The automaton* $\mathcal{A}(T, k)$ *for avoidance of a set of patterns* T *in words is defined by replacing* $\mathcal{E}(\tau, k)$ *with* $\mathcal{E}(T, k)$, *the set of equivalence classes of* $\sim_T$.

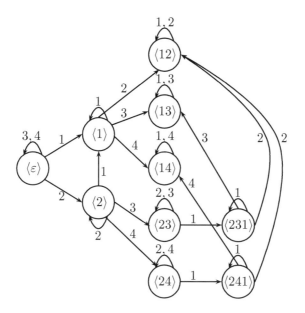

FIGURE 7.14: Automaton $\mathcal{A}(\{123, 132\}, 4)$.

Example 7.68 *Figure 7.14 shows the automaton for 4-ary words that avoid the pair of patterns* $\{123, 132\}$. *Note that for this automaton, there are edges from states that have labels with three letters to states that have two letter labels; nevertheless, there is no cycle in this graph. (We have seen a similar situation for the automaton of Example 7.32.) Using the order of states* $\langle \varepsilon \rangle$, $\langle 1 \rangle$, $\langle 2 \rangle$, $\langle 12 \rangle$, $\langle 13 \rangle$, $\langle 14 \rangle$, $\langle 23 \rangle$, $\langle 24 \rangle$, $\langle 213 \rangle$, *and* $\langle 241 \rangle$, *the adjacency matrix for this automaton is given by*

$$A(\{123, 132\}, 4) = \begin{bmatrix} 2 & 1 & 1 & 0 & 0 & 0 & 0 & 0 & 0 & 0 \\ 0 & 1 & 0 & 1 & 1 & 1 & 0 & 0 & 0 & 0 \\ 0 & 1 & 1 & 0 & 0 & 0 & 1 & 1 & 0 & 0 \\ 0 & 0 & 0 & 2 & 0 & 0 & 0 & 0 & 0 & 0 \\ 0 & 0 & 0 & 0 & 2 & 0 & 0 & 0 & 0 & 0 \\ 0 & 0 & 0 & 0 & 0 & 2 & 0 & 0 & 0 & 0 \\ 0 & 0 & 0 & 0 & 0 & 0 & 2 & 0 & 1 & 0 \\ 0 & 0 & 0 & 0 & 0 & 0 & 0 & 2 & 0 & 1 \\ 0 & 0 & 0 & 1 & 1 & 0 & 0 & 0 & 1 & 0 \\ 0 & 0 & 0 & 1 & 0 & 1 & 0 & 0 & 0 & 1 \end{bmatrix}.$$

Note that even though the matrix with this ordering of states is not a triangular matrix, there is an ordering that results in a triangular adjacency matrix (which one?). Using the Maple command

```
f:=simplify(evalm((1-x*A)^(-1)&*[1,1,1,1,1,1,1,1,1,1])[1]);
with(genfunc): rgf_expand(f,x,n);
```

with this adjacency matrix A gives the generating function

$$AW_{[4]}^{\{123,132\}}(x) = \frac{4x^2 - 2x + 1}{(1 - 2x)^3},$$

and $AW_{[4]}^{\{123,132\}}(n) = (n^2 + n + 2)2^{n-1}$. The corresponding Mathematica *commands are*

```
g[x_]:=Apply[Plus,Inverse[IdentityMatrix[10]-x A][[1]]]//Factor
SeriesCoefficient[g[x], {x,0,n}]
```

Now we will turn our attention to compositions.

Definition 7.69 *For pattern avoidance in compositions, we define the automaton $\mathcal{A}(T, A)$ for any (finite) ordered subset A of $\mathbb{N}$ with $\mathcal{E}(T, A)$ as the set of equivalence classes of $\sim_T$, and label the edges between the state $\langle \sigma \rangle$ and $\langle \sigma a_i \rangle$ by x^{a_i} instead of by a_i. In this case, the edge labels are weights, unlike in Definition 7.29, where the labels a_i were for identification only. The (i, j) entry of the associated weighted adjacency matrix now counts the total weight of the edges between states i and j. An edge with multiple labels $x^{i_1}, x^{i_2}, \ldots, x^{i_m}$ has weight $\sum_{j=1}^{m} x^{i_j}$.*

Note that the use of weights on the edges in the case of compositions leads to results in terms of generating functions rather than results on the number of compositions of n that avoid a pattern τ or set of patterns T. We will now present some results and examples on subsequence patterns, subword patterns and generalized patterns in compositions.

Example 7.70 *As indicated in Definition 7.69, we need to put weights on the edges when enumerating compositions. It is easy to draw the automaton for avoidance of the subsequence pattern $\tau = 123$ in compositions with parts in $A = [3]$ (do it) to obtain the associated adjacency matrix as*

$$A = \begin{bmatrix} x^2 + x^3 & x & 0 \\ 0 & x + x^3 & x^2 \\ 0 & 0 & x + x^2 \end{bmatrix}.$$

Using the Maple command

```
factor(evalm( (1-A)^(-1)&*[1,1,1] )[1]);
```

or the Mathematica *command*

```
Apply[Plus, Inverse[IdentityMatrix[3] - A][[1]]] // Factor
```

with the matrix A as defined above, we obtain the generating function

$$AC_{[3]}^{123}(x) = \frac{1 - x - x^2 + x^4 + x^5}{(1 - x^2 - x^3)(1 - x - x^3)(1 - x - x^2)}.$$

This example can be generalized. The fact that the set A consists exactly of the letters of the pattern forces a very specific structure on the automaton which allows us to give an explicit formula for the generating function.

Theorem 7.71 *Let $\tau = \tau_1 \tau_2 \cdots \tau_\ell$ be any subsequence pattern of length ℓ with d distinct letters. Then the generating function for the number of compositions of n with m parts in $[d]$ that avoid the pattern τ is given by*

$$AC_{[d]}^\tau(x, y) = \sum_{j=1}^{\ell} \frac{(-1)^{j-1} y^{j-1} x^{\tau_1 + \cdots + \tau_{j-1}}}{\left(1 - y \sum_{i \in [d] \setminus \{\tau_1\}} x^i\right) \cdots \left(1 - y \sum_{i \in [d] \setminus \{\tau_j\}} x^i\right)}.$$

In particular,

$$AC_{[d]}^\tau(x) = \sum_{j=1}^{\ell} \frac{(-1)^{j-1} x^{\tau_1 + \cdots + \tau_{j-1}}}{\left(1 - \sum_{i \in [d] \setminus \{\tau_1\}} x^i\right) \cdots \left(1 - \sum_{i \in [d] \setminus \{\tau_j\}} x^i\right)}.$$

Proof Let τ be any pattern of length ℓ with d distinct letters. Then the automaton that enumerates the number of compositions of n with parts in $A = [d]$ that avoid the pattern τ is given by Figure 7.15, where the state $\langle t_i \rangle = \langle \tau_1 \tau_2 \cdots \tau_i \rangle$.

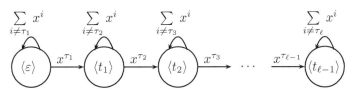

FIGURE 7.15: Automaton for avoidance of $\tau \in [d]^\ell$.

Thus, the adjacency matrix is of the form

$$A = \begin{bmatrix} \sum_{i \neq \tau_1} x^i & x^{\tau_1} & 0 & \cdots & 0 \\ 0 & \sum_{i \neq \tau_2} x^i & x^{\tau_2} & \cdots & 0 \\ & & \ddots & \ddots & \\ 0 & 0 & 0 & \ddots & x^{\tau_{\ell-1}} \\ 0 & 0 & 0 & & \sum_{i \neq \tau_\ell} x^i \end{bmatrix}.$$

We now use Equation (7.1) of Theorem 7.33 and compute $(I - yA)^{-1}$. Note that for compositions, the length of the walk is enumerated by y, the number of parts of the composition. Now let B be an $\ell \times \ell$ matrix of the form $B = (B_{ij})$ with $B_{ii} = \alpha_i$, $B_{i,i+1} = \beta_i$, and $B_{ij} = 0$ for all $j - i \neq 0, 1$. Then it can be shown by induction that $B^{-1} = (B^{-1})_{ij}$ is given by

$$
(B^{-1})_{ij} = \begin{cases} 0, & \text{if } i > j \\ (-1)^{i+j} \dfrac{\beta_i \beta_{i+1} \cdots \beta_{j-1}}{\alpha_i \alpha_{i+1} \cdots \alpha_j} & \text{if } i \leq j \end{cases}.
$$

Thus for $B = I - yA$ we have that

$$
\alpha_i = 1 - y \sum_{i \neq \tau_i} x^i \quad \text{and} \quad \beta_i = y x_i^\tau.
$$

This implies that the first row of B^{-1} is given by

$$
\left[\frac{1}{1 - y \sum_{i \neq \tau_1} x^i}, \frac{-y x^{\tau_1}}{\prod_{j=1}^2 \left(1 - y \sum_{i \neq \tau_j} x^i\right)}, \cdots, \frac{(-1)^{\ell-1} y^{\ell-1} x^{\sum_{j=1}^{\ell-1} \tau_j}}{\prod_{j=1}^{\ell} \left(1 - y \sum_{i \neq \tau_j} x^i\right)} \right].
$$

Adding these entries yields the desired result. $\qquad\square$

We can obtain a similar result for avoidance of subword patterns in compositions.

Theorem 7.72 *Let $\tau = \tau_1 \tau_2 \cdots \tau_\ell$ be any subword pattern of length ℓ with d distinct letters. Then the generating function for the number of compositions of n with m parts in $[d]$ that avoid the pattern τ is given by*

$$
AC_{[d]}^\tau(x, y) = \frac{1 + \sum_{j=1}^{\ell-1} x^{\tau_1 + \cdots + \tau_j} y^j}{1 - \sum_{j=1}^{\ell} x^{\tau_1 + \cdots + \tau_{j-1}} \left(\sum_{i \in [d] \setminus \{\tau_j\}} x^i\right) y^j}.
$$

In particular,

$$
AC_{[d]}^\tau(x) = \frac{1 + \sum_{j=1}^{\ell-1} x^{\tau_1 + \cdots + \tau_j}}{1 - \sum_{j=1}^{\ell} x^{\tau_1 + \cdots + \tau_{j-1}} \left(\sum_{i \in [d] \setminus \{\tau_j\}} x^i\right)}.
$$

Proof   See Exercise 7.8. $\qquad\square$

We conclude this section with an example for the generalized pattern 12–3.

Example 7.73 *Figure 7.16 shows the automaton for compositions in $A = [4]$ that avoid the pattern 12–3. Using the order of states*

$$
\langle \varepsilon \rangle, \ \langle 1 \rangle, \ \langle 2 \rangle, \ \langle 12 \rangle, \ \langle 13 \rangle, \ \langle 131 \rangle,
$$

the adjacency matrix for this automaton is given by

$$\begin{bmatrix} x^3 + x^4 & x^1 & x^2 & 0 & 0 & 0 \\ x^4 & x^1 & 0 & x^2 & x^3 & 0 \\ x^4 & x^1 & x^2 & 0 & x^3 & 0 \\ 0 & 0 & 0 & x^1 + x^2 & 0 & 0 \\ 0 & 0 & 0 & 0 & x^2 + x^3 & x^1 \\ 0 & 0 & 0 & x^2 & x^3 & x^1 \end{bmatrix}.$$

To see that $\Delta(\langle 13 \rangle 1) = \langle 131 \rangle$, *note that* $13123 \overset{A}{\approx} 1323$. *Using Maple or* Mathematica *as in Example 7.70, we can obtain the generating function* $AC^{12-3}_{[4]}(x)$ *as*

$$AC^{12-3}_{[4]}(x) = \frac{(1-x)^5}{(1 - 4x + 3x^2 - x^3)(1 - 2x)}.$$

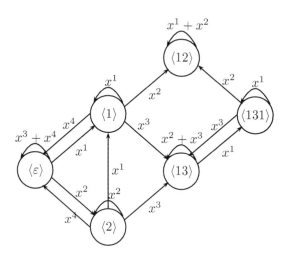

FIGURE 7.16: Automaton for compositions in [4] avoiding 12–3.

Note that for generalized patterns the automaton can have cycles, unlike in the case of subsequence patterns, where a cycle would have resulted in all equivalence classes in the cycle being in the same class. Here, however, the adjacency requirements prevent that the equivalence classes of a cycle are the same.

Finally a word on automata for avoiding substrings. We have not focused on these types of patterns but want to point out that automata are very well-suited for enumeration of words or compositions avoiding substrings. The main reason is that for a substring, we only have to consider the last letter, not a history.

Example 7.74 *We look at the automaton for avoidance of the substring 123 in words on the alphabet* [4]. *For avoidance of substrings, the states are of the form* $\langle s_1 \cdots s_j \rangle$ *for* $j = 1, 2, \ldots, \ell - 1$ *for a substring pattern* $s_1 s_2 \cdots s_\ell$. *So for the pattern 123 we have states* $\langle \varepsilon \rangle$, $\langle 1 \rangle$, *and* $\langle 12 \rangle$. *The transfer matrix is given by*

$$A = \begin{bmatrix} 3 & 1 & 0 \\ 2 & 1 & 1 \\ 2 & 1 & 0 \end{bmatrix}.$$

Using Mathematica *or* Maple *we can obtain the generating function as*

$$AW_{[4]}^{123}(x) = \frac{1}{1 - 4x + x^3}$$

and the number of words on [4] *avoiding the substring 123 for* $n = 0, 1, \ldots, 15$ *is given by* 1, 4, 16, 63, 248, 976, 3841, 15116, 59488, 234111, 921328, 3625824, 14269185, 56155412, 220995824, *and* 869714111.

7.4 Generating trees

Generating trees were introduced by West in [191] as a method to capture the structure of a recurrence. He coined the term *transfer matrix* for the adjacency matrix of the digraph associated with the generating tree. Since this has become standard terminology in any paper on generating trees, we will use the term transfer matrix instead of adjacency matrix. We will also see that we can translate any generating tree into an associated automaton.

Definition 7.75 *A generating tree is a rooted labeled tree such that if* v_1 *and* v_2 *are two nodes with the same label, and* ℓ *is any label, then* v_1 *and* v_2 *have exactly the same number of children with label* ℓ. *To specify a generating tree it therefore suffices to specify:*

1. the label of the root, and

2. a set of succession rules explaining how to derive from the label of a parent the labels of all of its children.

Example 7.76 (*The complete binary tree*) *The complete binary tree is a tree in which each vertex has exactly two children. Since all the nodes in the complete binary tree are identical, it is enough to use only one label, which we choose to be* (2). *So we get the following description:*

Root: (2)

Rule: $(2) \rightsquigarrow (2)(2)$.

Example 7.77 (*The Fibonacci tree*) *The Fibonacci tree has many uses in computer science applications. It is a variant of the binary tree and can be visualized as follows. The root is a red vertex. A red vertex has a blue offspring, and a blue vertex has a blue and a red offspring. The two different types of vertices are thus distinguished by the number of offspring, so we label the vertices according to their number of children and obtain these rules:*

Root: (1)

Rules: $(1) \rightsquigarrow (2)$, $(2) \rightsquigarrow (1)(2)$

We can encode the information from the generating tree as an automaton whose states correspond to the labels. Furthermore, we place an edge from vertex l_i to vertex l_j for every occurrence of l_j in the succession rule of l_i given by $(l_i) \rightarrow \cdots$. We use the integers from 1 to the maximum number of labels in all the succession rules as the "alphabet" for the edge labels. Formally, we have the following definition of the automaton.

Definition 7.78 *For the automaton of a generating tree, the set of states $\mathcal{E}$ consists of the labels of the generating tree. The alphabet is $\Sigma = \{1, 2, \ldots, m\}$, where m is the maximal number of labels in any of the succession rules. For any label l_i with succession rule $(l_i) \rightsquigarrow (l_{i_1})(l_{i_2}) \cdots (l_{i_j})$, the transition rule is defined by $\Delta((l_i), k) = (l_{i_k})$ for $k = 1, \ldots, j$. S_0 is the label of the root, and $F = \mathcal{E}$.*

Example 7.79 (*Continuation of Example 7.76*) *The automaton for the complete binary tree is given by*

$$\mathcal{E} = \{(2)\}, \ \Sigma = \{1, 2\}, \ \Delta((2), 1) = \Delta((2), 2) = (2), \ S_0 = (2),$$

and its graph is shown in Figure 7.17. The transfer matrix for the binary tree is $A = [2]$. Using (7.1) of Theorem 7.33 we obtain that the generating function for the number of vertices at height n (which correspond to paths of length n starting from the root) in the complete binary tree as $\frac{1}{1-2x} = \sum_{n \geq 0} 2^n x^n$, and we recover the well-known result that the number of vertices at height n in the complete binary tree is given by 2^n. This is a rather trivial application of the transfer matrix, but it shows the power of the approach.

Example 7.80 (*Continuation of Example 7.77*) *The automaton for the Fibonacci tree is given by*

$$\mathcal{E} = \{(1), (2)\}, \ \Sigma = \{1, 2\}, \ \Delta((1), 1) = (2), \ \Delta((2), 1) = (1), \ \Delta((2), 2) = (2),$$

with initial state $S_0 = (1)$. The corresponding transfer matrix is

$$A = \begin{bmatrix} 0 & 1 \\ 1 & 1 \end{bmatrix}.$$

To compute the generating function for the number of vertices at height n, we need to compute the sum of the entries in the first row of

$$(I - xA)^{-1} = \begin{bmatrix} 1 & -x \\ -x & (1-x) \end{bmatrix}^{-1} = \frac{1}{1-x-x^2} \begin{bmatrix} 1-x & x \\ x & 1 \end{bmatrix},$$

and we obtain the generating function as $1/(1-x-x^2)$. This is the generating function of the sequence $\{f_n\}_n$, where f_n is the n-th shifted Fibonacci number (see Table A.1).

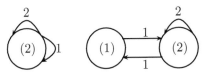

FIGURE 7.17: Automata for the complete binary and Fibonacci trees.

Generating trees were used by West in [191] to enumerate permutations avoiding a given subsequence pattern $\tau \in \mathcal{S}_n$. We start by giving the generating tree for the set of permutations $\mathcal{S}_n$.

Example 7.81 *The permutations in $\mathcal{S}_{n+1}$ can be created from the permutations in $\mathcal{S}_n$ by inserting the value $n+1$ at all possible positions in the permutations of n. Since each permutation in $\mathcal{S}_n$ creates exactly $n+1$ permutations of size $n+1$, the labels chosen for the generating tree give the number of children, with the empty permutation denoted by ε and given the label (1). Therefore, the succession rules are*

Root: (1)

Rules: $(n) \rightsquigarrow (n+1)^n,$

where $(i)^k$ is shorthand for k labels (i). The generating tree for the permutations is given in Figure 7.18, showing both the permutations and their respective labels.

The next step is to derive generating trees for permutations that avoid a pattern τ.

Example 7.82 *Given a pattern τ, one can define the rooted tree for permutations avoiding τ as follows. The nodes at level n are precisely the elements*

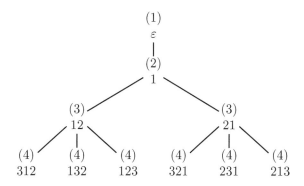

FIGURE 7.18: Generating tree for $\mathcal{S}_n$.

of $\mathcal{S}_n(\tau)$. *The parent of a permutation $\pi = \pi_1 \pi_2 \cdots \pi_{n+1} \in \mathcal{S}_{n+1}(\tau)$ is the unique permutation $\pi' = \pi_1 \cdots \pi_{j-1} \pi_{j+1} \cdots \pi_n$ such that $\pi_j = n+1$. Note that here we define the parent via the child, which ensures that the τ-avoidance is preserved. If we were to define the child via the parent, then we would have to delete those children for which the insertion of the value n creates an occurrence of the pattern τ.*

Definition 7.83 *The generating tree whose vertices at level n are the permutations of $\mathcal{S}_n(\tau)$ is denoted by $\mathcal{T}(\tau)$. Similarly, the tree corresponding to the set $\mathcal{S}_n(T)$ is denoted by $\mathcal{T}(T)$.*

The difficult work, which is truly an art, consists of translating this backward structural description of the generating tree $\mathcal{T}(T)$ into succession rules. The "translation" is known for some patterns, and not for others. We will give the rules for the generating trees $\mathcal{T}(123)$ and $\mathcal{T}(\{123, (k-1) \cdots 1k\})$. More examples of pattern avoidance in permutations can be found in [25] and the references cited therein.

Example 7.84 *West derived the generating tree $\mathcal{T}(123)$ [192, Example 4]. In this case the pattern to be avoided creates a very distinctive structure. If the permutation of n has at least one rise, with the leftmost rise $\pi_{i_1} \pi_{i_2}$ at position i_2, then we can insert the part $(n+1)$ only at positions $1, 2, \ldots, i_2$, where insertion at position i means the placement of the new part between π_{i-1} and π_i. If insertion is at position 1, the leftmost rise now occurs at position $i_2 + 1$; otherwise, the leftmost rise occurs at the insertion position i. If a permutation of n does not have any rise, then we can insert $(n+1)$ at any of $n+1$ positions, creating a rise at position i if the insertion is at position $i > 1$. Insertion at position 1 does not create a rise, which we can express as a rise at $n+2$ in order to combine the two cases. Overall, we have these* *rules, where the label indicate the position of the rise and the number of the children (why?):*

Root: (1)

Rules: $(p) \rightsquigarrow (p+1)(2)(3)\cdots(p)$.

Figure 7.19 gives the generating tree $T(123)$ showing both the permutations and the respective labels.

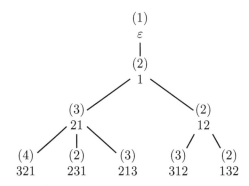

FIGURE 7.19:   Generating tree for 123-avoiding permutations.

Example 7.85 *Chow and West [53] proved that the succession rules for the tree $T(\{123, (k-1)\cdots 1k\})$ are*

Root: (2)

Rules: $(l) \rightsquigarrow (2)\cdots(l)(l+1)$, $l < k-1$
      $(k-1) \rightsquigarrow (2)\cdots(k-2)(k-1)(k-1)$.

Note that in this case, the label indicates the number of children of the vertex. The corresponding graph is shown in Figure 7.20. The transfer matrix is given by

$$
A_k = \begin{bmatrix}
1\,1\,0\,0 \cdots 0\,0\,0 \\
1\,1\,1\,0 \cdots 0\,0\,0 \\
1\,1\,1\,1 \cdots 0\,0\,0 \\
\vdots\,\vdots\,\vdots\,\vdots\,\ddots\,\vdots\,\vdots\,\vdots \\
1\,1\,1\,1 \cdots 1\,0\,0 \\
1\,1\,1\,1 \cdots 1\,1\,0 \\
1\,1\,1\,1 \cdots 1\,1\,1 \\
1\,1\,1\,1 \cdots 1\,1\,2
\end{bmatrix}
$$

It is well known (see [53, Thereom 3.1]) that the generating function for the number of permutations in $\mathcal{S}(\{123, (k-1)\cdots 21k)\})$ is given by

$$\frac{U_{k-1}\left(\frac{1}{2\sqrt{x}}\right)}{\sqrt{x}U_k\left(\frac{1}{2\sqrt{x}}\right)},$$

where U_k are the Chebyshev polynomials of the second kind (see Appendix C.2).

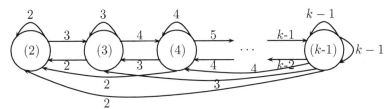

FIGURE 7.20: The automaton for $\mathcal{T}(123, (k-1)\cdots 21k)$.

We will now look at generating trees for words.

Example 7.86 *Every k-ary word of length $n+1$ can be created from a k-ary word of length n by adding one of the letters in $[k]$. Thus, we get a k-ary tree, that is, a tree where each vertex has exactly k children, and therefore, all vertices have the same label. If we let the label denote the number of children, then the succession rules are as follows:*

Root: (k)

Rules: $(k) \rightsquigarrow (k)^k$.

The generating tree for 2-ary words is shown in Figure 7.21.

Example 7.87 *The set $\mathcal{C}_n$ of compositions of n has almost the same succession rules as the binary tree, as each composition of n creates two compositions of $n+1$ by either appending a 1 on the right side or by increasing the rightmost part by 1. The only difference is the root, which corresponds to the empty composition. It has only one child, the composition 1. Therefore, the rules are:*

Root: (1)

Rules: $(1) \rightsquigarrow (2)$

$(2) \rightsquigarrow (2)$

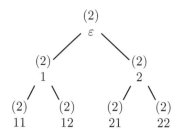

FIGURE 7.21: Generating tree for 2-ary words.

We now discuss the ECO method, a general framework for enumeration of combinatorial objects that is closely related to generating trees and therefore to automata.

7.5 The ECO method

We start by presenting the general framework for the ECO method. In essence, it is a road map (much like the Noonan-Zeilberger algorithm) in that it lays out the task to be done, but does not give a formulaic description of the steps.

Definition 7.88 *Let $\mathcal{O}$ be a class of objects which have an associated concept of size, that is, there exists a map $p : \mathcal{O} \to \mathbb{N}$, such that*

$$|\mathcal{O}_n| = |\{O \in \mathcal{O} \mid p(O) = n\}|$$

is finite. An operator *ϑ on the class $\mathcal{O}$ maps $\mathcal{O}_n$ to $2^{\mathcal{O}_{n+1}}$, where $2^{\mathcal{O}_{n+1}}$ is the power set of $\mathcal{O}_{n+1}$.*

Now we can present the main result for the ECO method, which asserts that there is a recurrence which allows the construction of all elements of size $n + 1$ from those of size n.

Theorem 7.89 *[16, Proposition 2.1] Let ϑ be an operator on $\mathcal{O}$. If ϑ satisfies the following conditions:*

1. *For each $O' \in \mathcal{O}_{n+1}$, there exists $O \in \mathcal{O}_n$ such that $O' \in \vartheta(O)$.*

2. *If $O, O' \in \mathcal{O}_n$ such that $O \neq O'$, then $\vartheta(O) \cap \vartheta(O') = \varnothing$.*

Then the family of sets $\{\vartheta(O) : O \in \mathcal{O}_n\}$ is a partition of $\mathcal{O}_{n+1}$.

If such an operator ϑ exists and it can be translated into a set of succession rules of the form

$$\textbf{Root: } (b)$$

$$\textbf{Rules: } (d) \rightsquigarrow (c_1)(c_2)\cdots(c_{q(d)}),$$

where $b, d, c_i \geq 0$, and $q : \mathbb{N} \to \mathbb{N}$, then the recursion can be described via a generating tree. Elements of $\mathcal{O}_n$ occur at the same level of the tree, and any element of $\mathcal{O}_n$ has as its children the elements of $\mathcal{O}_{n+1}$ created by the operator ϑ.

We illustrate the ECO method by enumerating several classes of words avoiding pairs of generalized patterns. The main idea in Bernini at al. [23] is to enumerate a special type of words avoiding a set of patterns via the ECO method, and then to obtain the total number of words avoiding the set of patterns from the count of the special words. We need some notation before describing the process.

Definition 7.90 *We call a word* reduced *if it is in reduced form (see Definition 1.20), and denote the set of reduced k-ary words of length n that avoid a set of patterns T by $\widetilde{\mathcal{AW}}_n^{[k]}(T)$. The number of such words will be denoted by $\widetilde{AW}_{[k]}^T(n)$.*

Note that each $w \in \widetilde{\mathcal{AW}}_n^{[m]}(T)$ is associated with exactly $\binom{k}{m}$ distinct words in $\mathcal{AW}_n^{[k]}(T)$ as we can replace the set of letters of w with any possible subset of $[k]$ having m elements, taking care of preserving the relative order of the letters. For instance, the reduced word $121321 \in \widetilde{\mathcal{AW}}_6^{[3]}(\{1\text{–}22, 2\text{–}12\})$ is associated with the following $\binom{5}{3} = 10$ words of $\mathcal{AW}_6^{[5]}(\{1\text{–}22, 2\text{–}12\})$:

$$\begin{array}{ccccc}
121321 & 121421 & 121521 & 131431 & 131531 \\
141541 & 232432 & 232532 & 242542 & 343543 \, \dot{}
\end{array}$$

Thus

$$AW_{[k]}^T(n) = \sum_{m=0}^{n} \binom{k}{m} \widetilde{AW}_{[m]}^T(n), \tag{7.7}$$

and we can restrict ourselves to finding an operator ϑ for the sets $\widetilde{\mathcal{AW}}_n^{[m]}(T)$. The ECO method now consists of the following steps:

1. Find a recursion for $\widetilde{\mathcal{AW}}_n^{[k]}(T)$, that is, identify the operator ϑ.

2. If possible, translate ϑ into a generating tree.

3. If possible, determine the generating function for $\widetilde{AW}_{[k]}^T(n)$, or even better, derive a closed formula.

Then apply (7.7) to obtain the corresponding results for $AW_{[k]}^T(n)$, assuming Steps 2 and 3 of the ECO method were successful.

So how can we create reduced words of length $n + 1$ from those of n? Usually, when recursively constructing words we append a letter to the right end of the word. We do the same thing here, except we have to ensure that the resulting word is again a reduced word and that none of the patterns in T are created. To assist with this task, we visualize the word as a path, where the height of the path corresponds to the letter. Since we are dealing with words (as opposed to permutations) we can append letters that have previously occurred (indicated by lines) or new letters (indicated by circles). To account for new letters we identify the regions above, between, and below the lines. These regions and the lines encode the potential insertion letters and we will call them *active sites*.

Now we identify for a word $w \in \widetilde{AW}^T_{[m]}(n)$ the *allowed active sites*, those that do not create any of the patterns in T. We mark regions and lines that are allowed by an empty circle. Obviously, this depends very much both on the word w and the patterns T to be avoided, which seems to make it hard to find a general recursion. For example, Figure 7.22 shows the word $121321 \in \widetilde{AW}^{[3]}_6(\{1\text{–}22, 2\text{–}12\})$ with the allowed active sites identified. Since 121321 ends in a 1, avoidance of 1–22 does not put any restrictions on the active sites. However, avoidance of 2–12 eliminates all previously occurring values bigger than the last one, that is, we cannot append the letters 2 or 3.

FIGURE 7.22:   Allowed active sites for $121321 \in \widetilde{AW}^{[3]}_6(\{1\text{–}22, 2\text{–}12\})$.

Once we have identified the allowed active sites, we have to describe what happens as the respective letter is appended. Appending a letter that has previously occurred to $w \in \widetilde{AW}^{[m]}_n(T)$ creates a word $w' \in \widetilde{AW}^{[m]}_{n+1}(T)$. If we insert a new value, that is, we choose an allowed active region, then we have to make some adjustments to the letters in w. If the region between i and $i + 1$ is chosen (for $i = 0, \ldots, m$), then the letter appended is $i + 1$, and all letters $j \geq i + 1$ in w get mapped to $j + 1$. This creates a word $w' \in \widetilde{AW}^{[m+1]}_{n+1}(T)$. Obviously, this case is only possible if $m < k$. Note that the process of creating words of length $n + 1$ satisfies the conditions for an operator ϑ stated in Theorem 7.89. Figure 7.23 shows the five words created from $121321 \in \widetilde{AW}^{[3]}_6(\{1\text{–}22, 2\text{–}12\})$ with their allowed active sites indicated. The allowed values were added in order from top to bottom. For example, insertion above the top line does not require relabeling of any letters in w.

However, insertion in the region between 2 and 3 requires that any letter 3 in w is renamed to a 4, thus mapping 121321 to $w = 1214213$. The only insertion on a line is the fourth case, which results in a reduced word on [3]. All together, starting from the word $121321 \in \widetilde{\mathcal{AW}}_6^{[3]}(\{1\text{--}22, 2\text{--}12\})$, we can construct five new reduced words of length 7, namely 1213214, 1214213, 1314312, 1213211, and 2324321.

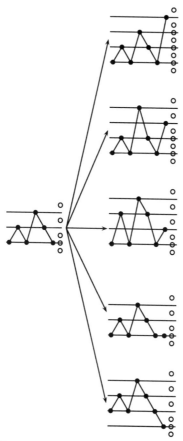

FIGURE 7.23: Generating the reduced words of length 7 from 121321.

After having explored a particular example, we need to find the general structure of the insertions for words $w \in \widetilde{\mathcal{AW}}_n^{[m]}(T)$ avoiding the pair of patterns $T = \{1\text{--}12, 2\text{--}21\}$. Suppose that the last letter of w is $h \leq m$. Since the patterns to be avoided have repeated letters, we distinguish two cases:

- h occurs for the first time in w.

- h has previously occurred in w.

To distinguish these two cases, we use the following notation: the label (m_h) denotes a word of $\widetilde{\mathcal{AW}}_n^{[m]}(T)$ ending with (the first occurrence of) the letter h; the label $(\bar{m}_h)$ denotes a word in $\widetilde{\mathcal{AW}}_n^{[m]}(T)$ ending with the letter h, where h also appears in some previous position. Now let's see what happens in each case. In the first case, h cannot be part of an occurrence of either 1–12 or 2–21, and we can add any letter on the right-hand side. Since w is a reduced word on $[m]$, the letters $1, \ldots, m$ have occurred before, and so insertion of any of these letters results in the labels $(\bar{m}_1) \cdots (\bar{m}_m)$. Insertion of any new letter i will create a word on the alphabet $[m+1]$ for which the letter i is the first occurrence, resulting in the labels

$$((m+1)_1) \cdots ((m+1)_{m+1}).$$

In the second case, h can play the role of 1 in the pattern 1–12 and therefore we cannot add any value that is bigger than h. For the pattern 2–21, the opposite is true, that is, we cannot add any value that is smaller than h, so h is the only allowed value. Thus, only one word gets created and the succession rule is given by

$$(\bar{m}_h) \rightsquigarrow (\bar{m}_h).$$

The smallest word is the word 1, which for obvious reasons avoids simultaneously 1–12 and 2–21, and therefore, the root is given by (1_1). All together we have the following succession rules:

Root: (1_1)

Rules: $(m_h) \rightsquigarrow (\bar{m}_1) \cdots (\bar{m}_m)((m+1)_1) \cdots ((m+1)_{m+1})$ for $m < k$
$(\bar{m}_h) \rightsquigarrow (\bar{m}_h)$.

These succession rules do not easily allow for enumeration, so let's look at them more closely. First of all, in the rule $(\bar{m}_h) \rightsquigarrow (\bar{m}_h)$, the value of h does not matter at all. What is important is the parameter m as it plays a significant role in the enumeration. Since each vertex with this label has exactly one child, we use the label (1_m) instead, and rewrite the rule as $(1_m) \rightsquigarrow (1_m)$, where (1_m) denotes a reduced word on the alphabet $[m]$ whose last letter appears somewhere else in the word. Now let's look at the rule for the first case. Here again, the value of h is unimportant. The vertex with label (m_h) has $m + (m+1) = 2m + 1$ children, so we should use the label $(2m+1)$ instead for a reduced word on the alphabet $[m]$ whose last letter does not appear anywhere else in the word. The rule thus translates to $(2m+1) \rightsquigarrow (1_m)^m (2m+3)^{m+1}$. Making the empty word the root with label (1) puts the words of length n at level n in the generating tree and we obtain this modified set of succession rules:

Root: (1)

Rules: $(2m+1) \rightsquigarrow (1_m)^m (2m+3)^{m+1}$ for $m < k$
$(1_m) \rightsquigarrow (1_m)$.

So what can we say about the new set of succession rules? From their definition, it is clear that

(i) each label $(2m+1)$ and (1_m) represents a word on the alphabet $[m]$;

(ii) the unique labels appearing at level n of the generating tree are (1_1), (1_2), $\cdots$, (1_{n-1}) and $(2n+1)$.

Therefore $\widetilde{AW}^{T}_{[m]}(n)$ is either the number of labels (1_m) at level n if $m < n$, or the number of labels $(2m+1)$ at level n if $m = n$. This results in the following explicit formula for $\widetilde{AW}^{T}_{[m]}(n)$ given by Bernini et al.

Theorem 7.91 *[23, Theorem 4.1] For fixed n and k and $m \le k$, we have*

$$\widetilde{AW}^{T}_{[m]}(n) = \begin{cases} m \cdot m! & \text{for } m < n \\ n! & \text{for } m = n \end{cases}.$$

Proof Let $a_{n,m} = \widetilde{AW}^{T}_{[m]}(n)$ be the number of reduced words of length n on the alphabet $[m]$. From the succession rules we obtain the following recursions:

$$a_{n,n} = n \cdot a_{n-1,n-1} \tag{7.8}$$

$$a_{n,n-1} = (n-1) \cdot a_{n-1,n-1} \tag{7.9}$$

$$a_{n,k} = a_{k+1,k} \tag{7.10}$$

where the last equation holds for $k < n - 1$. Since $a_{1,1} = 1$, we immediately obtain from (7.8) that $a_{n,n} = n!$. This result together with (7.9) gives that $a_{n,n-1} = (n-1)(n-1)!$. Finally, (7.10) yields $a_{n,k} = k \cdot k!$, for $k < n - 1$. Obviously, these formulas hold only for $m \le k$, and therefore we can use only a finite part of the generating tree. □

As an easy consequence of Theorem 7.91 we obtain the following proposition.

Proposition 7.92 *[23, Theorem 4.2] The number of k-ary words of length n that avoid the pair of generalized patterns $T = \{1\text{-}12, 2\text{-}21\}$ is given by*

$$AW^{T}_{[k]}(n) = \begin{cases} \sum_{m=0}^{n-1} m \cdot (k)_m + (k)_n & \text{for } n \le k \\ \sum_{m=0}^{k-1} m \cdot (k)_m & \text{for } n > k \end{cases},$$

where $(k)_m = k(k-1) \cdots (k-m+1) = k!/m!$.

Proof   The proof follows from Theorem 7.91 and (7.7). □

Bernini et al. [23] have used the ECO method for several pairs of generalized patterns of type $(2,1)$ and have obtained either an explicit formula or a generating function. We summarize their results in Table 7.3 and also indicate which cases are still missing for a complete classification of the pairs of generalized patterns of type $(2,1)$ with repeated letters. As before, the complement and reversal maps reduce the total number of different pairs to be considered. Bernini et al. pose the pairs with * as questions for further research.

TABLE 7.3:   Results for pairs of patterns of type $(1,2)$

1–11, 1–12	[23, Thm 6.3]
1–11, 1–21	[23, Thm 7.1]
1–11, 1–22	[23, Thm 7.2]
1–12, 1–21	*
1–12, 1–22	*
1–12, 2–11	*
1–12, 2–12	*
1–12, 2–21	Theorem 7.92
1–21, 1–22	*
1–21, 2–11	*
1–21, 2–12	[23, Thm 5.1]
2–11, 1–22	[23, Thm 7.3]

Note that we can obtain simplified expressions for some of the generating functions given in [23]. For instance, in Theorem 7.3 of [23] the generating function $AW_{[k]}^T(x)$ for $T = \{2\text{–}11, 1\text{–}22\}$ is given as

$$AW_{[k]}^{\{2\text{–}11,1\text{–}22\}}(x) = \sum_{m=0}^{k} \frac{(k)_m \cdot x^{k-1}}{\left(\prod_{i=1}^{m-1}(1 - ix)\right) \cdot (1 - x)}.$$

We will derive a simpler generating function for the Wilf-equivalent pair

$$T = \{11\text{–}2, 22\text{–}1\} \sim \{2\text{–}11, 1\text{–}22\}$$

using generating function techniques.

Theorem 7.93 *The generating function for k-ary words w of length n that avoid* $\{11\text{–}2, 22\text{–}1\}$ *is given by*

$$AW_{[k]}^T(x) = \frac{1 + (k - 1)x^2}{1 - kx + (k - 1)x^2}.$$

Proof Define $AW_{[k]}^T(w_1 w_2 \cdots w_s | x)$ to be the generating function for k-ary words w of length n that avoid T and start with $w_1 w_2 \cdots w_s$. Clearly,

$$AW_{[k]}^T(x) = 1 + \sum_{i=1}^{k} AW_{[k]}^T(i|x), \qquad (7.11)$$

where

$$AW_{[k]}^T(i|x) = x + \sum_{j \neq i} AW_{[k]}^T(ij|x) + AW_{[k]}^T(ii|x). \qquad (7.12)$$

Let w avoid T. If $w_1 = w_2 = i$, then avoiding 11–2 and 22–1 implies that we have to have $w_j = i$ for all j, and therefore, $AW_{[k]}^T(ii|x) = \frac{x^2}{1-x}$. If $w_1 \neq w_2$, then $AW_{[k]}^T(ij|x) = xAW_{[k]}^T(j|x)$ because both patterns start with repeated letters. All together, (7.12) can be written as

$$AW_{[k]}^T(i|x) = x + x\sum_{j \neq i} AW_{[k]}^T(j|x) + \frac{x^2}{1-x}$$

$$= x + x(AW_{[k]}^T(x) - AW_{[k]}^T(i|x) - 1) + \frac{x^2}{1-x},$$

where the second equality follows from (7.11). Thus,

$$(1+x)AW_{[k]}^T(i|x) = xAW_{[k]}^T(x) + \frac{x^2}{1-x}.$$

Summing over $i = 1, 2, \ldots, k$ leads to

$$(1+x)(AW_{[k]}^T(x) - 1) = kxAW_{[k]}^T(x) + \frac{kx^2}{1-x},$$

which gives the desired result after appropriate simplification. $\square$

We now obtain a formula for the generating function $AW_{[k]}^{\{11-2,12-2\}}(x)$, one of the cases listed as open in [23] by using the scanning-element algorithm.

Theorem 7.94 *Let* $T = \{11\text{–}2, 12\text{–}2\}$, $P(n, k) = AW_n^{[k]}(T)$,

$$\overline{P}(n, k) = \{w \in AW_n^{[k]}(T) \mid w_j \neq w_1 \text{ for } j = 2, \ldots, n\},$$

and let $P_k(x; v)$ *and* $\overline{P}_k(x; v)$ *be defined as in (6.23). Then*

$$P_k(x; v) = 1 + x + \frac{vx}{1-v}(P_k(x; v) - v^{k-1} P_k(x; 1))$$

$$+ \frac{x}{1-v}(\overline{P}_k(x; 1) - \overline{P}_k(x; v)) + x^2 \sum_{i=1}^{k} v^{i-1} P_i(x; 1),$$

and

$$\overline{P}_k(x; v) = x + \frac{xv}{1-v}(P_{k-1}(x; v) - v^{k-1}P_{k-1}(x; 1))$$
$$+ \frac{x}{1-v}(\overline{P}_{k-1}(x; 1) - v\overline{P}_{k-1}(x; v)),$$

where $P_0(x; v) = 1$ and $\overline{P}_0(x; v) = 0$.

Proof We use the scanning-element algorithm to find a formula for the generating function for the number of k-ary words of length n that avoid both 11–2 and 12–2. From the definition of the algorithm (see Section 6.5.3) we have that $p(n, k) = \sum_{i=1}^{k} p(n, k; i)$ and

$$p(n, k; i) = \sum_{j=1}^{i-1} p(n, k; i, j) + p(n, k; i, i) + \sum_{j=i+1}^{k} p(n, k; i, j)$$
$$= \sum_{j=1}^{i-1} p(n-1, k; j) + p(n-2, i) + \sum_{j=i+1}^{k} \overline{p}(n-1, k; j), \quad (7.13)$$

where $\overline{p}(n, k; j)$ is the number of k-ary words of length n that avoid both 11–2 and 12–2 such that the letter j occurs only at the beginning of the word. Applying the scanning-element algorithm now for $\overline{P}(n, k)$ we obtain

$$\overline{p}(n, k; i) = \sum_{j=1}^{i-1} p(n-1, k-1; j) + \sum_{j=i}^{k-1} \overline{p}(n-1, k-1; j), \quad (7.14)$$

as the letter i cannot occur anywhere else in the word, and we therefore can map the alphabet $[k] \setminus \{i\}$ to $[k-1]$. Let $P_{n,k}(v) = \sum_{i=1}^{k} p(n, k; i)v^{i-1}$ and $\overline{P}_{n,k}(v) = \sum_{i=1}^{k} \overline{p}(n, k; i)v^{i-1}$. Multiplying the recurrence relations (7.13) and (7.14) by v^{i-1} and summing over $i = 1, 2, \ldots, k$, we get that for all $n \geq 2$

$$P_{n,k}(v) = \sum_{i=1}^{k} v^{i-1} \sum_{j=1}^{i-1} p(n-1, k; j) + \sum_{i=1}^{k} v^{i-1} p(n-2, i)$$
$$+ \sum_{i=1}^{k} v^{i-1} \sum_{j=i+1}^{k} \overline{p}(n-1, k; j),$$

and

$$\overline{P}_{n,k}(v) = \sum_{i=1}^{k} v^{i-1} \sum_{j=1}^{i-1} p(n-1, k-1; j) + \sum_{i=1}^{k} v^{i-1} \sum_{j=i}^{k-1} \overline{p}(n-1, k-1; j).$$

Change of order of summation leads to

$$P_{n,k}(v) = \frac{v}{1-v}(P_{n-1,k}(v) - v^{k-1}P_{n-1,k}(1))$$

$$+ \frac{1}{1-v}(\overline{P}_{n-1,k}(1) - \overline{P}_{n-1,k}(v)) + \sum_{i=1}^{k} v^{i-1}p(n-2,i),$$

and

$$\overline{P}_{n,k}(v) = \frac{v}{1-v}(P_{n-1,k-1}(v) - v^{k-1}P_{n-1,k-1}(1))$$

$$+ \frac{1}{1-v}(\overline{P}_{n-1,k-1}(1) - v\overline{P}_{n-1,k-1}(v)).$$

Let $P_k(x;v) = \sum_{n\geq 0} P_{n,k}(v)x^n$ and $\overline{P}_k(x;v) = \sum_{n\geq 0} \overline{P}_{n,k}(v)x^n$. Using the initial conditions $P_{0,k}(v) = 1$, $\overline{P}_{0,k}(v) = 0$, and $P_{1,k}(v) = \overline{P}_{1,k}(v) = \frac{1-v^k}{1-v}$, we obtain that

$$P_k(x;v) = 1 + x + \frac{vx}{1-v}(P_k(x;v) - v^{k-1}P_k(x;1)) + x^2\sum_{i=1}^{k} v^{i-1}P_i(x;1)$$

$$+ \frac{x}{1-v}(\overline{P}_k(x;1) - \overline{P}_k(x;v)) \tag{7.15}$$

and

$$\overline{P}_k(x;v) = x + \frac{xv}{1-v}(P_{k-1}(x;v) - v^{k-1}P_{k-1}(x;1))$$

$$+ \frac{x}{1-v}(\overline{P}_{k-1}(x;1) - v\overline{P}_{k-1}(x;v)), \tag{7.16}$$

where $P_0(x;v) = 1$ and $\overline{P}_0(x;v) = 0$. □

The results of Theorem 7.94 allow us to iteratively compute $AW^T_{[k]}(x) = P_k(x;1)$. Rewriting (7.15) so that we can apply the kernel method leads to

$$P_k(x;v)\left(1 - \frac{vx}{1-v}\right) = 1 + x + \frac{x}{1-v}(\overline{P}_k(x;1) - \overline{P}_k(x;v))$$

$$+ \left(x^2v^{k-1} - \frac{v^k x}{1-v}\right)P_k(x;1) + x^2\sum_{i=1}^{k-1} v^{i-1}P_i(x;1). \tag{7.17}$$

Setting the factor of $P_k(x;v)$ to zero in (7.17) results in $v = \frac{1}{1+x}$ for all values of k. Substituting this value of v into (7.17) and solving for $P_k(x;1)$ leads to

$$P_k(x;1) = \frac{(1+x)^{k-1}}{1-x^2}[(1+x)(1+\overline{P}_k(x;1) - P_k(x;1/(1+x)))$$

$$+ x^2\sum_{i=1}^{k-1} v^{i-1}P_i(x;1)]. \tag{7.18}$$

We now compute $AW^T_{[k]} = P_k(x; 1)$ for the first few values of k. For $k = 1$, we obtain from the initial conditions that $\overline{P}_1(x; v) = x$, and from (7.18) that $P_1(x; 1) = \frac{1}{1-x}$. Substituting these functions into (7.17) yields that $P_1(x; v) = \frac{1}{1-x}$. We follow the same procedure for $k = 2, 3$, and 4, using (7.16) instead of the initial conditions to obtain that

$$\overline{P}_2(x; v) = x + x^2 + \frac{xv}{1-x} \quad \text{and} \quad P_2(x; 1) = \frac{1 + x^2}{(1-x)^2}.$$

Similarly, we can find that

$$P_3(x; 1) = \frac{1 + 3x^2 + 2x^3 + 3x^4 - x^6}{(1-x)^3}$$

and

$$P_4(x; 1) = \frac{1 + 6x^2 + 8x^3 + 21x^4 + 24x^5}{(1-x)^4}$$
$$+ \frac{20x^6 + x^7 - 14x^8 - 6x^9 + 2x^{10} + x^{11}}{(1-x)^4}.$$

Note that the scanning-element algorithm can also be applied to some of the other pairs of letters which are listed as open problems in Table 7.3.

We now return to the ECO method and its applications. So far we have seen the ECO method applied to pattern avoidance in words. A somewhat different focus was used by Baril and Do [17]. They used the ECO method to encode the generating trees for various types of restricted compositions via restricted permutations. For example, they consider $(1, k)$-compositions, compositions without the part k, and compositions whose largest part is p. The focus of their paper is generating these compositions via permutations for which efficient (that is, constant time) recursive algorithms are known. So the succession rules and the encodings are given without much detail. Since our focus is not on algorithmic creation but rather on the structure of the compositions, we will give an interpretation of the succession rules in terms of the recursive creation of the respective compositions. We start by deriving, the succession rules for $(1, k)$-compositions (see Example 3.14), which were studied by [7, 17, 51].

Theorem 7.95 (*compare to [17, Table 1]*) *The system of succession rules for the $(1, k)$-compositions is given by*

$$\textbf{Root: } (1_0)$$
$$\textbf{Rules: } (1_i) \rightsquigarrow (1_{i+1}), \textit{ for } 0 \le i < k - 2$$
$$(1_{k-2}) \rightsquigarrow (2)$$
$$(2) \rightsquigarrow (2)(1_0)$$

Proof We can generate the $(1, k)$-compositions of $n+1$ by either appending a 1 to a composition of n, or converting the $k-1$ rightmost consecutive 1s into a k. A composition with at least $k-1$ consecutive rightmost 1s therefore produces two new compositions (label (2)), while all other compositions produce one new composition. We use the label (1_i) to denote a composition that has i consecutive rightmost 1s. The root is the empty composition, with label (1_0). Note that the algorithm does not produce duplicates, and creates all $(1, k)$-composition of $n+1$ from those of n. Figure 7.24 shows the generating tree for $(1, 4)$-compositions. □

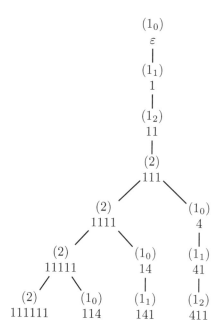

FIGURE 7.24: Generating tree for $(1, 4)$ compositions.

We give one more example of the compositions listed in [17, Table 1], namely the succession rules for the compositions with largest part p.

Example 7.96 *The succession rules for compositions in $[k]$ with largest part p where $k \geq p$ are given by*

Root: (0)

Rules: $(0) \rightsquigarrow (1)$

$\qquad\qquad (i) \rightsquigarrow (1)(i+1), \text{ for } 1 \leq i \leq p-1$

$\qquad\qquad (p) \rightsquigarrow (1)$

The derivation of these rules is left as Exercise 7.11.

7.6 Exercises

Exercise 7.1 *The adjacency matrix for the automaton for avoidance of the subsequence pattern 3241 is given in Example 7.32. Carefully explain the following edges:* $\langle 42 \rangle \rightarrow \langle 324 \rangle$, $\langle 432 \rangle \rightarrow \langle 3243 \rangle$, *and* $\langle 3243 \rangle \rightarrow \langle 435 \rangle$.

Exercise 7.2 *Determine the adjacency matrices for the automata* $\mathcal{A}(\tau, 4)$ *where* τ *is a permutation pattern in* S_3. *Several of the patterns in* S_3 *have isomorphic graphs, so their adjacency matrices will be identical for an appropriate ordering of states.*

(1) *Determine the number of nonisomorphic graphs. Justify your answer by stating the canonical ordering of states that show that certain graphs are isomorphic, and giving a reason why the other graphs are nonisomorphic.*

(2) *All permutation patterns of length three are in the same Wilf-equivalence class and thus have the same generating function. Show that each of the nonisomorphic adjacency matrices leads to the same generating function using the formula given in Theorem 7.34.*

Exercise 7.3 *Let* $\tau = 11$–2 *be a generalized pattern of type* $(2, 1)$. *Find the automata* $\mathcal{A}(\tau, k)$ *for the number of k-ary words that avoid* τ *for* $k = 2$ *and* $k = 3$.

Exercise 7.4 *Prove that the automaton given in Figure 7.25 counts the number of smooth k-ary words of length n (see Exercises 2.17 and 6.9).*

Exercise 7.5 *Find the simple paths associated with each of the Young tableaux of Figure 7.13 (see Example 7.53).*

Exercise 7.6 *Fill in the details of the proof of Corollary 7.65.*

Exercise 7.7 *Use the algorithm given in Lemma 7.40 to derive the automaton* $\mathcal{A}(1234, 5)$.

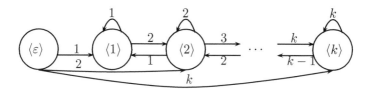

FIGURE 7.25: Automaton of smooth words.

Exercise 7.8 *Give the proof of Theorem 7.72.*

Exercise 7.9 *Let* $\tau = 11\text{--}2$ *be a generalized pattern of type* $(2, 1)$. *Find the automata* $\mathcal{A}(\tau, k)$ *for the number of compositions of* n *with parts in* $[k]$ *that avoid* τ *for* $k = 2$ *and* $k = 3$.

Exercise 7.10 *Derive the generating function for the number of compositions avoiding the set of patterns* $\{123, 132, 122\}$ *using any method discussed so far.*

Exercise 7.11 *Derive the succession rules given in Example 7.96 for the re-cursive creation of the compositions in* $[k]$ *with largest part* p.

Exercise 7.12 *Let* $N(\tau, k)$ *be the number of accept states in the automaton* $\mathcal{A}(\tau, k)$ *for pattern avoidance in words, that is,* $N(\tau, k) = |\mathcal{E}(\tau, k)| - 1$. *From Lemma 7.40 we obtain that* $N(1234, k) = \binom{k}{3}$. *For the pattern* 1324, *we obtain the first few values of* $N(1324, k)$ *by using the program TOU_AUTO (available on Mansour's home page [139]) as given in the table below.*

τ	$N(\tau, 4)$	$N(\tau, 5)$	$N(\tau, 6)$	$N(\tau, 7)$	$N(\tau, 8)$
1324	4	12	34	98	294

The values of $N(\tau, k)$ *for the pattern* $\tau = 1324$ *coincide with the values of the sequence A014143 in the Online Encyclopedia of Integer Sequences [180] (with a shift). Use the program TOU_AUTO to determine the next two (or possibly three) values in the sequence* $N(1324, k)$ *and check whether they still agree with the values of the sequence A014143.*

7.7 Research directions and open problems

We now suggest several research directions, which are motivated both by the results and exercises of this chapter. In later chapters we will revisit some of these research directions and pose additional questions.

Research Direction 7.1 *Table 6.3 shows that there are five Wilf-equivalence classes for permutation patterns of length four, namely* $1234 \sim 1243 \sim 1432 \sim 2143$, 1324, 1342, 1423 *and* 2413. *Theorem 7.54 gives the asymptotic expression for the number of words avoiding* 1234 *for all* $k \geq 3$ *as*

$$AW_{[k]}^{1234}(n) \approx \frac{2 \cdot (3(k-3)!)}{(k-3)!(k-2)!(k-1)!} n^{\ell(k-3)} 3^{n-3k+9} \qquad (n \to \infty).$$

A natural extension is to ask about finding the asymptotics for the number of k-ary words of length n that avoid a subsequence pattern τ *from the remaining four Wilf-equivalence classes, namely for* $\tau = 1324$, $\tau = 1342$, $\tau = 1423$ *and* $\tau = 2413$.

Research Direction 7.2 *We were able to obtain results for the increasing patterns since the automaton for those patterns had a very specific structure for the loops. This leads to the following questions:*

(1) *Are there subsequence permutation patterns* τ *that have an automaton* $\mathcal{A}(\tau, k)$ *where each state has exactly* $\ell - 1$ *loops for all* $k > k_0$, *where* ℓ *is the number of letters in* τ *and* k_0 *is fixed (to avoid the special cases when k is too small)?*

(2) *More generally, are there subsequence permutation patterns* τ *that have an automaton* $\mathcal{A}(\tau, k)$ *such that the set of the number of loops in the automaton is* L *for all* $k > k_0$, *for a given set* L *and fixed* k_0? *For example, if* $L = \{3\}$, $\tau = 1234$ *would be such a pattern. If* $L = \{2, 3\}$, *then from Example 7.32, the pattern 3241 could be a candidate (as we do not know that the loop pattern is the same for all k).*

Research Direction 7.3 *In Lemma 7.40 we showed that the number of accept states for the automaton* $\mathcal{A}(\tau_\ell, k)$ *is given by* $N(\tau_\ell, k) = \binom{k}{\ell}$. *Below is a table of the first few values of* $N(\tau, k)$ *for permutation patterns of length four from each of the five Wilf-equivalence classes (see Table G.3), as computed using the program TOU_AUTO (except for* $\tau = 1234$).

τ	$N(\tau, 4)$	$N(\tau, 5)$	$N(\tau, 6)$	$N(\tau, 7)$	$N(\tau, 8)$
1234	4	10	20	35	56
1324	4	12	34	98	294
1342	4	14	48	164	560
1423	4	14	45	136	396
2413	4	18	72	280	1072

The following questions arise:

(1) *Compute the first few values of* $N(\tau, k)$ *for the remaining patterns of length four, both permutation patterns and patterns with repeated letters. Remember that even though* τ *and* ρ *may belong to the same Wilf-equivalence class,* $N(\tau, k)$ *and* $N(\rho, k)$ *may be different (see Exercise*

7.2). *The only general rule we have is that* $N(\tau, k) = N(c(\tau), k)$, *thus reducing the number of patterns to be considered by a factor of one half.*

(2) *In Exercise 7.12 we saw that the values of* $N(1324, k)$ *for* $k \geq 3$ *coincide with the values of the sequence A014143, which enumerates the partial sums of the Catalan numbers. Give a combinatorial proof that* $N(1324, k) = \sum_{i=0}^{k-3} C_i$, *where* C_i *is the* i-*th Catalan number.*

(3) *Find an explicit formula for* $N(\tau, k)$, *where* τ *is any of the other subsequence pattern of length four.*

(4) *Compute values and find an explicit formula for* $N(\tau, k)$, *where* τ *is a subsequence pattern of length five.*

Research Direction 7.4 *We have derived formulas for* $\mathrm{pr}(\ell, i, r)$, *the number of primitive segmented tableaux in* $[\ell] \times [i]$ *with* r *different entries, for two special cases, namely* $i = r$ (*see Theorem 7.57*) *and* $\ell = 2$ (*see Theorem 7.58*). *Brändén and Mansour [27] derived a system of equations for the generating function of* $\mathrm{pr}(\ell, i, r)$, *but did not succeed in solving it. Either find a clever way of solving this system of equations, or find a different approach to derive explicit formulas for* $\mathrm{pr}(\ell, i, r)$ *or the corresponding generating functions. Of specific interest is* $\mathrm{pr}(3, i, r)$, *which would lead to a formula for the number of* k-*ary words avoiding the subsequence pattern* 1234.

Research Direction 7.5 *A* noncrossing (nonnesting) *word* w *is a word that avoids the subsequence pattern* 1212 (1221). *Find explicit formulas for the number of* k-*ary noncrossing* (nonnesting) *words of length* n. (*Note that the terminology comes from noncrossing and nonnesting partitions and matchings. For details see [47, 112, 113]*).

Chapter 8

Asymptotics for Compositions

8.1 History

Asymptotic analysis usually involves complex analysis or probability theory or both. The idea is to get a handle on the behavior of the number of compositions or words of interest, or of a statistic on these objects as $n \to \infty$. Applying results from complex analysis we can obtain growth rates for the statistics of interest. On the other hand, when considering compositions as randomly selected from all compositions (that is, all compositions of n are equally likely to occur), we can derive results on the average or variance of a statistic, which now is a random variable.

Bender [18] blended the two approaches and gave results on the asymptotic distribution for the quantity of interest based on a generating function that keeps track of the order of the composition and one other statistic, such as the number of parts. Later on, Bender and Richmond derived central and local limit theorems for multi-variate generating functions [21].

Major research activity in the area of asymptotics for compositions started with a paper by Hwang and Yeh [102] who looked at measures of distinctness in random partitions and compositions, where distinctness is defined as the number of distinct part sizes in the composition (or partition). Their results on the average number of distinct parts revealed an expression that grows at a rate that is essentially $\log n$ (where log without base refers to the natural logarithm), based on a mixture of complex analysis and probability theory. Hitczenko and Stengle [97] proved the asymptotic result on the average number of distinct parts given by Hwang and Yeh in a different way. (The approach of Hwang and Yeh was more general and thus required more machinery. Apparently, Hitczenko and Stengle were unaware of the results of Hwang and Yeh.) We will present the shorter proof of Hitczenko and Stengle as well as variations on the theme by Hitczenko and several collaborators. With Louchard [93], Hitczenko derived more detailed results on the distribution of the number of distinct part sizes and also considered the statistics largest part and first empty part (first part size that is missing in the composition). Hitczenko and Savage [95] refined these results by deriving results on the multiplicity of parts in a random composition. Finally, Louchard [133] provided further analysis on the multiplicity of parts, studying for example the maximum part size of multiplicity m.

Similar questions were studied for Carlitz compositions. Knopfmacher and Prodinger [119] studied asymptotics for the number of parts, occurrences of the (subword) pattern 11 (with Carlitz compositions being those that have a zero count of the pattern 11), largest part, and other restricted compositions. Their results were based solely on complex analysis tools. Louchard and Prodinger added tools from probability and derived results on the number of parts, the last part size, and correlation between successive parts in Carlitz compositions. They made use of the results of Bender [18] repeatedly to obtain asymptotic distribution results. Finally, Goh and Hitczenko [73] obtained results on the average number of distinct part sizes in Carlitz compositions.

Most recently, Mansour and Sirhan [145] have obtained generating functions and asymptotics for the number of compositions that avoid subword patterns of length three and longer using tools from complex analysis as well as central and local limit theorems based on Bender's results [18].

We present a sampling of these results in the following sections. In Section 8.2 we give the necessary background from probability theory and derive explicit results for some average statistics. Section 8.3 contains the necessary background from complex analysis. In the next two sections we give a sample of asymptotic results for compositions and Carlitz compositions and provide outlines of the proofs of these results. Note that there are very few asymptotic results for words. This is simply because the generating function for the number of words avoiding a pattern is rational, and therefore asymptotic results are easy to come by. We will present an example in Section 8.6.

As before, our focus is on compositions and words. However, the interested reader who wants to delve deeper into asymptotics for more general combinatorial enumeration is referred to the recent book of Flajolet and Sedgewick [68]. This book contains literally hundreds of examples and is a one-stop reference for techniques for asymptotic analysis.

8.2   Tools from probability theory

Oftentimes we are interested in the value of a statistic for a "typical" composition. We can compute this "typical" value by assuming that we are presented with a composition that is randomly selected from all compositions of n, making all compositions of n equally likely. Then the statistic becomes a random quantity and we can compute its average and its variance. If we know an explicit formula for the number of compositions for which the statistic has a given value, then we can compute the average of the statistic over all compositions, or equivalently, the "typical" value of the statistic for a randomly selected composition.

To describe this approach, we only need very basic definitions from probability theory. We restrict ourselves throughout to definitions for countable sample spaces, as those are the only ones to be considered here. The definitions and result we give, as well as the corresponding definitions and results for more general sample spaces, can be found for example in [24, 80].

Definition 8.1 *The* sample space Ω *is the set of all possible* outcomes *of an experiment. A subset* E *of* Ω *is called an* event. *An* event occurs *if any outcome* $\omega \in E$ *occurs. The* probability function $\mathbf{P} : \Omega \to [0, 1]$ *satisfies the following three axioms:*

1. *For any* $E \subseteq \Omega$, $\mathbf{P}(E) \geq 0$.

2. $\mathbf{P}(\Omega) = 1$.

3. *For a countable sequence* $E_1, E_2, \ldots$ *of pairwise disjoint sets (that is,* $E_i \neq E_j$ *for* $i \neq j$), $\mathbf{P}(E_1 \cup E_2 \cup \cdots) = \sum_i \mathbf{P}(E_i)$.

The value $\mathbf{P}(E)$ *gives the probability that the event* E *will occur.*

As a consequence of the third axiom it is enough to specify $\mathbf{P}(\{\omega\})$ for each of the outcomes $\omega \in \Omega$ when Ω is a countable set.

Definition 8.2 *If all outcomes in a (finite) sample space are equally likely, then we say that the sample space is equipped with the* uniform probability measure. *In this case,* $\mathbf{P}(\{\omega\}) = 1/|\Omega|$ *for every* $\omega \in \Omega$. *Consequently for any subset* E *of* Ω,

$$\mathbf{P}(E) = \frac{|E|}{|\Omega|}.$$

We will consider what happens to a statistic of a random composition with certain characteristics, for example, the number of parts (statistic) in Carlitz compositions (compositions with certain characteristics) of n when n is large. Thus our sample space is the set of all compositions with those characteristics, for example compositions with parts in $\mathbb{N}$, Carlitz compositions, and so on. Unless stated otherwise, we will assume that the sample space is equipped with the uniform probability measure.

Definition 8.3 *A* discrete random variable X *is a function from* Ω *to a countable subset of* $\mathbb{R}$. *Its* probability mass function *is defined by*

$$\mathbf{P}(X = m) = \mathbf{P}(\{\omega \mid X(\omega) = m\}).$$

The set of values together with their respective probabilities characterize the distribution of X. *If the set of values of* X *is given by* $\{x_1, x_2, \ldots\}$, *then the probability mass function satisfies the following properties:*

1. $\mathbf{P}(X = x) = 0$ *if* $x \notin \{x_1, x_2, \ldots\}$.

2. $\mathbf{P}(X = x_i) \geq 0$ *for* $i = 1, 2, \ldots$

3. $\sum_{i=1}^{\infty} \mathbf{P}(X = x_i) = 1$.

Note that it is customary to denote the random variable with a capital letter and its values with the corresponding lower case letter.

Example 8.4 *Let* $\Omega = C_5$ *be the set of compositions of 5, where we assume that all compositions occur equally likely. Then* $X = \mathrm{par}(\sigma)$ *for* $\sigma \in C_5$ *is a discrete random variable that takes values* $m = 1, 2, 3, 4, 5$. *The respective (nonzero) probabilities are computed via the compositions (outcomes) that have* m *parts.*

m	1	2	3	4	5
ω	5	14	113	1112	11111
		41	131	1121	
		23	311	1211	
		32	122	2111	
			212		
			221		
$\mathbf{P}(X = m)$	$\frac{1}{16}$	$\frac{4}{16}$	$\frac{6}{16}$	$\frac{4}{16}$	$\frac{1}{16}$

In this particular case, we could have computed the probabilities by using Theorem 1.3, as it gives the number of compositions of n with exactly m parts as $\binom{n-1}{m-1}$. For larger values of n and if we have no explicit formulas for the quantity of interest, we can use Mathematica or Maple to compute the probabilities. For example, in Mathematica, we can use the following code to compute the probability distribution for $X = \mathrm{par}(\sigma)$ and obtain the values for $n = 5$:

```
comps[n_]:=Flatten[Map[Permutations,IntegerPartitions[n]],1]
probs[n_]:=BinCounts[Map[Length,comps[n],1],{Range[1,n+1]}]
            /Length[comps[n]]
probs[5]
```

which gives the result $\left\{\frac{1}{16}, \frac{1}{4}, \frac{3}{8}, \frac{1}{4}, \frac{1}{16}\right\}$.

Other random variables on this sample space are $X =$ number of occurrences of a pattern τ or $X =$ largest part in the composition (see Exercise 8.1). Example 8.4 is an illustration of the following lemma.

Lemma 8.5 *Let* Ω *be finite, and* $|\Omega| = N$. *If* Ω *is equipped with the uniform probability measure and* X *is a discrete random variable on* Ω, *then*

$$\mathbf{P}(X = m) = |\{\omega \mid X(\omega) = m\}|/|\Omega|.$$

Definition 8.6 *For a discrete random variable X which takes values $i \in \mathbb{Z}$ with probability $\mathbf{P}(X = i)$, the expected value of $f(X)$ for a (measurable) function f is defined as*

$$\mathbb{E}(f(X)) = \sum_{i \in \mathbb{Z}} \mathbf{P}(X = i) f(i),$$

assuming that the sum converges. In particular, the mean *(or* average *or* expected value*) μ and the* variance $\mathrm{Var}(X)$ *of X are given by*

$$\mu = \mathbb{E}(X) = \sum_{i \in \mathbb{Z}} i \mathbf{P}(X = i)$$

and

$$\mathrm{Var}(X) = \sigma^2 = \mathbb{E}\left((X - \mu)^2\right) = \mathbb{E}(X^2) - \mu^2.$$

The quantity $\mathbb{E}(X^k)$ is called the k-th moment of X, and $\sigma = \sqrt{\mathrm{Var}(X)}$ is called the standard deviation *of X. Furthermore,*

$$\mathbb{E}(X(X - 1)(X - 2) \cdots (X - k + 1))$$

is called the k-th factorial moment of X.

Example 8.7 (*Continuation of Example 8.4*) *Using the distribution derived in Example 8.4 for the random variable $X = \mathrm{par}(\sigma)$, we have that*

$$\mathbb{E}(X) = \frac{1}{16} + 2\frac{4}{16} + 3\frac{6}{16} + 4\frac{4}{16} + 5\frac{1}{16} = 3$$

and

$$\mathrm{Var}(X) = \left(\frac{1}{16} + 2^2\frac{4}{16} + 3^2\frac{6}{16} + 4^2\frac{4}{16} + 5^2\frac{1}{16}\right) - 3^2 = 1.$$

We can easily derive a general formula for the mean and the variance of a statistic if all compositions under consideration are equally likely.

Lemma 8.8 *If Ω is equipped with the uniform probability measure, $|\Omega| = N$ and $a_{n,m}$ is the number of compositions of n with a given characteristic such that the statistic X has the value m, then*

$$\mathbb{E}(X) = \mathbb{E}(X(n)) = \frac{1}{N} \sum_m a_{n,m} \cdot m. \tag{8.1}$$

Proof Follows immediately from the definition of the expected value and Lemma 8.5. $\qquad \square$

Using Equation (8.1) it is straightforward to compute the average value of a statistic X if we have an explicit formula for the quantity $a_{n,m}$. To illustrate this, we will compute the average number of parts in all compositions of n and in all palindromic compositions of n.

Example 8.9 *We start by computing the average number of parts in a random composition of n. From Theorem 3.3, we have that the number of compositions of n is given by $C(n) = 2^{n-1}$ and the number of compositions of n with m parts is given by $C(m; n) = \binom{n-1}{m-1}$. Let $X_{par}(n)$ denote the number of parts in a random composition of n. Then the average number of parts is given by*

$$\mathbb{E}(X_{par}(n)) = \frac{1}{2^{n-1}} \sum_{m=1}^{n} m \binom{n-1}{m-1}$$

$$= \frac{1}{2^{n-1}} \left[\sum_{m=1}^{n} \frac{(m-1)(n-1)!}{(n-m)!(m-1)!} + \sum_{m=1}^{n} \binom{n-1}{m-1} \right]$$

$$= \frac{1}{2^{n-1}} \left[(n-1) \sum_{m=2}^{n} \binom{n-2}{m-2} + 2^{n-1} \right]$$

$$= \frac{1}{2^{n-1}} \left[(n-1)2^{n-2} + 2^{n-1} \right] = \frac{n-1}{2} + 1 = \frac{n+1}{2}.$$

This was not too hard! With a little bit more effort we can establish the corresponding result for palindromic compositions.

Example 8.10 *We now compute the average number of parts in a random palindromic composition of n where we have to distinguish between odd and even n. From Theorems 3.4 and 3.5 we have that $P(n) = 2^{\lfloor n/2 \rfloor}$ and that*

$$P(2k; 2n') = \binom{n'-1}{k-1} \text{ and } P(2k+1; 2n'-1) = P(2k+1; 2n') = \binom{n'-1}{k}.$$

Let $X_{par}(n)$ denote the number of parts in a random palindromic composition of n. Then

$$\mathbb{E}(X_{par}(2n')) = \frac{1}{2^{n'}} \left[\sum_{k=1}^{n'} 2k \binom{n'-1}{k-1} + \sum_{k=0}^{n'-1} (2k+1) \binom{n'-1}{k} \right]$$

$$= \frac{1}{2^{n'}} \left[\sum_{k=1}^{n'} 2(k-1) \binom{n'-1}{k-1} + 2 \binom{n'-1}{k-1} \right.$$

$$\left. + \sum_{k=0}^{n'-1} 2k \binom{n'-1}{k} + \binom{n'-1}{k} \right]$$

$$= \frac{1}{2^{n'}} \left(2(n'-1)2^{n'-2} + 2 \cdot 2^{n'-1} + 2(n'-1)2^{n'-2} + 2^{n'-1} \right)$$

$$= \frac{2n'+1}{2} = \frac{n+1}{2}.$$

Similarly, for odd $n = 2n' - 1$, we obtain

$$\mathbb{E}(X_{\text{par}}(2n' - 1)) = \frac{1}{2^{n'-1}} \sum_{k=0}^{n'-1} (2k+1) \binom{n'-1}{k}$$

$$= \frac{1}{2^{n'-1}} \left(2(n'-1)2^{n'-2} + 2^{n'-1} \right)$$

$$= n' - 1 + 1 = \frac{n+1}{2}.$$

How do we proceed if we do not have an explicit formula for the statistic of interest, but we do have the generating function? In this case, we can compute the mean and the variance of the statistic from a different but related generating function, namely the probability generating function.

Definition 8.11 *The* probability generating function *of a discrete random variable X is defined as $pg_X(u) = \sum_m \mathbf{P}(X = m)u^m = \mathbb{E}(u^X)$. The* moment generating function *of a random variable X is defined as $M_X(t) = \mathbb{E}(e^{tX}) = pg_X(e^t)$.*

Example 8.12 *Let $\Omega = C_n$ with $n \geq 1$ be the set of compositions of n where we assume that all compositions occur equally likely. Then $X_{\text{par}}(n) = \text{par}(\sigma)$ for $\sigma \in C_n$ is a discrete random variable that takes values $m = 1, 2, \ldots, n$. The respective (nonzero) probabilities are computed via the compositions that have m parts:*

$$\mathbf{P}(X_{\text{par}}(n) = m) = \frac{1}{2^{n-1}} \binom{n-1}{m-1}.$$

Thus, the probability generating function of X_{par} is given by

$$pg_X(u) = \sum_{m=1}^{n} \mathbf{P}(X_{\text{par}}(n) = m)u^m = \frac{u}{2^{n-1}} \sum_{m=0}^{n-1} \binom{n-1}{m} u^m$$

$$= u \left(\frac{1+u}{2} \right)^{n-1}.$$

Lemma 8.13 *We can obtain the k-th moment and the k-th factorial moment of a discrete random variable X from the probability generating function $pg_X(u)$ as follows:*

1. $\mathbb{E}(X^k) = \left. \dfrac{\partial^k}{\partial t^k} M_X(t) \right|_{t=0} = \left. \dfrac{\partial^k}{\partial t^k} pg_X(e^t) \right|_{t=0}.$

2. $\mathbb{E}(X(X-1)(X-2) \cdots (X-k+1)) = \left. \dfrac{\partial^k}{\partial u^k} pg_X(u) \right|_{u=1}.$

How is the generating function for $a_{n,m}$ related to the probability generating function from which we can obtain the moments via differentiation? Lemma 8.14 provides the answer.

Lemma 8.14 *For fixed n, let X be a statistic on the compositions of n, $A(x,q) = \sum_{n,m\geq 0} a_{n,m} x^n q^m$, and $N = [x^n]A(x,1)$. Then the mean μ and the variance σ^2 of X are given by*

$$\mu = \mathbb{E}(X) = \frac{1}{N}[x^n]\frac{\partial}{\partial q}A(x,q)\bigg|_{q=1} \tag{8.2}$$

and

$$\sigma^2 = \text{Var}(X) = \frac{1}{N}[x^n]\left(\frac{\partial^2}{\partial q^2}A(x,q) + \frac{\partial}{\partial q}A(x,q) - \left(\frac{\partial}{\partial q}A(x,q)\right)^2\right)\bigg|_{q=1}$$

$$= \frac{1}{N}[x^n]\frac{\partial^2}{\partial q^2}A(x,q)\bigg|_{q=1} + \mu - \mu^2. \tag{8.3}$$

Proof Follows immediately from Lemma 8.13 as we have that $pg_X(q) = \frac{1}{N}[x^n]A(x,q)$. □

We illustrate the use of this lemma by revisiting Example 8.9 and recomputing the average number of parts in a random composition from $\mathcal{C}_n$. In addition, we derive the variance of the number of parts.

Example 8.15 (*Example 8.9 revisited*) *From (4.4) we have that the generating function for the number of compositions of n with m parts is given by*

$$C(x,y) = \frac{1}{1 - y\frac{x}{1-x}} = \frac{1-x}{1-x-xy}.$$

It can be shown (see Exercise 8.3) that

$$[x^n]\frac{\partial^k}{\partial y^k}C(x,y)\bigg|_{y=1} = k! \cdot 2^{n-1-k}\left(2\binom{n}{k} - \binom{n-1}{k}\right). \tag{8.4}$$

Using this expression, we can now compute the mean and the variance. For $k=1$, we obtain that

$$\mathbb{E}(X_{\text{par}}(n)) = \frac{1}{2^{n-1}}2^{n-2}(2n - (n-1)) = \frac{n+1}{2}$$

as before. For the variance we use $k=2$ to compute

$$\mathbb{E}(X_{\text{par}}(n)^2) = \frac{1}{2^{n-1}}[x^n]\left(\frac{\partial^2}{\partial y^2}C(x,y)\bigg|_{y=1}\right) = \frac{n^2+n-2}{4}.$$

Therefore, the variance for the number of parts is given by

$$\sigma^2 = \frac{n^2 + n - 2}{4} + \frac{n+1}{2} - \frac{(n+1)^2}{4} = \frac{n-1}{4}.$$

As another example, we compute the average number of rises in compositions.

Example 8.16 *Using (4.8) in the proof of Corollary 4.5, we obtain that for* $r_A(x,y) = \sum_{n \geq 0} \sum_{\sigma \in \mathcal{C}_n^A} \text{ris}(\sigma) x^n y^{\text{par}(\sigma)},$

$$r_N(x,y) = \frac{y^2 \sum_{i \geq 1} \sum_{j \geq i+1} x^{i+j}}{\left(1 - y\dfrac{x}{1-x}\right)^2} = \frac{y^2 \sum_{i \geq 1} x^i \dfrac{x^{i+1}}{1-x}}{\left(1 - y\dfrac{x}{1-x}\right)^2} = \frac{\dfrac{y^2 x}{1-x} \cdot \dfrac{x^2}{1-x^2}}{\left(1 - y\dfrac{x}{1-x}\right)^2}$$

$$= \frac{y^2 x^3}{(1 - x - yx)(1+x)}.$$

Since we are interested in the number of rises, we set $y = 1$. *This leads to*

$$r_N(x,1) = \frac{x^3}{(1+x)(1-2x)^2} = \sum_{n \geq 1} \frac{3n \cdot 2^{n-2} - 5 \cdot 2^{n-2} - (-1)^n}{9} x^n.$$

Dividing by the total number of compositions and extracting the coefficient of x^n *we obtain that the average number of rises in a random composition of* n *is given by*

$$\mathbb{E}(X_{\text{ris}}(n)) = \frac{3n-5}{18} - \frac{(-1)^n}{9 \cdot 2^{n-1}} \qquad \text{for } n \geq 1.$$

Similarly, for $l_A(x,y) = \sum_{n \geq 0} \sum_{\sigma \in \mathcal{C}_n^A} \text{lev}(\sigma) x^n y^{\text{par}(\sigma)}$, *we obtain (by comparing the expressions given in Theorem 4.5 and the above derivation) that*

$$l_N(x,y) = \frac{y^2 \sum_{j \geq 1} x^{2j}}{\left(1 - y\dfrac{x}{1-x}\right)^2} = \frac{y^2 x^2 (1-x)}{(1+x)(1 - x - xy)}.$$

This implies that

$$l_N(x,1) = \frac{x^2(1-x)}{(1+x)(1-2x)^2} = \sum_{n \geq 2} \frac{3n \cdot 2^n + 2^n + 8(-1)^n}{36} x^n,$$

so that the average number of levels in a random composition of n *is given by*

$$\mathbb{E}(X_{\text{lev}}(n)) = \frac{3n+1}{18} + \frac{(-1)^n}{9 \cdot 2^{n-2}}.$$

Note that we have obtained the series expansions for $r_\mathbb{N}(x,y)$ and $l_\mathbb{N}(x,y)$ with the coefficients expressed as functions of n (as opposed to numerical values) by using the following Maple command. The function `rgf_expand` *takes as its input a rational function $f(x)$ and returns $[x^n]f(x)$ in terms of n.*

```
with(genfunc);
rgf_expand(f(x),x;n);
```

In Mathematica *we use the function* `SeriesCoefficient` *to compute*

$$[x^n]r_\mathbb{N}(x,1)$$

as a function of n:

```
SeriesCoefficient[x^3/((1 + x) (1 - 2 x)^2), {x, 0, n}]
```

However, knowing the generating function for the statistic of interest does not always guarantee success in computing the expected value. For example, if we want to compute the same averages for Carlitz compositions (see Definition 1.18), trouble arises when we try to extract the coefficient of x^n of the function

$$\frac{\partial}{\partial y}C_\mathbb{N}(x,y,1,0,1)\bigg|_{y=1} = \frac{\sum_{i\geq 1}\frac{x^i}{(1+x^i)^2}}{\left(1-\sum_{i\geq 1}\frac{x^i}{1+x^i}\right)^2}.$$

In this case, even though we have an explicit formula for the number of Carlitz compositions with respect to the statistic of interest, the expression is too complicated to compute an explicit result for the average. In such cases, we can obtain only asymptotic results (see Exercise 8.4) using tools from complex analysis that we will introduce and will discuss in the next section.

We conclude this section by giving some definitions that will play an important role for asymptotic results in terms of distributions.

Definition 8.17 *Let X be a random variable. If there exists a nonnegative, real-valued function $f : \mathbb{R} \to [0,\infty)$ such that*

$$\mathbf{P}(X \in A) = \int_A f(x)dx$$

for any measurable set $A \subset \mathbb{R}$ (a set that can be constructed from intervals by a countable number of unions and intersections), then X is called (absolutely) continuous and f is called its density function.

Definition 8.18 *A continuous random variable X is called a* normal random variable with mean μ and variance σ^2 *if its density function is of the form*

$$f(x) = \frac{1}{\sigma\sqrt{2\pi}}\exp\left[\frac{-(x-\mu)^2}{2\sigma^2}\right],$$

and we will write $X \sim \mathcal{N}(\mu, \sigma^2)$. *The parameters* μ *and* σ *that appear in the density are the mean and the standard deviation of* X. *If* $\mu = 0$ *and* $\sigma = 1$, *then* X *is called a* standard normal *random variable.*

It can be shown that if $X \sim \mathcal{N}(\mu, \sigma^2)$, then the standardized random variable $\frac{X-\mu}{\sigma}$ is a standard normal random variable, that is, $\frac{X-\mu}{\sigma} \sim \mathcal{N}(0,1)$. Following Bender [18], we define asymptotic normality for a sequence of two indices.

Definition 8.19 *To a given a sequence* $\{a_n(k)\}_{n,k\geq 0}$ *with* $a_n(k) \geq 0$ *for all* k, *we associate a normalized sequence*

$$p_n(k) = \frac{a_n(k)}{\sum_{j\geq 0} a_n(j)}.$$

We say that $\{a_n(k)\}_{n,k\geq 0}$ *is* asymptotically normal with mean μ_n *and variance* σ_n^2 *or* satisfies *a* central limit theorem *if*

$$\lim_{n\to\infty} \sup_x \left| \sum_{k\leq \sigma_n x + \mu_n} p_n(k) - \frac{1}{\sqrt{2\pi}} \int_{-\infty}^x e^{-t^2/2} dt \right| = 0,$$

and write $a_n(k) \overset{d}{\approx} \mathcal{N}(\mu_n, \sigma_n^2)$.

8.3 Tools from complex analysis

For the moment, we leave randomness and probabilities behind and focus on complex analysis instead, even though we will eventually combine the two areas in the quest for determining asymptotic behavior. The main result that we draw on is that for a complex function, the location of the singularities determines the growth behavior of the coefficients of the power series expansion of the function when n becomes large. Therefore, we will regard the generating function as a complex function instead of as a formal expression, and use variables z and w instead of variables x and y to remind ourselves of that viewpoint. We provide the necessary definitions for basic terminology and related results in Appendix E and start by giving the specific results (drawn primarily from Flajolet and Sedwick [68]) that will be used in the subsequent analysis of restricted compositions.

The typical generating function that we will consider satisfies the following theorem.

Theorem 8.20 *[68, Theorems IV.5 and IV.6] A function* $f(z)$ *that is analytic at the origin and whose expansion at the origin has a finite radius of* *convergence R necessarily has a singularity on the boundary of its disc of convergence, $|z| = R$. Furthermore, if the expansion of $f(z)$ at the origin has nonnegative coefficients, then the point $z = R$ is a singularity of $f(z)$, that is, there is at least one singularity of smallest modulus that is real-valued and positive.*

Note that the second part of Theorem 8.20 indicates that we can very easily find the dominant singularity for our generating functions – we just need to check the analyticity of the function along the positive real line and find the smallest value where the function ceases to be analytic. This singularity will determine the rate of growth, expressed using the notation $\approx$ (see Definition 7.36).

Theorem 8.21 (*Exponential Growth Formula*) *[68, Theorem IV.7] If $f(z)$ is analytic at the origin and R is the modulus of a singularity nearest to the origin in the sense that*

$$R = \sup\{r \geq 0 \mid f \text{ is analytic in } |z| < r\},$$

then the coefficient $f_n = [z^n]f(z)$ satisfies

$$f_n \approx \left(\frac{1}{R}\right)^n.$$

In the case of functions with nonnegative coefficients, specifically combinatorial generating functions, one can also define

$$R = \sup\{r \geq 0 \mid f \text{ is analytic for all } 0 \leq z < r\}.$$

Theorem 8.21 directly links the exponential growth of the coefficients to the location of the singularities nearest to the origin (called "First Principle of Coefficient Analysis" by Flajolet and Sedwick [68]). If the generating function is not only analytic, but also rational, then we can give a more detailed answer for the asymptotic behavior, which includes slower growth factors.

Theorem 8.22 (*Expansion of rational functions*) *[68, Theorem IV.9] If $f(z)$ is a rational function that is analytic at the origin with poles at $\rho_1, \rho_2, \ldots, \rho_m$ of orders $r_1, r_2, \ldots, r_m$, respectively, then there exist polynomials $\{P_j(z)\}_{j=1}^m$ such that for $n > n_0$,*

$$f_n = [z^n]f(z) = \sum_{j=1}^m P_j(n)\rho_j^{-n}. \tag{8.5}$$

Furthermore, the degree of P_j is one less than the multiplicity of ρ_j.

Proof Let $f(z) = N(z)/D(z)$ and let $n_0 = \deg(N) - \deg(D)$. Using partial fraction expansion, we can write

$$f(z) = Q(z) + \sum_{\rho,r} \frac{c_{\rho,r}}{(z - \rho)^r},$$

where $Q(z)$ is a polynomial of degree n_0, ρ ranges over the poles of $f(z)$ and r is bounded from above by the multiplicity of ρ as a pole of f. Thus, for $n > n_0$, only the terms in the sum contribute to the coefficient of z^n, and using (A.3) we obtain that

$$[z^n]\frac{1}{(z - \rho)^r} = \frac{(-1)^r}{\rho^r}[z^n]\frac{1}{\left(1 - \frac{z}{\rho}\right)^r} = \frac{(-1)^r}{\rho^r}\binom{n + r - 1}{n}\rho^{-n}. \qquad (8.6)$$

Since the binomial coefficient is a polynomial of degree $r - 1$ in n, the statement follows by collecting the terms associated with a given ρ_j. □

Theorem 8.22 gives a very precise representation of the coefficients. Sometimes all we care about is the asymptotic behavior, that is, we are only interested in the factor that dominates the growth. In such a case, we do not need to perform the partial fraction decomposition and explicit determination of the polynomials P_j.

Example 8.23 *Let*

$$f(z) = \frac{1}{(1 - z^4)^2(1 - z)^2(1 - z^2/4)^2}.$$

Then f can be rewritten as

$$f(z) = \frac{16}{(1 - z)^4(1 + z)^2(2 - z)^2(2 + z)^2(1 + z^2)^2}$$

to reveal that f has a pole of order four at $z = 1$ and poles of order two at $z = -1, 2, -2, i, -i$. Therefore

$$f_n = P_1(n) + P_2(n)2^{-n} + P_3(n)(-2)^{-n} + P_4(n)i^{-n}$$
$$+ P_5(n)(-i)^{-n} + P_6(n)(-1)^{-n},$$

where the degree of P_1 is three and all other polynomials have degree one. To derive how f_n behaves asymptotically, only the poles at roots of unity (those closest to the origin) need to be considered, as they correspond to the fastest exponential growth. Among those, it suffices to consider the root $\rho = 1$ because its order is bigger than the order of the other roots of unity. Substituting $z = 1$ everywhere except at the singularity gives that

$$f(z) \approx \frac{16}{2^2 \cdot 1^2 \cdot 3^2 \cdot 2^2(1 - z)^4}.$$

Now we can use (A.3) *to derive that*

$$f_n \approx \frac{1}{9}\binom{n+3}{3} \approx \frac{n^3}{54}.$$

If the generating function is not rational but only meromorphic, then we obtain an asymptotic result that resembles the structure of the exact result for rational functions.

Theorem 8.24 (*Expansion of meromorphic functions*) *[68, Theorem IV.10]* *Let $f(z)$ be a meromorphic function in the closed disk $\{z \mid |z| \leq R\}$ with poles at $\rho_1, \rho_2, \ldots, \rho_m$ of orders $r_1, r_2, \ldots, r_m$, respectively. If $f(z)$ is analytic on $|z| = R$ and at $z = 0$, then there exist m polynomials $\{P_j(z)\}_{j=1}^m$ such that*

$$f_n = [z^n]f(z) \approx \sum_{j=1}^{m} P_j(n)\rho_j^{-n}. \tag{8.7}$$

Furthermore, the degree of P_j is one less than the multiplicity of ρ_j.

Proof We provide a proof based on subtracting singularities. (A second, alternative proof based on contour integration can be found in [68].) If ρ is a pole, we can expand $f(z)$ locally as

$$f(z) = \sum_{k \geq -M} c_{\rho,k}(z - \rho)^k = S_\rho(z) + H_\rho(z), \tag{8.8}$$

where $S_\rho(z)$ consists of the "singular part," that is, the terms with negative indices, and $H_\rho(z)$ is analytic at ρ. Therefore we can express $S_\rho(z)$ as $N_\rho(z)/(z - \rho)^M$, where $N_\rho(z)$ is a polynomial of degree less than M. If we let $S(z) = \sum_j S_{\rho_j}(z)$, then it is easy to see that $f(z) - S(z)$ is analytic for $|z| \leq R$, because we have removed the singularities of $f(z)$ by subtracting them. We determine the asymptotic behavior of

$$[z^n]f(z) = [z^n]S(z) + [z^n](f(z) - S(z))$$

by analyzing the singular and the analytic part separately. The coefficient $[z^n]S(z)$ is obtained by applying Theorem 8.22. Using Cauchy's coefficient formula (E.7) with the contour $\gamma = \{z \mid |z| = R\}$ and bounding the integral leads to

$$\left| [z^n](f(z) - S(z)) \right| = \frac{1}{2\pi}\left| \int_{|z|=R} (f(z) - S(z))\frac{dz}{z^{n+1}} \right| \leq \frac{1}{2\pi}\frac{C}{R^{n+1}}2\pi R,$$

where $C = \sup\{|f(z) - S(z)| \mid |z| = R\}$. Thus, the contribution of the analytic part has a slower growth rate and can be neglected, which proves the statement. □

A special case that we will encounter in many applications is that of a meromorphic function with a single dominant pole. In this case, we can explicitly state the exponential growth rate.

Theorem 8.25 *Let $f(z) = 1/h(z)$ and let ρ^* be the positive zero with smallest modulus and multiplicity one. Then, in a neighborhood of ρ^*,*

$$f(z) \approx \frac{A}{1 - z/\rho^*}$$

and

$$[z^n]f(z) \approx A(\rho^*)^{-n} \text{ with } A = \frac{-1}{\rho^* \frac{d}{dz} h(z)|_{z=\rho^*}}.$$

Proof From the proof of Theorem 8.24 we have that $S_{\rho^*}(z) = \frac{c}{z-\rho^*}$, where $c = \lim_{z \to \rho^*} (z - \rho^*) f(z)$. Furthermore, using (8.6), we have that

$$[z^n]S_{\rho^*}(z) = \frac{-c}{\rho^*}(\rho^*)^{-n}.$$

We can determine c by using l'Hôpital's Rule to obtain that

$$c = \lim_{z \to \rho^*} \frac{z - \rho^*}{h(z)} = \lim_{z \to \rho^*} \frac{1}{h'(z)} = \frac{1}{h'(z)|_{z=\rho^*}},$$

which proves the statement. $\qquad \square$

Example 8.26 *We illustrate the use of Theorem 8.25 by applying it to the generating function for the number of compositions with parts in $A = \{1,2\}$ (see Example 3.14). In this case we actually know an explicit formula for the coefficients, so we can see how the asymptotic result compares to the explicit one. From Theorem 3.10, we know that*

$$C_A(n) = F_{n+1} = \frac{1}{\sqrt{5}}(\alpha^{n+1} - \beta^{n+1})$$

where $\alpha = (1 + \sqrt{5})/2$ and $\beta = (1 - \sqrt{5})/2$. Since $|\alpha| > 1$ and $|\beta| < 1$, the dominant term is the α term and we obtain that

$$C_A(n) \approx \frac{\alpha}{\sqrt{5}}\alpha^n = 0.723607(1.61803)^n.$$

Now we derive the asymptotics by applying Theorem 8.25. From Example 3.14 we know that the generating function for these compositions is given by

$$C_A(z) = \frac{1}{1 - z - z^2} = \frac{-1}{(z - (1 - \sqrt{5})/2)(z - (1 + \sqrt{5})/2)}.$$

Clearly, $C_A(z)$ is meromorphic having two poles

$$\rho_1 = (-1 + \sqrt{5})/2 = 0.618034 = -\beta$$

and

$$\rho_2 = (-1 - \sqrt{5})/2 = -1.61803 = -\alpha.$$

The pole of smaller modulus is $\rho^* = (-1 + \sqrt{5})/2$. Since $h(z) = 1 - z - z^2$, we obtain that

$$\left. \frac{-1}{\rho^* \cdot h'(z)} \right|_{z=\rho^*} = \frac{1}{\rho^*(1 + 2\rho^*)} = 0.723607.$$

Therefore,

$$C_A(n) \approx 0.723607(1.61803)^n.$$

Note that we give between six and ten significant digits in numerical approximations. So, for example, $\rho_1 = (-1 + \sqrt{5})/2 = 0.618034$ should be interpreted as $\rho_1 = 0.618034\cdots$.

A final result that is of great importance for establishing that a numerically computed root is indeed associated with the dominant singularity is Henrici's Principle of the Argument, which we will use in several instances.

Theorem 8.27 (*Henrici's Principle of the Argument*) *[83, Theorem 4.10a] Let $f(z)$ be analytic in a region $\Re$ and let γ be a simple closed curve in the interior of $\Re$, where $f(z) \neq 0$ on γ. Then the number of zeros of $f(z)$ (counted with multiplicities) inside γ equals the winding number of the transformed contour $f(\gamma)$ around the origin.*

Surprisingly, complex analysis also helps us establish asymptotic results in terms of convergence to a normal distribution (see Definition 8.19). The following proposition is a corollary to a more general result by Bender [18, Theorem 1].

Proposition 8.28 *[18, Section 3] Suppose that*

$$f(z, w) = \sum_{n,k \geq 0} a_n(k) z^n w^k = \frac{g(z, w)}{P(z, w)},$$

where

(i) *$P(z, w)$ is a polynomial in z with coefficients continuous in w,*

(ii) *$P(z, 1)$ has a simple root at ρ^* and all other roots have larger absolute value,*

(iii) *$g(z, w)$ is analytic for w near 1 and $z < \rho^* + \varepsilon$, and*

(iv) $g(\rho^*, 1) \neq 0$.

Then

$$\mu = -\frac{r(z,w)}{\rho^*}\Big|_{z=\rho^*,w=1} \quad and \quad \sigma^2 = \mu^2 - \frac{s(z,w)}{\rho^*}\Big|_{z=\rho^*,w=1},$$

where

$$r = r(z,w) = -\frac{\frac{\partial}{\partial w}P}{\frac{\partial}{\partial z}P} \tag{8.9}$$

and

$$s = s(z,w) = -\frac{r^2\frac{\partial^2}{\partial z^2}P + 2r\frac{\partial}{\partial z}\frac{\partial}{\partial w}P + \frac{\partial}{\partial w}P + \frac{\partial^2}{\partial w^2}P}{\frac{\partial}{\partial z}P}. \tag{8.10}$$

If $\sigma \neq 0$, then $a_n(k) \overset{d}{\approx} \mathcal{N}(n\mu, n\sigma^2)$.

A second, local limit theorem was also given by Bender.

Theorem 8.29 *[18, Theorem 3] Let $f(z,w)$ have power series expansion*

$$f(z,w) = \sum_{n,k\geq 0} a_n(k)z^n w^k$$

with nonnegative coefficients $a_n(k)$, and let $a < b$ be real numbers. Define

$$R(\varepsilon) = \{z \mid a \leq \mathrm{Re}(z) \leq b, |\mathrm{Im}(z)| \leq \varepsilon\}.$$

Suppose there exist $\varepsilon > 0$, $\delta > 0$, a nonnegative integer ℓ and functions $A(s)$ and $r(s)$ such that

1. *$A(s)$ is continuous and nonzero for $s \in R(\varepsilon)$;*

2. *$r(s)$ is nonzero and has bounded third derivative for $s \in R(\varepsilon)$;*

3. *for $s \in R(\varepsilon)$ and $|z| \leq |r(s)|(1 + \delta)$*

$$\left(1 - \frac{z}{r(s)}\right)^\ell f(z, e^s) - \frac{A(s)}{1 - z/r(s)}$$

   is analytic and bounded;

4.

$$\left(\frac{r'(\alpha)}{r(\alpha)}\right)^2 - \frac{r''(\alpha)}{r(\alpha)} \neq 0 \quad for \quad a \leq \alpha \leq b;$$

5. *$f(z, e^s)$ is analytic and bounded for*

$$|z| \leq |r(\mathrm{Re}(s))|(1 + \delta) \quad and \quad \varepsilon \leq |\mathrm{Im}(s)| \leq \pi.$$

Then we have that

$$a_n(k) \approx \frac{n^\ell e^{-\alpha k} A(\alpha)}{\ell! r(\alpha)^n \sigma_\alpha \sqrt{2\pi n}} \tag{8.11}$$

as $n \to \infty$, uniformly for $a \leq \alpha \leq b$, where

$$\frac{k}{n} = -\frac{r'(\alpha)}{r(\alpha)} \quad and \quad \sigma_\alpha^2 = \left(\frac{k}{n}\right)^2 - \frac{r''(\alpha)}{r(\alpha)}.$$

Finally, to be able to specify growth rates easily, we will use *Landau's big and little O notation.*

Definition 8.30 *We say that a function $f(x)$ is of the* order *of $g(x)$ or $O(g(x))$ as $x \to \infty$ if and only if there exists a value x_0 and a constant $M > 0$ such that $|f(x)| \leq M|g(x)|$ for all $x > x_0$ or, equivalently, if and only if $\limsup_{x \to \infty} |f(x)/g(x)| < \infty$. We say that $f(x)$ is $o(g(x))$ if and only if $|f(x)/g(x)| \to 0$ as $x \to \infty$, that is, $f(x)$ grows slower than $g(x)$ as $x \to \infty$.*

In general, $g(x)$ is a function that is simpler than $f(x)$ and consists of the terms that dominate the growth of $f(x)$ in the limit.

Example 8.31 *Let $f(x) = 3x^2 + x - 1$. Then $f(x) = O(x^2)$ and $f(x) = o(x^n)$ for $n \geq 3$.*

Having all the necessary tools available for the asymptotic analysis, we now consider various statistics for compositions and Carlitz compositions.

8.4   Asymptotics for compositions

In this section we derive various asymptotic results for compositions. In Section 8.4.1 we concentrate on compositions that avoid or contain a given subword pattern, and derive results on the number of such compositions, the average number of parts in such compositions, as well as asymptotic distributions for these compositions. In Sections 8.4.2 and 8.4.3 we focus on asymptotic results for the largest part and the number of distinct parts in compositions of n.

8.4.1   Asymptotics for subword pattern avoidance

Avoidance of subword patterns of length two reduces either to Carlitz compositions (avoidance of the subword pattern 11, which will be discussed in Section 8.5) or partitions (avoidance of the subword pattern 12). The asymptotic behavior for partitions is well-known and dates back to the early part of the last century. Hardy and Ramanujan in 1918, and independently Uspensky in 1920, derived the following asymptotic result for partitions (see for example [10] or [164]).

Theorem 8.32 *Let $p(n)$ be the number of partitions of n. Then*

$$AC^{12}(n) = p(n) \approx \frac{e^{\pi\sqrt{2n/3}}}{4n\sqrt{3}} \qquad \text{as } n \to \infty.$$

Subword patterns of length three were investigated by Heubach and Mansour [92]. We present the results for the eight single subword patterns 111, 112, 221, 212, 121, 123, 132 and 213, and for the patterns *valley* and *peak*.

In all cases, the process is the same, namely checking that the generating functions are meromorphic (see Definition E.8) in a disc around the origin, identifying the dominant singularity, checking that it is a pole of multiplicity one, and then using Theorem 8.25 for the rate of growth. Similar analyses for the longer subword patterns discussed in Sections 4.3.3, 4.3.4, and 4.3.5 together with results on the asymptotic distribution were given by Mansour and Sirhan [145].

Theorem 8.33 *For subword patterns τ of length three, we have that*

$$AC^{\tau}(n) \approx A_{\tau}(\rho_{\tau}^*)^{-n}, \quad n \to \infty, \qquad (8.12)$$

where the zeros ρ_{τ}^* are given by

$$
\begin{aligned}
\rho_{111}^* &= 0.5233508903 & \rho_{112}^* &= 0.5534397072 \\
\rho_{221}^* &= 0.5133872872 & \rho_{212}^* &= 0.5120333765 \\
\rho_{121}^* &= 0.5386079645 & \rho_{123}^* &= 0.5132858388 \\
\rho_{132}^* &= 0.5055555128 & \rho_{213}^* &= 0.5220916260 \\
\rho_{valley}^* &= 0.543206932 & \rho_{peak}^* &= 0.6135900256
\end{aligned}
$$

and the constants A_{τ} are given by

$$
\begin{aligned}
A_{111} &= 0.4993008432 & A_{112} &= 0.6920054131 \\
A_{221} &= 0.5453621593 & A_{212} &= 0.5372331234 \\
A_{121} &= 0.5838933009 & A_{123} &= 0.5760960195 \\
A_{132} &= 0.5353097161 & A_{213} &= 0.6121896742 \\
A_{valley}^* &= 0.7282070589 & A_{peak} &= 1.3945612176
\end{aligned}
$$

Note that Theorem 8.33 for $\tau = 111$ gives the asymptotics for the number of 2-Carlitz compositions.

Proof Let τ be any of the ten subword patterns. Using results from Chapter 4 it can be shown that the functions $AC^{\tau}(z, w)$ are meromorphic in the domain $|z| < 1$ if $w \in [0, 1]$ and meromorphic in the domain $|z| < 1/w$ if

$w \in [1, \infty]$. Thus the asymptotic behavior of $AC^\tau(n)$ is determined by the dominant pole of the function $AC^\tau(z, 1) = 1/h_\tau(z)$. By Theorem 8.20 this pole occurs at the smallest positive root ρ_τ^* of $h_\tau(z)$. We numerically compute the roots ρ_τ^* by converting $h_\tau(z)$ into a series up to the z^{30} term. Since all the zeros have modulus less than 0.7, we use the curve $\gamma = \{z \mid |z| = 0.7\}$ when applying Theorem 8.27. Figure 8.1 shows the curves $h_\tau(\gamma)$ (in the same order as the zeros given in Theorem 8.33) where we have used the expansion of $h_\tau(z)$ up to the z^{30} term.

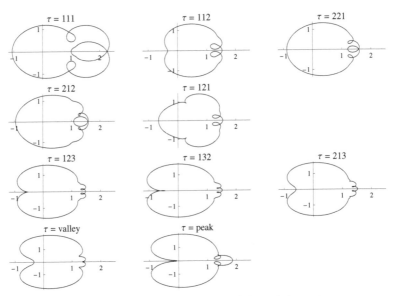

FIGURE 8.1: Image of $h_\tau(z)$ for $|z| = 0.7$

In each case, the winding number is one, so that the function $h_\tau(z)$ has exactly one root ρ_τ^* in the domain $|z| < 0.7$. Thus, Theorem 8.25 gives (8.12) where the values A_τ are computed using the expansion of $h(z)$ to the z^{30} term. $\qquad \square$

We can also obtain results on the average number of parts in a random composition avoiding a given subword pattern. Knopfmacher and Prodinger [119, Section 2] derived asymptotics on the average number of parts in a random Carlitz composition of n, thus covering the case of the subword pattern 11. More precisely, if we let $X_{\text{par}}^{11}(n)$ denote the number of parts in a random Carlitz composition of n, then Knopfmacher and Prodinger showed that

$$\mathbb{E}(X_{\text{par}}^{11}(n)) \approx 0.350571n.$$

We derive the corresponding results for random compositions that avoid the patterns of length three given in Table 4.2.

Theorem 8.34 *Let $X_{par}^\tau(n)$ denote the number of parts in a random composition of n that avoids the pattern τ. Then we have that*

$$\mathbb{E}(X_{par}^\tau(n)) \approx C_\tau \cdot n,$$

where

$$
\begin{array}{ll}
C_{111} = 1.715995518 & C_{112} = 1.915578938 \\
C_{221} = 1.990357910 & C_{212} = 1.953193954 \\
C_{121} = 1.312572461 & C_{123} = 1.744084928 \\
C_{132} = 1.784275109 & C_{213} = 1.342234855 \\
C_{valley} = 1.447227208 & C_{peak} = 0.651749
\end{array}
$$

Proof We give the proof for the pattern *peak*. The proofs for the other patterns follow along the same lines of reasoning. Applying (8.2) in Lemma 8.14 we obtain the desired expected value by differentiating the functions

$$AC^\tau(z, w) = C^\tau(z, w, 0)$$

given in Section 4.3. The value for $N = C_\tau(n) = [z^n]C^\tau(z, 1, 0)$ can be approximated by using the results of Theorem 8.33. Specifically, for the pattern *peak* we obtain from Example 4.27 that

$$AC^{peak}(z, w) = \frac{1 + \sum_{j \geq 1} \dfrac{z^{j(j+2)}w^{2j}}{\prod_{i=1}^{2j}(1 - z^i)}}{1 + \sum_{j \geq 1} \dfrac{z^{j(j+2)}w^{2j}}{\prod_{i=1}^{2j}(1 - z^i)} - \sum_{j \geq 0} \dfrac{z^{j^2+3j+1}w^{2j+1}}{\prod_{i=1}^{2j+1}(1 - z^i)}} = \frac{n(z, w)}{h(z, w)},$$

and therefore,

$$
\begin{aligned}
F_{peak}(z) &= \frac{\partial}{\partial w} AC^{peak}(z, w)\Big|_{w=1} \\
&= \frac{\frac{\partial}{\partial w}n(z, w)\big|_{w=1}h(z, 1) - n(z, 1)\frac{\partial}{\partial w}h(z, w)\big|_{w=1}}{h(z, 1)^2}.
\end{aligned}
$$

Since it can be shown numerically that the numerator of $F_{peak}(\rho_{peak}^*) \neq 0$ and $h(\rho_{peak}^*, 1) = 0$ (that is how ρ_{peak}^* was computed), we have that for z close to the dominant singularity ρ_{peak}^*

$$F_{peak}(z) \approx \frac{B_{peak}}{(1 - z/\rho_{peak}^*)^2},$$

where $B_{peak} = \lim_{z \to \rho_{peak}^*}(1 - z/\rho_{peak}^*)^2 F_{peak}(z) = 0.908904458$ (using Maple with an approximation to the z^{30} term). From (8.6) we then obtain that

$$[x^n]F_{peak}(z) \approx B_{peak} \cdot n \cdot (\rho_{peak}^*)^{-n},$$

and therefore, $\mathbb{E}(X_{\text{par}}^{peak}(n)) \approx \frac{B_{peak}}{A_{peak}} \cdot n$ where A_{peak} is the value given in Theorem 8.33. □

Using Theorems 8.28 and 8.29 we can obtain a more detailed picture of the asymptotic distribution function for the number of parts in compositions of n (as $n \to \infty$). In fact, Flajolet and Sedgewick [68, Chapter IX.6] show that one can obtain Gaussian limit laws for many statistics that have bivariate meromorphic generating functions by using singularity perturbation on the secondary variable. Mansour and Sirhan gave results for all three-letter patterns and for selected patterns of length ℓ [145, Theorem 3.1].

Theorem 8.35 *For large n, the number of parts M in a composition of n that avoids τ is asymptotically Gaussian with mean $n\mu_\tau$ and standard deviation $\sqrt{n}\sigma_\tau$, that is,*

$$\frac{M - \mu_\tau n}{\sqrt{n}\sigma_\tau} \overset{d}{\approx} \mathcal{N}(0, 1).$$

Furthermore,

$$AC_{\mathbb{N}}^\tau(m; n) \approx \frac{A_\tau}{(\rho_\tau^*)^n \cdot \sqrt{2\pi n} \cdot \sigma_\tau} e^{-(m - n\mu_\tau)^2/(2n\sigma_\tau^2)},$$

for $m - n\mu_\tau = O(\sqrt{n})$, where A_τ and ρ_τ^* are given in Theorem 8.33 and μ_τ and σ_τ^2 are given by

τ	μ_τ	σ_τ^2
111	0.4277999660	0.1654603703
112	0.4465604097	0.2471428219
221	0.4866127277	0.2498205897
212	0.4879666379	0.2498550090
121	0.4613949299	0.2484778134
123	0.4854585926	0.2606813427
132	0.5112956159	0.2517270954
213	0.5030376743	0.2777839362

Proof Let $h_\tau(z, w) = 1/AC^\tau(z, w)$. Then we can use Theorem 8.28 to compute the values of $r_\tau(z, w)$ and $s_\tau(z, w)$ given in (8.9) and (8.10). Setting $w = 1$ and $z = \rho_\tau^*$ we derive $\mu_\tau = -r_\tau/\rho_\tau^*$ and $\sigma_\tau^2 = \mu_\tau^2 - s_\tau/\rho_\tau^*$, where r_τ and s_τ given below are numerically computed by developing $h_\tau(z, w)$ into a series up to the z^{30} term and using the values of ρ_τ^* given in Theorem 8.33.

τ	r_τ	s_τ
111	-0.2238894933	0.00918608541
112	-0.2471442624	-0.02641380791
221	-0.2498207882	-0.00668873965
212	-0.2498552052	-0.00601309945 .
121	-0.2485109840	-0.01917042125
123	-0.2491790209	-0.01283794488
132	-0.2584883173	0.00490192258
213	-0.2626317573	-0.01291499850

The statement to be proven now follows from Theorem 8.29 with $r(s) = r(s, 1)$ as defined in Theorem 8.28, $A(s) = A_\tau$, and $\ell = 1$. Furthermore, if $m - n\mu_\tau = O(\sqrt{n})$, then the assumptions of Theorem 8.29 are satisfied and the statement about the asymptotics of $AC_{\mathbb{N}}(m; n)$ follows. $\qquad\square$

Remark 8.36 *Even though both Theorem 8.34 and 8.35 give results on the asymptotics for the number of parts, one has to be careful to distinguish between them. Theorem 8.34 gives a result for a fixed large n, giving the average number of parts over all compositions of n. Theorem 8.35 gives an asymptotic distribution for the number of parts for cases where the order of the composition and the number of parts are of given relative size as expressed by the big O requirement.*

Results similar to those of Theorem 8.35 can also be obtained for the patterns *peak* and *valley* (see Exercise 8.5) and for longer patterns (see [145]).

We conclude this section with some results on the asymptotics for the number of compositions of n containing a subword pattern τ a given number of times. Knopfmacher and Prodinger [119] showed that the number of compositions of n with exactly r occurrences of the subword pattern 11 has the asymptotic behavior

$$C^{11}(n, r) \approx C_r n^r (\rho^*)^{-n},$$

where $\rho^* = 0.571349$ is the single positive real pole of $\mathcal{CC}(z)$ in the interval $[0, 1]$ and the constants C_r are explicitly computable. For example, the constants for $r = 0, 1, 2$ are given by $C_0 = 0.4563$, $C_1 = 0.0482$, and $C_2 = 0.0025$.

Remark 8.37 *In general, if $C^\tau(z, w, q) = 1/h_\tau(z, w, q)$ is the generating function for the number of compositions of n with m parts according to the number of occurrences of the subword pattern τ, then applying Taylor's formula results in*

$$[q^r]C^\tau(z, 1, q) = \frac{1}{r!}\frac{\partial^r}{\partial q^r}C^\tau(z, 1, q)\Big|_{q=0} = \frac{G_\tau(z)}{h_\tau(z, 1, 0)^{r+1}},$$

where $G_\tau(z)$ is a function consisting of derivatives of $h_\tau(z, 1, q)$ evaluated at $q = 0$. Consequently, for fixed r and $n \to \infty$, the number of compositions with exactly r occurrences of the subword pattern τ has asymptotic behavior

$$C^\tau(n, r) \approx C_r^\tau n^{s(r+1)-1} (\rho_\tau^*)^{-n},$$

where ρ_τ^* has multiplicity s and is the smallest positive real root of $h_\tau(z, 1, 0)$.

Example 8.38 *For instance, if $\tau = 111$, then we obtain from Theorem 4.32 and Remark 8.37 that the number of compositions with exactly r occurrences of the subword pattern 111 has the following asymptotic behavior (see Exercise 8.6):*

$$C^{111}(n, 0) \approx 0.4993008432 \cdot 0.5233508903^{-n},$$
$$C^{111}(n, 1) \approx 0.05827141770 \cdot n \cdot 0.5233508903^{-n},$$
$$C^{111}(n, 2) \approx 0.00339555803 \cdot n^2 \cdot 0.5233508903^{-n}.$$

We now leave pattern avoidance and enumeration behind and consider the asymptotics for statistics on unrestricted compositions.

8.4.2 Asymptotics for the largest part

In the analysis so far we either computed the average of a statistic from the generating function using Lemma 8.14, or obtained asymptotic results using Theorem 8.25. We now show an example where we compute the average using a different method and in the process encounter an estimate that contains a small periodic part, very different from the results discussed so far.

Theorem 8.39 *Let $X_{\max}(n)$ denote the largest part in a random composition of n. Then,*

$$\mathbb{E}(X_{\max}(n)) \approx \log_2 n + 0.3327461773 + d(\log_2 n),$$

where $d(x)$ is a function that has period one, mean zero, and small amplitude.

Proof Since the random variable $X_{\max}(n)$ is nonnegative, we have from Theorem D.8 that

$$\mathbb{E}(X_{\max}(n)) = \sum_{h=0}^{\infty} P(X_{\max}(n) > h) = \sum_{h=0}^{\infty} \left(1 - \frac{C_{[h]}(n)}{C(n)}\right), \qquad (8.13)$$

where $C_{[h]}(n)$ denotes the number of compositions of n with parts in $[h] = \{1, 2, \ldots, h\}$ (all parts are less than or equal h) and $C(n)$ is the number of compositions of n with parts in $\mathbb{N}$. So we need to derive asymptotic results for both these quantities. From Example 2.35 we have that

$$C(z) = \frac{1}{1 - g(z)} = \frac{1}{1 - \frac{z}{1-z}},$$

and from Theorem 3.13 we obtain that

$$C_{[h]}(z) = \frac{1}{1 - (z + z^2 + \cdots + z^h)} = \frac{1}{1 - \frac{z(1-z^h)}{1-z}} = \frac{1}{1 - g_h(z)}.$$

It is easy to show that $C(z)$ has a dominant pole at $\rho^* = 1/2$ of multiplicity one, so applying Theorem 8.25 we obtain that $C(n) \approx \frac{1}{\rho^* g'(\rho^*)}(\rho^*)^{-n}$. Now let ρ_h^* be the smallest positive real solution of the equation $g_h(z) = 1$ (and therefore, the dominant pole of $C_{[h]}(z)$). It can be shown that ρ_h^* is a simple pole (see Exercise 8.7) and therefore,

$$C_{[h]}(n) \approx \frac{1}{\rho_h^* g_h'(\rho_h^*)}(\rho_h^*)^{-n}.$$

This expression is not very useful as it depends on h, so we will express ρ_h^* in terms of ρ^* and $g_h(z)$ in terms of $g(z)$. Clearly, in the disk $|z| < 1$, $\lim_{h\to\infty} C_{[h]}(z) = C(z)$ and $\lim_{h\to\infty} \rho_h^* = \rho^*$. Following the "bootstrapping method" described in [124], we obtain that around $z = \rho^*$,

$$g_h(z) \approx g(z) - \frac{(\rho^*)^{h+1}}{1 - \rho^*},$$

and therefore

$$C_{[h]}(z) \approx \frac{1}{1 - g(z) + \frac{(\rho^*)^{h+1}}{1-\rho^*}} \tag{8.14}$$

near $z = \rho^*$. Using $g(\rho^*) = 1$ and Taylor's theorem for $g(\rho_h^*)$ with $\rho_h^* = \rho^*(1 + \varepsilon_h)$ we obtain that

$$0 \approx \rho^* \cdot \varepsilon_h \cdot g'(\rho^*) - \frac{(\rho^*)^{h+1}}{1 - \rho^*},$$

and therefore

$$\varepsilon_h \approx \frac{1}{(1 - \rho^*)g'(\rho^*)}(\rho^*)^h.$$

Applying Theorem 8.25 to (8.14) results in

$$C_{[h]}(n) \approx \frac{1}{\rho_h^* g'(\rho_h^*)}(\rho_h^*)^{-n} \approx \frac{1}{\rho^* g'(\rho^*)}(\rho^*)^{-n}(1 + \varepsilon_h)^{-n},$$

where we are using that $\rho_h^* \approx \rho^*$ for the fraction and substitute $\rho_h^* = \rho^*(1+\varepsilon_h)$ in the power term for the second approximation. Thus, the probability that the largest part is greater than h is approximated by

$$1 - \frac{C_{[h]}(n)}{C(n)} \approx 1 - (1 + \varepsilon_h)^{-n} \approx 1 - (1 - \varepsilon_h)^n$$

$$\approx 1 - \left(1 - \frac{1}{(1 - \rho^*)g'(\rho^*)}(\rho^*)^h\right)^n. \tag{8.15}$$

Using the fact that $(1 - a)^n \approx e^{-a \cdot n}$ we obtain from (8.13) that

$$\mathbb{E}(X_{\max}(n)) \approx \sum_{h \geq 1}(1 - \exp(-Mn(\rho^*)^h)) \quad \text{with } M = \frac{1}{(1 - \rho^*)g'(\rho^*)}.$$

The expression that approximates $\mathbb{E}(X_{\max}(n))$ is quite well studied (see for example [67]). Using the results given in Flajolet et al. [67], we obtain that

$$\mathbb{E}(X_{\max}(n)) \approx \log_{1/\rho^*} n + \log_{1/\rho^*} M - \frac{\gamma}{\log \rho^*} + \frac{1}{2} + d(\log_{1/\rho^*} nM), \quad (8.16)$$

where $\gamma = \lim_{n \to \infty} \sum_{i=1}^{n} \frac{1}{i} - \log n = 0.57721566490 \cdots$ and the function $d(x)$ has period one, mean zero, and small amplitude. For $\rho^* = 1/2$ and $g(z) = z/(1 - z)$ we obtain that $\log_{1/\rho^*} M - \frac{\gamma}{\log \rho^*} + \frac{1}{2} = 0.3327461773 \cdots$, giving the desired result. □

8.4.3   Asymptotics for the number of distinct parts

We now look at a different statistic, namely the number of distinct parts in a random composition.

Definition 8.40 *For a given random composition $\sigma = \sigma_1 \sigma_2 \cdots \sigma_m$ of n with m parts, we denote the number of distinct parts by $D_n(\sigma)$. That is,*

$$D_n(\sigma) = 1 + \sum_{i=2}^{m} I_{\{\sigma_i \neq \sigma_j, \, j=1,2,\ldots,i-1\}}, \quad (8.17)$$

where I_A, the indicator function of the set A equals 1 if A occurs and 0 otherwise.

We present a rather precise result on the asymptotics of $\mathbb{E}(D_n)$ that was originally obtained by Hwang and Yeh [102] using generating function techniques, but was also derived by Hitczenko and Stengle [97] using a probabilistic approach. We will summarize the key steps in the probabilistic approach as these techniques can be used to study the asymptotics of other statistics. The analysis is based on a reformulation of the notion of a random composition by exploiting a bijection (mentioned in [10]) between compositions of n and binary words of length n that end in 1. Specifically, a composition of n with m parts is associated with the binary word $w_1 w_2 \cdots w_n$ of length n where $w_i = 1$ for all $i \in \{\sigma_1, \sigma_1 + \sigma_2, \ldots, \sigma_1 + \sigma_2 + \cdots + \sigma_m\}$ and $w_i = 0$ otherwise.

Example 8.41 *The composition 213114 of 12 with six parts corresponds to the word 011001110001, while the composition 321 of 6 with three parts corresponds to the word 001011. The correspondence is easily visualized as follows:*

$$\underbrace{01}_{2} \mid \underbrace{1}_{1} \mid \underbrace{001}_{3} \mid \underbrace{1}_{1} \mid \underbrace{1}_{1} \mid \underbrace{0001}_{4} \quad and \quad \underbrace{001}_{3} \mid \underbrace{01}_{2} \mid \underbrace{1}_{1}.$$

Note that the number of parts in the composition corresponds to the number of 1s in the binary word.

With this bijection it is easy to see that choosing a composition at random is equivalent to having a 0 or 1 occur with probability $1/2$ at each of the first $n-1$ positions, with the occurrences at different positions being independent (see Definition D.5) of each other. In other words, the number of 1s in the first $n-1$ positions is a binomial random variable (see Definition D.11) $\mathcal{B}(n-1, 1/2)$ with $n-1$ trials and success probability $1/2$. Thus, the total number of parts $X_{\text{tot}}(n)$ of a random composition of n equals the number of 1s (including the 1 at the rightmost position) in the binary word and therefore,

$$X_{\text{tot}}(n) \sim \mathcal{B}(n-1, 1/2) + 1.$$

The numbers $\sigma_1, \ldots, \sigma_m$ are "waiting times" for the first, second,...., and m-th appearance of a 1. It is well-known and easy to check that in an infinite sequence of independent Bernoulli trials with success probability p, the waiting times for successes are independent and identically distributed (i.i.d.) random variables whose common distribution is that of a geometric random variable with parameter p (see Definition D.12). Since we are considering only $n-1$ trials, this is no longer true. However, we have the following characterization.

Proposition 8.42 *[96, Proposition 1] Let $\Gamma_1, \Gamma_2, \ldots$, be i.i.d. geometric random variables with parameter $1/2$ (that is, $\mathbf{P}(\Gamma_1 = j) = 2^{-j}$, for all $j \geq 1$) and define*

$$\eta = \inf\{k \geq 1 \mid \Gamma_1 + \Gamma_2 + \cdots + \Gamma_k \geq n\}.$$

Then, the distribution of a randomly chosen composition in $\mathcal{C}_n$ is given by

$$\left(\Gamma_1, \Gamma_2, \ldots, \Gamma_{\eta-1}, n - \sum_{j=1}^{\eta-1} \Gamma_j \right).$$

Proposition 8.42 provides an alternative way to create a random composition, namely sampling geometric random variables, with an adjustment for the last part (see Appendix F and Exercise 8.11). Using this fact, Hitczenko and Stengle [97] showed the following result using probabilistic estimates.

Theorem 8.43 *[97, page 520] As $n \to \infty$,*

$$\mathbb{E}(D_n) \approx \log_2 n - 0.6672538227 + d(\log_2 n),$$

where d is a function with mean zero and period one satisfying $|d| \leq 0.0000016$.

Thus, the expected number of distinct parts in a random composition of n asymptotically behaves like $\log_2 n$ plus a constant plus a small but periodic oscillation. This behavior is similar to the asymptotic behavior for the average size of the maximal part given in Theorem 8.39. Clearly, the maximal part is always greater than or equal to the number of distinct parts.

The proof of Theorem 8.43 is split into two parts. It consists of first estimating the expected value as an infinite sum of a particular type and then showing that this type of infinite sum has a certain rate of growth.

Proposition 8.44 *[97, Proposition 2.1] As $n \to \infty$,*

$$\mathbb{E}(D_n) = \sum_{m \geq 1} \left[1 - \left(1 - \frac{1}{2^m} \right)^{\frac{n}{2}(1+o(1))} \right] + o(1).$$

Proof We sketch the proof by stating the main steps. Let $\tilde{\Gamma}_i(\sigma) = \Gamma_i(\sigma)$ for all $i < \eta(\sigma)$, and $\tilde{\Gamma}_{\eta(\sigma)}(\sigma) = n - \sum_{i=1}^{\eta(\sigma)-1} \Gamma_i(\sigma)$ denote the parts of a randomly chosen composition σ. Clearly, $\tilde{\Gamma}_\eta \leq \Gamma_\eta$. Rewriting (8.17) in terms of the binary words model leads to

$$\mathbb{E}(D_n) = 1 + \mathbb{E}\left(\sum_{i=2}^{\eta} I_{\{\tilde{\Gamma}_i \neq \tilde{\Gamma}_j, \, j=1,2,\ldots,i-1\}} \right),$$

or equivalently

$$\mathbb{E}(D_n) = 1 + \mathbb{E}\left(\sum_{i=2}^{\eta-1} I_{\{\Gamma_i \neq \Gamma_j, \, j=1,2,\ldots,i-1\}} \right) + \mathbf{P}(\tilde{\Gamma}_\eta \neq \Gamma_1, \ldots, \Gamma_{\eta-1}). \quad (8.18)$$

The remainder of the proof consists of estimating each term in (8.18). To show that the last probability term is negligible we break up the values for η into those where $|\eta - \mathbb{E}(\eta)| \leq t_n$ and those where $|\eta - \mathbb{E}(\eta)| > t_n$, where $t_n = \sqrt{(n-1)\log n}$. To estimate the latter probability, we use that $\eta = X_{\text{tot}}(n)$ is a $1 + \mathcal{B}(n-1, 1/2)$ random variable and therefore satisfies the bound (using a large deviation estimate, see Theorem D.14)

$$\mathbf{P}(|\eta - \mathbb{E}(\eta)| \geq t) \leq 2 \exp\left(\frac{-2t^2}{n-1} \right). \quad (8.19)$$

Specifically,

$$\mathbf{P}\left(|\eta - \mathbb{E}(\eta)| \geq \sqrt{(n-1)\log n} \right) \leq 2\exp(-2\log n) = \frac{2}{n^2}. \quad (8.20)$$

In order to deal with the values of η where $|\eta - \mathbb{E}(\eta)| \leq t_n$, we define

$$n_0 = \mathbb{E}(\eta) - t_n = (n+1)/2 - \sqrt{(n-1)\log n} \quad (8.21)$$

and

$$n_1 = \mathbb{E}(\eta) + t_n = (n+1)/2 + \sqrt{(n-1)\log n}. \quad (8.22)$$

With this definition of n_0 and n_1 we have that

$$|\eta - \mathbb{E}(\eta)| \leq t_n \iff n_0 \leq \eta \leq n_1,$$

which we will use repeatedly. Using basic rules of probability and changing the order of summation results in

$$\mathbf{P}(\tilde{\Gamma}_\eta \neq \Gamma_1, \ldots, \Gamma_{\eta-1}) \leq \sum_{j=1}^{n} \mathbf{P}(\tilde{\Gamma}_\eta = j, \Gamma_1, \ldots, \Gamma_{\eta-1} \neq j)$$

$$\leq \sum_{j=1}^{n} \mathbf{P}(\Gamma_\eta \geq j, \Gamma_1, \ldots, \Gamma_{\eta-1} \neq j, |\eta - \mathbb{E}(\eta)| \leq t_n)$$

$$+ n\mathbf{P}(|\eta - \mathbb{E}(\eta)| > t_n)$$

$$\leq \sum_{j=1}^{n} \sum_{k=n_0}^{n_1} \mathbf{P}(\Gamma_k \geq j, \Gamma_1, \ldots, \Gamma_{k-1} \neq j, \eta = k) + \frac{2}{n}$$

$$\leq \sum_{k=n_0}^{n_1} \sum_{j \geq 1} \frac{1}{2^{j-1}} \left(1 - \frac{1}{2^j}\right)^{k-1} + \frac{2}{n}$$

$$\leq \sum_{k=n_0}^{n_1} C \int_0^\infty \frac{1}{2^x} \left(1 - \frac{1}{2^{x+1}}\right)^{k-1} dx + \frac{2}{n}$$

$$\leq \sum_{k=n_0}^{n_1} \frac{\tilde{C}}{k} + \frac{2}{n} \leq \tilde{C} \frac{\sqrt{n \log n}}{n} \to 0,$$

where the last inequality follows from Exercise 8.10. Now we estimate the term involving the expectation in (8.18) as

$$\mathbb{E}\left[\sum_{i=2}^{\eta-1} I_{\{\Gamma_i \neq \Gamma_j, j=1,2,\ldots,i-1\}}\right]$$

$$\leq \mathbb{E}\left[\sum_{i=2}^{\eta} I_{\{\Gamma_i \neq \Gamma_j, j=1,2,\ldots,i-1\}} I_{\{|\eta - \mathbb{E}(\eta)| \leq t_n\}}\right] + (n-2)\mathbf{P}(|\eta - \mathbb{E}(\eta)| > t_n)$$

$$\leq \mathbb{E}\left[\sum_{i=2}^{\eta-1} I_{\{\Gamma_i \neq \Gamma_j, j=1,2,\ldots,i-1\}} I_{\{|\eta - \mathbb{E}(\eta)| \leq t_n\}}\right] + \frac{2(n-2)}{n^2}, \tag{8.23}$$

where we have used that $I_{A \cap B} = I_A \cdot I_B$, $\mathbb{E}(I_A) = \mathbf{P}(A)$, and $\eta \leq n$. Since the second term in (8.23) is of order $1/n$, we only need to estimate the first term. Translating the condition $|\eta - \mathbb{E}(\eta)| \leq t_n$ into values for η, we obtain that

$$\mathbb{E}\left[\sum_{i=2}^{\eta-1} I_{\{\Gamma_i \neq \Gamma_j, j=1,2,\ldots,i-1\}} I_{\{|\eta - \mathbb{E}(\eta)| \leq t_n\}}\right]$$

$$\leq \mathbb{E}\left[\sum_{i=2}^{n_1} I_{\{\Gamma_i \neq \Gamma_j, j=1,2,\ldots,i-1\}}\right] = \sum_{i=2}^{n_1} \mathbf{P}(\Gamma_i \neq \Gamma_j, j = 1, 2, \ldots, i-1),$$

and similarly,

$$
\mathbb{E}\Big[\sum_{i=2}^{\eta-1} I_{\{\Gamma_i \neq \Gamma_j,\, j=1,2,\dots,i-1\}} I_{\{|\eta-\mathbb{E}(\eta)|\le t_n\}}\Big]
$$

$$
\ge \mathbb{E}\Big[\sum_{i=2}^{n_0-1} I_{\{\Gamma_i \neq \Gamma_j,\, j=1,2,\dots,i-1\}} I_{\{|\eta-\mathbb{E}(\eta)|\le t_n\}}\Big]
$$

$$
= \mathbb{E}\Big[\sum_{i=2}^{n_0-1} I_{\{\Gamma_i \neq \Gamma_j,\, j=1,2,\dots,i-1\}}\Big] - \mathbb{E}\Big[\sum_{i=2}^{n_0-1} I_{\{\Gamma_i \neq \Gamma_j,\, j=1,2,\dots,i-1\}} I_{\{|\eta-\mathbb{E}(\eta)|>t_n\}}\Big]
$$

$$
\ge \sum_{i=2}^{n_0-1} \mathbf{P}(\Gamma_i \neq \Gamma_j,\, j=1,2,\dots,i-1) - (n_0-2)\mathbf{P}(|\eta-\mathbb{E}(\eta)| > t_n)
$$

$$
\ge \sum_{i=2}^{n_0-1} \mathbf{P}(\Gamma_i \neq \Gamma_j,\, j=1,2,\dots,i-1) - \frac{n-3-2\sqrt{(n-1)\log n}}{n^2}.
$$

Both bounds have the same type of sum which we now estimate. For fixed i,

$$
\mathbf{P}(\Gamma_i \neq \Gamma_j,\, j=1,2,\dots,i-1) = \sum_{m=1}^{\infty} \mathbf{P}(\Gamma_i = m, \Gamma_j \neq m,\, j=1,2,\dots,i-1)
$$

$$
= \sum_{m=1}^{\infty} \frac{1}{2^m}\left(1 - \frac{1}{2^m}\right)^{i-1}
$$

because the Γ_i are geometric random variables. Summing over i and performing some algebraic manipulations leads to

$$
\sum_{i=2}^{k} \mathbf{P}(\Gamma_i \neq \Gamma_j,\, j=1,2,\dots,i-1) = \sum_{m\ge 1}\left[1 - \left(1 - \tfrac{1}{2^m}\right)^{k}\right] - 1.
$$

After ignoring terms of order $o(1)$ we obtain that

$$
\sum_{m\ge 1}\left[1 - \left(1 - \frac{1}{2^m}\right)^{n_0-1}\right] \le \mathbb{E}D_n \le \sum_{m\ge 1}\left[1 - \left(1 - \frac{1}{2^m}\right)^{n_1}\right].
$$

Both bounding sums have powers of the form $\frac{n}{2}(1 + o(1))$. Thus the second, purely analytical step, consists of analyzing the asymptotic behavior of the infinite sum

$$
\sum_{m\ge 1}\left[1 - \left(1 - \frac{1}{2^m}\right)^{\frac{n}{2}(1+o(1))}\right].
$$

Defining

$$
f(x) = \sum_{m\ge 1}\left[1 - \left(1 - \frac{1}{2^m}\right)^{2^x}\right]
$$

we can apply Proposition A.1 with $x = \log_2(\frac{n}{2})$ and $k = \log_2(\frac{n}{2}o(1))$ to obtain the result of Theorem 8.43. $\qquad\square$

In fact, Hitczenko and Louchard [93] extended Theorem 8.43 by studying the poissonized generating function for the probability distribution function of D_n, that is,

$$G(z,q) = \sum_{n\geq 0}\sum_{k\geq 0} \mathbf{P}(D_n = k)q^k \frac{z^n e^{-z}}{n!}.$$

Conditioning on the number of Γ_j that are equal to 1 they obtained that

$$G_n(q) = \mathbf{P}(D_n = k)q^k = \begin{cases} 1 & \text{if } n = 0 \\ q\sum_{j=0}^{n-1}\binom{n}{j}\frac{G_j(q)}{2^n} + \frac{G_n(q)}{2^n} & \text{if } n \geq 1 \end{cases}.$$

Using simple generating function techniques they showed that $G(z,q)$ satisfies the following functional equation:

$$G(z,q) = G(z/2,q)(e^{-z/2}(1-q)+q).$$

Then, by nontrivial analysis involving Mellin transforms (see Definition E.12), the authors found the variance and higher moments of the random variable D_n as

$$\text{Var}(D_n) \approx 1$$
$$\mu_3(D_n) \approx -3 + 2\log 3 = 0.1699250014,$$
$$\mu_4(D_n) \approx 10 - 12\log 3 + 6\log 4 = 2.980449991, \text{ and}$$
$$\mu_5(D_n) \approx -45\log 2 + 70\log 3 - 60\log 4 + 24\log 5 = 1.673649353$$

as $n \to \infty$, where the variance does not have a periodic contribution, but the other moments do. Hitczenko and Louchard also conducted simulations and investigated additional properties for compositions and Carlitz compositions.

Another generalization of the results given in Proposition 8.44 was investigated by Hitczenko and Savage [95] who studied the probability that a randomly chosen part in a random composition of n has multiplicity k. We present their result and an outline of the proof.

Definition 8.45 *For a composition $\sigma \in \mathcal{C}_n$, we define $\mathcal{U}_n^k = \mathcal{U}_n^k(\sigma)$ to be the set of parts in σ that have multiplicity k, and let $U_n^k = U_n^k(\sigma) = |\mathcal{U}_n^k(\sigma)|$ be the number of such parts, given by $U_n^k(\sigma) = \sum_{i=1}^m I_{B_i}$, where*

$$B_i = \{\sigma_i \neq \sigma_j, j < i \text{ and } |\{\ell > i : \sigma_\ell = \sigma_i\}| = k - 1\}$$

is the event that the i-th part is the first occurrence of a new part size with multiplicity k. Furthermore, let E_n^k denote the event that a randomly chosen part size of a randomly chosen composition $\sigma \in \mathcal{C}_n$ has multiplicity k.

Example 8.46 *For $\sigma = 11321325 \in \mathcal{C}_{18}$ the distinct part sizes are $\{1, 2, 3, 5\}$ with respective multiplicities three, two, two, and one and $\mathcal{U}_{18}^2(\sigma) = \{2, 3\}$. The probability that a randomly chosen part size has multiplicity two would be given by $2/4 = U_{18}^2/D_{18}$. If we wanted to compute $\mathbf{P}(E_{18}^2)$ then we would have to take the average over all such probabilities for $\sigma \in \mathcal{C}_{18}$. Thus, in general*

$$\mathbf{P}(E_n^k) = \mathbb{E}\left(\frac{U_n^k}{D_n}\right).$$

With this notation, Hitczenko and Savage [95] showed the following result.

Theorem 8.47 *[96, Theorem 1] Let k be a fixed integer and let $\mathcal{U}_n^k$ and E_n^k be defined as above. Then*

$$(\log n)\mathbf{P}(E_n^k) = (\log n)\mathbb{E}\left(\frac{U_n^k}{D_n}\right) = O(1),$$

that is, there exist positive constants $c_1(k)$ and $c_2(k)$ such that for all $n \geq 2$,

$$c_1(k) \leq (\log n)\mathbf{P}(E_n^k) \leq c_2(k).$$

More precisely, as $n \to \infty$,

$$(\log n)\mathbf{P}(E_n^k) = \frac{1}{k} + H^k(c \log n) + o(1),$$

where H^k is a function of period one with mean zero whose Fourier coefficients are given by

$$\Phi_\ell = \frac{1}{k!}\Gamma\left(k - \frac{2\ell\pi\imath}{\log 2}\right)$$

where $\ell \neq 0$ and $\imath = \sqrt{-1}$.

Proof The proof consists of several parts. The first part concerns the derivation of two sequences $\{\ell_n\}$ and $\{k_n\}$ such that both sequences increase to infinity with $\lim_{n \to \infty}(\ell_n/k_n) = 1$, and both $\mathbf{P}(D_n \leq \ell_n)$ and $\mathbf{P}(D_n \geq k_n)$ tend to zero at a rate faster than $\log_2 n$. This allows for substitution of either sequence in the denominator of the expectation that is to be estimated.

Let $S_n(\sigma)$ denote the number of consecutive part sizes (starting with 1) in σ. For example, if $\sigma = 12135612$, then $S_{21}(\sigma) = 3$, as the value 4 does not occur in σ. Clearly, $S_n(\sigma) \leq D_n(\sigma)$, and using (8.20) we obtain for any sequence $\{\ell_n\}$ that

$$\mathbf{P}(D_n \leq \ell_n) \leq \mathbf{P}(S_n \leq \ell_n) \leq \mathbf{P}(\exists j \leq \ell_n, \forall i < \eta : \Gamma_i \neq j)$$

$$\leq \frac{2}{n^2} + \mathbf{P}(\{\exists j \leq \ell_n, \forall i < \eta : \Gamma_i \neq j\} \cap \{|\eta - \mathbb{E}(\eta)| < t_n\})$$

$$\leq \frac{2}{n^2} + 2\exp\left(-\frac{n_0}{2^{\ell_n}}\right),$$

where t_n and n_0 are defined in (8.21). Choosing $\ell_n = \log_2(n_0/\log_2 n_0)$ gives that $\mathbf{P}(D_n) = O(1/n)$. We obtain an upper bound by conditioning on whether $D_n \leq \ell_n$ or $D_n > \ell_n$ and using that $0 \leq U_n^k/D_n \leq 1$:

$$\mathbb{E}\left(\frac{U_n^k}{D_n}\right) = \mathbb{E}\left(\frac{U_n^k}{D_n}I_{\{D_n \leq \ell_n\}}\right) + \mathbb{E}\left(\frac{U_n^k}{D_n}I_{\{D_n > \ell_n\}}\right)$$

$$\leq \mathbf{P}(D_n \leq \ell_n) + \frac{\mathbb{E}(U_n^k)}{\ell_n}.$$

On the other hand,

$$\mathbb{E}\left(\frac{U_n^k}{D_n}\right) \geq \mathbb{E}\left(\frac{U_n^k}{D_n}I_{\{D_n \leq k_n\}}\right) \geq \frac{1}{k_n}\mathbb{E}\left(U_n^k I_{\{D_n \leq k_n\}}\right)$$

$$\geq \frac{1}{k_n}[\mathbb{E}(U_n^k) - \mathbb{E}(U_n^k I_{\{D_n > k_n\}})].$$

Choosing $k_n = \log_2(n(\log_2 n)^2)$ ensures that $\mathbb{E}(U_n^k I_{\{D_n > k_n\}}) \to 0$ and we have that

$$\mathbb{E}\left(\frac{U_n^k}{D_n}\right) = \mathbb{E}\left(\frac{U_n^k}{D_n}I_{\{\ell_n \leq D_n \leq k_n\}}\right) + \mathbb{E}\left(\frac{U_n^k}{D_n}I_{\{\ell_n \leq D_n \leq k_n\}^c}\right)$$

$$= \mathbb{E}\left(\frac{U_n^k}{\log_2 n \pm o(\log_2 n)}\right) + \mathbb{E}\left(\frac{U_n^k}{D_n}I_{\{D_n < \ell_n\}}\right) + \mathbb{E}\left(\frac{U_n^k}{D_n}I_{\{D_n > k_n\}}\right)$$

$$\leq \mathbb{E}\left(\frac{U_n^k}{\log_2 n \pm o(\log_2 n)}\right) + \mathbf{P}(D_n < \ell_n) + \mathbf{P}(D_n > k_n).$$

Since the number of distinct parts is less than or equal to the largest part, we can bound the last two terms as follows:

$$\mathbf{P}(D_n < \ell_n) + \mathbf{P}(D_n > k_n) \leq O\left(\frac{1}{n}\right) + \mathbf{P}(\max \Gamma_i > k_n)$$

$$\leq O\left(\frac{1}{n}\right) + n\mathbf{P}(\Gamma_1 > k_n) \leq O\left(\frac{1}{n}\right) + \frac{n}{2^{k_n}} \leq O\left(\frac{1}{(\log_2 n)^2}\right)$$

and therefore,

$$(\log n)\mathbf{P}(E_n^k) = \mathbb{E}\left(\frac{U_n^k}{\log_2 e \pm o(1)}\right) + o(1)$$

provided that $\mathbb{E}(U_n^k)$ is bounded. The last part of the proof consists of estimating the behavior of $\mathbb{E}(U_n^k)$ as $n \to \infty$. Note that

$$\mathbb{E}(U_n^k) = \mathbb{E}\left(\sum_{j \leq n/k} I_{\{j \in U_n^k\}}\right) = \sum_{j \leq n/k} \mathbf{P}(j \in U_n^k).$$

With $\tilde{\Gamma}_i$ defined as in the proof of Proposition 8.44,

$$\mathbf{P}(j \in \mathcal{U}_n^k) = \mathbf{P}\left(\sum_{i=1}^{\eta-1} I_{\{\Gamma_i = j\}} = k\right) - \mathbf{P}\left(\{\tilde{\Gamma}_\eta = j\} \cap \left\{\sum_{i=1}^{\eta-1} I_{\{\Gamma_i = j\}} = k\right\}\right)$$

$$+ \mathbf{P}\left(\{\tilde{\Gamma}_\eta = j\} \cap \left\{\sum_{i=1}^{\eta-1} I_{\{\Gamma_i = j\}} = k - 1\right\}\right). \tag{8.24}$$

Defining $q_n^\pm = (n+1)/2 \pm \sqrt{2(n-1)\log n}$, it can be shown that

$$\binom{q_n^-}{k}\frac{1}{2^{jk}}(1 - 2^{-j})^{q_n^+ - k} - O\left(\frac{1}{n^4}\right) \leq \mathbf{P}(\sum_{i=1}^{\eta-1} I_{\{\Gamma_i = j\}} = k)$$

$$\leq \binom{q_n^+}{k}\frac{1}{2^{jk}}(1 - 2^{-j})^{q_n^- - 1 - k} + O\left(\frac{1}{n^4}\right).$$

Note that for $q \approx n/2$, a comparison to an appropriate integral yields that

$$\binom{q}{k}\sum_{j=1}^{\infty} 2^{-jk}(1 - 2^{-j})^{q-k} = O(1).$$

Using that q_n^+ and q_n^- are asymptotically the same, and that

$$\sum_{j>n} 2^{-jk}(1 - 2^{-j})^{q_n^\pm - k} \leq \sum_{j>n} 2^{-j} = \frac{1}{2^n},$$

we obtain that

$$\mathbb{E}(\mathcal{U}_n^k) \approx \binom{q}{k}\sum_{j=1}^{\infty} 2^{-jk}(1 - 2^{-j})^{q-k} \quad \text{as } n \to \infty.$$

The remainder of the proof obtains that the right-hand side does not have a limit, but oscillates (see [95]).   □

For further extensions of this work, see [133]. We now turn to results about Carlitz compositions.

8.5   Asymptotics for Carlitz compositions of n

We now analyze the asymptotics for the total number of Carlitz compositions and for certain statistics on these compositions. Recall that $\mathcal{CC}(n)$ (respectively $\mathcal{CC}(m; n)$) denotes the number of Carlitz compositions of n with parts in $\mathbb{N}$ (respectively with m parts in $\mathbb{N}$). Theorem 3.7 gives that the generating function $CC(z, w) = \sum_{n,m \geq 0} CC(m; n) x^n y^m$ is given by

$$CC(z, w) = \cfrac{1}{1 - \sum_{j \geq 1} \frac{z^j w}{1 + z^j w}} = \cfrac{1}{1 - \sum_{j \geq 1} \sum_{i \geq 0} (-1)^i z^{j+ij} w^{i+1}}$$

$$= \cfrac{1}{1 - \sum_{i \geq 1} \sum_{j \geq 1} (-1)^{i-1} z^{ij} w^i} = \cfrac{1}{1 - \sum_{i \geq 1} (-1)^{i-1} \frac{z^i w^i}{1 - z^i}}$$

$$= \cfrac{1}{1 - g(z, w)}. \tag{8.25}$$

Note that $CC(z) = CC(z, 1)$ and that we will denote $g(z, 1)$ by $g(z)$.

Again, we consider the generating function as a function of one or more complex variables. Carlitz [38] noted that the radius of convergence of $CC(z)$ is at least $\frac{1}{2}$. Knopfmacher and Prodinger [119] found that the generating function $CC(z)$ has a dominant pole ρ^*, which is the unique real solution in the interval $[0, 1]$ of the equation

$$g(z) = \sum_{j \geq 1} \frac{z^j}{1 + z^j} = 1.$$

Numerically, $\rho^* = 0.571349$. The other poles are further away from the origin which can be proved using Henrichi's Principle of the Argument (see Theorem 8.27). Figure 8.2 shows the the curve $1 - g(z)$ for $|z| = 0.6$, where $1 - g(z)$ is converted into a series up to the z^{60} term. The winding number of this curve is one, therefore $1 - g(z)$ has only the root ρ^* inside the disc $|z| < 0.6$. Thus

$$CC(z) \approx \frac{A}{1 - z/\rho^*}$$

with

$$A = \frac{1}{\rho^* g'(\rho^*)} = 0.456363474.$$

Therefore,

$$CC(n) \approx A \cdot (\rho^*)^{-n} = 0.456363474 \cdot (1.750243)^n. \tag{8.26}$$

Louchard and Prodinger [134] added to the picture by giving the asymptotic distribution of the number of parts in a random Carlitz composition, similar to Theorem 8.35 for unrestricted compositions.

Theorem 8.48 *[134, Theorem 2.1] The number M of parts in a random Carlitz composition of large given n is asymptotically Gaussian with mean $n\mu$ and standard deviation $\sqrt{n}\sigma$, that is,*

$$\frac{M - n\mu}{\sqrt{n}\sigma} \overset{d}{\approx} \mathcal{N}(0, 1)$$

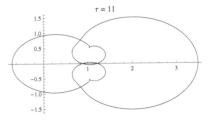

FIGURE 8.2:   The curve $1 - g(z)$ with $|z| = 0.6$.

as $n \to \infty$, where $\mu = 0.3506012746$ and $\sigma^2 = 0.1339166786$. *Moreover,*

$$\mathcal{CC}(m; n) \approx \frac{A}{(\rho^*)^n} \frac{\exp\left(-\frac{(m-n\mu)^2}{2n\sigma^2}\right)}{\sqrt{2\pi n}\sigma},$$

as $n \to \infty$ and $m - n\mu = O(\sqrt{n})$, with A and ρ^* as in (8.26).

Proof   The result follows immediately from Theorems 8.28 and 8.29, as in the proof of Theorem 8.35.   □

8.5.1   Asymptotics for the rightmost part

We now consider the rightmost (or last) part in Carlitz compositions. Louchard and Prodinger [134] gave the following result.

Theorem 8.49 *[134, Section 2.2] Let $X_L(n)$ denote the last part in a random Carlitz composition of n. Then the asymptotic distribution for $X_L(n)$ is given by*

$$\mathbf{P}(X_L(n) = k) \approx \frac{(\rho^*)^k}{1 + (\rho^*)^k}$$

uniformly for all k as $n \to \infty$. Thus,

$$\mathbb{E}(X_L(n)) \approx \sum_{k \geq 1} \frac{k(\rho^*)^k}{1 + (\rho^*)^k} = 2.5833049669.$$

Note that this is a very different result from the ones we have derived so far. There is no dependence on n in the average value! We will explore this phenomenon in Exercise 8.12.

Proof   Let $F(z, w, q)$ be the generating function for the number of Carlitz compositions of $n \geq 1$ according to order n, number of parts m and rightmost (or last) part σ_m, that is,

$$F(z, w, q) = \sum_{n \geq 1} \sum_{m \geq 1} \sum_{\sigma_1 \sigma_2 \cdots \sigma_m \in \mathcal{C}_{n,m}} z^n w^m q^{\sigma_m}.$$

Letting $F(z, w, q|j)$ denote the generating function for the number of Carlitz compositions of $n \geq 1$ with m parts whose rightmost part is j, we obtain that

$$F(z, w, q) = \sum_{j \geq 1} F(z, w, q|j)$$

$$= w(F(z, w, 1) + 1) \sum_{j \geq 1} z^j q^j - w \sum_{j \geq 1} F(z, w, 1|j) z^j q^j$$

$$= \frac{zwq}{1 - zq} \mathcal{C}(z, w) - wF(z, w, zq), \tag{8.27}$$

where we have expressed the compositions with m parts that end with j as compositions with one fewer part and any last part, subtracting off those that end in jj. Iterating (8.27) gives

$$F(z, w, q) = \sum_{j \geq 1} (-1)^{j-1} \frac{z^j w^j q}{1 - z^j q} \mathcal{C}(z, w) =: G(z, w, q) \mathcal{C}(z, w) \tag{8.28}$$

since $w^n F(z, w, z^n q) \to 0$ as $n \to \infty$ (why?). Setting $w = 1$ in (8.28) we derive from Theorem 8.25 that

$$[z^n] F(z, 1, q) \approx \frac{G(\rho^*, 1, q)}{(\rho^*)^{n+1} \frac{\partial}{\partial z} g(z, 1)\big|_{z = \rho^*}}, \tag{8.29}$$

as $n \to \infty$, uniformly for q in some complex neighborhood of the origin. Using Lemma 8.14 to compute the expected value proves difficult, so we will use (8.29) instead. We obtain that

$$\mathbf{P}(X_L = k) \approx \frac{[q^k]([z^n] F(z, 1, q))}{[z^n] F(z, 1, 1)} = \frac{[q^k] G(\rho^*, 1, q)}{G(\rho^*, 1, 1)}.$$

The last equality in (8.25) yields that $G(z, w, 1) = g(z, w)$, and in particular, $G(\rho^*, 1, 1) = g(\rho^*) = 1$. Now let's determine the numerator of $\mathbf{P}(X_L = k)$. Since

$$G(z, 1, q) = \sum_{j \geq 1} \sum_{i \geq 1} q^i z^{ij} (-1)^{j-1} = \sum_{i \geq 1} q^i \frac{z^i}{1 + z^i},$$

we obtain that

$$[q^k] G(\rho^*, 1, q) = \frac{(\rho^*)^k}{1 + (\rho^*)^k},$$

which completes the proof of the result for $\mathbf{P}(X_L = k)$. The expected value follows from Definition 8.6 and the numerical value is obtained by approximating the infinite sum by a finite one. $\quad\square$

We leave it to the interested reader to prove a similar result for unrestricted compositions.

8.5.2   Asymptotics for the largest part

Knopfmacher and Prodinger [119] also derived asymptotics for the average size of the largest part in Carlitz compositions.

Theorem 8.50 *[119, Section 4] Let $X_{\max}(n)$ denote the largest part in a random Carlitz composition of n. Then,*

$$\mathbb{E}(X_{\max}(n)) \approx \log_{1/\rho^*} n + 0.64311 + d(\log_{1/\rho^*} 0.60829n),$$

where $\rho^* = 0.571349$ and $d(x)$ is a function that has period one, mean zero, and small amplitude.

Comparing this result to Theorem 8.39 we see that the average largest parts in compositions and Carlitz compositions, respectively, tend to infinity at the same rate (up to a multiplicative constant).

Proof   We know that

$$\mathcal{CC}(z) = \frac{1}{1 - \sum_{j \geq 1} \frac{(-1)^{j-1} z^j}{1 - z^j}} = \frac{1}{1 - g(z)},$$

with a dominant simple pole at $\rho^* = 0.571349$. Next, we derive the generating function $\mathcal{CC}_{[h]}(z)$ for the number of Carlitz compositions of n with parts less than or equal to h, and let $\mathcal{CC}_{[h]}(z|j)$ be the generating function for those compositions that end with j. Then

$$\mathcal{CC}_{[h]}(z) = 1 + \sum_{j=1}^{h} \mathcal{CC}_{[h]}(z|j). \tag{8.30}$$

Clearly,

$$\mathcal{CC}_{[h]}(z|j) = z^j + z^j \sum_{i=1,\, i \neq j}^{h} \mathcal{CC}_{[h]}(z|i) = z^j \mathcal{CC}_{[h]}(z) - z^j \mathcal{CC}_{[h]}(z|j)$$

which yields

$$\mathcal{CC}_{[h]}(z|j) = \frac{z^j}{1 + z^j} \mathcal{CC}_{[h]}(z). \tag{8.31}$$

Combining (8.30) and (8.31) results in

$$\mathcal{CC}_{[h]}(z) = 1 + \sum_{j=1}^{h} \frac{z^j}{1 + z^j} \mathcal{CC}_{[h]}(z),$$

which can be solved for $\mathcal{C}_{[h]}(z)$:

$$\mathcal{C}_{[h]}(z) = \frac{1}{1 - \sum_{j=1}^{h}\frac{z^j}{1+z^j}} = \frac{1}{1 - \sum_{i \geq 1}(-1)^{i-1}\sum_{j=1}^{h}(z^i)^j}$$

$$= \frac{1}{1 - \sum_{i \geq 1}(-1)^{i-1}\frac{z^i - z^{i(h+1)}}{1-z^i}} = \frac{1}{1 - g_h(z)}.$$

Following the steps in the proof of Theorem 8.39 we obtain that the probability for a random Carlitz composition to have largest part less than or equal to h is approximated by

$$(1 + \varepsilon_h)^{-n} \approx \left(1 - \frac{1}{(1-\rho^*)g'(\rho^*)}(\rho^*)^h\right)^n$$

with $\rho^* = 0.571349$ and $g'(\rho^*) = 3.83517756$. Substituting these values into (8.16) gives the desired result. $\qquad\square$

8.5.3 Asymptotics for the number of distinct parts

We now present results on the average number of distinct parts in a random Carlitz composition obtained by Goh and Hitczenko [73]. Their proof is different from the proof of Theorem 8.43 for the same statistic for unrestricted compositions. We will give an outline of the proof by Goh and Hitczenko which draws upon some of the ideas in the proofs of Theorems 8.39 and 8.50.

Theorem 8.51 *[73, Theorem 1] The expected value of the number of distinct parts in a random Carlitz composition of n as $n \to \infty$ is given by*

$$\mathbb{E}(D_n) = 1.786495 \log n - 2.932545 + d(\rho^* \log(n/g'(\rho^*))) + o(1),$$

where the amplitude of d is bounded by $0.588234 \cdot 10^{-7}$, and $\rho^* = 0.571349$ is the pole of $\mathcal{C}(z)$.

Proof We start by rewriting the expression for D_n. Let B_j be the event that a composition has at least one part of size j. Then we can write

$$D_n = \sum_{j=1}^{n} I_{B_j}$$

and therefore

$$\mathbb{E}(D_n) = \sum_{j=1}^{n}\mathbf{P}(B_j) = \sum_{j=1}^{n}(1 - \mathbf{P}(B_j^c)) = \sum_{j=1}^{n}\left(1 - \frac{\mathcal{C}_{\mathbb{N}\setminus\{j\}}(n)}{\mathcal{C}(n)}\right), \qquad (8.32)$$

where B_j^c denotes the complement of B_j, that is, the event that the composition does not contain the part j. Let $CC_{\bar{j}}(n)$ be the number of Carlitz compositions of n that do not use the part j. Then Example 4.8 with $A = \mathbb{N}\backslash\{j\}$ yields that the corresponding generating function is given by

$$CC_{\bar{j}}(z) = \frac{1}{1 + \frac{z^j}{1+z^j} - \sum_{i \geq 1} \frac{z^i}{1+z^i}} = \frac{1}{1 - g_j(z)},$$

where

$$g_j(z) = g(z) - \frac{z^j}{1 + z^j}. \tag{8.33}$$

We now express the asymptotics for $CC_{\bar{j}}(n)$ in terms of those for $CC(n)$. In Section 8.5 we showed that $CC(z) = \frac{1}{1-g(z)}$ has a unique singularity in the disc $|z| \leq 0.6$ which occurs at $\rho^* = 0.571349$. (Actually, this can be shown to be true also for the larger disk $|z| < 0.663$.) Since $g(z)$ and $g_j(z)$ do not differ by too much, the functions $g_j(z)$ will also have a unique singularity, at least for values of j that are sufficiently large. In fact, on the disc $|z| < 0.663$ and for $j \geq 6$, the equation $g_j(z) = 1$ has a unique real simple root ρ_j in the interval $[0, 1]$. Moreover, there exists $\delta > 0$ such that for all $j \geq 6$ the following properties hold (see [73, Appendix]):

(1) $0 < \rho_j \leq \rho^* + \delta$,

(2) all roots ζ of the equation $g_j(z) = 1$ other than ρ_j satisfy $|\zeta| \geq \rho^* + 2\delta$,

(3) the ρ_j are strictly decreasing and $\rho_j \to \rho^*$ as $j \to \infty$, that is, $\rho_j = \rho^* + \varepsilon_j$ where $\varepsilon_j = o(1)$.

Since $g_j(\rho_j) = 1$ by definition of ρ_j, (8.33) together with $\rho_j = \rho^* + \varepsilon_j$ implies that

$$1 = g(\rho^* + \varepsilon_j) - \frac{(\rho^* + \varepsilon_j)^j}{1 + (\rho^* + \varepsilon_j)^j} = g(\rho^* + \varepsilon_j) - \frac{(\rho^*)^j(1 + \varepsilon_j/\rho^*)^j}{1 + (\rho^*)^j(1 + \varepsilon_j/\rho^*)^j}$$

$$= g(\rho^* + \varepsilon_j) + \sum_{i=1}^{\infty} \left(-(\rho^*)^j(1 + \varepsilon_j/\rho^*)^j\right)^i.$$

Expanding the right-hand side into its Taylor series results in

$$1 = g(\rho^*) + g'(\rho^*)\varepsilon_j + O(\varepsilon_j^2) - (\rho^*)^j(1 + \varepsilon_j/\rho^*)^j + O((\rho^*)^{2j}). \tag{8.34}$$

Using $(1 + \varepsilon_j/\rho^*)^j \approx 1$ and $g(\rho^*) = 1$, we solve (8.34) for ε_j to obtain that

$$\rho_j = \rho^* + \varepsilon_j = \rho^* + \frac{(\rho^*)^j}{g'(\rho^*)} + o((\rho^*)^j) \tag{8.35}$$

which is sufficient for our needs. Let $A_j = \frac{-1}{g'_j(\rho_j)}$ be the residue of $\frac{1}{1-g_j(z)}$ at $z = \rho_j$. Then the function $\frac{1}{1-g_j(z)} - \frac{A_j}{z-\rho_j}$ is analytical for $|z| \leq \rho^* + \delta$ and employing Theorem E.7 (the Cauchy coefficient formula) we get that

$$CC_{\bar{j}}(n) = \frac{-A_j}{\rho_j}(\rho_j)^{-n} + O((\rho^* + \delta)^{-n}) \tag{8.36}$$

similar to the proof of Theorem 8.24. Since $g_j(z) = g(z) - \dfrac{z^j}{1+z^j}$ we obtain

that $g'_j(z) = g'(z) - \dfrac{jz^{j-1}}{(1+z^j)^2}$, and together with (8.35) this results in

$$A_j = -\frac{1}{g'(\rho^*)} + O(j(\rho^*)^j).$$

Substituting this value of A_j and (8.35) into (8.36), we obtain that

$$CC_{\bar{j}}(n) = \left(\frac{1}{g'(\rho^*)} + O(j(\rho^*)^j)\right) \frac{1}{\left(\rho^* + (\rho^*)^j \frac{1+o(1)}{g'(\rho^*)}\right)^{n+1}} + O((\rho^* + \delta)^{-n})$$

for some $\delta > 0$ (universally for $j \geq 6$). This expression can now be used in (8.32). Goh and Hitczenko established that (see [73, Lemma 3])

$$\mathbb{E}(D_n) = \sum_{j=1}^{n}\left(1 - \frac{CC_{\mathbb{N}\setminus\{j\}}(n)}{CC(n)}\right) = \sum_{j=0}^{\infty}(1 - (1 - (\rho^*)^j M)^n) + o(1) \tag{8.37}$$

as $n \to \infty$, with $M = 1/g'(\rho^*)$. The proof of this lemma is rather technical, but the benefit is that we are back to a sum that we have encountered previously, namely in (8.15). Therefore,

$$\mathbb{E}(D_n) \approx \log_{1/\rho^*} n + \log_{1/\rho^*} M - \frac{\gamma}{\log \rho^*} + \frac{1}{2} + d(\log_{1/\rho^*} nM).$$

Substituting the values for ρ^* and M and simplifying gives the statement of the theorem. For the more detailed bounds on the function d see [73]. $\qquad\square$

8.6 A word on the asymptotics for words

The astute reader may wonder why there is no chapter on the asymptotics for words. The simple answer is that for words, the asymptotics are much easier to obtain or are not very interesting. For example, the average number of parts in a k-ary word of length n is always n. Some of the other statistics, for example the largest part and the rightmost part, are very easy to obtain.

Example 8.52 *Let $X_L(n)$ be the last part in a k-ary word of length n. Then the number of words for which the last part is j is given by k^{n-1}, while the total number of words is given by k^n, and therefore,*

$$\mathbf{P}(X_L(n) = j) = \frac{1}{k} \quad \text{with} \quad \mathbb{E}(X_L(n)) = \sum_{j=1}^{k} j \frac{1}{k} = \frac{k+1}{2}.$$

Almost as easy is the determination of the average size of the largest part. Let $X_{\max}(n)$ be the largest part in a k-ary word of length n. Then

$$\mathbf{P}(X_{\max}(n) = i) = \frac{k^i - k^{i-1}}{k^n},$$

and

$$\mathbb{E}(X_{\max}(n)) = \sum_{i=1}^{n} i \frac{k^i - k^{i-1}}{k^n} = \frac{1 - (n+1)k^n + nk^{n+1}}{k^n(k-1)}.$$

8.7   Exercises

Exercise 8.1 *Let $X_{\max}(n)$ denote the largest part in a random composition of n. Use* Mathematica *or* Maple *to compute the probability distribution of $X_{\max}(n)$ for $n = 1, 2, \ldots, 10$ and then use the distributions to compute the expected value and the variance of $X_{\max}(n)$ for $n = 1, 2, \ldots, 10$.*

Exercise 8.2 *Derive the asymptotic behavior for the average number of rises in all palindromic compositions of n.*

Exercise 8.3 *Derive Equation (8.4), namely that*

$$[x^n] \frac{\partial^k}{\partial y^k} C(x, y) \bigg|_{y=1} = k! \cdot 2^{n-1-k} \left(2 \binom{n}{k} - \binom{n-1}{k} \right)$$

for $C(x, y) = (1 - x)/(1 - x - xy)$.

Exercise 8.4 *Derive the asymptotic behavior for the average number of rises in all Carlitz compositions of n.*

Exercise 8.5 *Derive results analogous to those in Theorem 8.35 for the patterns peak and valley.*

Exercise 8.6 *Derive the asymptotic behavior for the number of compositions with r occurrences of the pattern 111 given in Example 8.38.*

Exercise 8.7 *Show that the function*

$$f_h(z) = \sum_{m \geq 0} (z + z^2 + \cdots + z^h)^m = \frac{1}{1 - \frac{z(1-z^h)}{1-z}} = \frac{1 - z}{1 - 2z + z^{h+1}}$$

has a simple pole as its dominant singularity. (Hint: use Rouché's Theorem.)

Exercise 8.8 *Let $\bar{c}(n)$ be the number of Carlitz compositions of n with parts in $\mathbb{N} \cup \{0\}$.*

(1) *Find an explicit formula for the generating function $\sum_{n \geq 0} \bar{c}(n) z^n$.*

(2) *Prove that there is a dominant singularity for the generating function at $\bar{\rho} = 0.386960$.*

(3) *Show that $\bar{c}(n) \approx 1.337604 \cdot (2.584243)^n$.*

Exercise 8.9 *Let $C_P(n)$ be the number of compositions of n whose parts are prime numbers (that is, the parts are in $P = \{2, 3, 5, 7, 11, 13, 17, 19, \ldots\}$).*

(1) *Find an explicit formula for the generating function $\sum_{n \geq 0} C_P(n) z^n$.*

(2) *Prove that $C_P(n) \approx 3.293208186 (1.476228784)^n$.*

Exercise 8.10 *Show that $\sum_{k=n_0}^{n_1} \frac{C}{k} + \frac{2}{n} \leq \tilde{C} \dfrac{\sqrt{n \log n}}{n} \to 0$ where n_0 and n_1 are defined by (8.21) and (8.22).*

Exercise 8.11 *This exercise explores how to create random compositions using* Mathematica *or* Maple.

(1) *Write a function that creates the compositions of n. (In* Mathematica, *you may use the function* comps[n] *defined in Example 8.4.) Use this function to create the compositions of $n = 10$ and then create a random sample of 200 compositions of $n = 10$ (by randomly selecting from all compositions of $n = 10$).*

(2) *Write a function that creates a random composition of n using the method described in Proposition 8.42. (Hint: create a sequence of geometric random variables until their sum becomes greater or equal to n and then adjust the last value.) Now create 200 (random) compositions of $n = 10$.*

(3) *For each of the samples created in Parts (1) and (2) and the complete set of compositions of $n = 10$, enumerate the number of compositions according to the largest part. Draw a bar chart for each of the three sets of data and compare.*

Exercise 8.12 *This exercise explores the surprising result that the average size of the last part of a Carlitz composition of n does not depend on n.*

(1) *Write a program in Maple or* Mathematica *to recursively create the Carlitz compositions of n.*

(2) *Use this program to compute the probability distribution for the random variable $X_L(n)$, the last part of a random Carlitz composition. That is, compute $P(X_L(n) = k)$ for $k = 1, 2, \ldots, n$ by directly enumerating the Carlitz compositions according to the last part.*

(3) *Use Definition 8.6 and the probabilities computed in Part (2) to compute the average size of the last part. Compare the expectation that you have computed to the approximation given in Theorem 8.49.*

Exercise 8.13* *In Exercises 6.9 and 6.10 we gave the generating functions for the number of k-ary smooth and k-ary smooth cyclic words of length n. Compute the asymptotic behavior for the number of k-ary smooth and k-ary smooth cyclic words of length n as $n \to \infty$.*

Exercise 8.14* *In Exercise 6.13 we gave the generating function for the number of smooth partition words of $[n]$. Show that the number of smooth partition words of $[n]$ with exactly k blocks is asymptotically given by*

$$\frac{4}{k+1} \cos^2 \frac{\pi}{2(k+1)} \left(1 + 2\cos \frac{\pi}{k+1} \right)^{n-1}$$

as $n \to \infty$.

Exercise 8.15* *A strong record in a word $w_1 \cdots w_n$ is an element w_i such that $w_i > w_j$ for all $j = 1, 2, \ldots, i-1$ (that is, w_i is strictly larger than all the letters to the left of it). A weak record is an element w_i such that $w_i \geq w_j$ for all $j = 1, 2, \ldots, i-1$ (that is, w_i is greater than or equal to the letters to the left of it). Furthermore, the position i is called the* position of the strong record (weak record). *We denote the sum of the positions of all strong (respectively, weak) records in a word w by $\mathrm{srec}(w)$ (respectively, $\mathrm{wrec}(w)$). Prodinger [165] originally studied the number of strong and weak records in samples of geometrically distributed random variables. More recently, he gave results on the expected value of the sum of the positions of strong records in random geometrically distributed words of length n [166]. Prove that*

(1) *the average sum of the positions of the strong records $\mathbb{E}(X_{\mathrm{srec}}(n))$ in compositions of n has the asymptotic expansion*

$$\mathbb{E}(X_{\mathrm{srec}}(n)) = \frac{n}{4\log 2} \left(1 + \delta\left(\log_2 n\right) \right) + o(n);$$

(2) *the average sum of the positions of the weak records* $\mathbb{E}(X_{\mathrm{wrec}}(n))$ *in compositions of n has the asymptotic expansion*

$$\mathbb{E}(X_{\mathrm{wrec}}(n)) = \frac{n}{2\log 2}\left(1 + \delta\left(\log_2 n\right)\right) + o(n),$$

where $\delta(x)$ is a periodic function of period one, mean zero and small amplitude, given by the Fourier series

$$\delta(x) = \sum_{k\neq 0} \frac{2k\pi\imath}{\log 2}\Gamma\left(-1 - \frac{2k\pi\imath}{\log 2}\right)e^{2k\pi\imath x}.$$

Exercise 8.16* *We say that a word $w = w_1 w_2 \cdots w_n$ has an ascent of size d or more if $w_{i+1} \geq w_i + d$ for some i. For example, there are three ascents of size two or more in the 4-ary words of length two, namely 13, 14, and 24. Let $f_d(n, j, s)$ be the number of compositions of n with last part j and s ascents of size d or more. Define $F_d(z, u, v) = \sum_{n,j\geq 1}\sum_{s\geq 0} f_d(n, j, s)z^n u^j v^s$.*

(1) *Prove that the generating function $F_d(z, u, v)$ satisfies*

$$F_d(z, u, v) = \frac{zu}{1 - zu}\left(F_d(z, 1, v) + 1\right) - \frac{(1 - v)z^d u^d}{1 - zu}F_d(z, zu, v).$$

(2) *Iterate the recurrence in (1) to obtain that*

$$F_d(z, v) := 1 + F_d(z, 1, v) = \frac{1}{1 - G_d(z, v)}$$

where

$$G_d(z, v) = \sum_{i\geq 0}\frac{(-1)^i z^{i+1}(1 - v)^i z^{d(i+1)i/2}}{\prod_{k=1}^{i}(1 - z^k)}.$$

(3) *Let $X_{\mathrm{asc}}^d(n)$ be the number of ascents of size d or more in a random composition of n. Derive from (2) that for $n \geq d$,*

$$\mathbb{E}(X_{\mathrm{asc}}^d(n)) = \frac{2^{-d}}{9}(3n - 3d - 2) + \frac{2}{9}\frac{(-1)^{n-d}}{2^n}$$

and

$$\mathrm{Var}(X_{\mathrm{asc}}^d(n)) = \frac{2^{-d}}{3}n\left(1 - \frac{2^{-d}}{9}(13 + 6d) + \frac{8\cdot 2^{-2d}}{7}\right).$$

(4) *Derive from (2) that the distribution of $X_{\mathrm{asc}}^d(n)$ converges to a Gaussian distribution with a rate of convergence of $O(1/\sqrt{n})$ with mean and variance given in (3).*

Exercise 8.17* *Let $w = w_1 w_2 \cdots w_n$ be a k-ary word of length n. We define the* total variation (*or just* variation) $u(w) = u_{n,k}(w)$ *of w as the sum of the differences of adjacent letters of w, that is,*

$$u(w) = \sum_{i=1}^{n-1} |w_i - w_{i+1}|.$$

For example, the total variation of the word $w = 1121415221 \in [5]^{10}$ equals $u(w) = 0 + 1 + 1 + 3 + 3 + 4 + 3 + 0 + 1 = 16$.

(1) *Use the scanning-element algorithm to obtain that the generating function*

$$F_k(x, q) = \sum_{n \geq 0} x^n \sum_{w \in [k]^n} q^{u(w)}$$

is given by

$$1 + \frac{k(1-q)x}{1 - q - x(1+q)} - \frac{2x^2 q \frac{1-t^k}{1-t(q)}}{(1 - q - x(1+q))\left(1 - x\frac{1-qt^k}{1-qt}\right)}$$

where

$$t = \frac{1 - x + q^2(1+x) - \sqrt{(1-q^2)(1+q-x(1-q))(1-q-x(1+q))}}{2q}.$$

(2) *Prove that the generating function $F_k(x, q)$ can be expressed as*

$$F_k(x, q) = 1 + \sum_{i=1}^{k} (x_i - q x_{i-1}),$$

where

$$x_i = \frac{x(1-q)}{q} \sum_{j=i}^{k-1} \frac{U_{i-1}(s/2) \sum_{s=0}^{j-1} U_s(s/2)}{U_{j-1}(s/2) U_j(s/2)}$$

$$+ \frac{x U_{i-1}(s/2)}{U_{k-1}(s/2)} \left(\frac{U_{k-1}(s/2) + (1-q)/q \sum_{j=0}^{k-1} U_j(s/2)}{U_k(s/2) - q(1+x) U_{k-1}(s/2)} \right),$$

U_i is the i-th Chebychev polynomial of the second kind (see Definition C.1), and $s = \dfrac{1 + q^2 - x(1 - q^2)}{q}$.

(3) *Derive from (1) that the mean and the variance of the total variation of a randomly chosen k-ary word of length n are given by*

$$\frac{(n-1)(k+1)(k-1)}{3k}$$

and

$$\frac{(k-1)(k+1)(k^2(6n-7) + 2(3n-1))}{90k^2},$$

respectively.

(4) *Prove (3) directly without using (1) and (2).*

8.8 Research directions and open problems

We now suggest several research directions, which are motivated both by the results and exercises of this and earlier chapter(s).

Research Direction 8.1 *As shown in the previous chapter, finding either the number of k-ary words of length n avoiding a fixed pattern or the number of compositions of n avoiding a fixed pattern is a hard problem. A potentially easier problem is to find the asymptotics for either the number of k-ary words of length n avoiding a fixed pattern or the number of compositions of n that avoid a fixed pattern. This question has been answered for words avoiding a subsequence pattern of length three by Brändén and Masour [27], and for compositions avoiding a subsequence pattern of length three by Savage and Wilf [176].*

Find either the asymptotics for the number of words (compositions) avoiding a fixed subsequence pattern of length four or five, or find an upper bound for the rate of growth for the number of these words (compositions). Obviously, we are interested in the smallest upper bound. (The trivial bound for compositions is 2^n, and for words on the alphabet $[k]$ the trivial bound is k^n).

Research Direction 8.2 *Baik, Deift, and Johansson [12] solved a problem about the asymptotic behavior of the length $X_{\text{lis}}(n)$ of the longest increasing subsequence for random permutations of order n as $n \to \infty$ (with the uniform distribution on the set of permutations S_n). They proved that the sequence*

$$\frac{X_{\text{lis}}(n) - 2\sqrt{n}}{n^{1/6}}$$

converges in distribution, as $n \to \infty$, to a certain random variable whose distribution function can be expressed via a solution of the Painlevé II equation (see [12] for details).

More recently, the case of the longest (strongly and weakly) increasing subsequence in k-ary words has been studied by Tracy and Widom [187]. For example, the word $w = 1123254$ has longest strongly (=strictly) increasing subsequence of length 4, namely $w_1 w_3 w_4 w_6 = w_2 w_3 w_4 w_6 = 1235$, while the longest (weakly) increasing subsequence is of length 5, namely either $w_1 w_2 w_3 w_4 w_6 =$

11235, $w_1w_2w_3w_4w_7 = 11234$, $w_1w_2w_3w_5w_6 = 11225$, or $w_1w_2w_3w_5w_7 = 11224$. *The natural extension is to study the same statistic, longest (strongly or weakly) increasing subsequence, in random compositions of n with parts in* $\mathbb{N}$.

Appendix A

Useful Identities and Generating Functions

A.1 Useful formulas

The Binomial theorem[1]

$$(x + y)^n = \sum_{i \geq 0} \binom{n}{i} x^i y^{n-i} \tag{A.1}$$

is often attributed to Blaise Pascal (who described it in the 17th century) but it was known to many mathematicians who preceded him. Around 1665, Isaac Newton generalized (A.1) to allow exponents other than nonnegative integers. In this generalization the finite sum is replaced by an infinite series. Namely, if x and y are real numbers with $x > |y|$ and r is any complex number, then

$$(x + y)^r = \sum_{j \geq 0} \binom{r}{j} x^{r-j} y^j = 1 + \frac{r}{1!} x^{r-1} y + \frac{r(r-1)}{2!} x^{r-2} y^2 + \cdots .$$

In the special case where $x = 1$ and $r = 1/2$, we obtain that

$$\sqrt{1 + y} = \sum_{m \geq 0} \binom{1/2}{m} y^m = 1 + \sum_{m \geq 1} (-1)^{m-1} \frac{(2m - 2)!}{2^{2m-1} m! (m-1)!} y^m. \tag{A.2}$$

Choosing $r = -(k + 1)$ yields another handy formula which we will use over and over:

$$\frac{1}{(1 - x)^{k+1}} = \sum_{i \geq 0} \binom{k + i}{k} x^i. \tag{A.3}$$

Equation (A.3) has also a nice combinatorial proof. On the left-hand side of the equation we see a convolution of $k + 1$ geometric series, that is, a sum of $k + 1$ nonnegative values. In how many ways can these $k + 1$ values add up to n? Well, dividing n items into $k + 1$ groups is the same as placing k dividers

---

[1] http://en.wikipedia.org/wiki/Binomial_theorem

in between the n items, that is, choosing k of the possible $n + k$ positions for the dividers. For example,

$$\bullet\bullet\bullet\,|\,\bullet\,|\,|\,\bullet\bullet\bullet\bullet\,|\,\bullet \qquad \leftrightarrow \qquad 3 + 1 + 0 + 4 + 1$$

Equation (A.3) has two important special cases for $k = 0$ and $k = 1$, namely the geometric series and its derivative.

$$\frac{1}{1 - x} = \sum_{i \geq 0} x^i \tag{A.4}$$

$$\frac{1}{(1 - x)^2} = \sum_{i \geq 0} (i + 1) x^i \tag{A.5}$$

Another handy equation that involves binomial coefficients can be easily proved by induction on b.

$$\binom{a}{a} + \binom{a + 1}{a} + \cdots + \binom{b}{a} = \binom{b + 1}{a + 1} \tag{A.6}$$

Taylor series expansions of specific functions can be utilized for estimates and asymptotic considerations. The approximations and the bound in the equations below are valid for small values of x.

$$\left(\frac{1}{1 + x}\right)^n = (1 - x + x^2 - x^3 + \cdots)^n \approx (1 - x)^n \tag{A.7}$$

$$e^{(-x \cdot n)} = \left(1 - x + \frac{x^2}{2} - \frac{x^3}{3!} + \cdots\right)^n \approx (1 - x)^n \tag{A.8}$$

$$\log(1 + x) = \sum_{n=1}^{\infty} (-1)^{n-1} \frac{x^n}{n} < x \tag{A.9}$$

Another useful identity for asymptotic estimation was proved by Hitczenko and Stengle.

Proposition A.1 *[97, Proposition 2.2] Let*

$$f(x) = \sum_{m=1}^{\infty} \left\{ 1 - \left(1 - \frac{1}{2^m}\right)^{2^x} \right\}.$$

Then, for large positive k,

$$f(x + k) = x + k + \frac{\gamma}{\log 2} - \frac{1}{2} + g(x) + o(2^{-x-k}),$$

where $\gamma = \lim_{n \to \infty} \sum_{i=1}^{n} \frac{1}{i} - \log n = 0.57721566490\cdots$ is Euler's constant and

$$g(x) = -x - \frac{\gamma}{\log 2} + \frac{1}{2} - \sum_{m=-\infty}^{0} e^{-2^{-m+x}} + \sum_{m=1}^{\infty} (1 - e^{-26-m+x})$$

is a nonconstant, mean-zero function of period 1 satisfying $|g(x)| \le 0.0000016$.

A.2 Generating functions of important sequences

Table A.1 displays the recurrence relations and generating functions for important sequences discussed in this text.

TABLE A.1: Generating functions of important sequences

Fibonacci	$F_n = F_{n-1} + F_{n-2}$ $F_0 = 0, F_1 = 1$	$\dfrac{x}{1 - x - x^2}$
Shifted Fibonacci	$f_n = F_{n+1}$	$\dfrac{1}{1 - x - x^2}$
Modified Fibonacci	$\hat{F}_n = F_n$ for $n \ge 1$ $\hat{F}_0 = 1$	$\dfrac{1 - x^2}{1 - x - x^2}$
Lucas	$L_n = L_{n-1} + L_{n-2}$ $L_0 = 2, L_1 = 1$	$\dfrac{2 - x}{1 - x - x^2}$
Catalan	$C_{n+1} = \sum_{i=0}^{n} C_i C_{n-i}$ $C_0 = 1$	$\dfrac{1 - \sqrt{1 - 4x}}{2x}$
Compositions in $\mathbb{N}$	$C(n) = 2C(n-1)$ $C(0) = C(1) = 1$	$\dfrac{1 - x}{1 - 2x}$
Modified compositions	$\tilde{C}(n) = C(n)$ for $n \ge 1$ $\tilde{C}_0 = 0$	$\dfrac{x}{1 - 2x}$
Stirling numbers of the first kind (unsigned)	$\begin{bmatrix} n+1 \\ j \end{bmatrix} = \begin{bmatrix} n \\ j-1 \end{bmatrix} + n\begin{bmatrix} n \\ j \end{bmatrix}$ $\begin{bmatrix} n \\ 0 \end{bmatrix} = \delta_{n0}, \begin{bmatrix} 0 \\ 1 \end{bmatrix} = 0$	$\prod_{i=0}^{n-1}(x+i)$
Stirling numbers of the second kind	$\begin{Bmatrix} n \\ k \end{Bmatrix} = \begin{Bmatrix} n-1 \\ k-1 \end{Bmatrix} + k\begin{Bmatrix} n-1 \\ k \end{Bmatrix}$ $\begin{Bmatrix} n \\ 1 \end{Bmatrix} = \begin{Bmatrix} n \\ n \end{Bmatrix} = 1$	$\dfrac{x^k}{\prod_{i=0}^{k}(1 - ix)}$

Appendix B

Linear Algebra and Algebra Review

B.1 Linear algebra review

Definition B.1 *An $n \times m$ matrix*

$$A = (a_{ij}) = \begin{bmatrix} a_{11} & a_{12} & \cdots & a_{1m} \\ a_{21} & a_{22} & \cdots & a_{2m} \\ \vdots & \vdots & \cdots & \vdots \\ a_{n1} & a_{n2} & \cdots & a_{nm} \end{bmatrix}$$

is a rectangular array of numbers consisting of n rows and m columns. An $n \times n$ matrix is called a square *matrix. The entries $a_{11}, a_{22}, \ldots, a_{nn}$ are said to be on the* main diagonal *of A. The sum of the elements on the main diagonal of an $n \times n$ matrix is called the* trace *of the matrix, that is*

$$\operatorname{tr}(A) = a_{11} + a_{22} + \cdots + a_{nn} = \sum_{i=1}^{n} a_{ii}.$$

A matrix whose entries are all zero is called the zero matrix *and denoted by O. A square matrix whose entries on the main diagonal are 1 and whose off-diagonal entries are 0 is called the* identity matrix *and denoted by I. A matrix whose only nonzero entries are on the main diagonal is called a* diagonal *matrix.*

Matrix addition and scalar multiplication of matrices are very straightforward, while matrix multiplication is a little more involved.

Definition B.2   *1. For any matrix A and scalar k, $k \cdot A = (k\,a_{ij})$.*

2. For two matrices A and B of the same dimensions, $A \pm B = (a_{ij} \pm b_{ij})$.

3. Let A be an $n \times r$ matrix and B an $r \times m$ matrix. Then $C = A \cdot B$ is an $n \times m$ matrix, where $c_{ij} = \sum_{k=1}^{r} a_{ir} b_{rj}$.

Definition B.3 *If A is an n×n matrix, then the (unique) n×n matrix B such that*

$$AB = BA = I = \begin{bmatrix} 1 & 0 & 0 & \cdots & 0 \\ 0 & 1 & 0 & \cdots & 0 \\ \vdots & \vdots & \vdots & \ddots & \vdots \\ 0 & 0 & 0 & \cdots & 1 \end{bmatrix}$$

is called the inverse of A and denoted by A^{-1}.

Definition B.4 *If A is any $n \times m$ matrix, then the transpose of A, denoted by A^T, is defined to be the $m \times n$ matrix that results from interchanging the rows and columns of A, that is, $A^T = (a_{ji})$.*

Definition B.5 *The determinant $\det(A)$ of an $n \times n$ matrix $A = (a_{ij})$ is a scalar which can be computed using Leibnitz' formula:*

$$\det(A) = \sum_{\pi \in \mathcal{S}_n} sgn(\pi) \prod_{i=1}^{n} a_{i,\pi_i},$$

where π is a permutation, and $sgn(\pi) = 1$ if π is an even permutation, and $sgn(\pi) = -1$ if π is an odd permutation. (A permutation is even if it can be obtained from $123\cdots n$ by an even number of exchanges of two numbers, and odd otherwise.)

Example B.6 *The permutation $\pi = 321$ is an odd permutation as there is one pairwise exchange, namely switching 1 and 3 to obtain π from 123.*

The above definition of the determinant is mainly used for 2×2 and 3×3 matrices. In the case of a 2×2 matrix, the general formula reduces to

$$\det(A) = \det \begin{bmatrix} a_{11} & a_{12} \\ a_{21} & a_{22} \end{bmatrix} = a_{11}a_{22} - a_{21}a_{12} \tag{B.1}$$

since there are two permutations 12 and 21 in $\mathcal{S}_2$. For 3×3 matrices, summing over the six permutations of 123 (of which three are odd and three are even), we obtain that

$$\det(A) = a_{11}a_{22}a_{33} + a_{12}a_{23}a_{31} + a_{13}a_{21}a_{32}$$
$$- a_{11}a_{23}a_{32} - a_{12}a_{21}a_{33} - a_{13}a_{22}a_{31}. \tag{B.2}$$

This formula is easily remembered as follows: extend the 3×3 matrix by appending the first two columns at the right end. Then the positive factors are exactly the products of the entries on the three diagonals from top left to bottom right, and the negative factors are the products of the entries on the three diagonals from bottom left to top right, as shown in Figure B.1.

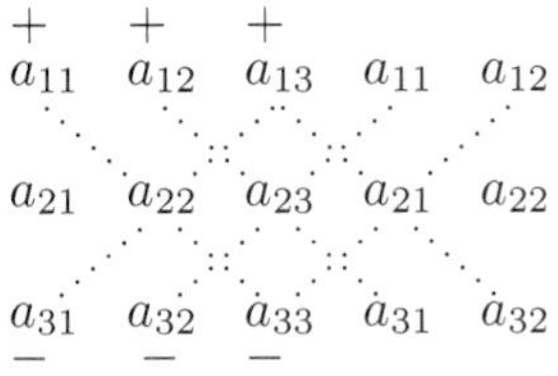

$$
\begin{array}{ccccc}
+ & + & + & & \\
a_{11} & a_{12} & a_{13} & a_{11} & a_{12} \\
a_{21} & a_{22} & a_{23} & a_{21} & a_{22} \\
a_{31} & a_{32} & a_{33} & a_{31} & a_{32} \\
- & - & - & &
\end{array}
$$

FIGURE B.1: Determinant computation for 3×3 matrices.

Example B.7 *Let* $A = \begin{bmatrix} 1\,2\,3 \\ 2\,5\,3 \\ 1\,0\,8 \end{bmatrix}$. *Then*

$$
\begin{aligned}
\det(A) &= 1 \cdot 5 \cdot 8 + 2 \cdot 3 \cdot 1 + 3 \cdot 2 \cdot 0 - (1 \cdot 5 \cdot 3 + 0 \cdot 3 \cdot 1 + 8 \cdot 2 \cdot 2) \\
&= 46 - 47 = -1.
\end{aligned}
$$

A more commonly used way to compute the determinant of a matrix, especially for large matrices, is the use of cofactors.

Definition B.8 *If A is a square $n \times n$ matrix, then the* minor *of entry a_{ij} is denoted by M_{ij} and is defined to be the determinant of the submatrix that remains after the i-th row and the j-th column are deleted from A. The* cofactor *of entry a_{ij} is denoted by C_{ij} and is computed as $C_{ij} = (-1)^{i+j} M_{ij}$. The matrix*

$$
\begin{bmatrix}
C_{11} & C_{12} & \cdots & C_{1n} \\
C_{21} & C_{22} & \cdots & C_{2n} \\
\vdots & \vdots & & \vdots \\
C_{n1} & C_{n2} & \cdots & C_{nn}
\end{bmatrix}
$$

is called the matrix of cofactors *from A. The transpose of the cofactor matrix is called the* adjoint *of A and is denoted by* $\mathrm{adj}(A)$, *that is*

$$
\mathrm{adj}(A) = \begin{bmatrix}
C_{11} & C_{21} & \cdots & C_{n1} \\
C_{12} & C_{22} & \cdots & C_{n2} \\
\vdots & \vdots & & \vdots \\
C_{1n} & C_{2n} & \cdots & C_{nn}
\end{bmatrix}.
$$

Example B.9 *For the matrix A given in Example B.7, the cofactors are* *given by*

$$C_{11} = (-1)^{1+1} \det \begin{bmatrix} 5 & 3 \\ 0 & 8 \end{bmatrix} = 40, \qquad C_{12} = (-1)^{1+2} \det \begin{bmatrix} 2 & 3 \\ 1 & 8 \end{bmatrix} = -13,$$

$$C_{13} = (-1)^{1+3} \det \begin{bmatrix} 2 & 5 \\ 1 & 0 \end{bmatrix} = -5, \qquad C_{21} = (-1)^{2+1} \det \begin{bmatrix} 2 & 3 \\ 0 & 8 \end{bmatrix} = -16,$$

$$C_{22} = (-1)^{2+2} \det \begin{bmatrix} 1 & 3 \\ 1 & 8 \end{bmatrix} = 5, \qquad C_{23} = (-1)^{2+3} \det \begin{bmatrix} 1 & 2 \\ 1 & 0 \end{bmatrix} = 2,$$

$$C_{31} = (-1)^{3+1} \det \begin{bmatrix} 2 & 3 \\ 5 & 3 \end{bmatrix} = -9, \qquad C_{32} = (-1)^{3+2} \det \begin{bmatrix} 1 & 3 \\ 2 & 3 \end{bmatrix} = 3,$$

$$C_{33} = (-1)^{3+3} \det \begin{bmatrix} 1 & 2 \\ 2 & 5 \end{bmatrix} = 1.$$

Thus,

$$\mathrm{adj}(A) = \begin{bmatrix} C_{11} & C_{21} & C_{31} \\ C_{12} & C_{22} & C_{32} \\ C_{13} & C_{23} & C_{33} \end{bmatrix} = \begin{bmatrix} 40 & -16 & -9 \\ -13 & 5 & 3 \\ -5 & 2 & 1 \end{bmatrix}.$$

Theorem B.10 *The determinant of an $n \times n$ matrix A can be computed by multiplying the entries in any row (or column) by their cofactors and adding the resulting products. That is, for each $1 \le i \le n$ and $1 \le j \le n$, we can compute the determinant via the cofactor expansion along the i-th row:*

$$\det(A) = a_{i1}C_{i1} + a_{i2}C_{i2} + \cdots + a_{in}C_{in}$$

or via the cofactor expansion along the j-th column:

$$\det(A) = a_{1j}C_{1j} + a_{2j}C_{2j} + \cdots + a_{nj}C_{nj}.$$

Example B.11 (*Continuation of Example B.9*). *We can use the cofactors computed in Example B.9 to compute the determinant of the matrix A given in Example B.7 in a different way. For example, using the cofactor expansion along the first row, we obtain*

$$\det(A) = a_{11}C_{11} + a_{12}C_{12} + a_{13}C_{13} = 1 \cdot 40 + 2 \cdot (-13) + 3 \cdot (-5) = -1.$$

Using the cofactor expansion along the second column gives

$$\det(A) = a_{12}C_{12} + a_{22}C_{22} + a_{32}C_{32} = 2 \cdot (-13) + 5 \cdot 5 + 0 \cdot 3 = -1.$$

Note that it is advantageous to use rows or columns of A that have zero entries (requires computation of fewer cofactors).

Another use of cofactors and the adjoint matrix is the computation of the inverse of a matrix.

Theorem B.12 *If A is an invertible matrix, then*

$$A^{-1} = \frac{1}{\det(A)} \mathrm{adj}(A).$$

Example B.13 (*Continuation of Example B.9*). *For the matrix A given in Example B.7, the inverse matrix is computed as*

$$A^{-1} = \frac{1}{\det(A)} \mathrm{adj}(A) = (-1) \begin{bmatrix} 40 & -16 & -9 \\ -13 & 5 & 3 \\ -5 & 2 & 1 \end{bmatrix} = \begin{bmatrix} -40 & 16 & 9 \\ 13 & -5 & -3 \\ 5 & -2 & -1 \end{bmatrix}.$$

For special types of matrices, the determinant can be computed very easily.

Theorem B.14 *If M is a diagonal matrix with diagonal elements $d_i = m_{ii}$, then $\det(M) = \prod_{i=1}^{n} d_i$.*

Theorem B.15 *For a* block matrix $M = \begin{pmatrix} A & O \\ C & D \end{pmatrix}$, *where A and D are square matrices and O is the zero matrix, $\det(M) = \det(A)\det(D)$.*

Oftentimes, we like to manipulate a matrix to make computation of the determinant easier. For elementary row (and the corresponding elementary column) operations, the effect on the value of the determinant is well understood.

Definition B.16 *The following operations are called* elementary row *operations:*

1. *Multiply each entry in a row by a nonzero constant.*

2. *Interchange two rows.*

3. *Add a multiple of one row to another row.*

Theorem B.17 *Let A be an $n \times n$ matrix.*

1. *If B is the matrix that results when a single row or column of A is multiplied by a scalar k, then $\det(B) = k \det(A)$.*

2. *If B is the matrix that results when two rows or two columns of A are interchanged, then $\det(B) = -\det(A)$.*

3. *If B is the matrix that results when a multiple of one row of A is added to another row or when a multiple of one column of A is added to another column, then $\det(B) = \det(A)$.*

The determinant and inverse of a matrix and its transpose are closely related.

Theorem B.18 *Let A be a square matrix. Then*

1. $\det(A^T) = \det(A)$.

2. *If A is invertible, that is, $\det(A) \neq 0$, then $(A^T)^{-1} = (A^{-1})^T$*

Matrices arise naturally in connection with systems of equations. The following theorem provides one way to compute the solution of a system of equations.

Theorem B.19 (*Cramer's Rule*) *If $A\vec{x} = \vec{b}$ is a system of n linear equations in n unknowns such that $\det(A) \neq 0$, then the system has a unique solution $x_i = \frac{\det(A_i)}{\det(A)}$ for $i = 1, 2, \ldots, n$, where A_i is the matrix obtained by replacing the entries in the i-th column of A by the entries in the vector $\vec{b}$ on the right-hand side of the system of equations.*

An important characterization of a matrix is its eigenvalues and the associated eigenvectors.

Definition B.20 *A value λ for which the system of equations*

$$A\vec{x} = \lambda\vec{x} \iff (\lambda I - A)\vec{x} = \vec{0} \tag{B.3}$$

has a nontrivial solution $\vec{x} \neq \vec{0}$ is called an eigenvalue *of A. The nontrivial solutions of* (B.3) *are called the* eigenvectors *of A corresponding to λ. The eigenvalues of A are the roots of the* characteristic polynomial $\det(\lambda I - A)$ *of A.*

Example B.21 *Let $A = \begin{bmatrix} 1 & 3 \\ 4 & 2 \end{bmatrix}$. To find the eigenvalues of A we solve the equation*

$$\det(\lambda I - A) = \begin{bmatrix} \lambda - 1 & -3 \\ -4 & \lambda - 2 \end{bmatrix} = 0$$

which by (B.1) *is equivalent to*

$$(\lambda - 1)(\lambda - 2) - 12 = \lambda^2 - 3\lambda - 10 = (\lambda + 2)(\lambda - 5) = 0.$$

Thus the two eigenvalues are $\lambda_1 = -2$ and $\lambda_2 = 5$. To find the eigenvectors of A that correspond to λ_1, we need to find nontrivial solutions for the system of equations $(\lambda_1 I - A)\vec{x} = 0$, which reduces to solving $-x_1 - x_2 = 0$. Thus any vector of the form $[-t, t]^T$ for any $t \in \mathbb{R}$ is an eigenvector for $\lambda_1 = -2$. Similarly, the eigenvectors for $\lambda_2 = 5$ are of the form $[3t, 4t]^T$.

There is a strong relation between the eigenvalues of a matrix A and its determinant and trace.

Theorem B.22 *If A is an $n \times n$ matrix and if $\lambda_1, \lambda_2, \ldots, \lambda_n$ are the eigenvalues of A listed according to their algebraic multiplicities, then*

$$\det(A) = \lambda_1 \lambda_2 \cdots \lambda_n.$$

Furthermore,

$$\operatorname{tr}(A) = \lambda_1 + \lambda_2 + \cdots + \lambda_n.$$

Example B.23 *(Continuation of Example B.21). For the matrix given in Example B.21,*

$$\det(A) = 2 - 12 = -10 = (-2) \cdot 5 \quad and \quad \operatorname{tr}(A) = 1 + 2 = (-2) + 5.$$

B.2 Algebra review

Definition B.24 *A group $(G, +)$ is a set G with a binary operation $+$ on G that satisfies:*

1. *Closure: For all $a, b \in G$, $a + b \in G$.*

2. *Associativity: For all $a, b, c \in G$, $(a + b) + c = a + (b + c)$.*

3. *Identity element: There exists a (unique) element $e \in G$ such that for all $a \in G$, $e + a = a + e = a$.*

4. *Inverse element: For each $a \in G$, there exists an element $b \in G$ such that $a + b = b + a = e$, where e is the identity element.*

If the group operation $+$ is commutative so that for all $a, b \in G$, $a + b = b + a$, then G is said to be an abelian *or* commutative *group.*

Example B.25 *The set of integers $\mathbb{Z}$ together with the usual addition is a commutative group with identity element $e = 0$, and inverse elements $-a$. $\mathbb{Z}$ together with the usual multiplication is not a group, as there is no inverse element in $\mathbb{Z}$ for every $a \in \mathbb{Z}$.*

Example B.26 *The set of square matrices A with $\det(A) \neq 0$ of size $n \times n$ together with the usual matrix multiplication is a group. However, the group is not commutative because for the matrices*

$$A = \begin{bmatrix} 1 & 2 \\ -3 & 4 \end{bmatrix} \quad and \quad B = \begin{bmatrix} 0 & 2 \\ -1 & 1 \end{bmatrix}$$

we have that

$$A \cdot B = \begin{bmatrix} -2 & 4 \\ 4 & -2 \end{bmatrix} \neq B \cdot A = \begin{bmatrix} -6 & 8 \\ -4 & 2 \end{bmatrix}.$$

Definition B.27 *A* ring *is an abelian group* $(R, +)$ *equipped with a second binary operation* $*$ *such that for all* $a, b, c \in R$,

1. $a * (b * c) = (a * b) * c$.

2. $a * (b + c) = (a * b) + (a * c)$.

3. $(a + b) * c = (a * c) + (b * c)$.

If there also exists a multiplicative identity in the ring, that is, an element e' *such that for all* $a \in R$, $a * e' = e' * a = a$, *then the ring is said to be a* ring with unity.

Example B.28 *The set of rational numbers* $\mathbb{Q}$ *together with regular addition and multiplication is a ring, but it does not have a unity.*

Appendix C

Chebychev Polynomials of the Second Kind

Definition C.1 Chebyshev polynomials of the second kind *are defined by*

$$U_j(\cos\theta) = \frac{\sin((j+1)\theta)}{\sin\theta}$$

for $j \geq 0$.

Proposition C.2 *The j-th Chebyshev polynomial of the second kind, $U_j(x)$, is a polynomial of degree j with integer coefficients that satisfies the recurrence relation*

$$U_{j+1}(x) = 2xU_j(x) - U_{j-1}(x) \qquad (C.1)$$

with initial conditions $U_0(x) = 1$ and $U_1(x) = 2x$.

Proof From Definition C.1 we have that $U_0(\cos\theta) = 1$ and $U_1(\cos\theta) = \sin(2\theta)/\sin(\theta) = 2\cos\theta$, which implies that $U_0(x) = 1$ and $U_1(x) = 2x$. Using the trigonometric identity

$$\sin((j+2)\theta) = \sin((j+1)\theta)\cos\theta + \cos((j+1)\theta)\sin\theta$$

we obtain that

$$U_{j+1}(x) - xU_j(x) = \cos((j+1)\theta). \qquad (C.2)$$

On the other hand, using the trigonometric identity

$$\cos((j+1)\theta) = \cos(j\theta)\cos\theta - \sin(j\theta)\sin\theta$$

together with (C.2), we obtain that

$$U_{j+1}(x) - xU_j(x) = x(U_j(x) - xU_{j-1}(x)) - U_{j-1}(x)(1 - x^2),$$

which implies that $U_j(x)$ satisfies the recurrence relation (C.1). Using induction we obtain that $U_j(x)$ is a polynomial of degree j with integer coefficients, as required. □

Example C.3 *The first terms of the sequence $\{U_j(x)\}_{j\geq 0}$ are*

$$U_0(x) = 1,$$
$$U_1(x) = 2x,$$
$$U_2(x) = 4x^2 - 1,$$
$$U_3(x) = 8x^3 - 4x,$$
$$U_4(x) = 16x^4 - 12x^2 + 1, \text{ and}$$
$$U_5(x) = 32x^5 - 32x^4 + 6x.$$

Using (C.1) one can extend the definition of the Chebyshev polynomials of the second kind to all $j \in \mathbb{Z}$ to obtain $U_{-1}(x) = 0$, $U_{-2}(x) = -1$, and so on.

Chebyshev polynomials were invented for the needs of approximation theory but are also widely used in various other branches of mathematics including algebra, combinatorics, and number theory (see Rivlin [173]).

Definition C.4 *For $j \geq 0$ we define $R_j(x)$ by*

$$R_j(x) = \frac{U_{j-1}\left(\frac{1}{2\sqrt{x}}\right)}{\sqrt{x}U_j\left(\frac{1}{2\sqrt{x}}\right)}.$$

Example C.5 *Using Definition C.4, we can compute that*

$$R_0(x) = 0,$$
$$R_1(x) = 1,$$
$$R_2(x) = \frac{1}{1-x}, \text{ and}$$
$$R_3(x) = \frac{1-x}{1-2x}.$$

Actually, $R_j(x)$ is the generating function for the number of Dyck paths (see Example 2.50) of height at most $j - 1$. To prove this claim we use the first return decomposition of a Dyck path P as $P = UP'DP''$, where D is the first return to the x-axis, P' and P'' are shorter Dyck paths, and P' is reduced in height by one. Therefore we can get a recurrence relation for $R_j(x)$ as $R_j(x) = 1 + xR_{j-1}(x)R_j(x)$, which implies that $R_j(x) = \frac{1}{1-xR_{j-1}(x)}$ with initial condition $R_0(x) = 0$.

From Proposition C.2 we see that for any $j \geq 0$, $R_j(x)$ is rational in x. These rational functions arise in some interesting results, see for example [126, 147, 148].

Proposition C.6 *Let* $f_j(x) = \dfrac{1}{1 - x f_{j-1}(x)}$ *for all* $j \geq 1$ *with initial condition* $f_0(x) = 0$. *Then*

$$f_j(x) = R_j(x) = \frac{U_{j-1}\left(\frac{1}{2\sqrt{x}}\right)}{\sqrt{x} U_j\left(\frac{1}{2\sqrt{x}}\right)}.$$

Proof We proceed by induction on j. For $j = 0$ and $j = 1$ the claim is true since $U_{-1}(t) = 0$, $U_0(t) = 1$, and $U_1(t) = 2t$. Assuming that the claim holds for j, we prove it for $j + 1$:

$$f_{j+1}(x) = \frac{1}{1 - x f_j(x)} = \frac{1}{1 - x \dfrac{U_{j-1}\left(\frac{1}{2\sqrt{x}}\right)}{\sqrt{x} U_j\left(\frac{1}{2\sqrt{x}}\right)}}$$

$$= \frac{U_j\left(\frac{1}{2\sqrt{x}}\right)}{\sqrt{x}\left(\frac{1}{\sqrt{x}} U_j\left(\frac{1}{2\sqrt{x}}\right) - U_{j-1}\left(\frac{1}{2\sqrt{x}}\right)\right)},$$

which by (C.1) implies that

$$f_{j+1}(x) = \frac{U_j\left(\frac{1}{2\sqrt{x}}\right)}{\sqrt{x} U_{j+1}\left(\frac{1}{2\sqrt{x}}\right)} = R_{j+1}(x),$$

as required. □

Example C.7 *Using the recurrence relation for* $f_j(x) = R_j(x)$ *given in Proposition C.6, we obtain that* $R_j(x)$ *can be expressed as the following finite continued fraction:*

$$R_j(x) = \cfrac{1}{1 - \cfrac{x}{\ddots \atop 1 - \cfrac{x}{1 - x}}}.$$

Thus, we have found three different expressions for the generating function of the Dyck paths of height at most j *— an explicit formula (see Definition C.4), a recursion (see Proposition C.6), and finally the continued fraction. Moreover, from Example 2.62 we obtain that*

$$\lim_{j \to \infty} R_j(x) = \frac{1 - \sqrt{1 - 4x}}{2x},$$

the generating function for the Catalan sequence, which counts Dyck paths (of any height).

Proposition C.6 can be generalized.

Proposition C.8 *Let* $f_j(x) = \dfrac{a(x)}{1 - b(x) - c(x)f_{j-1}(x)}$, $j \geq 1$, *with initial condition* $f_0(x)$. *Then*

$$f_j(x) = \frac{\sqrt{a(x)}\left[\sqrt{\frac{a(x)}{c(x)}}U_{j-1}\left(\frac{1-b(x)}{2\sqrt{c(x)a(x)}}\right) - U_{j-2}\left(\frac{1-b(x)}{2\sqrt{c(x)a(x)}}\right)f_0(x)\right]}{\sqrt{c(x)}\left[\sqrt{\frac{a(x)}{c(x)}}U_j\left(\frac{1-b(x)}{2\sqrt{c(x)a(x)}}\right) - U_{j-1}\left(\frac{1-b(x)}{2\sqrt{c(x)a(x)}}\right)f_0(x)\right]}.$$

Proof Again, we proceed by induction on j. Since $U_{-2}(t) = -1$, $U_{-1}(t) = 0$ and $U_0(t) = 1$, we get that the claim holds for $j = 0$. Assuming that the claim holds for j, we prove it for $j + 1$. Then, with $t = \dfrac{1-b(x)}{2\sqrt{c(x)a(x)}}$,

$$f_{j+1}(x) = \frac{a(x)}{1 - b(x) - c(x)f_j(x)}$$

$$= \frac{a(x)}{1 - b(x) - c(x)\dfrac{\sqrt{a(x)}\left[\sqrt{\frac{a(x)}{c(x)}}U_{j-1}(t) - U_{j-2}(t)f_0(x)\right]}{\sqrt{c(x)}\left[\sqrt{\frac{a(x)}{c(x)}}U_j(t) - U_{j-1}(t)f_0(x)\right]}}$$

$$= \frac{\sqrt{a(x)}\left[\sqrt{\frac{a(x)}{c(x)}}U_j(t) - U_{j-1}(t)f_0(x)\right]}{\sqrt{c(x)}\left[\sqrt{\frac{a(x)}{c(x)}}(2tU_j(t) - U_{j-1}(t)) - (2tU_{j-1}(t) - U_{j-2}(t))f_0(x)\right]},$$

which, using (C.1), gives the desired result. □

Appendix D

Probability Theory

From the axioms given in Definition 8.1 we can derive the following results.

Proposition D.1 *Let A, B, and E_i, $i = 1, \ldots, n$ be events in some sample space Ω. Then*

1. $\mathbf{P}(A \cup B) = \mathbf{P}(A) + \mathbf{P}(B) - \mathbf{P}(A \cap B)$, *and more generally, the* inclusion-exclusion principle *holds:*

$$\mathbf{P}\left(\bigcup_{i=1}^{n} E_i\right) = \sum_{i=1}^{n} \mathbf{P}(E_i) - \sum_{i<j} \mathbf{P}(E_i E_j) + \sum_{i<j<k} \mathbf{P}(E_i E_j E_k)$$
$$- \cdots + (-1)^{n-1} \mathbf{P}(E_1 E_2 \cdots E_n).$$

2. $\mathbf{P}(A^c) = 1 - \mathbf{P}(A)$, *where A^c denotes the complement of A.*

Definition D.2 *If $\mathbf{P}(B) > 0$, then the* conditional probability of A given B, *denoted by $\mathbf{P}(A \mid B)$, is defined as*

$$\mathbf{P}(A \mid B) = \frac{\mathbf{P}(A \cap B)}{\mathbf{P}(B)}.$$

Rearranging the terms in the definition of conditional probability gives a way of computing intersection probabilities as

$$\mathbf{P}(A \cap B) = \mathbf{P}(A \mid B)\mathbf{P}(B) = \mathbf{P}(B \mid A)\mathbf{P}(A). \tag{D.1}$$

Definition D.3 *Let Ω be a sample space. A finite or countably infinite sequence $\{B_n\}$ is a* partition *of Ω if $B_n \subseteq \Omega$, $B_i \cap B_j = \varnothing$ for $i \neq j$ and $\bigcup_n B_n = \Omega$.*

Theorem D.4 *If $\{B_n\}$ is a partition of Ω, then*

$$\mathbf{P}(A) = \sum_n \mathbf{P}(A \cap B_n) = \sum_n \mathbf{P}(A \mid B_n)\mathbf{P}(B_n).$$

An important concept in probability theory is the independence of events.

Definition D.5 *Two events A and B are called* independent *if*

$$\mathbf{P}(AB) = \mathbf{P}(A)\mathbf{P}(B).$$

If A and B are not independent, then they are called dependent. *More generally, the events $E_1, E_2, \ldots, E_n$ are called* independent *if for every subset $\{E_{i_1}, E_{i_2}, \ldots, E_{i_k}\}$ of $\{E_1, E_2, \ldots, E_n\}$ where $2 \le k \le n$,*

$$\mathbf{P}(E_{i_1} E_{i_2} \cdots E_{i_k}) = \mathbf{P}(E_{i_1})\mathbf{P}(E_{i_2}) \cdots \mathbf{P}(E_{i_k}).$$

Definition D.6 *Two discrete random variables X and Y are called* independent *if for any countable sets $A, B \subset \mathbb{R}$,*

$$\mathbf{P}(X \in A, Y \in B) = \mathbf{P}(X \in A)\mathbf{P}(Y \in B).$$

If X and Y are not independent, then they are called dependent. *The definition extends to more than two random variables in the obvious manner.*

Theorem D.7 *If random variables X and Y are independent, and g and f are measurable functions, then $f(X)$ and $g(Y)$ are also independent. Furthermore,*

$$\mathbb{E}(X \cdot Y) = \mathbb{E}(X)\mathbb{E}(Y).$$

We will utilize this theorem only for continuous (and therefore, measurable) functions f and g.

Theorem D.8 *If X is a discrete random variable taking values in $\mathbb{N}_0$, then*

$$\mathbb{E}(X) = \sum_{k=0}^{\infty} \mathbf{P}(X > k).$$

Proof By definition,

$$\mathbb{E}(X) = \sum_{k=0}^{\infty} k\mathbf{P}(X = k) = \sum_{k=0}^{\infty} \left(\sum_{i=1}^{k} \mathbf{P}(X = k) \right)$$

$$= \sum_{i=1}^{\infty} \sum_{k=i}^{\infty} \mathbf{P}(X = k) = \sum_{i=0}^{\infty} \mathbf{P}(X > i),$$

as claimed. □

Definition D.9 *A* Bernoulli trial *is an experiment which has exactly two outcomes,* success *and* failure. *The associated random variable, which takes values 1 (success has occurred) and 0 (failure has occurred) with probability p and $1 - p$, respectively, is called a* Bernoulli random variable with parameter p.

If X is a Bernoulli random variable with parameter p then we obtain from Definition 8.6 that

$$\mathbb{E}(X) = p, \quad \mathrm{Var}(X) = p(1-p), \text{ and } \sigma = \sqrt{p(1-p)}.$$

Example D.10 *The indicator function of a set A, denoted by I_A, is a Bernoulli random variable with $p = \mathbf{P}(success) = \mathbf{P}(A)$. Therefore, $\mathbb{E}(I_A) = \mathbf{P}(A)$.*

Definition D.11 *Let X be a random variable which counts the number of successes in n independent Bernoulli trials, each having success probability p. Then X is called a Binomial random variable with parameters n and p and we write $X \sim \mathcal{B}(n,p)$. Its probability function is given by*

$$\mathbf{P}(X = k) = \binom{n}{k} p^k (1-p)^{n-k} \qquad for\ k = 0, 1, \ldots, n.$$

If X is a Binomial random variable with parameters n and p then

$$\mathbb{E}(X) = np, \quad \mathrm{Var}(X) = np(1-p), \text{ and } \sigma = \sqrt{np(1-p)}.$$

Definition D.12 *Let X be a random variable which counts the number of trials needed until the first success occurs in a sequence of independent Bernoulli trials each having success probability p. Then X is called a* geometric random variable with parameter p. *Its probability function is given by*

$$\mathbf{P}(X = k) = (1-p)^{k-1} p \qquad for\ k = 1, 2, \ldots.$$

If X is a geometric random variable with parameter p then

$$\mathbb{E}(X) = \frac{1}{p}, \quad \mathrm{Var}(X) = \frac{1-p}{p^2}, \text{ and } \sigma = \frac{\sqrt{1-p}}{p}.$$

Theorem D.13 (*Markov and Chebyshev inequalities*) *Let X be a nonnegative random variable and let Y be an arbitrary random variable. Then for any $t > 0$,*

$$\mathbf{P}(X \geq t) \leq \frac{\mathbb{E}(X)}{t} \qquad \text{(Markov inequality)} \qquad (\text{D.2})$$

$$\mathbf{P}(|Y - \mathbb{E}(Y)| \geq t) \leq \frac{\mathrm{Var}(Y)}{t^2} \qquad \text{(Chebyshev inequality)} \qquad (\text{D.3})$$

Proof We prove the theorem for discrete random variables. Let A be the set of values of X, and let $B = \{x \in A \mid x \geq t\}$. Then for $t > 0$

$$\mathbb{E}(X) = \sum_{x \in A} x \mathbf{P}(X = x) \geq \sum_{x \in B} x \mathbf{P}(X = x)$$

$$\geq t \sum_{x \in B} \mathbf{P}(X = x) = t\mathbf{P}(X \geq t),$$

from which the Markov inequality follows. The proof of the Chebychev inequality follows directly from the Markov inequality, since

$$\mathbf{P}(|Y - \mathbb{E}(Y)| \geq t) = \mathbf{P}((Y - \mathbb{E}(Y))^2 \geq t^2) \leq \frac{\mathrm{Var}(Y)}{t^2},$$

as claimed. For (absolutely) continuous random variables the proof follows with sums replaced by integrals. □

Note that these two inequalities tell us that the probability of being much larger than the mean must decay. Specifically, an upper bound on the decay is given in terms of the standard deviation of the random variable. These two inequalities are examples of moment inequalities which are discussed for example in Billingsley [24]. In the area of discrete mathematics, these types of inequalities have been used to show the existence of certain combinatorial objects in an approach introduced by Erdős and referred to as the *probabilistic method* (in combinatorics). More details on the probabilistic method can be found in Alon and Spencer [8].

Besides showing existence of combinatorial objects, so-called large deviation estimates can be used to derive asymptotic behavior by showing that the probability for large deviations of the random variable from its mean decays exponentially.

Theorem D.14 (*Large Deviation estimate*) Let $X \sim \mathcal{B}(n, 1/2)$, then

$$\mathbf{P}(|X - \mathbb{E}(X)| > a) \leq 2 \exp(-2a^2/n).$$

Proof Let X_i be a Bernoulli random variable with success probability $1/2$ for $i = 1, \ldots, n$. Then

$$X - \mathbb{E}(X) = \sum_{i=1}^{n} \left(X_i - \frac{1}{2} \right) = \sum_{i=1}^{n} Y_i = Y,$$

where the Y_i are independent symmetric random variables taking values $1/2$ and $-1/2$ with equal probability. For any $t > 0$, using the Markov inequality and Theorem D.7, we have that

$$\mathbf{P}(Y > a) = \mathbf{P}(e^{tY} > e^{ta}) \leq \frac{\mathbb{E}(e^{tY})}{e^{ta}} = \frac{\prod_{i=1}^{n} \mathbb{E}(e^{tY_i})}{e^{ta}}.$$

Using Definition 8.6 we compute

$$\mathbb{E}(e^{tY_i}) = \frac{1}{2} \left(\exp\left(\frac{1}{2}t \right) + \exp\left(-\frac{1}{2}t \right) \right) = \cosh\left(\frac{1}{2}t \right).$$

We can show that $\cosh(x) \leq \exp(x^2/2)$ for $x > 0$ by comparing their respective Taylor expansions. Thus,

$$\mathbf{P}(Y > a) \leq \exp\left(\frac{1}{8}t^2 n - at \right) \quad \text{for } t > 0.$$

Minimizing the right-hand side over all $t > 0$ yields $t = 4a/n$. Using this value and the symmetry of Y completes the proof. $\qquad\square$

Alon and Spencer showed a more general result which we state without proof.

Theorem D.15 *[8, Theorem A.1.4] Let X_i be mutually independent random variables with*

$$\mathbf{P}(X_i = 1 - p_i) = p_i \quad and \quad \mathbf{P}(X_i = -p_i) = 1 - p_i \quad for \; p_i \in [0, 1].$$

Then for $X = \sum_{i=1}^{n} X_i$ and $a > 0$,

$$\mathbf{P}(X > a) < \exp(-2a^2/n).$$

Appendix E

Complex Analysis Review

This section provides the basic definitions and theorems that are used in the asymptotic analysis. They can be found in any reference text (see for example [5, 150]).

Definition E.1 *A function $f(z)$ defined over a region $\Re$ is* analytic *at a point $Z_0 \in \Re$ if, for some open disc centered at z_0 and contained in $\Re$, $f(z)$ can be represented by a convergent power series expansion*

$$f(z) = \sum_{n \geq 0} c_n (z - z_0)^n.$$

A function is analytic *in a region $\Re$ if and only if it is analytic at every point of $\Re$. If f is analytic at z_0, then there is a disc (of possibly infinite radius) such that $f(z)$ converges for z inside the disc and is divergent for z outside the disc. This disc is called the* disc of convergence *and its radius is the* radius of convergence, *denoted by $R(f; z_0)$.*

Example E.2 *Consider the function $f(z) = 1/(1 - z)$ which is defined in $\mathbb{C} \setminus \{1\}$. It is analytic at $z = 0$ since it represents the geometric series which converges for $|z| < 1$. For $z_0 \neq 1$, we can obtain the radius of convergence by rewriting $f(z)$ as follows:*

$$\frac{1}{1 - z} = \frac{1}{1 - z_0 - (z - z_0)} = \frac{1}{1 - z_0} \frac{1}{1 - \frac{z - z_0}{1 - z_0}}$$

$$= \sum_{n \geq 0} \left(\frac{1}{1 - z_0} \right)^{n+1} (z - z_0)^n.$$

From the last equation we can read off that $f(z)$ is analytic in the disc centered at z_0 with radius of convergence $|1 - z_0|$, and therefore $R(f; z_0) = |1 - z_0|$.

Definition E.3 *A function $f(z)$ defined over a region $\Re$ is called* complex-differentiable *or* holomorphic *at z_0 if*

$$\lim_{\delta \to 0} \frac{f(z_0 + \delta) - f(z_0)}{\delta}$$

exists for $\delta \in \mathbb{C}$. In particular, the limit is independent of the way in which δ tends to 0 in $\mathbb{C}$. A function is holomorphic *in the region $\mathfrak{R}$ if and only if it is holomorphic for every $z_0 \in \mathfrak{R}$.*

Theorem E.4 (*Basic Equivalence Theorem*) *[68, Theorem IV.1] A function is analytic in a region $\mathfrak{R}$ if and only if it is holomorphic on $\mathfrak{R}$.*

Theorem E.5 *Let $f(z)$ be an analytic function in $\mathfrak{R}$. Then $f(z)$ has derivatives of all orders and the derivatives can be obtained through term-by-term differentiation of the series expansion of the function.*

Definition E.6 *A* curve *γ in a region $\mathfrak{R}$ is a continuous function which maps $[0,1]$ into $\mathfrak{R}$. A curve is called* simple *if the mapping γ is one-to-one, and* closed *if $\gamma(0) = \gamma(1)$. A closed curve is a* loop *if γ can be continuously deformed to a single point within $\mathfrak{R}$.*

Theorem E.7 (*Cauchy's Coefficient Formula*) *[68, Theorem IV.4] Let $f(z)$ be analytic in a region $\mathfrak{R}$ containing 0 and let γ be a simple loop around 0 in $\mathfrak{R}$ that is positively oriented. Then*

$$f_n = [z^n]f(z) = \frac{1}{2\pi} \int_\gamma f(z) \frac{dz}{z^{n+1}}.$$

Definition E.8 *A function $h(z)$ is* meromorphic *at z_0 if and only if it can be represented as $f(z)/g(z)$ for z in a neighborhood of z_0 with $z \neq z_0$, where $f(z)$ and $g(z)$ are analytic at z_0. If $h(z)$ is meromorphic, then it has an expansion of the form*

$$h(z) = \sum_{n \geq -M} h_n(z - z_0)^n \qquad (E.1)$$

near z_0. If $h_{-M} \neq 0$ and $M \geq 1$, then we say that $h(z)$ has a pole *of order M at $z = z_0$. A pole of order one is called a* simple *pole. The coefficient h_{-1} is called the* residue *of $h(z)$ at $z = z_0$ and is written as $\text{Res}[h(z); z = z_0]$. A function is* meromorphic *in a region $\mathfrak{R}$ if and only if it is meromorphic at every point of the region.*

One of the most important theorems in complex analysis is Cauchy's Residue Theorem. It relates global properties of a function (the integral along closed curves) to local properties (residues at poles).

Theorem E.9 (*Cauchy's Residue Theorem*) *[68, Theorem IV.3] Let $h(z)$ be meromorphic in the region $\mathfrak{R}$ and let γ be a simple loop in $\mathfrak{R}$ along which the function $h(z)$ is analytic. Then*

$$\frac{1}{2\pi} \int_\gamma h(z)dz = \sum_s \text{Res}[h(z); z = s],$$

where the sum is over all poles s of $h(z)$ enclosed by γ.

Definition E.10 *Let* $f(z)$ *be an analytic function defined over the interior region determined by a simple closed curve* γ, *and let* z_0 *be a point of the bounding curve* γ. *If there exists an analytic function* $f^*(z)$ *defined over some open set* $\Re^*$ *containing* z_0 *such that* $f^*(z) = f(z)$ *in* $\Re \cap \Re^*$ *then* f *is analytically continuable at* z_0. *If* f *is not analytically continuable at* z_0 *then* z_0 *is called a* singularity *or* singular point. *If* $f(z)$ *is analytic at* 0, *then a singularity that lies on the boundary of the disc of convergence is called a* dominant singularity.

Theorem E.11 (*Rouché's theorem*) *Let the functions* $f(z)$ *and* $g(z)$ *be analytic in a region containing the interior of the simple closed curve* γ. *Assume that* $|g(z)| < |f(z)|$ *on* γ. *Then* $f(z)$ *and* $f(z) + g(z)$ *have the same number of zeros inside the interior domain bounded by* γ. *Equivalently,* $f(\gamma)$ *and* $(f + g)(\gamma)$ *have the same winding number.*

Definition E.12 *Let* $f(x)$ *be a function defined over the positive real numbers. Then the* Mellin transform *of* f *is defined as the complex function* $f^*(s)$ *where*

$$f^*(s) = \int_0^\infty f(x)x^{s-1}dx.$$

The Mellin transform is used in many areas. We will use it for the asymptotic analysis of sums of the form

$$G(x) = \sum_k \lambda_k g(\mu_k x).$$

These types of sums are often called *harmonic sums*. In this context, the λ_k are the *amplitudes*, the μ_k are the *frequencies*, and $g(x)$ is the *base function*.

Example E.13 *[67, page 4] The expected height of a Catalan tree involves the quantity*

$$T_n = \sum_{k=0}^n d(k)\frac{\binom{2n}{n-k}}{\binom{2n}{n}},$$

where $d(k)$ *denotes the number of divisors of* k. *The continuous analog to* T_n *is given by*

$$H(x) = \sum_{k=0}^\infty d(k)e^{-k^2x^2},$$

which is a harmonic sum with $\mu_k = k$, $\lambda_k = d(k)$ *and* $g(x) = e^{-x^2}$. *With elementary analysis it can be established that* $T_n \approx H(n)$. *In this case, the asymptotic analysis is made difficult by the divisor function which fluctuates heavily and in an irregular manner.*

Appendix F

Using Mathematica® and Maple™

Throughout this book we have encountered several problems that are time consuming at best and impossible to solve by hand at worst. Solving systems of equations or finding partial derivatives of complicated expressions are best done with the assistance of computer algebra systems such as Maple and *Mathematica*. The advantage of a computer algebra system is that it can do symbolic algebra and has many specialized functions already built-in or available as packages. *Mathematica* and Maple can be used for different aspects of research: to do the drudge work that is tedious by hand, to check theoretically derived results quickly for algebraic mistakes, to simplify complicated expressions, to find patterns, and to guess generating functions. For example, we can use *Mathematica* or Maple to compute a few derivatives or a few iterations of a recurrence relation to detect a pattern in the results. Such a pattern might then be proved by induction.

F.1   *Mathematica*® functions

In this section we focus on *Mathematica* functions that are useful for or specific to combinatorial enumeration and provide a compact overview of these functions and how they are applied. Note that all *Mathematica* functions start with capital letters and use square brackets, and that code followed by ; does not return any output.

F.1.1   General functions

Here we list a number of general functions that are not specific to combinatorics but play an important supportive role. The functions `Sum[f,{i,n,m}]` and `Product[f,{i,n,m}]` compute the sum $f(n)+\cdots+f(m)$ and the product $f(n)\cdots f(m)$, respectively. To compute the sum and the product of the first five square numbers we use

```
Sum[x^2,{x,1,5}]
Product[x^2,{x,1,5}]
```

which returns 55 for the sum and 14400 for the product.

Two very useful functions are `Simplify[expr]` and `FullSimplify[expr]` which apply various simplification rules to `expr`, with `FullSimplify` trying a wider range of transformations. For example,

```
Simplify[1/(1-x)/(1-2*x)+x/(1-2*x)]
```

returns

```
(1+x-x^2)/(1-3x+2x^2)
```

Solving systems of equations is usually tedious by hand and prone to algebraic error. We can use `Solve[eqns, vars]` to solve the system `eqns` in the variables `vars`. To solve the system of three equations $u + v + w = 1$, $3u + v = 3$, and $u - 2v - w = 0$ in terms of u, v, and w, we use

```
Solve[{u+v+w==1,3u+v==3,u-2v-w==0},{u,v,w}]
```

which results in

```
{{u->4/5,v->3/5,w->-(2/5)}}
```

Note that in *Mathematica*, equality in an equation (as opposed to an assignment) is expressed using double equal signs.

Finally, there are functions to create sequences of values, usually the values of a sequence whose coefficients are given by an explicit formula. The function `Table[expr, {i,n,m}]` generates the sequence created by evaluating `expr` at the values of i from n to m. For example, the sequence of the third to tenth powers of two is computed as

```
Table[2^i,{i,3,10}]
```

with output

```
{8,16,32,64,128,256,512,1024}
```

To create sequences of consecutive integers we use the function `Range[n,m]` which creates the list `{n,...,m}`.

```
Range[3,12]
```

produces the output

```
{3,4,5,6,7,8,9,10,11,12}
```

F.1.2 Special sequences and combinatorial formulas

We will frequently encounter the binomial coefficients $\binom{n}{m}$. The relevant function is `Binomial[n,m]`. For example

```
Binomial [10,5]
```

returns 252. There is also a function that computes the list of all possible permutations of the elements in a list, `Permutations[list]`. For example, to create the permutations of [3], we use

```
Permutations[Range[1,3]]
```

which results in

```
{{1,2,3},{1,3,2},{2,1,3},{2,3,1},{3,1,2},{3,2,1}}
```

Mathematica also has functions for the Fibonacci and Lucas sequences. Fibonacci[n] and LucasL[n] compute the n-th Fibonacci and Lucas number, respectively. Combined with Table we can create the sequences of the first ten Fibonacci and Lucas numbers:

```
Table[Fibonacci[n],{n,1,10}]
Table[LucasL[n],{n,1,10}]
```

which returns

```
{1,1,2,3,5,8,13,21,34,55}
{1,3,4,7,11,18,29,47,76,123}
```

The partitions of n are computed by the function IntegerPartitions. For example, the partitions of 4 are computed as

```
IntegerPartitions[4]
```

which gives

```
{{4},{3,1},{2,2},{2,1,1},{1,1,1,1}}
```

So how can we compute the compositions of n? There is no built-in function, so we apply the function Permutations to each of the partitions of n. For example, the compositions of 3 are created by

```
Flatten[Map[Permutations, IntegerPartitions[3]], 1]
```

which produces the output

```
{{3},{2,1},{1,2},{1,1,1}}
```

Note that Map applies the function Permutations to each of the partitions of [3]. Flatten is used to get the appropriate list structure, so that each composition is in its own list.

F.1.3 Series manipulation

The function Series computes power series expansions in one or more variables. Series[f, {x,x0,n}] generates the power series expansion of f about the point $x = x0$ to order $(x - x0)^n$. For functions of several variables, Series[f, {x,x0,nx},{y,y0,ny},...] successively finds series expansions with respect to x, then y, and so on. To see the first six terms of the power series of the shifted Fibonacci sequence $f(x) = \frac{1}{1-x-x^2}$ (see Table A.1) about $x = 0$ we use

```
Series[1/(1-x-x^2),{x,0,5}]
```

with output

```
1+x+2x^2+3x^3+5x^4+8x^5+O[x]^6
```

The multivariate series expansion of $f(x) = 1/(1-x-y)$ about $(x,y) = (0,0)$ to order 3 for x and order 2 for y is computed as

```
Series[1/(1 - x - y), {x, 0, 3}, {y, 0, 2}]
```

and returns

```
(1+y+y^2+O[y]^3)+(1+2y+3y^2+O[y]^3)x+(1+3y+6y^2+O[y]^3)x^2
      +(1+4y+10y^2+O[y]^3)x^3+O[x]^4
```

Usually we are interested in the coefficients of the generating functions. We can obtain these by using the function `Coefficient[expr, form, n]` which gives the coefficient of `form^n` in `expr`. For example, if we are interested in finding

$$[x^5]\left(\frac{1 - x - \sqrt{1 - 2x - 3x^2}}{2x^2}\right),$$

then we use

```
Series[(1-x-Sqrt[1-2x-3x^2])/(2x^2),{x,0,7}]
Coefficient[%,x,5]
```

which gives

```
1+x+2x^2+4x^3+9x^4+21x^5+51x^6+127x^7+O[x]^8
21
```

Note that % refers to the immediate last output (which may be nonsense if there was a syntax error in the code, so beware). The two commands may be combined as

```
Coefficient[Series[(1-x-Sqrt[1-2x-3x^2])/(2x^2),{x,0,7}],x,5]
```

and now the output consists of only the value 21. If we want coefficients of an exponential generating function then we need to take into account the factorial factor. Say the function to be considered is $f(x) = e^{e^x - 1}$ and we would like to find $[x^5]f(x)$. Then we enter

```
5! Coefficient[Series[Exp[Exp[x]-1],{x,0,6}],x,5]
```

to obtain the answer 52. If instead of a single coefficient value we want the sequence of coefficients (like those we have presented in many of the examples) then the function `CoefficientList[poly,var]` gives the coefficients of powers of var, starting with power 0. For example,

```
CoefficientList[Series[(1-x-Sqrt[1-2x-3x^2])/(2x^2),{x,0,7}],x]
```

returns

```
{1,1,2,4,9,21,51,127}
```

F.1.4 Solving recurrence relations

The function `RSolve[eqn,a[n],n]` solves a recurrence relation for `a[n]`. A more general version solves a system of recurrence relations or a partial recurrence relation. We illustrate the function `RSolve` with the recurrence for the Fibonacci sequence:

```
RSolve[{f[n+2]==f[n+1]+f[n],f[0]==0,f[1]==1},f[n],n]
```

This results in

```
{{f[n] -> Fibonacci[n]}}
```

that is, *Mathematica* recognizes this as the recurrence for the Fibonacci sequence. In order to obtain a generating function from a recurrence relation, we apply the function `GeneratingFunction` to the result of `RSolve`. `GeneratingFunction[expr,n,x]` gives the generating function in x for the sequence whose n-th series coefficient is given by the expression `expr`. For example, to obtain the generating function from the recurrence of the Fibonacci sequence, we use

```
sol=RSolve[{f[n+2]==f[n+1]+f[n],f[0]==0,f[1]==1},f[n],n];
GeneratingFunction[sol[[1,1,2]],n,x]//Simplify
```

This results in

```
-(x/(-1+x+x^2))
```

If we have already derived a generating function for a sequence then we can try to obtain an explicit formula for the coefficients a_n in terms of n by using the function `SeriesCoefficient`. Its syntax is basically the same as `Series`, except we enter `n` instead of an integer value indicating the order of the expansion. For the Fibonacci sequence we use

```
SeriesCoefficient[x/(1-x-x^2),{x,0,n}]
```

This returns the familiar formula for the Fibonacci sequence.

$$\begin{cases} \dfrac{-\left(\frac{1}{2}\left(1-\sqrt{5}\right)\right)^n+\left(\frac{1}{2}\left(1+\sqrt{5}\right)\right)^n}{\sqrt{5}} & n \geq 0 \\ 0 & \text{True} \end{cases}$$

F.1.5 Obtaining results from sequence values

In Version 7.0 of *Mathematica*, several new functions were introduced. They all take as input a list of sequence values and then try to find either an explicit formula, a recurrence relation, or a generating function. We give some examples of these functions using either the first few values of the Fibonacci sequence or the first few values of the Catalan sequence as our starting point. First let's find a recurrence relation:

```
FindLinearRecurrence[{1,1,2,3,5,8,13,21}]
```

which returns the kernel (the list of coefficients) of the shortest recurrence relation that creates these values:

```
{1,1}
```

This means that the recurrence relation is of the form $a_n = 1 \cdot a_{n-1} + 1 \cdot a_{n-2}$. We can recreate the sequence by using `LinearRecurrence[ker, init, n]` which gives the sequence of length n obtained by iterating the linear recurrence with kernel `ker` and initial values `init`:

```
LinearRecurrence[{1,1},{1,1}, 8]
```

The output is our original sequence:

```
{1,1,2,3,5,8,13,21}
```

Next we look at finding the generating function from a sequence of values. `FindGeneratingFunction[{a1,a2,...}, x]` attempts to find a simple generating function in x whose n-th series coefficient is a_n.

```
FindGeneratingFunction[{1,1,2,5,14,42,132,429,1430},x]
```

which returns

```
2/(1+Sqrt[1-4x])
```

And finally there is the function `FindSequenceFunction[list,n]` which attempts to find a formula for the n-th entry in the list.

```
FindSequenceFunction[{1,1,2,3,5,8,13,21},n]
```

which returns once more

```
Fibonacci[n]
```

F.1.6 Pattern matching

Mathematica has good capabilities for pattern matching. All we need to do is to translate the structure of the pattern that we want to find in either compositions or words into *Mathematica* patterns. For example, the subword pattern 112 has structure ?xxy? where ? can be any integer(s) and x needs to be less than y. In *Mathematica* terminology this becomes

$$\{\_\_\_,x\_,x\_,y\_,\_\_\_\}/;\ x < y$$

where the ___ stands for any (possibly empty) pattern, and x_ and y_ are patterns that need to be in the relation specified by the condition that follows /;. For a subsequence pattern there are no adjacency requirements, and therefore, the *Mathematica* pattern corresponding to the subsequence pattern 112 would be

$$\{\_\_\_,x\_,\_\_\_,x\_,\_\_\_,y\_,\_\_\_\}/;\ x < y$$

This idea can be extended to patterns on a larger alphabet. For example, the subsequence pattern 132 is coded as follows (where && denotes the logical AND):

$$\{\_\_\_, x\_, \_\_\_, y\_, \_\_\_, z\_, \_\_\_\}/;\ x\ <\ z\ \&\&\ z{<}y$$

We use the function `Position[list,pattern]` to find where in a list of objects the pattern occurs. For example, if we want to find the subsequence pattern 132 in the list $\{\{1,2,3,4\},\{1,4,2,3,5\},\{1,5,4\},\{1,1,1,1\}\}$ we use:

```
Position[{{1,2,3,4},{1,4,2,3,5},{1,5,4},{1,1,1,1}},
    {___,x_,___,y_,___,z_,___} /; x < z && z < y]
```

which returns as answer the list

```
{{2},{3}}
```

This means that the pattern was detected in the second and third element of the list. Since the result is a list, we can count the number of elements in the list by using the function `Length`:

```
Length[{{2},{3}}]
```

which returns 2 as an answer. Thus two of the four entries of the original list contained the subsequence pattern 132. Note that each word or composition has to be expressed as a list, the the dominant structure in *Mathematica*. However, the function used to create compositions renders them exactly in this form. Now we can obtain the number of compositions that avoid a given pattern by taking the total number of compositions and subtracting from it the number of compositions that contain the pattern. For an example of enumeration by direct pattern matching check out Example 5.14.

F.1.7   Random compositions

Mathematica has several built-in functions that are useful for creating random compositions and computing probabilities. To select an element at random from a list of values or list of compositions, use `RandomChoice`. `RandomChoice[vals,n]` gives a list of n pseudorandom choices from the list `vals`. Note that each value is equally likely to be chosen and a value can be chosen more than once. For example, to obtain 10 random compositions of $n = 5$, we use

```
comps[n_] := Flatten[Map[Permutations,IntegerPartitions[n]],1]
RandomChoice[comps[5],10]
```

which produces an answer similar to $\{\{3,2\},\ \{3,1,1\},\ \{2,1,1,1\},\ \{2,1,1,1\},$ $\{2,1,2\},\ \{1,2,1,1\},\ \{3,1,1\},\ \{3,1,1\},\ \{1,2,2\},\ \{1,1,2,1\}\}$. Each time this command is executed, a different answer will result.

Mathematica has also many functions relating to discrete and continuous random variables. Probability mass functions and density functions are computed using `PDF[dist,x]`. To see a list of all the built-in distributions, use

?*Distribution. For our purposes, the Binomial and geometric distributions are the most relevant, and they are named `BinomialDistribution` and `GeometricDistribution`. To create random values from a distribution, the function `RandomInteger` is used. For example, to create ten random values from a geometric random variable (which counts the number of trials until the first success) with parameter $p = 1/3$, we use

```
Table[RandomInteger[GeometricDistribution[1/3]]+1,{10}]
```

which produces an answer similar to $\{1, 1, 5, 6, 1, 3, 7, 3, 4, 1\}$. Note that the function `GeometricDistribution` produces values that count the number of failures before the first success, so one needs to add a 1 to obtain the number of trials until the first success used in the model for the random compositions.

F.2   Maple™ functions

In this section we focus on Maple functions that are useful for or specific to combinatorial enumeration. We provide a compact overview of these functions and how they are applied. A few general remarks: code that is followed by a : does not return its output, while code followed by ; does return the output. Functions that exist in packages can only be used after first loading the package using the function `with(package)`.

F.2.1   General functions

Here we list a number of general functions that are not specific to combinatorics but play an important supportive role. The functions `sum(f, k=m..n)` and `product(f, k=m..n)` compute the sum $f(m) + \cdots + f(n)$ and the product $f(m) \cdots f(n)$, respectively. To compute the sum and the product of the first five square numbers we use

```
sum(k^2, k=1..5);
product(k^2, k=1..5);
```

which returns 55 for the sum and 14400 for the product. Another very useful function is `simplify(expr)` which applies various simplification rules to `expr`. For example,

```
simplify(1/(1-x)/(1-2*x)+x/(1-2*x));
```

returns

```
-(-1-x+x^2)/(-1+x)/(-1+2*x)
```

Solving systems of equations is usually tedious by hand and prone to algebraic error. We can use `solve(eqns,vars)` to solve the system `eqns` in the variables `vars`. To solve the system of three equations $u + v + w = 1$, $3u + v = 3$, and $u - 2v - w = 0$ in terms of u, v, and w, we use

```
eqns:= {u+v+w=1,3*u+v=3,u-2*v-w=0}: vars:={u,v,w}:
solve(eqns,vars);
```

which results in

```
{v=3/5, u=4/5, w=-2/5}
```

We will also frequently encounter the binomial coefficients $\binom{n}{m}$. The relevant Maple function is `binomial(n,m)`. For example

```
binomial(10,5);
```

returns 252. Finally, there is a function to create the values of a sequence if an explicit formula for its coefficients is known. The function `seq(f,i=m..n)` generates the sequence $f(m), f(m+1), \ldots, f(n)$. For example, the sequence of the third to tenth powers of two are computed as

```
seq(2^i,i=3..10);
```

with output

```
8,16,32,64,128,256,512,1024
```

F.2.2   Series manipulation

Maple has a number of functions that deal with series expansions. The most general of these is the function `series(f,x=a,m+1)` which generates the power series expansion of f about the point $x = a$ to order $(x - a)^m$. To see the first six terms of the power series of the shifted Fibonacci sequence $f(x) = \frac{1}{1-x-x^2}$ (see Table A.1) about $x = 0$ we use

```
series(1/(1-x-x^2),x=0,6);
```

resulting in the output

```
1+x+2*x^2+3*x^3+5*x^4+8*x^5+0[x]^6
```

More specific series expansions are the Taylor series expansions for one or more variables. For a single variable we use `taylor(f,x=a,n)` which gives the Taylor series expansion of the function f around $x = a$ up to order n. For several variables we use `mtaylor(f, v, n)` which computes a truncated multivariate Taylor series expansion of the function f with respect to the variables v to order n. By default, the expansion is about the origin, but a different point for expansion can be specified as an optional argument. For example, the Taylor series of $f(x) = 1/(1 - x - y)$ about $x = 0$ to order 10 is computed as

```
taylor(1/(1-x),x=0,10);
```

with output

```
1+1*x+1*x^2+1*x^3+1*x^4+1*x^5+1*x^6+1*x^7+1*x^8+1*x^9+0(x^10)
```

The multivariate Taylor expansion of $f(x) = 1/(1 - x - y)$ about $(x, y) = (0, 0)$ to order 3 is computed as

```
mtaylor(1/(1-x-y), [x,y], 4);
```

which results in

```
1+x+y+x^2+2*y*x+y^2+x^3+3*y*x^2+3*y^2*x+y^3
```

Usually we are interested in the coefficients of the generating functions. We can obtain these by using the function `coeff(p,x,n)` which extracts the coefficient of x^n in the polynomial p. For example, if we are interested in finding

$$[x^5]\left(\frac{1 - x - \sqrt{1 - 2x - 3x^2}}{2x^2}\right),$$

then we use

```
series((1-x-sqrt(1-2*x-3*x^2))/2/x^2,x=0,8);
coeff(%,x,5);
```

to obtain the output

```
1+x+2*x^2+4*x^3+9*x^4+21*x^5+51*x^6+127*x^7+0[x]^8
21
```

Note that the % sign refers to the immediate last output (which may be nonsense if there was a syntax error in the code, so beware). The two commands may be combined as

```
coeff(series((1-x-sqrt(1-2*x-3*x^2))/2/x^2,x=0,8),x,5);
```

and now the output consists of only the value 21.

If we want to get the coefficients of an exponential generating function then we need to take into account the factorial factor. Say the function to be considered is $f(x) = e^{e^x - 1}$ and we would like to find $[x^5]f(x)$. Then we enter

```
5!*coeff(series(exp(exp(x)-1),x=0,6),x,5);
```

to obtain the answer 52.

F.2.3 Solving recurrence relations

The function `rsolve(eqns,fcns)` attempts to solve the recurrence relation(s) specified in `eqns` for the functions in `fcns`, returning an expression for the general term of the function. The first argument should be a single recurrence relation or a set of recurrence relations and boundary conditions. Any expressions in `eqns` which are not equations will be understood to be equal to zero. The second argument `fcns` indicates the functions that `rsolve` should solve for. We use the recurrence relation of the Fibonacci sequence to check out this function. Note that the recurrence and the initial conditions have to be entered as a list.

```
rsolve({f(n+2)=f(n+1)+f(n),f(0)=0,f(1)=1},f);
```

This results in

$$\frac{\sqrt{5}\left(\frac{1}{2}+\frac{\sqrt{5}}{2}\right)^n}{5} - \frac{\sqrt{5}\left(\frac{1}{2}-\frac{\sqrt{5}}{2}\right)^n}{5},$$

which is of course the explicit formula for the n-th Fibonacci number. To obtain the generating function from the recurrence relation, we use a more general version of `rsolve`.

```
rsolve({f(n) = f(n-1)+f(n-2), f(0)=1,(1)=1},f,'genfunc'(x));
```

The answer is what we expect:

$$-\frac{x}{-1+x+x^2}.$$

If we have derived a generating function for a sequence then we can try to obtain an explicit formula for the coefficients a_n in terms of n by using the function `rgf_expand` in the package `genfunc`. We use again the Fibonacci sequence for illustration.

```
with(genfunc): rgf_expand(x/(1-x-x^2), x, n);
```

This returns

$$\frac{\sqrt{5}\left(-\frac{2}{\sqrt{5}+1}\right)^n}{5} - \frac{(\sqrt{5}-1)\sqrt{5}\left(-\frac{2}{-\sqrt{5}+1}\right)^n}{5(-\sqrt{5}+1)},$$

which needs to be manipulated a bit with the function `simplify` to obtain the familiar expression for the n-th term of the Fibonacci sequence.

F.2.4 Asymptotic behavior

The function `asympt(f,x,n)` computes the asymptotic expansion of f with respect to the variable x as $x \to \infty$. The third argument n specifies the truncation order of the series expansion. For instance, if we enter

```
asympt(n!,n,2);
```

we will get the output

$$\frac{\frac{\sqrt{2}\sqrt{\pi}}{\sqrt{\frac{1}{n}}} + \frac{\sqrt{2}\sqrt{\pi}\sqrt{\frac{1}{n}}}{12} + O\left(\left(\frac{1}{n}\right)^{\frac{3}{2}}\right)}{\left(\frac{1}{n}\right)^n e^n}.$$

As another example of the use of `asympt` we look at the well-known limit $\sin(x)/x \to 1$ as $x \to 0$. How quick is this convergence? Since `asympt` only works when the variable tends to infinity, we rewrite the function in terms of $1/x$ and compute

```
asympt(sin(1/x)x,x,10);
```

The answer returned is

$$1 - \frac{1}{6x^2} + \frac{1}{120x^4} - \frac{1}{5040x^6} + \frac{1}{362880x^8} + O\left(\frac{1}{x^9}\right),$$

This implies that the function $\frac{\sin x}{x}$ has asymptotic behavior

$$1 - \frac{1}{6}x^2 + \frac{1}{120}x^4 - \frac{1}{5040}x^6 + \frac{1}{362880}x^8 + O(x^9)$$

when $x \to 0$.

F.2.5　Gfun Maple™ package

The package Gfun was created to assist in finding generating functions. Given the first few coefficients of a sequence, the package helps us to conjecture the corresponding generating function. In some cases the answer will be an explicit formula. However, most of the time such a formula does not exist and the answer will be an equation satisfied by the generating function in question. Full details on the package Gfun can be found at

http://hal.inria.fr/docs/00/07/00/25/PDF/RT-0143.pdf.

We show some examples to give the reader an idea of what is to be found in the Gfun package. First we load the package

```
with(gfun);
```

Now let's explore what we get if we use the first few terms of the Fibonacci sequence, namely $0, 1, 1, 2, 3, 5, 8, 13$ as our starting point. The simplest task is to convert the sequence into a series:

```
F1:=listtoseries([0,1,1,2,5,8,13],x,ogf);
```

which results in

```
F1:=x+x^2+2x^3+5x^4+8x^5+13x^6+O(x^7)
```

This is not a hard task, so let's do something more challenging. To find the ordinary generating function for this sequence we enter

```
listtoratpoly([0,1,1,2,3,5,8,13],x);
```

and we obtain the generating function for the Fibonacci sequence

```
[-\frac{x}{-1+x+x^2}, ogf]
```

Another approach is to look for a differential equation that is satisfied by the generating function for the sequence. In this case we use

```
listtodiffeq([0,1,1,2,3,5,8,13],y(x));
```

which returns

```
[x+(-1+x+x^2)y(x), ogf]
```

That is, the ordinary generating function satisfies the differential equation $x + (-1 + x + x^2)y(x) = 0$ or equivalently, $y(x) = \frac{x}{1-x-x^2}$. This differential equation is particularly simple (no derivatives involved!), so we present a more typical example. We use the first few values of the Catalan sequence as the input to listtodiffeq, that is

```
listtodiffeq([1,1,2,5,14,42,132,429,1430],y(x));
```

The result

```
[{-1+(1-2x)y(x)+(x-4x^2)diff(y(x),x), y(0) = 1, ogf]
```

suggests that the Catalan sequence may satisfy the differential equation

$$1 - (1 - 2x)y(x) - (x - 4x^2)y'(x) = 0$$

with initial condition $y(0) = 1$. Finally, we can ask for a recurrence relation for the generating function for the Catalan sequence:

```
listtorec([1,1,2,5,14,42,132,429,1430],a(n));
```

We obtain that

```
[{a(0) = 1, (-4n-2)a(n)+(n+2)a(n+1)}, ogf]
```

which gives a conjecture that the Catalan numbers satisfy the recurrence relation $(n + 2)a(n + 1) - (4n + 2)a(n) = 0$ with initial condition $a(0) = 1$ (Prove it!!).

F.2.6 Random compositions

Maple has a package that can create all the compositions of n, namely combinat. The function composition(n,i) produces the compositions of n with i parts. The function rand(1..k) produces a pseudorandom integer in the range $1, \ldots, k$. For example, to obtain 10 random compositions of $n = 5$, we use

```
with(combinat): n:=5:
    a:=composition(n,1):
    for i from 2 to n do
        a:=a union composition(n,i):
    od:
    nn:=rand(1..2^(n-1)):
    for s from 1 to 10 do
        a[nn()];
    od;
```

which produces an answer similar to $[1, 3, 1]$, $[2, 3]$, $[1, 2, 2]$, $[1, 1, 1, 1, 1]$, $[1, 1, 3]$, $[1, 1, 1, 1, 1]$, $[2, 1, 2]$, $[1, 1, 1, 1, 1]$, $[1, 1, 1, 2]$, $[1, 2, 1, 1]$. Each time this command is executed, a different answer will result.

Maple has also many functions relating to discrete and continuous random variables. To access these, load the package `Statistics`. For our purposes, the Binomial and geometric distributions are the most relevant, and they are named `Binomial` and `Geometric`. To create random values from a distribution, the function `RandomVariable` is used. For example, to create ten random values from a geometric random variable (which counts the number of trials until the first success) with parameter $p = 1/3$, we use

```
with(Statistics):
Y := RandomVariable(Geometric(1/3))+1;
Sample(Y,10);
```

which produces for example the sequence 2, 5, 1, 2, 10, 6, 2, 3, 5, 3. Note that the function `Geometric` in Maple counts the number of failures before the first success, so one needs to add a 1 to obtain the number of trials until the first success used in the model for the random compositions.

Appendix G

C++ and Maple™ Programs

Sections G.1 through G.4 contain C++ programs to count the number of compositions and words, respectively, that avoid either a single pattern entered by the user or all subsequence patterns τ of a fixed length. Section G.5 contains a C++ and a Maple program. The C++ program TOU_AUTO computes the states and the transfer matrix of the automaton for avoidance of subsequence patterns in words. The Maple program TOU_FORMULA computes the first few values and an explicit formula for the sequence $\{AW_{[k]}^{\tau}(n)\}$ from the transfer matrix obtained as output of the C++ program. These programs, as well as Java versions of the programs in Sections G.1 and G.3 can be downloaded from the author's website `http://math.haifa.ac.il/toufik/manbook.html`. The Java versions are provided for readers that do not have access to a C++ compiler, but they are slower than the C++ versions and are not suitable for massive computation.

G.1   Program to compute $AC_{\mathbb{N}}^{\tau}(n)$ for given τ

This program computes the number of compositions of n for $n = 1, 2, \ldots, 20$ that avoid a pattern entered by the user. If the pattern is a subsequence pattern of length three, four, five, or six then the functions `vpatterns3()`, `vpatterns4()`, `vpatterns5()`, and `vpatterns6()` are to be used. For a subword pattern of length dp the relevant function is `swpatterns(dp)`, while for generalized patterns of length four we use `gpatterns112()`, `gpatterns121()`, and `gpatterns22(dp)`. Note that the routine `gpatterns22` is more general as it enumerates the compositions that avoid the generalized pattern $(dp - 2, 2)$ for $dp \geq 4$. The routine `gpatterns22(dp)` illustrates how the user might modify the other routines for generalized patterns of length four to deal with generalized patterns of length greater than four. The program also contains two routines that check whether a pattern occurs: `testK` for subsequence patterns and `testW` for the other two types of patterns.

The routine `builtcomps` recursively creates all compositions of n and increases the count if the composition avoids the pattern p. The user can adjust the type of patterns for which the program checks by making the relevant routines active. As presented below, the program checks for subsequence patterns. The user is prompted for the length of the pattern and the specific pattern to be checked. After the computation the output is presented on the screen and also written into a file named data.dat.

```c
#include <stdio.h>

#define MAXN 40 /*max number of parts in the composition*/

int n; /* composition of n*/

int comp[MAXN],    /* composition */
    p[MAXN],    /* pattern */
    lcomp;      /* number of parts of comp */

long numbercomp=0;
/*number of compositions of n that avoid the pattern p */

FILE *fout; /*output file */

/* comparison between two elements */
int order(int a,int b)
{
  if (a==b) return 0; else if(a>b) return 1; else return -1;
}

/* testing whether two sequences are order-isomorphic */
/* k = length of sequences */
int testK(int k,int * p1,int * p2)
{
    int i;
    for (i=0;i<k;i++)
        if (order(p1[i],p1[k])!=order(p2[i],p2[k])) return 1;
    return 0;
}

int testW(int k,int * p1,int * p2)
{
    int i,j;
    for (i=0;i<k-1;i++)
      for(j=i+1;j<k;j++)
        if (order(p1[i],p1[j])!=order(p2[i],p2[j])) return 1;
    return 0;
}

/* check whether comp avoids subsequence pattern p of length 3 */
int vpatterns3()
{
    int curword[3];  //subword of comp[] currently being tested
```

```
    int nn=lcomp;
    int i1,i2,i3;
    for(i3=0;i3<nn-2;i3++){
      curword[0]=comp[i3];
      for(i2=i3+1;i2<nn-1;i2++){
        curword[1]=comp[i2];
        if (testK(1,curword,p)) continue;
        for(i1=i2+1;i1<nn;i1++){
          curword[2]=comp[i1];
          if (!testK(2,curword,p))
            return 0;
    }}}
    return 1;//no pattern
}


/* check whether comp avoids subsequence pattern p of length 4 */
int vpatterns4() {
    int curword[4]; //subword of comp[] currently being tested
    int nn=lcomp;
    int i1,i2,i3,i4;
    for(i4=0;i4<nn-3;i4++){
      curword[0]=comp[i4];
      for(i3=i4+1;i3<nn-2;i3++){
        curword[1]=comp[i3];
        if (testK(1,curword,p)) continue;
        for(i2=i3+1;i2<nn-1;i2++){
          curword[2]=comp[i2];
          if (testK(2,curword,p)) continue;
          for(i1=i2+1;i1<nn;i1++){
            curword[3]=comp[i1];
            if (!testK(3,curword,p)
            return 0;
    }}}}
    return 1;//no pattern
}


/* check whether comp avoids subsequence pattern p of length 5 */
int vpatterns5() {
    int curword[5]; //subword of comp[] currently being tested
    int nn=lcomp;
    int i1,i2,i3,i4,i5;
    for(i5=0;i5<nn-4;i5++){
      curword[0]=comp[i5];
      for(i4=i5+1;i4<nn-3;i4++){
        curword[1]=comp[i4];
        if (testK(1,curword,p)) continue;
        for(i3=i4+1;i3<nn-2;i3++){
          curword[2]=comp[i3];
          if (testK(2,curword,p)) continue;
```

```
        for(i2=i3+1;i2<nn-1;i2++){
          curword[3]=comp[i2];
          if (testK(3,curword,p)) continue;
          for(i1=i2+1;i1<nn;i1++){
            curword[4]=comp[i1];
            if (!testK(4,curword,p))
              return 0;
  }}}}}
  return 1;//no pattern
}


/* check whether comp avoids subsequence pattern p of length 6 */
int vpatterns6() {
    int curword[6]; //subword of comp[] currently being tested
    int nn=lcomp;
    int i1,i2,i3,i4,i5,i6;
    for(i6=0;i6<nn-5;i6++){
      curword[0]=comp[i6];
      for(i5=i6+1;i5<nn-4;i5++){
        curword[1]=comp[i5];
        if (testK(1,curword,p)) continue;
        for(i4=i5+1;i4<nn-3;i4++){
          curword[2]=comp[i4];
          if (testK(2,curword,p)) continue;
          for(i3=i4+1;i3<nn-2;i3++){
            curword[3]=comp[i3];
            if (testK(3,curword,p)) continue;
            for(i2=i3+1;i2<nn-1;i2++){
              curword[4]=comp[i2];
              if (testK(4,curword,p)) continue;
              for(i1=i2+1;i1<nn;i1++){
                curword[5]=comp[i1];
                if (!testK(5,curword,p))
                  return 0;
    }}}}}}
    return 1;//no pattern
}


/* check whether comp avoids subword pattern p of length dp */
int swpatterns(int dp) {
    int curword[10]; //subword of comp[] currently being tested
    int nn=lcomp,j,i1;
    for(i1=0;i1<nn+1-dp;i1++)
    {
      for(j=0;j<dp;j++) curword[j]=comp[i1+j];
      if (!testW(dp,curword,p)) return 0;
    }
    return 1;//no pattern
}
```

```
/*  check whether  comp avoids the pattern p of type (1,1,2) */
int gpatterns112() {
    int curword[10]; //subword of comp[] currently being tested
    int nn=lcomp,i1,i2,i3;
    for(i1=0;i1<nn-3;i1++)
     for(i2=i1+1;i2<nn-2;i2++)
      for(i3=i2+1;i3<nn-1;i3++)
      {
         curword[0]=comp[i1]; curword[1]=comp[i2];
         curword[2]=comp[i3]; curword[3]=comp[i3+1];
         if (!testW(4,curword,p)) return 0;
      }
    return 1;//no pattern
}

/* check whether comp avoids the pattern p of type (1,2,1) */
int gpatterns121() {
    int curword[10]; //subword of comp[] currently being tested
    int nn=lcomp;
    int i1,i2,i3,i4;
    for(i1=0;i1<nn-3;i1++)
     for(i2=i1+1;i2<nn-2;i2++)
     {
        i3=i2+1;
        for(i4=i3+1;i4<nn;i4++)
        {
           curword[0]=comp[i1]; curword[1]=comp[i2];
           curword[2]=comp[i3]; curword[3]=comp[i4];
           if (!testW(4,curword,p)) return 0;
        }
     }
    return 1;//no pattern
}

/* check whether comp avoids the pattern p of type (2,2) */
int gpatterns22(int dp) {
    int curword[10]; //subword of comp[] currently being tested
    int nn=lcomp;
    int j,i1,i2;
    for(i1=0;i1<nn+1-dp;i1++)
     for(i2=i1+dp-2;i2<nn-1;i2++)
     {
        for(j=0;j<dp-2;j++) curword[j]=comp[i1+j];
        curword[dp-2]=comp[i2]; curword[dp-1]=comp[i2+1];
        if (!testW(dp,curword,p)) return 0;
     }
    return 1;//no pattern
}
```

```
/* recursively generating all the compositions comp of n */
/* if comp avoids the pattern p then count it */
/* l = parts considered, nnn= weight of comp, lenp = length of p */
void builtcomps(int l,int nnn,int lenp) {
    int j;
    if (nnn==0)
    {
        lcomp=l;
/* subsequence */
        if(lenp==3) numbercomp=numbercomp+vpatterns3();
        if(lenp==4) numbercomp=numbercomp+vpatterns4();
        if(lenp==5) numbercomp=numbercomp+vpatterns5();
        if(lenp==6) numbercomp=numbercomp+vpatterns6();
/* generalized patterns */
        //if (lenp==4) numbercomp=numbercomp+gpatterns112();
        //if (lenp==4) numbercomp=numbercomp+gpatterns121();
        //if (lenp==4) numbercomp=numbercomp+gpatterns22(4);
                /* subword pattern of length dp */
        //if (lenp==4) numbercomp=numbercomp+swpatterns(4);

    }
    else
      for(j=1;j<=nnn;j++)
        { comp[l]=j;  builtcomps(l+1,nnn-j,lenp);  }
}

 /* User inputs length and pattern  */
void main() {
  int i,lenp;

  printf("Please enter the length of the
          subsequence pattern (3-6): ");
  scanf("%d",&lenp);

  printf("Please enter the pattern: ");
  for(i=0;i<lenp;i++) scanf("%d",&p[i]);

  fout=fopen("data.dat","w");
  fprintf(fout,"The number of compositions of n
                that avoid the pattern ");
  for(i=0;i<lenp;i++) fprintf(fout,"%d",p[i]);
  fprintf(fout," is given by\n");

  printf("The number of compositions of n
          that avoid the pattern ");
  for(i=0;i<lenp;i++) printf("%d",p[i]);
  printf(" is given by\n");
```

```
    for(n=1;n<21;n++)
    {
numbercomp=0;
    for(i=0;i<n;i++) comp[i]=0;
lcomp=0;
    builtcomps(0,n,lenp);
    printf("Number of compositions of %2d that
            avoid the pattern is %d\n",n,numbercomp);
    fprintf(fout,"Number of compositions of %2d that
                avoid the pattern is %d\n",n,numbercomp);
    }
    fclose(fout);
}
```

We illustrate the use of the program for the pattern 132.

Example G.1 *To compute the number of compositions of n that avoid the subsequence pattern* $\tau = 132$ *for* $n = 1, 2, \ldots, 20$, *we enter* 3 *at the first prompt, and* 1 3 2 *(with spaces between the numbers) at the second prompt. For subsequence patterns, the length can be between* 3 *and* 6.

```
    Please enter the length of the subsequence pattern (3-6): 3
    Please enter the pattern: 1 3 2
```

The resulting output is

```
    The number of compositions of n that avoid the pattern 132 is given
    by
    Number of compositions of  1 that avoid the pattern is 1
    Number of compositions of  2 that avoid the pattern is 2
    Number of compositions of  3 that avoid the pattern is 4
    Number of compositions of  4 that avoid the pattern is 8
    Number of compositions of  5 that avoid the pattern is 16
    Number of compositions of  6 that avoid the pattern is 31
    Number of compositions of  7 that avoid the pattern is 60
    Number of compositions of  8 that avoid the pattern is 114
    Number of compositions of  9 that avoid the pattern is 214
    Number of compositions of 10 that avoid the pattern is 398
    Number of compositions of 11 that avoid the pattern is 732
    Number of compositions of 12 that avoid the pattern is 1334
    Number of compositions of 13 that avoid the pattern is 2410
    Number of compositions of 14 that avoid the pattern is 4321
    Number of compositions of 15 that avoid the pattern is 7688
    Number of compositions of 16 that avoid the pattern is 13590
    Number of compositions of 17 that avoid the pattern is 23869
    Number of compositions of 18 that avoid the pattern is 41686
    Number of compositions of 19 that avoid the pattern is 72405
    Number of compositions of 20 that avoid the pattern is 125144
    Press any key to continue
```

G.2   Program to compute $AC_{\mathbb{N}}^{\tau}(n)$ for all subsequence patterns τ of fixed length

Unlike the program given in Section G.1, the following program computes the sequence $AC_{\mathbb{N}}^{\tau}(n)$ for **all** subsequence patterns τ of a fixed length, that is, the program will create those patterns and then basically run the program given in Section G.1. Note that the program below only contains routines that pertain to subsequence patterns, but can easily be adjusted by utilizing the routines for subword or generalized patterns of the program listed in Section G.1.

The additional routines are `isword`, `reversal`, and `builtpatterns`. The first one checks whether a word of length *lenp* is indeed a pattern, that is, in reduced form, while the second one checks whether the reversal of the current pattern has already been considered. The third function recursively builds all words of length *lenp*, then checks that the current word is a new pattern. If so, the pattern is added to the list and the program computes $AC_{\mathbb{N}}^{\tau}(n)$ for $n = 1, 2, \ldots, 24$. The output is once more printed to the screen and to a file named data.dat.

```
#include <stdio.h>

#define MAXN 40
#define MAXP 10000
int n;
int comp[MAXN],p[MAXN],lcomp;
long numbercomp=0;
int listp[MAXP][MAXN], /*list containing the patterns p, one per row*/
    nlistp=0;

FILE *fout;

int order(int a,int b)
{ if (a==b) return 0; else if(a>b) return 1; else return -1; }

int testK(int k,int * p1,int * p2)
{
    int i;
    for (i=0;i<k;i++)
      if (order(p1[i],p1[k])!=order(p2[i],p2[k])) return 1;
    return 0;
}

int vpatterns3()
{
    int curword[3]; //subword of comp[] currently being tested
    int nn=lcomp,i1,i2,i3;
```

```
    for(i3=0;i3<nn-2;i3++){
      curword[0]=comp[i3];
      for(i2=i3+1;i2<nn-1;i2++){
        curword[1]=comp[i2];
        if (testK(1,curword,p)) continue;
        for(i1=i2+1;i1<nn;i1++){
          curword[2]=comp[i1];
          if (!testK(2,curword,p))
            return 0;
}}}
    return 1;//no pattern
}


int vpatterns4()
{
    int curword[4]; //subword of word[] currently being tested
    int i1,i2,i3,i4;
    for(i4=0;i4<n-3;i4++){
      curword[0]=word[i4];
      for(i3=i4+1;i3<n-2;i3++){
        curword[1]=word[i3];
        if (testK(1,curword,p)) continue;
        for(i2=i3+1;i2<n-1;i2++){
          curword[2]=word[i2];
          if (testK(2,curword,p)) continue;
          for(i1=i2+1;i1<n;i1++){
            curword[3]=word[i1];
            if (!testK(3,curword,p))
            return 0;
}}}}
    return 1;//no pattern
}


int vpatterns5()
{
    int curword[5]; //subword of word[] currently being tested
    int i1,i2,i3,i4,i5;
    for(i5=0;i5<n-4;i5++){
      curword[0]=word[i5];
      for(i4=i5+1;i4<n-3;i4++){
        curword[1]=word[i4];
        if (testK(1,curword,p)) continue;
        for(i3=i4+1;i3<n-2;i3++){
          curword[2]=word[i3];
          if (testK(2,curword,p)) continue;
          for(i2=i3+1;i2<n-1;i2++){
            curword[3]=word[i2];
            if (testK(3,curword,p)) continue;
            for(i1=i2+1;i1<n;i1++){
```

```
                curword[4]=word[i1];
                if (!testK(4,curword,p))
                return 0;
    }}}}}
    return 1;//no pattern
}

int vpatterns6()
{
    int curword[6];
    int i1,i2,i3,i4,i5,i6;
    for(i6=0;i6<n-5;i6++){
      curword[0]=word[i6];
      for(i5=i6+1;i5<n-4;i5++){
        curword[1]=word[i5];
        if (testK(1,curword,p)) continue;
        for(i4=i5+1;i4<n-3;i4++){
          curword[2]=word[i4];
          if (testK(2,curword,p)) continue;
          for(i3=i4+1;i3<n-2;i3++){
            curword[3]=word[i3];
            if (testK(3,curword,p)) continue;
            for(i2=i3+1;i2<n-1;i2++){
              curword[4]=word[i2];
              if (testK(4,curword,p)) continue;
              for(i1=i2+1;i1<n;i1++){
                curword[5]=word[i1];
                if (!testK(5,curword,p))
                return 0;
    }}}}}}
    return 1;
}

void builtcomps(int l,int nnn,int lenp)
{
    int j;
    if (nnn==0){
      lcomp=l;
      if(lenp==3) numbercomp=numbercomp+vpatterns3();
      if(lenp==4) numbercomp=numbercomp+vpatterns4();
      if(lenp==5) numbercomp=numbercomp+vpatterns5();
      if(lenp==6) numbercomp=numbercomp+vpatterns6();
    }
    else
      for(j=1;j<=nnn;j++){
        comp[l]=j; builtcomps(l+1,nnn-j,lenp);
      }
}
```

```
/* check whether p is a  word in reduced form */
int isword(int lenp)
{
    int i,j,fg,m=0;
    for(i=0;i<lenp;i++)
      if(m<p[i]) m=p[i];
    for(i=1;i<=m;i++){
      fg=1;
      for(j=0;j<lenp;j++)
        if (p[j]==i) fg=0;
      if(fg==1) return(0);
    }
    return(1);
}


/* check whether the reverse of p is already in the list of patterns */
int reversal(int lenp)
{
    int i,j,fg;
    for(i=0;i<nlistp;i++){
      fg=1;
      for(j=0;j<lenp;j++)
        if (listp[i][j]!=p[lenp-1-j]) fg=0;
        if (fg==1) return(0);
    }
    return(1);
}


/* check whether p is a new pattern */
int checkp(int lenp)
{
    if (isword(lenp)==0) return(0);
    if (reversal(lenp)==0) return(0);
    return(1);
}


/* recursively builds patterns of length lenp */
void builtpatterns(int l,int lenp)
{
    int i,j;
    if (l==lenp){
      if (checkp(lenp)==1)  /* if a new pattern */
      {
        /* add to list */
        for(i=0;i<lenp;i++) listp[nlistp][i]=p[i];
        nlistp++;
        printf("%4d) ",nlistp);
        fprintf(fout,"%4d) ",nlistp);
        for(i=0;i<lenp;i++) printf("%d",p[i]);
```

```
    printf(" : ");
    for(i=0;i<lenp;i++) fprintf(fout,"%d",p[i]);
    fprintf(fout," :");

    for(n=1;n<25;n++){
      numbercomp=0;
      for(i=0;i<n;i++) comp[i]=0;
      lcomp=0;
      builtcomps(0,n,lenp);
      printf("%8d,",numbercomp);
      fprintf(fout,"%8d,",numbercomp);
    }
    printf("\n");
    fprintf(fout,"\n");
    }
  }
  else
    for(j=1;j<=lenp;j++){
      p[l]=j; builtpatterns(l+1,lenp);
    }
}

void main()
{
  int lenp=4;
  fout=fopen("data.dat","w");
  builtpatterns(0,lenp);
  fclose(fout);
}
```

In Chapter 6, we saw that Propositions 6.18 and 6.19 together with Theorem 6.20 completely classify the Wilf-equivalence classes for subsequence patterns of lengths four and five. Tables G.1 and G.2 provide the complete listing of all Wilf-equivalence classes, including the ones that consist of a single pattern. The program above was used to compute values of the sequence $\{AC_{\mathbb{N}}^\tau(n)\}$. Since large values of n are time-consuming to compute, we display the values of $AC_{\mathbb{N}}^\tau(n)$ for $n = 17, 18, \ldots, 24$ for the patterns of length four.

TABLE G.1: Subsequence patterns of length four

τ	The sequence $\{AC_{\mathbb{N}}^\tau(n)\}_{n=17}^{24}$
1112,1121	16374, 27655, 46240, 77075, 127434, 209453, 341764, 555698
1111	16377, 28764, 48518, 82889, 137161, 237502, 390084, 646347
2112	31811, 57081, 101691, 180111, 316951, 554561, 964485, 1668026
1212	31791, 57043, 101638, 180057, 316955, 554777, 965283, 1670341
1221	31909, 57319, 102253, 181381, 319715, 560407, 976597, 1692732
1122	32105, 57767, 103235, 183465, 324009, 569113, 993995, 1726932
1312	37161, 67418, 121524, 217661, 387562, 686136, 1208267, 2116810

τ	The sequence $\{AC_{\mathbb{N}}^{\tau}(n)\}_{n=17}^{24}$
1231	37315, 67813, 122485, 219897, 392573, 697020, 1231297, 2164445
2113	37378, 67960, 122823, 220639, 394149, 700295, 1237957, 2177740
1213	37392, 68004, 122946, 220961, 394948, 702193, 1242302, 2187393
1123	37427, 68096, 123170, 221485, 396128, 704754, 1247712, 2198554
1222	44909, 84261, 157331, 292357, 540667, 995338, 1824181, 3328584
1223,1232 1322,2123 2132,2213	48219, 90722, 169789, 316127, 585660, 1079755, 1981407, 3619549
2313	59019, 115202, 224253, 435348, 842864, 1627482, 3134190, 6020078
1332	59021, 115210, 224285, 435447, 843157, 1628279, 3136275, 6025299
1323	59021, 115210, 224285, 435449, 843165, 1628312, 3136381, 6025619
3123	59024, 115222, 224319, 435546, 843417, 1628950, 3137925, 6029274
1233,2133	59025, 115226, 224334, 435591, 843544, 1629284, 3138770, 6031329
2413	61181, 120014, 234818, 458273, 892134, 1732500, 3356468, 6487681
1423	61181, 120014, 234819, 458278, 892154, 1732568, 3356678, 6488279
1342	61181, 120014, 234819, 458278, 892156, 1732578, 3356721, 6488429
2314	61182, 120019, 234839, 458346, 892365, 1733176, 3358335, 6492597
3124	61182, 120019, 234839, 458346, 892366, 1733182, 3358362, 6492696
1234,1243 1432,2134 2143,3214	61182, 120019, 234840, 458351, 892386, 1733249, 3358567, 6493277
1324	61182, 120019, 234840, 458351, 892387, 1733255, 3358594, 6493377

For subword patterns of length five, we provide the values $AC_{\mathbb{N}}^{\tau}(n)$ for $n = 17, 18, \ldots, 21$ for all 279 patterns to be considered (roughly a seven-fold increase from the number of patterns of length four). For some patterns, we have computed additional values to show that the specific patterns are indeed in a Wilf-equivalence class of their own.

TABLE G.2: Subsequence patterns of length five

τ	The sequence $\{AC_{\mathbb{N}}^{\tau}(n)\}_{n=17}^{\geq 21}$
11112,11121,11211	170944, 299744, 522829, 905728, 1561794
11111	170015, 303356 536877 932679, 1637383
12112	277374, 506651 920031 1661532, 2984319
21112	277480, 506816 920330 1661916, 2984760
11212	278222, 508621 924526 1671457, 3005810
12121	278276, 508768 924892 1672346, 3007872
11221	280120, 512922 933962 1691750, 3048552
11122	280794, 514454, 937414 , 1699250, 3064692
13112	317119, 585160, 1073587, 1958783, 3554467
11312	317346, 585816, 1075372, 1963346, 3565653
12131	317628, 586559, 1077226, 1967786, 3575872
11231	318174, 587969, 1080694, 1975987, 3594640
21113	318202, 588047, 1080918, 1976550, 3595962
11321	318264, 588201, 1081275, 1977378, 3597877
12113	318265, 588243, 1081479, 1978055, 3599799

τ	The sequence $\{AC_{\mathbb{N}}^{\tau}(n)\}_{n=17}^{\geq 21}$
11213	318412, 588642, 1082499, 1980553, 3605700
11123,11132	318804, 589632, 1084872, 1986054, 3618082
21212	369350, 698204, 1314047, 2462440, 4594680
12122	369389, 698312, 1314338, 2463181, 4596534
12212	369381, 698303, 1314327, 2463196, 4596646
12221	369544, 698720, 1315359, 2465690, 4602480
21122	369782, 699234, 1316421, 2467784, 4606410
11222	370136, 700162, 1318781, 2473590, 4620294
21312	392795, 744739, 1405410, 2639921, 4936338
13212	392822, 744825, 1405658, 2640589, 4938041
13122	393012, 745249, 1406555, 2642401, 4941526
21132	392964, 745170, 1406468, 2642420, 4942050
12312	392999, 745302, 1406880, 2643593, 4945188
21213	393008, 745334, 1406977, 2643861, 4945872
12132	393010, 745344, 1407015, 2643987, 4946251, 9212699
12123	393010, 745344, 1407015, 2643987, 4946251, 9212703
21123	393055, 745435, 1407185, 2644256, 4946564
12231	393092, 745510, 1407324, 2644500, 4946943
12213	393154, 745715, 1407936, 2646195, 4951406
12321	393159, 745731, 1407980, 2646307, 4951672
22113	393413, 746377, 1409556, 2650021, 4960144
11223,11232,11322	393429, 746443, 1409778, 2650699, 4962054
12222,21222,22122	428952, 827634, 1591632, 3050965, 5829342
12223,12232,12322 13222,21223,21232 21322,22123,22132 22213	445647, 861514, 1659551, 3185466, 6092743
13132	472809, 924977, 1805122, 3514011, 6823654
21313	472818, 925013, 1805231, 3514337, 6824536
12313	472823, 925037, 1805331, 3514671, 6825565
13123	472823, 925038, 1805341, 3514713, 6825727
13213	472825, 925047, 1805376, 3514829, 6826077
13231	472847, 925107, 1805539, 3515235, 6827061
13312	472854, 925129, 1805596, 3515373, 6827360
12331	472857, 925146, 1805668, 3515622, 6828139
23113	472903, 925247, 1805875, 3515993, 6828696
31123	472918, 925309, 1806086, 3516651, 6830599
31213	472929, 925355, 1806229, 3517085, 6831797
11332	472947, 925405, 1806383, 3517501, 6832927
11323	472947, 925405, 1806382, 3517506, 6832945
21133	472963, 925467, 1806569, 3518047, 6834381
12133	472963, 925467, 1806574, 3518069, 6834471
11233	472963, 925471, 1806593, 3518139, 6834703
13142	488636, 960244, 1882770, 3683216, 7189065
14213	488638, 960253, 1882806, 3683339, 7189448
21413	488638, 960254, 1882810, 3683354, 7189492
12413	488638, 960255, 1882816, 3683383, 7189608
13412	488639, 960260, 1882839, 3683461, 7189854
14123	488643, 960272, 1882872, 3683543, 7190038
14312	488640, 960265, 1882861, 3683542, 7190126
21314	488640, 960267, 1882874, 3683602, 7190350
12431	488644, 960278, 1882898, 3683640, 7190366

τ	The sequence $\{AC_{\mathbb{N}}^{\tau}(n)\}_{n=17}^{\geq 21}$
13124	488640, 960267, 1882874, 3683604, 7190369
13241	488644, 960279, 1882902, 3683653, 7190403
14132	488645, 960283, 1882916, 3683698, 7190536
24113	488674, 960372, 1883160, 3684321, 7192033
12314	488665, 960355, 1883152, 3684414, 7192588
13214	488665, 960355, 1883154, 3684425, 7192639
12341	488666, 960358, 1883161, 3684437, 7192641
31124	488679, 960398, 1883271, 3684719, 7193333
11423	488682, 960408, 1883297, 3684785, 7193473
11342	488682, 960408, 1883297, 3684784, 7193477
31214	488680, 960405, 1883302, 3684837, 7193723
21134,21143	488680, 960405, 1883304, 3684847, 7193765
12134,12143	488681, 960410, 1883324, 3684917, 7193987
23114	488685, 960422, 1883355, 3684984, 7194103
32114	488685, 960422, 1883358, 3685000, 7194171
11234,11243,11432	488686, 960428, 1883383, 3685094, 7194484
11324	488686, 960428, 1883383, 3685094, 7194488
23213	494582, 974957, 1918100, 3765918, 7378434
23132	494584, 974968, 1918147, 3766082, 7378953
23123	494587, 974977, 1918166, 3766118, 7378999
12323	494587, 974979, 1918177, 3766166, 7379166, 14427727, 28148807, 54800750, 106456739, 206356820
21323	494587, 974979, 1918177, 3766166, 7379166, 14427727, 28148807, 54800750, 106456739, 206356816
13232	494587, 974979, 1918177, 3766166, 7379168
31223	494594, 975010, 1918282, 3766474, 7379986
13223	494594, 975009, 1918280, 3766474, 7380013
32123	494594, 975010, 1918289, 3766518, 7380183
21332	494594, 975012, 1918300, 3766567, 7380361
12332	494594, 975012, 1918302, 3766581, 7380425
22313	494598, 975033, 1918388, 3766862, 7381265
13322	494608, 975078, 1918547, 3767380, 7382805
12233,21233,22133	494608, 975080, 1918559, 3767434, 7382995
24213	504899, 998671, 1971813, 3886039, 7644090
24123	504899, 998671, 1971813, 3886039, 7644095
13242	504899, 998671, 1971813, 3886042, 7644116, 15007564, 29406726, 57508444
12423	504899, 998671, 1971813, 3886042, 7644116, 15007566, 29406736, 57508486, 112244393, 218649086, 425095725, 824879229
21423	504899, 998671, 1971813, 3886042, 7644116, 15007566, 29406736, 57508486, 112244393, 218649086, 425095725, 824879228
21342	504899, 998673, 1971830, 3886132, 7644491
23142	504900, 998679, 1971856, 3886222, 7644769
12342	504900, 998680, 1971862, 3886250, 7644877
21432	504900, 998680, 1971864, 3886263, 7644938
12432	504900, 998680, 1971864, 3886263, 7644939
14223	504901, 998687, 1971893, 3886361, 7645226
22413	504901, 998687, 1971895, 3886375, 7645294
14232	504901, 998687, 1971895, 3886375, 7645298

τ	The sequence $\{AC_{\mathbb{N}}^{\tau}(n)\}_{n=17}^{\geq 21}$
31224	504901, 998687, 1971897, 3886387, 7645347
32214	504901, 998687, 1971897, 3886387, 7645350
32124	504901, 998687, 1971897, 3886387, 7645352
23124	504901, 998687, 1971897, 3886387, 7645353, 15011590
23214	504901, 998687, 1971897, 3886387, 7645353, 15011589
13224	504901, 998687, 1971897, 3886388, 7645357
12324	504901, 998687, 1971897, 3886389, 7645366, 15011651, 29419180, 57544352, 112343292, 218911964, 425773182, 826579262
21324	504901, 998687, 1971897, 3886389, 7645366, 15011651, 29419180, 57544352, 112343292, 218911964, 425773182, 826579261
13422	504901, 998688, 1971901, 3886403, 7645401
22314	504901, 998687, 1971899, 3886402, 7645425
14322	504901, 998688, 1971904, 3886424, 7645504
12234,12243,21234 21243,22134,22143	504901, 998688, 1971905, 3886430, 7645532
23313	511953, 1017130, 2018679, 4002039, 7924995
23133	511953, 1017130, 2018679, 4002039, 7924996
31323	511953, 1017130, 2018679, 4002039, 7924998
13323	511953, 1017130, 2018679, 4002039, 7925003, 15674892, 30965800
13332	511953, 1017130, 2018679, 4002039, 7925003, 15674892, 30965797
13233	511953, 1017130, 2018679, 4002039, 7925004
31233	511954, 1017135, 2018700, 4002114, 7925241
32133	511954, 1017135, 2018700, 4002114, 7925242
12333,21333	511954, 1017135, 2018700, 4002114, 7925246
24313	516589, 1027944, 2043430, 4057772, 8048745
23413	516589, 1027944, 2043430, 4057772, 8048750, 15946331
31324	516589, 1027944, 2043430, 4057772, 8048750, 15946334, 31554879
23134	516589, 1027944, 2043430, 4057772, 8048750, 15946334, 31554899, 62363616, 123095495, 242655378, 477713441, 939225481, 1844141265
23143	516589, 1027944, 2043430, 4057772, 8048750, 15946334, 31554899, 62363616, 123095495, 242655378, 477713441, 939225481, 1844141264
31423	516589, 1027944, 2043431, 4057777, 8048767
24133	516589, 1027944, 2043432, 4057784, 8048801, 15946514, 31555463
13342	516589, 1027944, 2043432, 4057784, 8048801, 15946514, 31555465
14233	516589, 1027944, 2043432, 4057784, 8048801, 15946514, 31555467
13423	516589, 1027944, 2043432, 4057784, 8048804, 15946534, 31555563, 62365649, 123101367, 242671568
13432	516589, 1027944, 2043432, 4057784, 8048804, 15946534, 31555563, 62365649, 123101367, 242671567, 477720039
14332	516589, 1027944, 2043432, 4057784, 8048804, 15946534, 31555563, 62365649, 123101367, 242671567, 477756437

τ	The sequence $\{AC_{\mathbb{N}}^{\tau}(n)\}_{n=17}^{\geq 21}$
23314	516589, 1027944, 2043432, 4057784, 8048804, 15946535, 31555566
14323	516589, 1027944, 2043432, 4057784, 8048804, 15946535, 31555569
13324	516589, 1027944, 2043432, 4057784, 8048804, 15946535, 31555570
32314	516589, 1027944, 2043432, 4057784, 8048804, 15946535, 31555580
13234	516589, 1027944, 2043432, 4057784, 8048804, 15946535, 31555582, 62365740, 123101723, 242672811, 477760435, 939348262, 1844453674
13243	516589, 1027944, 2043432, 4057784, 8048804, 15946535, 31555582, 62365740, 123101723, 242672811, 477760435, 939348262, 1844453673
31243	516589, 1027945, 2043436, 4057798, 8048843
32143	516589, 1027945, 2043437, 4057804, 8048869
31234	516589, 1027945, 2043437, 4057805, 8048877, 15946767, 31556254
32134	516589, 1027945, 2043437, 4057805, 8048877, 15946767, 31556256
33124	516589, 1027945, 2043438, 4057811, 8048903, 15946861, 31556255
33214	516589, 1027945, 2043438, 4057811, 8048903, 15946861, 31556258
12334,12343,12433 21334,21343,21433	516589, 1027945, 2043438, 4057811, 8048903, 15946861, 31556259
14243	522145, 1042632, 2081196, 4152622, 8282198, 16510861, 32899031
13424	522145, 1042632, 2081196, 4152622, 8282198, 16510863, 32899043, 65520227
31424	522145, 1042632, 2081196, 4152622, 8282198, 16510863, 32899043, 65520226
24143	522145, 1042632, 2081196, 4152622, 8282198, 16510865, 32899053
24134	522145, 1042632, 2081196, 4152622, 8282198, 16510867, 32899066
24314	522145, 1042632, 2081196, 4152622, 8282198, 16510867, 32899067
14342	522145, 1042632, 2081196, 4152622, 8282200, 16510875, 32899100
13442	522145, 1042632, 2081196, 4152622, 8282200, 16510877, 32899111, 65520497
24413	522145, 1042632, 2081196, 4152622, 8282200, 16510877, 32899111, 65520496
32414	522145, 1042632, 2081196, 4152623, 8282205, 16510899, 32899190
23414	522145, 1042632, 2081196, 4152623, 8282205, 16510899, 32899195
14234	522145, 1042632, 2081196, 4152623, 8282206, 16510902, 32899203
14324	522145, 1042632, 2081196, 4152623, 8282206, 16510902, 32899205
14423	522145, 1042632, 2081196, 4152623, 8282207, 16510909, 32899236
14432	522145, 1042632, 2081196, 4152623, 8282207, 16510909, 32899239

τ	The sequence $\{AC_{\mathbb{N}}^{\tau}(n)\}_{n=17}^{\geq 21}$
34124	522145, 1042632, 2081197, 4152627, 8282220, 16510941, 32899295, 65520977
34214	522145, 1042632, 2081197, 4152627, 8282220, 16510941, 32899295, 65520976
12443,21443	522145, 1042632, 2081197, 4152627, 8282222, 16510955, 32899367, 65521272, 130420893
12434,21434	522145, 1042632, 2081197, 4152627, 8282222, 16510955, 32899367, 65521272, 130420894
41324	522145, 1042632, 2081197, 4152628, 8282227, 16510978
42134	522145, 1042632, 2081197, 4152628, 8282227, 16510979
41234	522145, 1042632, 2081197, 4152628, 8282228
23144	522145, 1042632, 2081197, 4152628, 8282229, 16510991, 32899515, 65521811, 130422680, 259467469, 515903007
31244	522145, 1042632, 2081197, 4152628, 8282229, 16510991, 32899515, 65521811, 130422680, 259467469, 515903012
12344,21344,32144	522145, 1042632, 2081197, 4152628, 8282229, 16510991, 32899515, 65521811, 130422684, 259467492, 515903117
13244	522145, 1042632, 2081197, 4152628, 8282229, 16510991, 32899515, 65521811, 130422684, 259467492, 515903121
24153	523194, 1045321, 2087895, 4168924, 8321100, 16602179, 33110392
31524	523194, 1045321, 2087895, 4168924, 8321100, 16602179, 33110393
25314	523194, 1045321, 2087895, 4168924, 8321100, 16602179, 33110394
24513	523194, 1045321, 2087895, 4168924, 8321100, 16602179, 33110395
14253	523194, 1045321, 2087895, 4168924, 8321100, 16602180, 33110398
13524	523194, 1045321, 2087895, 4168924, 8321100, 16602180, 33110400, 66003540, 131511362
25134	523194, 1045321, 2087895, 4168924, 8321100, 16602180, 33110400, 66003540, 131511361
13542	523194, 1045321, 2087895, 4168924, 8321100, 16602180, 33110401, 66003546, 131511390
14352	523194, 1045321, 2087895, 4168924, 8321100, 16602180, 33110401, 66003546, 131511391
25413	523194, 1045321, 2087895, 4168924, 8321100, 16602180, 33110404
23514	523194, 1045321, 2087895, 4168924, 8321101, 16602187, 33110434
31425	523194, 1045321, 2087895, 4168924, 8321101, 16602187, 33110436, 66003689
24135	523194, 1045321, 2087895, 4168924, 8321101, 16602187, 33110436, 66003691
15324	523194, 1045321, 2087895, 4168924, 8321101, 16602188, 33110440, 66003703
32514	523194, 1045321, 2087895, 4168924, 8321101, 16602188, 33110440, 66003705

τ	The sequence $\{AC_{\mathbb{N}}^{\tau}(n)\}_{n=17}^{\geq 21}$
15234	523194, 1045321, 2087895, 4168924, 8321101, 16602188, 33110441, 66003711
14523	523194, 1045321, 2087895, 4168924, 8321101, 16602188, 33110441, 66003712, 131511996
14532	523194, 1045321, 2087895, 4168924, 8321101, 16602188, 33110441, 66003712, 131511998
15342	523194, 1045321, 2087895, 4168924, 8321101, 16602188, 33110441, 66003713
15243	523194, 1045321, 2087895, 4168924, 8321101, 16602188, 33110442, 66003716
14235	523194, 1045321, 2087895, 4168924, 8321101, 16602188, 33110442, 66003717
25143	523194, 1045321, 2087895, 4168924, 8321101, 16602188, 33110442, 66003719, 131512031
15423	523194, 1045321, 2087895, 4168924, 8321101, 16602188, 33110442, 66003719, 131512032
13425	523194, 1045321, 2087895, 4168924, 8321101, 16602188, 33110442, 66003719, 131512033
32415	523194, 1045321, 2087895, 4168924, 8321101, 16602188, 33110442, 66003719, 131512034
24315	523194, 1045321, 2087895, 4168924, 8321101, 16602188, 33110442, 66003721
13452	523194, 1045321, 2087895, 4168924, 8321101, 16602188, 33110443
12534,21534	523194, 1045321, 2087895, 4168925, 8321106, 16602208, 33110507, 66003908, 131512528, 261908366
12453,21453	523194, 1045321, 2087895, 4168925, 8321106, 16602208, 33110507, 66003908, 131512528, 261908367
35124	523194, 1045321, 2087895, 4168925, 8321106, 16602207, 33110499, 66003865, 131512342
35214	523194, 1045321, 2087895, 4168925, 8321106, 16602207, 33110499, 66003865, 131512341
42135	523194, 1045321, 2087895, 4168925, 8321107, 16602215, 33110544, 66004065, 131513117
41325	523194, 1045321, 2087895, 4168925, 8321107, 16602215, 33110544, 66004065, 131513119
23145,23154	523194, 1045321, 2087895, 4168925, 8321107, 16602215, 33110545, 66004072
31245,31254	523194, 1045321, 2087895, 4168925, 8321107, 16602215, 33110545, 66004073
34125	523194, 1045321, 2087895, 4168925, 8321107, 16602216, 33110550, 66004092, 131513218, 261910724,521332661
34215	523194, 1045321, 2087895, 4168925, 8321107, 16602216, 33110550, 66004092, 131513218, 261910724, 521332662
41235	523194, 1045321, 2087895, 4168925, 8321107, 16602216, 33110550, 66004092, 131513218, 261910721
42315	523194, 1045321, 2087895, 4168925, 8321107, 16602216, 33110550, 66004093
23415	523194, 1045321, 2087895, 4168925, 8321107, 16602216, 33110550, 66004094, 131513234
43125	523194, 1045321, 2087895, 4168925, 8321107, 16602216, 33110550, 66004094, 131513235

τ	The sequence $\{AC_{\mathbb{N}}^{\tau}(n)\}_{n=17}^{\geq 21}$
12345,12354,12543 15432,21345,21354 21543,32145,32154 43215	523194, 1045321, 2087895, 4168925, 8321107, 16602216,33110551, 66004100, 131513262, 261910916
12435,21435	523194, 1045321, 2087895, 4168925, 8321107, 16602216, 33110551, 66004100, 131513262, 261910917
14325	523194, 1045321, 2087895, 4168925, 8321107, 16602216, 33110551, 66004100, 131513263
13245,13254	523194, 1045321, 2087895, 4168925, 8321107, 16602216, 33110551, 66004101

G.3   Program to compute $AW_{[k]}^{\tau}(n)$ for given τ

We now present the corresponding programs for words. The routines are very similar, the only difference being that the user now also enters the number of letters of the alphabet. Also, since there is a much larger number of words of length n than there are compositions of n, we reset the counter at 10^7 and use the variable num2 to keep track of how often we have to reset.

```c
#include <stdio.h>

#define MAXN 40

int n,  /*length of word */
NK;   /* number of letters in the alphabet */
int word[MAXN],p[MAXN],numdp;

/* used in the counting of the number of words */
long numberword=0,num2=0;

FILE *fout;

int order(int a,int b)
{ if (a==b) return 0; else if(a>b) return 1; else return -1; }

int testK(int k,int *p1,int *p2)
{
    int i;
    for (i=0;i<k;i++)
     if (order(p1[i],p1[k])!=order(p2[i],p2[k])) return 1;
    return 0;
}
```

```
vpatterns3()
{
    int curword[3];   //subword of word[] currently being tested
    int i1,i2,i3;
    for(i3=0;i3<n-2;i3++){
      curword[0]=word[i3];
      for(i2=i3+1;i2<n-1;i2++){
        curword[1]=word[i2];
        if (testK(1,curword,p)) continue;
        for(i1=i2+1;i1<n;i1++){
          curword[2]=word[i1];
          if (!testK(2,curword,p))
          return 0;
    }}}
    return 1;//no pattern
}

int vpatterns4()
{
    int curword[4];  //subword of word[] currently being tested
    int i1,i2,i3,i4;
    for(i4=0;i4<n-3;i4++){
      curword[0]=word[i4];
      for(i3=i4+1;i3<n-2;i3++){
        curword[1]=word[i3];
        if (testK(1,curword,p)) continue;
        for(i2=i3+1;i2<n-1;i2++){
          curword[2]=word[i2];
          if (testK(2,curword,p)) continue;
          for(i1=i2+1;i1<n;i1++){
            curword[3]=word[i1];
            if (!testK(3,curword,p))
            return 0;
    }}}}
    return 1;//no pattern
}

int vpatterns5()
{
    int curword[5]; //subword of word[] currently being tested
    int i1,i2,i3,i4,i5;
    for(i5=0;i5<n-4;i5++){
      curword[0]=word[i5];
      for(i4=i5+1;i4<n-3;i4++){
        curword[1]=word[i4];
        if (testK(1,curword,p)) continue;
        for(i3=i4+1;i3<n-2;i3++){
          curword[2]=word[i3];
          if (testK(2,curword,p)) continue;
```

```
            for(i2=i3+1;i2<n-1;i2++){
              curword[3]=word[i2];
              if (testK(3,curword,p)) continue;
              for(i1=i2+1;i1<n;i1++){
                curword[4]=word[i1];
                if (!testK(4,curword,p))
                return 0;
    }}}}}
    return 1;//no pattern
}

int vpatterns6()
{
    int curword[6];
    int i1,i2,i3,i4,i5,i6;
    for(i6=0;i6<n-5;i6++){
      curword[0]=word[i6];
      for(i5=i6+1;i5<n-4;i5++){
        curword[1]=word[i5];
        if (testK(1,curword,p)) continue;
        for(i4=i5+1;i4<n-3;i4++){
          curword[2]=word[i4];
          if (testK(2,curword,p)) continue;
          for(i3=i4+1;i3<n-2;i3++){
            curword[3]=word[i3];
            if (testK(3,curword,p)) continue;
            for(i2=i3+1;i2<n-1;i2++){
              curword[4]=word[i2];
              if (testK(4,curword,p)) continue;
              for(i1=i2+1;i1<n;i1++){
                curword[5]=word[i1];
                if (!testK(5,curword,p))
                return 0;
    }}}}}}
    return 1;
}

int vpatterns7()
{
    int curword[7];
    int i1,i2,i3,i4,i5,i6,i7;
    for(i7=0;i7<n-6;i7++){
      curword[0]=word[i7];
      for(i6=i7+1;i6<n-5;i6++){
        curword[1]=word[i6];
        if (testK(1,curword,p)) continue;
        for(i5=i6+1;i5<n-4;i5++){
         curword[2]=word[i5];
          if (testK(2,curword,p)) continue;
```

```
        for(i4=i5+1;i4<n-3;i4++){
          curword[3]=word[i4];
          if (testK(3,curword,p)) continue;
          for(i3=i4+1;i3<n-2;i3++){
            curword[4]=word[i3];
            if (testK(4,curword,p)) continue;
            for(i2=i3+1;i2<n-1;i2++){
              curword[5]=word[i2];
              if (testK(5,curword,p)) continue;
              for(i1=i2+1;i1<n;i1++){
                curword[6]=word[i1];
                if (!testK(6,curword,p))
                return 0;
    }}}}}}
      return 1;
}

void builtwords(int l,int lenp)
{
    int j;
    if (l==n)
    {
      if(lenp==3){
        numberword=numberword+vpatterns3();
        if (numberword==10000000) {num2++; numberword=0;}
      }
      if(lenp==4){
        numberword=numberword+vpatterns4();
        if (numberword==10000000) {num2++; numberword=0;}
      }
      if(lenp==5){
        numberword=numberword+vpatterns5();
        if (numberword==10000000) {num2++; numberword=0;}
      }
      if(lenp==6){
        numberword=numberword+vpatterns6();
        if (numberword==10000000) {num2++; numberword=0;}
      }
      if(lenp==7){
        numberword=numberword+vpatterns7();
        if (numberword==10000000) {num2++; numberword=0;}
      }
    }
    else
      for(j=0;j<NK;j++){
        word[l]=j; builtwords(l+1,lenp);
      }
}
```

```
void main()
{
    int i,lenp;
    printf("Please enter the length of the subsequence
            pattern (3-7): ");
    scanf("%d",&lenp);
    printf("Please enter the pattern : ");
    for(i=0;i<lenp;i++) scanf("%d",&p[i]);

/*output file*/
    fout=fopen("data.dat","w");

    printf("Please enter the number of letters in your alphabet, k=");
    scanf("%d",&NK);

    for(n=1;n<10;n++)   /*the length n*/
    {
      for(i=0;i<n;i++) word[i]=0;
      builtwords(0,lenp);
      printf("The number of words in [%d]^%d that avoid the
              pattern is 10^7*%d+%d \n",NK,n,num2,numberword);
      fprintf(fout,"The number of words in [%d]^%d that avoid the
                    pattern is 10^7*%d+%d \n",NK,n,num2,numberword);
    }
  fclose(fout);
}
```

G.4   Program to compute $AW_{[k]}^{\tau}(n)$ for all subsequence patterns τ of fixed length

This is the program for words that corresponds to the one in Section G.2 for compositions. For words, there are two additional routines that check whether the complement and the reversal of the complement of the current pattern is already in the list, namely `compl1` and `reversalcompl`.

```
#include <stdio.h>

#define MAXN 40

int n,NK;
int word[MAXN],p[MAXN], numdp;
long numberword=0,num2=0;
int listp[100000][10],nlistp=0;

FILE *fout;
```

```
int order(int a,int b)
{ if (a==b) return 0; else if(a>b) return 1; else return -1; }

int testK(int k,int *p1,int *p2)
{
    int i;
    for (i=0;i<k;i++)
      if (order(p1[i],p1[k])!=order(p2[i],p2[k])) return 1;
    return 0;
}

vpatterns3()
{
    int curword[3];   //subword of word[] currently being tested
    int i1,i2,i3;
    for(i3=0;i3<n-2;i3++){
      curword[0]=word[i3];
      for(i2=i3+1;i2<n-1;i2++){
        curword[1]=word[i2];
        if (testK(1,curword,p)) continue;
        for(i1=i2+1;i1<n;i1++){
          curword[2]=word[i1];
          if (!testK(2,curword,p))
          return 0;
    }}}
    return 1;//no pattern
}

int vpatterns4()
{
    int curword[4]; //subword of word[] currently being tested
    int i1,i2,i3,i4;
    for(i4=0;i4<n-3;i4++){
      curword[0]=word[i4];
      for(i3=i4+1;i3<n-2;i3++){
        curword[1]=word[i3];
        if (testK(1,curword,p)) continue;
        for(i2=i3+1;i2<n-1;i2++){
          curword[2]=word[i2];
          if (testK(2,curword,p)) continue;
          for(i1=i2+1;i1<n;i1++){
            curword[3]=word[i1];
            if (!testK(3,curword,p))
            return 0;
    }}}}
    return 1;//no pattern
}
```

```
int vpatterns5()
{
    int curword[5]; //subword of word[] currently being tested
    int i1,i2,i3,i4,i5;
    for(i5=0;i5<n-4;i5++){
      curword[0]=word[i5];
      for(i4=i5+1;i4<n-3;i4++){
        curword[1]=word[i4];
        if (testK(1,curword,p)) continue;
        for(i3=i4+1;i3<n-2;i3++){
          curword[2]=word[i3];
          if (testK(2,curword,p)) continue;
          for(i2=i3+1;i2<n-1;i2++){
            curword[3]=word[i2];
            if (testK(3,curword,p)) continue;
            for(i1=i2+1;i1<n;i1++){
              curword[4]=word[i1];
              if (!testK(4,curword,p))
              return 0;
    }}}}}
    return 1;//no pattern
}

int vpatterns6()
{
    int curword[6];
    int i1,i2,i3,i4,i5,i6;
    for(i6=0;i6<n-5;i6++){
      curword[0]=word[i6];
      for(i5=i6+1;i5<n-4;i5++){
        curword[1]=word[i5];
        if (testK(1,curword,p)) continue;
        for(i4=i5+1;i4<n-3;i4++){
          curword[2]=word[i4];
          if (testK(2,curword,p)) continue;
          for(i3=i4+1;i3<n-2;i3++){
            curword[3]=word[i3];
            if (testK(3,curword,p)) continue;
            for(i2=i3+1;i2<n-1;i2++){
              curword[4]=word[i2];
              if (testK(4,curword,p)) continue;
              for(i1=i2+1;i1<n;i1++){
                curword[5]=word[i1];
                if (!testK(5,curword,p))
                return 0;
    }}}}}}
    return 1;
}
```

```
int vpatterns7()
{
    int curword[7];
    int i1,i2,i3,i4,i5,i6,i7;
    for(i7=0;i7<n-6;i7++){
      curword[0]=word[i7];
      for(i6=i7+1;i6<n-5;i6++){
        curword[1]=word[i6];
        if (testK(1,curword,p)) continue;
        for(i5=i6+1;i5<n-4;i5++){
         curword[2]=word[i5];
         if (testK(2,curword,p)) continue;
         for(i4=i5+1;i4<n-3;i4++){
          curword[3]=word[i4];
          if (testK(3,curword,p)) continue;
          for(i3=i4+1;i3<n-2;i3++){
           curword[4]=word[i3];
           if (testK(4,curword,p)) continue;
           for(i2=i3+1;i2<n-1;i2++){
            curword[5]=word[i2];
            if (testK(5,curword,p)) continue;
            for(i1=i2+1;i1<n;i1++){
             curword[6]=word[i1];
             if (!testK(6,curword,p))
             return 0;
    }}}}}}}
    return 1;
}

void builtwords(int l,int lenp)
{
   int j;
   if (l==n)
   {
     if(lenp==3){
       numberword=numberword+vpatterns3();
       if (numberword==10000000) {num2++; numberword=0;}
     }
     if(lenp==4){
       numberword=numberword+vpatterns4();
       if (numberword==10000000) {num2++; numberword=0;}
     }
  if(lenp==5){
       numberword=numberword+vpatterns5();
       if (numberword==10000000) {num2++; numberword=0;}
     }
     if(lenp==6){
       numberword=numberword+vpatterns6();
       if (numberword==10000000) {num2++; numberword=0;}
```

```
          }
      if(lenp==7){
        numberword=numberword+vpatterns7();
        if (numberword==10000000) {num2++; numberword=0;}
      }
  }
      else
        for(j=0;j<NK;j++){
          word[l]=j; builtwords(l+1,lenp);
        }
  }

int isword(int lenp)
{
      int i,j,m,fg;
      m=0;
      for(i=0;i<lenp;i++) if(m<p[i]) m=p[i];
for(i=1;i<=m;i++){
        fg=1;
        for(j=0;j<lenp;j++) if (p[j]==i) fg=0;
        if(fg==1) return(0);
      }
      return(1);
}

int reversal(int lenp)
{
      int i,j,fg;
      for(i=0;i<nlistp;i++){
        fg=1;
        for(j=0;j<lenp;j++)
          if (listp[i][j]!=p[lenp-1-j]) fg=0;
          if (fg==1) return(0);
      }
      return(1);
}
/* check whether the complement of p is already in list */
int compl1(int lenp)
{
      int i,j,fg,m;
      m=0;
      for(i=0;i<lenp;i++)
        if (m<p[i]) m=p[i];

      for(i=0;i<nlistp;i++){
        fg=1;
        for(j=0;j<lenp;j++)
          if(listp[i][j]!=(m+1-p[j])) fg=0;
          if (fg==1) return(0);
```

```
      }
    return(1);
}

/* check whether the reverse of the complement of p */
/* is already in list */
int reversalcompl(int lenp)
{
    int i,j,fg,m;
    m=0;
    for(i=0;i<lenp;i++)
      if (m<p[i]) m=p[i];
    for(i=0;i<nlistp;i++){
      fg=1;
      for(j=0;j<lenp;j++)
        if(listp[i][j]!=(m+1-p[lenp-1-j])) fg=0;
        if (fg==1) return(0);
    }
    return(1);
}

int checkp(int lenp)
{
    if (isword(lenp)==0) return(0);
    if (reversal(lenp)==0) return(0);
    if (compl1(lenp)==0) return(0);
    if (reversalcompl(lenp)==0) return(0);
    return(1);
}

void builtpatterns(int l,int lenp)
{
    int i,j;
    if (l==lenp)
    {
      if (checkp(lenp)==1)
  {
        for(i=0;i<lenp;i++) listp[nlistp][i]=p[i];
        nlistp++;
        numdp=0;
        for(i=0;i<lenp;i++) if (numdp<p[i]) numdp=p[i];
        NK=numdp+2;

        printf("%5d) pattern=",nlistp);
        fprintf(fout,"%5d) pattern=",nlistp);
        for(i=0;i<lenp;i++) printf("%d",p[i]);
        printf("  :\n");
        for(i=0;i<lenp;i++) fprintf(fout,"%d",p[i]);
        fprintf(fout,"  :\n");
```

```
    for(n=1;n<11;n++){
      numberword=0; num2=0;
      for(i=0;i<n;i++) word[i]=0;
      builtwords(0,lenp);
      printf("The number of words in [%d]^%d that avoid the
              pattern is 10^7*%d+%d\n",NK,n,num2,numberword);
      fprintf(fout,"The number of words in [%d]^%d that avoid the
              pattern is 10^7*%d+%d\n",NK,n,num2,numberword);
    }
    printf("\n");
    fprintf(fout,"\n");
    }
  }
  else
    for(j=1;j<=lenp;j++){
      p[l]=j; builtpatterns(l+1,lenp);
    }
}

void main()
{
    int lenp=3; /* length of the subsequence patterns */
    fout=fopen("data.dat","w");
    builtpatterns(0,lenp);
    fclose(fout);
}
```

We have listed the nontrivial Wilf-equivalence classes for subsequence patterns of length four and five for words in Tables 6.3 and 6.4. Propositions 6.18 and 6.19 together with Theorem 6.20 completely classify the Wilf-equivalence classes for these patterns. Tables G.3 and G.4 contain all Wilf-equivalence classes, including the ones that consist of a single pattern. Due to the much larger number of words of length n (as opposed to compositions of n), we present the values $AW^{\tau}_{[5]}(n)$ for $n = 4, \ldots, 9$ and $AW^{\tau}_{[6]}(n)$ for $n = 5, 6, \ldots, 9$, respectively. If these values do not distinguish between two patterns, then we provide values obtained using the above program for larger n from a convenient alphabet to show that the patterns are indeed in different classes.

TABLE G.3: Subsequence patterns of length four

τ	The sequence $\{|AW^{\tau}_{[5]}(n)|\}^9_{n=4}$	$|AW^{\tau}_{[6]}(9)|$
1111	620, 3020, 14300, 65100, 281400, 1138200	
1112,1121	615, 2935, 13425, 58259, 237798, 906722	
1122	615, 2915, 13095, 55235, 217710, 799910	
1123,1132	615, 2935, 13475, 59319, 250306, 1013790	

τ	The sequence $\{|AW_{[5]}^{\tau}(n)|\}_{n=4}^{9}$	$|AW_{[6]}^{\tau}(9)|$
1212	615, 2915, 13065, 54615, 211165, 752875	
1213	615, 2935, 13469, 59215, 249298, 1006608	
1221	615, 2915, 13075, 54815, 213180, 766450	
1223,1232 1322,2132	615, 2935, 13469, 59209, 249173, 1005169	5149600
1332	615, 2935, 13469, 59209, 249173, 1005169	5149598
1231	615, 2935, 13470, 59225, 249319, 1006165	
1234,1243 1432,2143	620, 3028, 14495, 67875, 310815, 1392795	
1312	615, 2935, 13459, 59019, 247148, 989288	
1324	620, 3028, 14495, 67876, 310839, 1393119	
1342	620, 3028, 14494, 67854, 310563, 1390527	
1423	620, 3028, 14494, 67853, 310540, 1390229	
2413	620, 3028, 14493, 67830, 310245, 1387439	

TABLE G.4: Subsequence patterns of length five

τ	The sequence $\{|AW_{[6]}^{\tau}(n)|\}_{n=5}^{9}$	$|AW_{[k]}^{\tau}(n)|$
13132	7756 46066 269850 1549246 8668438	
12313	7756 46066 269850 1549357 8672194	
21312	7756 46066 269850 1549366 8672416	
13212	7756 46066 269850 1549369 8672514	
13213	7756 46066 269850 1549387 8673242	
12312	7756 46066 269865 1549822 8680105	
12132	7756 46066 269865 1549846 8680923	
12123	7756 46066 269865 1549846 8680947	$|[6]^{10}(\tau)| = 47240506$
13232	7756 46066 269865 1549846 8680947	$|[6]^{10}(\tau)| = 47240485$
13312	7756 46066 269880 1550224 8685684	
21132	7756 46066 269880 1550269 8687310	
12231	7756 46066 269880 1550281 8687574	
13122	7756 46066 269880 1550281 8687576	
12213	7756 46066 269880 1550293 8688164	
12332	7756 46066 269880 1550314 8688780	
12321	7756 46066 269880 1550326 8689188	
13231	7756 46066 269880 1550326 8689200	
12331	7756 46066 269880 1550338 8689424	
13322	7756 46066 269910 1551280 8705626	
11223,11232 11322	7756 46066 269910 1551322 8707026	$|[5]^{11}(\tau)| = 32695299$
11332	7756 46066 269910 1551322 8707026	$|[5]^{11}(\tau)| = 32695309$
11323	7756 46066 269910 1551322 8707110	
11233	7756 46066 269910 1551406 8709812	
12112	7761 46191 271601 1566981 8808636	
11212	7761 46191 271601 1567021 8810056	$|[4]^{10}(\tau)| = 630518$
12121	7761 46191 271601 1567021 8810056	$|[4]^{10}(\tau)| = 630510$
12221	7761 46191 271601 1567061 8811456	
11221	7761 46191 271641 1568381 8835086	
11122	7761 46191 271641 1568501 8839406	
13112	7756 46096 270720 1563457 8839536	
11312	7756 46096 270720 1563478 8840287	
12131	7756 46096 270720 1563508 8841205	
12223,12232 12322,13222		

τ	The sequence $\{|AW_{[6]}^{\tau}(n)|\}_{n=5}^{9}$	$|AW_{[k]}^{\tau}(n)|$
21232,21322	7756 46096 270720 1563556 8842756	
13332	7756 46096 270720 1563577 8843421	
12113	7756 46096 270720 1563577 8843505	
11213	7756 46096 270720 1563598 8844100	
11231	7756 46096 270741 1564146 8852076	
11321	7756 46096 270741 1564161 8852553	
11123,11132	7756 46096 270741 1564221 8854416	
11112,11121 11211	7761 46231 272831 1588265 9080246	
24213	7761 46227 272775 1588571 9102298	
21413	7761 46227 272775 1588594 9102967	
23413	7761 46227 272775 1588606 9103392	
13142	7761 46227 272775 1588612 9103499	
14213	7761 46227 272775 1588621 9103889	
21423	7761 46227 272775 1588626 9104029	
12423	7761 46227 272775 1588626 9104030	
13242	7761 46227 272775 1588626 9104042	
14243	7761 46227 272775 1588626 9104049	
12413	7761 46227 272775 1588630 9104200	
13124	7761 46227 272775 1588644 9104648	
14312	7761 46227 272782 1588798 9106336	
13412	7761 46227 272782 1588802 9106427	
23142	7761 46227 272781 1588793 9106631	
14132	7761 46227 272782 1588812 9106775	
14123	7761 46227 272782 1588833 9107422	
21342	7761 46227 272782 1588843 9107811	
24113	7761 46227 272787 1588964 9109294	
22413	7761 46227 272787 1588979 9109802	
14223	7761 46227 272787 1588985 9110008	
13342	7761 46227 272787 1588985 9110048	
13442	7761 46227 272787 1589003 9110617	
14233	7761 46227 272787 1589003 9110623	
14332	7761 46227 272787 1589006 9110734	
14232	7761 46227 272787 1589006 9110739	
13432	7761 46227 272787 1589006 9110755	
14323	7761 46227 272787 1589006 9110764	
13423	7761 46227 272787 1589006 9110775	
21432	7761 46227 272788 1589026 9110925	
12342	7761 46227 272788 1589027 9110961	
14342	7761 46227 272787 1589013 9110997	
13224	7761 46227 272787 1589013 9111007	
13241	7761 46227 272789 1589046 9111087	
12431	7761 46227 272789 1589050 9111199	
12314	7761 46227 272788 1589034 9111208	
13243	7761 46227 272787 1589020 9111228	
12324	7761 46227 272787 1589020 9111229	
12432	7761 46227 272788 1589041 9111413	
13214	7761 46227 272788 1589048 9111684	
13422	7761 46227 272794 1589198 9113691	
14423	7761 46227 272794 1589205 9113892	
14322	7761 46227 272794 1589219 9114366	
12341	7761 46227 272795 1589241 9114562	
14432	7761 46227 272794 1589226 9114595	
12234,12243 12343,12433 21243,21433	7761 46227 272794 1589226 9114597	$|[6]^{10}(\tau)|=51363833$

τ	The sequence $\{|AW_{[6]}^{\tau}(n)|\}_{n=5}^{9}$	$|AW_{[k]}^{\tau}(n)|$
12443,21143	7761 46227 272794 1589226 9114597	$|[6]^{10}(\tau)| = 51363835$
12134,12143	7761 46227 272794 1589233 9114853	
11423	7761 46227 272801 1589429 9117845	
11342	7761 46227 272801 1589429 9117873	
11234,11243 11432	7761 46227 272801 1589457 9118797	
11324	7761 46227 272801 1589457 9118829	
11111	7770 46470 276570 1633170 9536520	
24153	7770 46480 276920 1640192 9642494	
25314	7770 46480 276920 1640195 9642602	
24513	7770 46480 276921 1640225 9643100	
25413	7770 46480 276921 1640227 9643168	
14253	7770 46480 276921 1640228 9643213	
23514	7770 46480 276921 1640228 9643215	
13524	7770 46480 276921 1640232 9643365	
25143	7770 46480 276922 1640262 9643856	
13542	7770 46480 276922 1640264 9643933	
15324	7770 46480 276922 1640264 9643936	
15243	7770 46480 276922 1640265 9643969	
14352	7770 46480 276922 1640265 9643973	
13425	7770 46480 276922 1640267 9644046	
14523	7770 46480 276923 1640296 9644505	
13452	7770 46480 276923 1640297 9644543	
15423	7770 46480 276923 1640297 9644545	
15234	7770 46480 276923 1640298 9644578	
12534,21534	7770 46480 276923 1640298 9644582	
15342	7770 46480 276923 1640298 9644583	
12453,21453	7770 46480 276923 1640299 9644620	
14532	7770 46480 276923 1640300 9644655	
12345,12354 12543,15432 21354,21543	7770 46480 276924 1640334 9645304	
12435,13254	7770 46480 276924 1640335 9645343	
14325	7770 46480 276924 1640335 9645344	

G.5 Programs TOU_AUTO and TOU_FORMULA

We now present the programs that were mentioned in Subsection 7.3.1. The C++ program TOU_AUTO has as input the alphabet and the subsequence pattern to be avoided, and returns the states of the automaton for pattern avoidance in words and the associated transfer matrix. The Maple program TOU_FORMULA takes as input the transfer matrix from the C++ program and returns the first terms of the sequence $\{AW_{[k]}^{\tau}(n)\}$, as well as an explicit formula for $AW_{[k]}^{\tau}(n)$.

G.5.1   C++ program TOU_AUTO

```c
#include <stdio.h>

/* maximal number of states of the automaton */
#define MAXCLASSES 2000

/* MAXBIN=max(# of states, k+1) */
#define MAXBIN 2000

/* maximal length of state name  */
#define MAXLCLASS 20

/* each pattern consists of exactly one row and at most 20 columns */
#define MAXROWS 1
#define MAXCOLS 20

/* number of letters k in the alphabet */
int NUMBERLETTER;

/* list of all the possible extensions for a state needed to  */
/* check equivalence (see  Lemma 7.27) */
int mat[MAXBIN][MAXROWS][MAXCOLS];

/* length of the extensions in mat[MAXBIN] */
int lmat[MAXBIN];

/* number of all the possible extensions listed in mat[MAXBIN]*/
int nallmat;

int pattern[MAXROWS][MAXCOLS];    /* pattern to be avoided */
int numbercolumn,numberrow,lenpattern;   /* lenpattern=3,4,5,6*/

/* three different states and their respective length */
int state0[MAXROWS][MAXLCLASS], lstate0;
int state1[MAXROWS][MAXLCLASS], lstate1;
int state2[MAXROWS][MAXLCLASS], lstate2;

/* the transfer matrix Au(p,k) and its dimension */
int Au[MAXCLASSES][MAXCLASSES], sizeAu=0;

/* the set E(p,k) of equivalence classes for a given pattern p */
int setclass[MAXCLASSES][MAXROWS][MAXLCLASS];

/* length of  a state in E(p,k) */
int lsetclass[MAXCLASSES];

/* cardinality of |E(p,k)| */
int nclass;
```

```
/* output file */
FILE *fout;

/* builds all the possible extensions for a state for the given  */
/* alphabet, that is, the list of words r in Lemma 7.27 */
void extendbymat()
{
 int nc,i,j,i1,j1,ii,nu;
 nallmat=0;
 for(nc=1;nc<1+numbercolumn;nc++)
 {
  nu=1;
  for(i=0;i<nc*(numberrow);i++) nu=nu*NUMBERLETTER;
  lmat[nallmat]=nc;
  for(i=0;i<numberrow;i++)
    for(j=0;j<nc;j++) mat[nallmat][i][j]=0;
  nallmat++;
  for(ii=0;ii<nu-1;ii++)
  {
   for(i=0;i<numberrow;i++)
     for(j=0;j<nc;j++)
       mat[nallmat][i][j]=mat[nallmat-1][i][j];
   i=0; j=0;
   while(mat[nallmat][i][j]==(NUMBERLETTER-1))
     {i++; if(i==numberrow) {j++; i=0;} }
   mat[nallmat][i][j]++;
   for(j1=0;j1<j;j1++)
     for(i1=0;i1<numberrow;i1++) mat[nallmat][i1][j1]=0;
   for(i1=0;i1<i;i1++)
     mat[nallmat][i1][j]=0;
   lmat[nallmat]=nc; nallmat++;
  }
 }
}


int ordereq(int a, int b)
{ if(a<b) return(1); else if(a>b) return(-1); else return(0); }

/* check if two sequences of length 3 are order-isomorphic*/
int equiv3(int a, int b, int c,
           int a1, int b1, int c1)
{
    int i,v[3],u[3];
    v[0]=ordereq(a,b);  v[1]=ordereq(a,c);  v[2]=ordereq(b,c);
    u[0]=ordereq(a1,b1);u[1]=ordereq(a1,c1);u[2]=ordereq(b1,c1);
    for(i=0;i<3;i++)
      if(v[i]!=u[i]) return(0);
    return(1);
}
```

```
/* check if two sequences of length 4 are order-isomorphic*/
int equiv4(int a, int b, int c, int d,
           int a1, int b1, int c1, int d1)
{
    int i,v[6],u[6];
    v[0]=ordereq(a,b);   v[1]=ordereq(a,c);   v[2]=ordereq(a,d);
    v[3]=ordereq(b,c);   v[4]=ordereq(b,d);   v[5]=ordereq(c,d);
    u[0]=ordereq(a1,b1);u[1]=ordereq(a1,c1);u[2]=ordereq(a1,d1);
    u[3]=ordereq(b1,c1);u[4]=ordereq(b1,d1);u[5]=ordereq(c1,d1);
    for(i=0;i<6;i++)
      if(v[i]!=u[i]) return(0);
    return(1);
}


/* check if two sequences of length 5 are order-isomorphic*/
int equiv5(int a, int b, int c, int d, int e,
           int a1, int b1, int c1, int d1,int e1)
{
    int i,v[10],u[10];
    v[0]=ordereq(a,b);   v[1]=ordereq(a,c);   v[2]=ordereq(a,d);
    v[3]=ordereq(a,e);   v[4]=ordereq(b,c);   v[5]=ordereq(b,d);
    v[6]=ordereq(b,e);   v[7]=ordereq(c,d);   v[8]=ordereq(c,e);
    v[9]=ordereq(d,e);
    u[0]=ordereq(a1,b1);u[1]=ordereq(a1,c1);u[2]=ordereq(a1,d1);
    u[3]=ordereq(a1,e1);u[4]=ordereq(b1,c1);u[5]=ordereq(b1,d1);
    u[6]=ordereq(b1,e1);u[7]=ordereq(c1,d1);u[8]=ordereq(c1,e1);
    u[9]=ordereq(d1,e1);
    for(i=0;i<10;i++)
      if(v[i]!=u[i]) return(0);
    return(1);
}


/* check if two sequences of length 6 are order-isomorphic*/
int equiv6(int a, int b, int c, int d, int e,int f,
           int a1, int b1, int c1, int d1,int e1,int f1)
{
    int i,v[10],u[10];
    v[0]=ordereq(a,b);   v[1]=ordereq(a,c);   v[2]=ordereq(a,d);
    v[3]=ordereq(a,e);   v[4]=ordereq(a,f);   v[5]=ordereq(b,c);
    v[6]=ordereq(b,d);   v[7]=ordereq(b,e);   v[8]=ordereq(b,f);
    v[9]=ordereq(c,d);   v[10]=ordereq(c,e);  v[11]=ordereq(c,f);
    v[12]=ordereq(d,e);  v[13]=ordereq(d,f);  v[14]=ordereq(e,f);
    u[0]=ordereq(a1,b1);  u[1]=ordereq(a1,c1);  u[2]=ordereq(a1,d1);
    u[3]=ordereq(a1,e1);  u[4]=ordereq(a1,f1);  u[5]=ordereq(b1,c1);
    u[6]=ordereq(b1,d1);  u[7]=ordereq(b1,e1);  u[8]=ordereq(b1,f1);
    u[9]=ordereq(c1,d1);  u[10]=ordereq(c1,e1);u[11]=ordereq(c1,f1);
    u[12]=ordereq(d1,e1);u[13]=ordereq(d1,f1);u[14]=ordereq(e1,f1);
    for(i=0;i<15;i++)
```

```
      if(v[i]!=u[i]) return(0);
    return(1);
}

/* check if state0 avoids pattern of length 3 */
int avoid3()
{
 int i1,i2,i3;
 for(i1=0;i1<lstate0-2;i1++)
 for(i2=i1+1;i2<lstate0-1;i2++)
 for(i3=i2+1;i3<lstate0;i3++)
  if(equiv3(state0[0][i1],state0[0][i2],state0[0][i3],
    pattern[0][0],pattern[0][1],pattern[0][2])==1)
        return(0);
 return(1);
}

/* check if state0 avoids pattern of length 4 */
int avoid4()
{
 int i1,i2,i3,i4;
 for(i1=0;i1<lstate0-3;i1++)
 for(i2=i1+1;i2<lstate0-2;i2++)
 for(i3=i2+1;i3<lstate0-1;i3++)
 for(i4=i3+1;i4<lstate0;i4++)
  if(equiv4(state0[0][i1],state0[0][i2],state0[0][i3],state0[0][i4],
    pattern[0][0],pattern[0][1],pattern[0][2],pattern[0][3])==1)
     return(0);
 return(1);
}

/* check if state0 avoids pattern of length 5 */
int avoid5()
{
 int i1,i2,i3,i4,i5;
 for(i1=0;i1<lstate0-4;i1++)
 for(i2=i1+1;i2<lstate0-3;i2++)
 for(i3=i2+1;i3<lstate0-2;i3++)
 for(i4=i3+1;i4<lstate0-1;i4++)
 for(i5=i4+1;i5<lstate0;i5++)
  if(equiv5(state0[0][i1],state0[0][i2],state0[0][i3],
     state0[0][i4],state0[0][i5],pattern[0][0],pattern[0][1],
     pattern[0][2],pattern[0][3],pattern[0][4])==1)
        return(0);
 return(1);
}

/* check if state0 avoids pattern of length 6 */
int avoid6()
```

```
{
 int i1,i2,i3,i4,i5,i6;
 for(i1=0;i1<lstate0-5;i1++)
 for(i2=i1+1;i2<lstate0-4;i2++)
 for(i3=i2+1;i3<lstate0-3;i3++)
 for(i4=i3+1;i4<lstate0-2;i4++)
 for(i5=i4+1;i5<lstate0-1;i5++)
 for(i6=i5+1;i6<lstate0;i6++)
   if(equiv6(state0[0][i1],state0[0][i2],state0[0][i3],
     state0[0][i4],state0[0][i5],state0[0][i6],pattern[0][0],
     pattern[0][1],pattern[0][2],pattern[0][3],pattern[0][4],
     pattern[0][5])==1) return(0);
 return(1);
}


/* check if a pattern of length lenpattern is avoided */
int avoiding()
{
 if (lenpattern==3) return(avoid3());
 if (lenpattern==4) return(avoid4());
 if (lenpattern==5) return(avoid5());
 if (lenpattern==6) return(avoid6());
 return(-1);
}
/* check if state1 and state2 are equivalent */
int equivstates()
{
 int i,j,i1,j1,ans1,ans2;
 int statea[MAXROWS][MAXLCLASS], lstatea,
       stateb[MAXROWS][MAXLCLASS], lstateb;

 lstate0=lstate1; lstatea=lstate1;
 for(i=0;i<numberrow;i++)
   for(j=0;j<lstate1;j++)
   {
     state0[i][j]=state1[i][j];
     statea[i][j]=state1[i][j];
   }
 ans1=avoiding();      /* check if state1 avoids pattern */
 lstate0=lstate2; lstateb=lstate2;
 for(i=0;i<numberrow;i++)
   for(j=0;j<lstate2;j++)
   {
     state0[i][j]=state2[i][j];
     stateb[i][j]=state2[i][j];
   }
 ans2=avoiding();      /* check if state2 avoids pattern */
 if(ans1!=ans2) return(0);   /* not equivalent since one state */
     /* avoids p and the other does not */
```

```
for(j=0;j<nallmat;j++)
{
 lstatea=lstate1; lstateb=lstate2;
 for(i1=0;i1<numberrow;i1++)
   for(j1=0;j1<lmat[j];j1++)
   {
     statea[i1][j1+lstatea]=mat[j][i1][j1];
     stateb[i1][j1+lstateb]=mat[j][i1][j1];
   }
 lstatea=lstate1+lmat[j];
 lstateb=lstate2+lmat[j];
 lstate0=lstatea;
 for(i1=0;i1<numberrow;i1++)
   for(j1=0;j1<lstatea;j1++) state0[i1][j1]=statea[i1][j1];
 ans1=avoiding();
 lstate0=lstateb;
 for(i1=0;i1<numberrow;i1++)
   for(j1=0;j1<lstateb;j1++) state0[i1][j1]=stateb[i1][j1];
 ans2=avoiding();
 if(ans1!=ans2) return(0);
 }
 return(1);
}

/* check if there is a state in E(p,k) that is  */
/* equivalent to an already existing state (state1) */
int checkequivstates()
{
 int i,i1,j1,ans;
 for(i=0;i<nclass;i++)
 {
  lstate2=lsetclass[i];
  for(i1=0;i1<numberrow;i1++)
    for(j1=0;j1<lstate2;j1++) state2[i1][j1]=setclass[i][i1][j1];
  ans=equivstates();
  if (ans==1) return(i);
 }
 return(nclass);
}

/* transfer matrix Au(p,k) and  set of  classes in E(p,k) */
void automat()
{
 int i,j,flag=1,nst=1,
     i1,j1,ans,pns,
     ans1,numvec;

 numvec=1;
 for(i=0;i<numberrow;i++) numvec=numvec*NUMBERLETTER;
```

```
/* first state, the empty state */

lsetclass[0]=0;   nclass=1;
printf("%d) empty state\n",nclass);
fprintf(fout,"%d) empty state\n",nclass);
sizeAu=1;  Au[0][0]=0;
pns=1;

/* find another state, if there is one */

while(flag==1)
{
flag=0;
for(i=nst-1;i<pns;i++)
{
 lstate1=lsetclass[i];     /* take old state */
 for(i1=0;i1<numberrow;i1++)
   for(j1=0;j1<lstate1;j1++)
     state1[i1][j1]=setclass[i][i1][j1];

 lstate1++;  /* extend the state in all possible ways */
 for(j=0;j<numvec;j++)
 {
  for(i1=0;i1<numberrow;i1++)
    state1[i1][lstate1-1]=mat[j][i1][0];

  ans=checkequivstates();    /* check after the extension if */
                             /* there is an old state equivalent to it */

  if (ans==nclass)  /* if not */
  {
   lstate0=lstate1;  /*check if state is avoiding the pattern */
   for(i1=0;i1<numberrow;i1++)
     for(j1=0;j1<lstate0;j1++)
       state0[i1][j1]=state1[i1][j1];
   ans1=avoiding();

   if(ans1==1)       /* if the new state avoids the pattern */
   {
    /* put it as a new state in E(p,k) */
    /* and extend the transfer matrix */
    flag=1;

    printf("%d) ",nclass+1);
    fprintf(fout,"%d) ",nclass+1);
    for(i1=0;i1<numberrow;i1++)
    {
        for(j1=0;j1<lstate1;j1++)
```

```
{
  setclass[nclass][i1][j1]=state1[i1][j1];
  printf("%d",state1[i1][j1]+1);
  fprintf(fout,"%d",state1[i1][j1]+1);
}
        printf(" , ");
        fprintf(fout," , ");
    }
    printf("\n");
    fprintf(fout,"\n");
    lsetclass[nclass]=lstate1;nclass++;
    for(i1=0;i1<sizeAu+1;i1++)
 {
   Au[sizeAu][i1]=0;
   Au[i1][sizeAu]=0;
 }
    Au[i][sizeAu]++; sizeAu++;
   }
   }
   else  Au[i][ans]++;
 }
}
nst=pns+1;
pns=nclass;
 }
}

/* print the transfer matrix Au(p,k) */
void printautomaton()
{
 int i,j;
 printf("\n\nThe matrix is given by
        (maple form)\n\nAu:=matrix([[");
 fprintf(fout,"\n\nThe matrix is given by
                (maple form)\n\nAu:=matrix([[");
 for(i=0;i<sizeAu-1;i++)
 {
  for(j=0;j<sizeAu-1;j++)
  {
  printf("%d,",Au[i][j]);
  fprintf(fout,"%d,",Au[i][j]);
  }
  printf("%d],\n[",Au[i][sizeAu-1]);
  fprintf(fout,"%d],\n[",Au[i][sizeAu-1]);
 }
 for(j=0;j<sizeAu-1;j++)
 {
 printf("%d,",Au[sizeAu-1][j]);
 fprintf(fout,"%d,",Au[sizeAu-1][j]);
```

```
  }
  printf("%d]]);\n",Au[sizeAu-1][sizeAu-1]);
  fprintf(fout,"%d]]);\n",Au[sizeAu-1][sizeAu-1]);
}

/* main procedure */
void main()
{
  int i;
  printf("Enter the number k of letters in your alphabet: ");
  scanf("%d",&NUMBERLETTER);
  numberrow=1;
  printf("Enter the length of the pattern (3-6): ");
  scanf("%d",&lenpattern);
  numbercolumn=lenpattern;
  printf("Enter the pattern (space between the letters): ");
  for(i=0;i<lenpattern;i++)
scanf("%d",&pattern[0][i]);

  fout=fopen("data.dat","w");
  printf("The automaton for the number
          of %d-ary words of length n\n",NUMBERLETTER);
  printf("that avoid the pattern ");
  fprintf(fout,"The automaton for the number
                of %d-ary words of length n\n",NUMBERLETTER);
  fprintf(fout,"that avoid the pattern ");
  for(i=0;i<lenpattern;i++)
  {
printf("%d ",pattern[0][i]);
    fprintf(fout,"%d ",pattern[0][i]);
  }
  printf(" is given by:\n\n");
  fprintf(fout," is given by:\n\n");
  printf("The states are \n");
  fprintf(fout,"The states are \n");
  extendbymat();
  nclass=0; sizeAu=0; automat();
  printautomaton();
  fclose(fout);
}
```

We show how to use the program for the pattern 112.

Example G.2 *To compute the automaton for avoidance of the subsequence pattern 112 in 3-ary words using the program TOU_AUTO, the user inputs the alphabet, the length of the pattern, and the pattern:*

```
Enter the number k of letters in your alphabet: 3
Enter the length of the pattern (3-6): 3
Enter the pattern (space between the letters): 1 1 2
```

This results in the output

```
The automaton for the number of 3-ary words of length n
that avoid the pattern 1 1 2  is given by:

The states are
1) empty state
2) 1 ,
3) 2 ,
4) 11 ,
5) 12 ,
6) 22 ,
7) 122 ,


The matrix is given by (maple form)

Au:=matrix([[1,1,1,0,0,0,0],
[0,1,0,1,1,0,0],
[0,0,1,0,1,1,0],
[0,0,0,1,0,0,0],
[0,0,0,1,1,0,1],
[0,0,0,0,0,1,1],
[0,0,0,1,0,0,1]]);
Press any key to continue
```

G.5.2 Maple™ program TOU_FORMULA

The core of this program is the implementation of the formula given in Theorem 7.34. Using partial fraction decomposition and geometric series allows us to obtain an explicit formula for $AW_{[k]}^\tau(n)$. The program has as its output this explicit formula and the values of $AW_{[k]}^\tau(n)$ for $n = 0, \ldots, 12$.

```
formula:=proc()
local S,Au,sizeAu,J,i,j,ii,jj,nn,Values,A,size,B,b,w,form,M:
Au:=args[1]: with(linalg):
J:=factor(minpoly(Au,x)):
S:=roots(J): sizeAu:=rowdim(Au): nn:=nops(S):
Values:={}: size:=0:
for i from 1 to nn do
   Values:=Values union {S[i][1]}: size:=size+S[i][2]:
od:
A:=matrix(size,size,[]): w:=linalg[vector](sizeAu,[]):
b:=vector(size,[]): B:=linalg[vector](size,[]):
for i from 1 to sizeAu do w[i]:=1: od:
b[1]:=1: B[2]:=evalm(Au): b[2]:=evalm(B[2]&*w)[1]:
for i from 2 to size-1 do
   B[i+1]:=evalm(B[i]&*Au): b[i+1]:=evalm(B[i+1]&*w)[1]:
od:
```

```
for i from 1 to size do jj:=0:
    for ii from 1 to nops(Values) do
        for j from 0 to S[ii][2]-1 do
            jj:=jj+1:
            if i-1>=j then
                A[i,jj]:=(S[ii][1])^(i-1-j)*binomial(i-1,j):
            else
                A[i,jj]:=0:
            fi:
od: od: od:
w:=linsolve(A,b);
printf("First values n=0...12");
M:=vector(13,[]):
for i from 0 to 12 do
    form:=0: jj:=0:
    for ii from 1 to nops(Values) do
        for j from 0 to S[ii][2]-1 do
            jj:=jj+1:
            form:=form+w[jj]*(S[ii][1])^(i-j)*binomial(i,j):
    od: od:
    M[i+1]:=form:
od:
print(M):
printf("The formula is");
form:=0: jj:=0:
for ii from 1 to nops(Values) do
    for j from 0 to S[ii][2]-1 do
        jj:=jj+1:
        form:=form+w[jj]*(S[ii][1])^(n-j)*binomial(n,j):
od: od:
print(form):
end:
```

Using the output of the program TOU_AUTO as input for the program TOU_FORMULA allows us to compute the values $\{AW_{[3]}^{112}(n)\}_{n=0}^{12}$ and an explicit formula for $AW_{[3]}^{112}(n)$.

Example G.3 (*Continutaion of Example G.2*) *The transfer matrix for the pattern 112 is given by*

```
Au:=matrix([[1,1,1,0,0,0,0],
[0,1,0,1,1,0,0],
[0,0,1,0,1,1,0],
[0,0,0,1,0,0,0],
[0,0,0,1,1,0,1],
[0,0,0,0,0,1,1],
[0,0,0,1,0,0,1]]);
```

This matrix is in the form needed for the argument of the Maple program. Using

```
formula(Au)
```

results in the following output:

```
First values n=0...12
    [1, 3, 9, 24, 56, 116, 218, 379, 619, 961, 1431, 2058, 2874]
The formula is
    1 + 2 n + 4 binomial(n, 2) + 5 binomial(n, 3) + 3 binomial(n, 4)
```

Appendix H

Notation

The table below lists the notation. The first part of the table contains sets, structures and equivalences, followed by an alphabetical listing of named quantities. Greek letters occur according to their sound. We use "gf" as an abbreviation for generating function. Note that in rare instances we use the same symbol for two different objects. This is a result of following the common notation in the literature. In each case, the two different notations do not occur in the same chapter.

Notation	Definition
$(a;q)_n$	$\prod_{j=0}^{n-1}(1 - aq^j)$
$[a_0, a_1, \ldots]$	infinite simple continued fraction
$[a_0, a_1, \ldots, a_k]$	finite simple continued fraction
$[d]$	set of first d integers, $\{1, 2, \ldots, d\}$
$[k]^*$	set of all finite words on the alphabet $[k]$
$[i, j]$	$\{i, i+1, \ldots, j\}$
$(j_1, j_2, \ldots, j_\ell)$	type of generalized pattern
k^j	shorthand for j letters $k \cdots k$
$[x^n]$	coefficient of x^n
$\langle x \rangle$	equivalence class of x for the automaton $\mathcal{A}$
$a_n \approx b_n$	a_n and b_n are of the same order
$a_n(k) \overset{d}{\approx} \mathcal{N}(\mu_n, \sigma_n^2)$	a_n is approximately normally distributed
$K \lesssim K'$	block K is to the left of block K'
$K \gtrsim K'$	block K is to the right of block K'
$\tau \sim \nu$	τ and ν are Wilf-equivalent
$\tau \overset{s}{\sim} \nu$	τ and ν are strongly equivalent
$\tau \overset{st}{\sim} \nu$	τ and ν are strongly tight Wilf-equivalent
$\tau \overset{t}{\sim} \nu$	τ and ν are tight Wilf-equivalent
$v \overset{\mathcal{A}}{\sim} w$	v and w are equivalent on $[k]^*$ with regard to avoidance of τ
$v \sim_T w$	v and w are equivalent on $[k]^*$ with regard to avoidance of T

Notation	Definition
$A \overset{\text{ops}}{\leftrightarrow} \{a_n\}_{n \geq 0}$	A is the ordinary gf of $\{a_n\}_{n \geq 0}$
$E \overset{\text{egf}}{\leftrightarrow} \{a_n\}_{n \geq 0}$	E is the exponential gf of $\{a_n\}_{n \geq 0}$
$\mathcal{A}$	finite automaton
$\mathcal{A}(\tau, k)$	automaton for avoidance of pattern τ for words on alphabet $[k]$
$A(\tau, k)$	adjacency matrix of $\mathcal{A}(\tau, k)$
$A(x), A(\vec{x})$	ordinary gf
$A(\vec{x}, \vec{y})$	ordinary gf with respect to order and statistic
A_k	$\{a_1, \ldots, a_k\}$
$\bar{A}_k$	$\{a_{k+1}, a_{k+2}, \ldots, a_d\}$
$\mathcal{AC}_n^A(\tau)$	set of compositions of n with parts in A that avoid the pattern τ
$AC_A^\tau(n)$	$|\mathcal{AC}_n^A(\tau)|$
$AC_A^\tau(x)$	gf for $AC_A^\tau(n)$
$AC_A^\tau(\sigma_1 \cdots \sigma_\ell | n)$	number of compositions in $\mathcal{AC}_n^A(\tau)$ that start with $\sigma_1 \cdots \sigma_\ell$
$AC_A^\tau(\sigma_1 \cdots \sigma_\ell | x)$	gf for $AC_A^\tau(\sigma_1 \cdots \sigma_\ell | n)$
$\mathcal{AC}_{n,m}^A(\tau)$	set of compositions of n with m parts in A that avoid the pattern τ
$AC_A^\tau(m; n)$	$|\mathcal{AC}_{n,m}^A(\tau)|$
$AC_A^\tau(x, y)$	gf for $AC_A^\tau(m; n)$
$AC_A^\tau(\sigma_1 \cdots \sigma_\ell | m; n)$	number of compositions in $\mathcal{AC}_{n,m}^A(\tau)$ that start with $\sigma_1 \cdots \sigma_\ell$
$AC_A^\tau(\sigma_1 \cdots \sigma_\ell | x, y)$	gf for $AC_A^\tau(\sigma_1 \cdots \sigma_\ell | m; n)$
$\text{adj}(A)$	adjoint of matrix A
$AT_\kappa^B(\tau)$	set of compositions in $\mathcal{T}_\kappa^B$ that avoid τ
$\mathcal{AW}^{[k]^*}(\tau)$	set of words in $[k]^*$ that avoid τ
$\mathcal{AW}_n^{[k]}(\tau)$	set of k-ary words of length n that avoid τ
$AW_{[k]}^\tau(n)$	$|\mathcal{AW}_n^{[k]}(\tau)|$
$AW_{[k]}^\tau(x)$	gf for $AW_{[k]}^\tau(n)$
$AW_{[k]}^\tau(w_1 \cdots w_\ell | x)$	gf for words in $AW_{[k]}^\tau(n)$ that start with $w_1 \cdots w_\ell$
$AW_\tau(x, y)$	gf for $AW_{[k]}^\tau(x)$
$\widetilde{\mathcal{AW}}_n^{[k]}(T)$	set of reduced k-ary words of length n that avoid T
$\widetilde{AW}_{[k]}^T(n)$	$|\widetilde{\mathcal{AW}}_n^{[k]}(T)|$
$B(\sigma)$	set of parts of σ

Notation	Definition
$\mathcal{B}(n,p)$	Binomial random variable with parameters n, p
$\mathbb{C}$	set of complex numbers
$c(\sigma)$	complement of σ with respect to A
$C(n)$	number of compositions of n in $\mathbb{N}$
$C(x)$	gf for $C(n)$
$C(m;n)$	number of compositions of n with m parts in $\mathbb{N}$
$C(m;x)$	gf for $C(m;n)$
$\tilde{C}(x)$	gf for number of nonempty compositions
$\bar{C}(n)$	number of n-color compositions of n with parts in $\mathbb{N}$
$\bar{C}(x)$	gf for $\bar{C}(n)$
$\bar{C}(m;n)$	number of n-color compositions of n with m parts in $\mathbb{N}$
$\bar{C}(m;x)$	gf for $\bar{C}(m;n)$
$\mathcal{C}^A$	set of compositions (of any order) with parts in A
$\mathcal{C}_n^A$	set of compositions of n with parts in A
$C_A(n)$	$|\mathcal{C}_n^A|$
$C_A(x)$	gf for $C_A(n)$
$\mathcal{C}_{n,m}^A$	set of compositions of n with m parts in A
$C_A(m;n)$	$|\mathcal{C}_{n,m}^A|$
$C_A(m;x)$	gf for $C_A(m;n)$
$\mathcal{C}_n^A(\tau;r)$	set of compositions of n with parts in A that contain τ exactly r times
$C_A^\tau(n,r)$	$|\mathcal{C}_n^A(\tau;r)|$
$C_A^\tau(x,q)$	gf for $C_A^\tau(n,r)$
$C_A^\tau(\sigma_1 \cdots \sigma_\ell|n,r)$	number of compositions in $\mathcal{C}_n^A(\tau;r)$ that start with $\sigma_1 \cdots \sigma_\ell$
$C_A^\tau(\sigma_1 \cdots \sigma_\ell|x,q)$	gf for $C_A^\tau(\sigma_1 \cdots \sigma_\ell|n,r)$
$\mathcal{C}_{n,m}^A(\tau;r)$	set of compositions of n with m parts in A that contain τ exactly r times
$C_A^\tau(m;n,r)$	$|\mathcal{C}_{n,m}^A(\tau;r)|$
$C_A^\tau(x,y,q)$	gf for $C_A^\tau(m;n,r)$
$C_A^\tau(\sigma_1 \cdots \sigma_\ell|m;n,r)$	number of compositions in $\mathcal{C}_{n,m}^A(\tau;r)$ that start with $\sigma_1 \cdots \sigma_\ell$
$C_A^\tau(\sigma_1 \cdots \sigma_\ell|x,y,q)$	gf for $C_A^\tau(\sigma_1 \cdots \sigma_\ell|m;n,r)$
$C_A(x,y,r,\ell,d)$	gf for the number of compositions of n with parts in A according to order and statistic $\overrightarrow{s}_\sigma$

Notation	Definition
$C_A(\sigma_1 \cdots \sigma_\ell \lvert x, y, r, \ell, d)$	gf for the number of compositions of n with parts in A according to order and statistic $\overrightarrow{s}_\sigma$ that start with $\sigma_1 \cdots \sigma_\ell$
$C_m(x)$	gf for number of compositions with at most m cracks per part
$C_{n,k}$	number of k-element combinations of $[n]$
$C\!C_A(n)$	number of Carlitz compositions of n in A
$C\!C_A(x)$	gf for $C\!C_A(n)$
$C\!C_A(\sigma_1 \cdots \sigma_\ell \lvert x)$	gf for the number of Carlitz compositions that start with $\sigma_1 \cdots \sigma_\ell$
$C\!C_A(m; n)$	number of Carlitz compositions of n with m parts in A
$C\!C_A(x, y)$	gf for $C\!C_A(m; n)$
$C\!P_A(n)$	number of palindromic Carlitz compositions of n in A
$C\!P_A(x)$	gf for $C\!P_A(n)$
$C\!P_A(\sigma_1 \cdots \sigma_\ell \lvert x)$	gf for the number of palindromic Carlitz compositions that start with $\sigma_1 \cdots \sigma_\ell$
$\overline{C\!P}(m; n)$	number of n-color palindromic compositions of n with m parts in $\mathbb{N}$
D	differential operator
$\mathrm{dro}(\sigma)$	number of drops in σ
$D_A^\tau(x, y, q)$	$\sum_{\sigma \in \mathcal{C}_n^A} x^{\mathrm{ord}(\sigma)} y^{\mathrm{par}(\sigma)} q^{\mathrm{occ}_\tau(\hat{\sigma})}$, $\hat{\sigma} = a\sigma$ or $\hat{\sigma} = \sigma b$
$\deg(i)$	degree of vertex i
$\deg^+(i)$	out-degree of vertex i
$\deg^-(i)$	in-degree of vertex i
$\det(A)$	determinant of matrix A
$\Delta(x)$	characteristic polynomial
$\Delta(x)$	transition function of automaton
$\mathcal{E}$	set of states for automaton
$\mathcal{E}(\tau, k)$	set of states for automaton for avoidance of pattern τ for words on alphabet $[k]$
$\vec{e}_i$	i-th unit vector
$E(x), E(\vec{x})$	exponential gf
$E(\vec{x}, \vec{y})$	exponential gf according to order and statistic $\overrightarrow{s}_\sigma$
$\mathbb{E}(X)$	expected value or mean of random variable X
$\langle \varepsilon \rangle$	equivalence class of empty word in automaton
f_n	shifted Fibonacci sequence

Notation	Definition
F_n	Fibonacci sequence
$F(x)$	gf of the Fibonacci sequence
$\mathcal{F}(T)$	set of sequences in $\mathcal{F}$ that avoid patterns from set T
G	graph
$J(P)$	set of order ideals of P
$\kappa(\sigma)$	core of σ
ξ	partially ordered generalized pattern
$\mathbb{L}[\overrightarrow{x}]$	set of polynomials in $\overrightarrow{x}$ with coefficients in $\mathbb{L}$
$\mathbb{L}[[\overrightarrow{x}]]$	set of formal power series in $\overrightarrow{x}$
L_n	Lucas sequence
$l_A(x, y)$	gf for the number of rises in the (palindromic) compositions with parts in A
$\lambda_\tau(n)$	number of compositions that are order-isomorphic to τ
$\Lambda_\tau(x)$	gf for $\lambda_\tau(n)$
$\lambda_{\lambda_1, \lambda_2, \cdots, \lambda_k}$	poset of Young diagrams
$\mathrm{lev}(\sigma)$	number of levels in σ
μ	mean of random variable X
μ_τ	content vector of τ
$\mathcal{N}(\mu, \sigma^2)$	Normal random variable with parameters μ, σ^2
$\mathcal{N}(0, 1)$	standard normal random variable
$\mathbb{N}$	set of natural numbers $= \{1, 2, 3, \ldots\}$
$\mathbb{N}_0$	$\mathbb{N} \cup \{0\}$
$N(i)$	neighborhood of vertex i
$N^+(i)$	set of out-neighbors of vertex i
$N^-(i)$	set of in-neighbors of vertex i
$\mathrm{occ}_\tau(\sigma)$	number of occurrences of τ in σ
$\mathrm{occ}_\tau(w)$	number of occurrences of τ in w
$\mathrm{ord}(\sigma)$	order of σ
Ω	sample space
$\omega_\xi(n, \ell, s)$	number of occurrences of ξ among the compositions of n with $\ell + k$ parts such that the sum of the parts preceding the occurrence is s
$\Omega_\xi(x, y, u)$	gf for $\omega_\xi(n, \ell, s)$
$\mathbf{P}(E)$	probability that event E occurs
$P(n)$	number of palindromic compositions of n in $\mathbb{N}$
$P(x)$	gf for $P(n)$

Notation	Definition
$P(m; n)$	number of palindromic compositions of n with m parts in $\mathbb{N}$
$pg_X(u)$	probability generating function
$\mathcal{P}_n^A$	set of palindromic compositions of n with parts in A
$P_A(n)$	$\lvert \mathcal{P}_n^A \rvert$
$P_A(x)$	gf for $P_A(n)$
$\mathcal{P}_{n,m}^A$	set of palindromic compositions of n with m parts in A
$P_A(m; n)$	$\lvert \mathcal{P}_{n,m}^A \rvert$
$P_A(x, y)$	gf for $P_A(m; n)$
$P_A(x, y, r, \ell, d)$	gf for the number of palindromic compositions of n with parts in A according to order and statistic $\overrightarrow{s}_\sigma$
$P_A(\sigma_1 \cdots \sigma_\ell \vert x, y, r, \ell, d)$	gf for the number of palindromic compositions of n with parts in A according to order and statistic $\overrightarrow{s}_\sigma$ that start with $\sigma_1 \cdots \sigma_\ell$
$\mathrm{par}(\sigma)$	number of parts in σ
π	permutation
$\mathcal{PR}_{\ell,i,r}$	set of primitive segmented tableaux of size $[\ell] \times [i]$ with r different entries
$\mathrm{pr}(\ell, i, r)$	$\lvert \mathcal{PR}_{\ell,i,r} \rvert$
$QC_A^\tau(x, y)$	gf for the number of compositions in $\mathcal{C}_{n,m}^A$ that quasi-avoid τ
$r_A(x, y)$	gf for the number of rises in the (palindromic) compositions with parts in A
$\mathrm{ris}(\sigma)$	number of rises in σ
$\Re$	region in $\mathbb{C}$
$R(\tau), R(\sigma)$	reverse of τ, σ
$\mathrm{Res}[f(z); z = z_0]$	residue of $f(z)$ at $z = z_0$
ρ^*	smallest positive pole, with multiplicity one
σ	composition
σ^2	variance of random variable X
Σ	alphabet for automaton
S_n	set of permutations of $[n]$
$S_{n,k}$	number of k-element permutations of $[n]$
$\overrightarrow{s}_\sigma$	$(\mathrm{par}(\sigma), \mathrm{ris}(\sigma), \mathrm{lev}(\sigma), \mathrm{dro}(\sigma))$
τ	pattern
$\tau = \tau_1 - \tau_2 - \cdots - \tau_s$	multi-pattern

Notation	Definition
$\tau = \tau_0-a_1-\cdots-a_s-\tau_s$	shuffle-pattern
$\tau-\mathrm{nlap}(\sigma)$	maximum number of nonoverlapping occurrences of τ in σ
$\mathcal{T}$	partially ordered alphabet
T	tree
$\mathcal{T}(\tau)$	generating tree for permutations of $[n]$ avoiding the pattern τ
$\mathcal{T}(T)$	generating tree for permutations of $[n]$ avoiding patterns in T
T_n	number of tilings of $1 \times n$ board with squares and dominoes
$T(x)$	gf for T_n
T_κ^B	set of compositions with core κ and set of values B
$\mathrm{Var}(X)$	variance of random variable X
w	word
$\mathcal{W}_n^{[k]}(\tau;r)$	set of k-ary words of length n with r occurrences of τ
$W_{[k]}^\tau(n,r)$	$|\mathcal{W}_n^{[k]}(\tau;r)|$
$W_{[k]}^\tau(x,q)$	gf for $W_{[k]}^\tau(n,r)$
X	random variable
$X_{\mathrm{lev}}(n)$	number of levels in a random composition of n
$X_{\mathrm{par}}(n)$	number of parts in a random composition of n
$X_{\mathrm{ris}}(n)$	number of rises in a random composition of n
$\mathbb{Z}$	set of integers $= \{\ldots,-3,-2,-1,0,1,2,3,\ldots\}$

References

[1] M. Abramson. Enumeration of sequences by levels and rises. *Discrete Math.*, 12:101–112, 1975.

[2] M. Abramson. Restricted combinations and compositions. *Fibonacci Quart.*, 14(5):439–452, 1976.

[3] A.K. Agarwal. *n*-colour compositions. *Indian J. Pure Appl. Math.*, 31(11):1421–1427, 2000.

[4] A.K. Agarwal. An analogue of Euler's identity and new combinatorial properties of *n*-colour compositions. *J. Comput. Appl. Math.*, 160:9–15, 2003.

[5] L.V. Ahlfors. *Complex Analysis*. McGraw-Hill Book Co., New York, third edition, 1978.

[6] M.H. Albert, R.E.L. Aldred, M.D. Atkinson, C. Handley, and D. Holton. Permutations of a multiset avoiding permutations of length 3. *European J. Combin.*, 22(8):1021–1031, 2001.

[7] K. Alladi and V.E. Hoggatt, Jr. Compositions with ones and twos. *Fibonacci Quart.*, 13(3):233–239, 1975.

[8] N. Alon and J.H. Spencer. *The Probabilistic Method*. Wiley-Interscience Series in Discrete Mathematics and Optimization. Wiley-Interscience [John Wiley & Sons], New York, second edition, 2000.

[9] G.E. Andrews. Ramanujan's "lost" notebook. IV. Stacks and alternating parity in partitions. *Adv. in Math.*, 53(1):55–74, 1984.

[10] G.E. Andrews. *The Theory of Partitions*. Cambridge Mathematical Library. Cambridge University Press, Cambridge, 1998.

[11] E. Babson and E. Steingrímsson. Generalized permutation patterns and a classification of the Mahonian statistics. *Sém. Lothar. Combin.*, 44:Art. B44b, 2000.

[12] J. Baik, P. Deift, and K. Johansson. On the distribution of the length of the longest increasing subsequence of random permutations. *J. Amer. Math. Soc.*, 12(4):1119–1178, 1999.

[13] C. Banderier, M. Bousquet-Mélou, A. Denise, P. Flajolet, D. Gardy, and D. Gouyou-Beauchamps. Generating functions for generating trees. *Discrete Math.*, 246(1-3):29–55, 2002.

[14] E. Barcucci, A. Del Lungo, E. Pergola, and R. Pinzani. A construction for enumerating k-coloured Motzkin paths. In *Computing and combinatorics (Xi'an, 1995)*, volume 959 of *Lecture Notes in Comput. Sci.*, pages 254–263. Springer, Berlin, 1995.

[15] E. Barcucci, A. Del Lungo, E. Pergola, and R. Pinzani. A methodology for plane tree enumeration. *Discrete Math.*, 180:45–64, 1998.

[16] E. Barcucci, A. Del Lungo, E. Pergola, and R. Pinzani. ECO: a methodology for the enumeration of combinatorial objects. *J. Differ. Equations Appl.*, 5(4-5):435–490, 1999.

[17] J.-L. Baril and P.-T. Do. ECO-generation for compositions and their restrictions. Proceedings of PP'08, 2008.

[18] E.A. Bender. Central and local limit theorems applied to asymptotic enumeration. *J. Combin. Theory Ser. A*, 15:91–111, 1973.

[19] E.A. Bender and E.R. Canfield. Locally restricted compositions. I. Restricted adjacent differences. *Electron. J. Combin.*, 12:#R57, 2005.

[20] E.A. Bender, G.F. Lawler, R. Pemantle, and H.S. Wilf. Irreducible compositions and the first return to the origin of a random walk. *Sém. Lothar. Combin.*, 50:Art. B50h, 2003/04.

[21] E.A. Bender and L.B. Richmond. Central and local limit theorems applied to asymptotic enumeration. II. multivariate generating functions. *J. Combin. Theory Ser. A*, 34(3):255–265, 1983.

[22] A.T. Benjamin and J.J. Quinn. *Proofs that really count*, volume 27 of *The Dolciani Mathematical Expositions*. Mathematical Association of America, Washington, DC, 2003.

[23] A. Bernini, L. Ferrari, and R. Pinzani. Enumeration of some classes of words avoiding two, generalized patterns of length three. *arXiv:0711.3387v1*, 2007.

[24] P. Billingsley. *Probability and Measure*. Wiley Series in Probability and Mathematical Statistics. John Wiley & Sons Inc., New York, third edition, 1995. A Wiley-Interscience Publication.

[25] M. Bóna. *Combinatorics of Permutations*. Discrete Mathematics and its Applications (Boca Raton). Chapman & Hall/CRC, Boca Raton, FL, 2004.

[26] M. Bousquet-Mélou and A. Rechnitzer. The site-perimeter of bargraphs. *Adv. in Appl. Math.*, 31:86–112, 2003.

[27] P. Brändén and T. Mansour. Finite automata and pattern avoidance in words. *J. Combin. Theory Ser. A*, 110(1):127–145, 2005.

[28] C. Brennan and A. Knopfmacher. The distribution of ascents of size d or more in compositions. *Discrete Math. Theor. Comput. Sci.*, 11(1):1–10, 2009.

[29] R. Brigham, R. Caron, P. Chinn, and R.P. Grimaldi. A tiling scheme for the Fibonacci numbers. *J. Recreational Mathematics*, 28(1):10–17, 1996.

[30] R.C. Brigham, P.Z. Chinn, L.M. Holt, and R.S. Wilson. Finding the recurrence relation for tiling $2 \times n$ rectangles. *Congr. Numer.*, 105:134–138, 1994.

[31] A. Burstein. *Enumeration of words with forbidden patterns*. Ph.D. thesis, University of Pennsylvania, 1998.

[32] A. Burstein, P. Hästö, and T. Mansour. Packing patterns into words. *Electron. J. Combin.*, 9(2):#R20, 2002/03.

[33] A. Burstein and T. Mansour. Words restricted by patterns with at most 2 distinct letters. *Electron. J. Combin.*, 9(2):#R3, 2002/03.

[34] A. Burstein and T. Mansour. Counting occurrences of some subword patterns. *Discrete Math. Theor. Comput. Sci.*, 6(1):1–11, 2003.

[35] A. Burstein and T. Mansour. Words restricted by 3-letter generalized multipermutation patterns. *Ann. Comb.*, 7(1):1–14, 2003.

[36] L. Carlitz. Enumeration of sequences by rises and falls: a refinement of the Simon Newcomb problem. *Duke Math. J.*, 39:267–280, 1972.

[37] L. Carlitz. Enumeration of up-down sequences. *Discrete Math.*, 4:273–286, 1973.

[38] L. Carlitz. Restricted compositions. *Fibonacci Quart.*, 14(3):254–264, 1976.

[39] L. Carlitz. Enumeration of compositions by rises, falls and levels. *Math. Nachr.*, 77:361–371, 1977.

[40] L. Carlitz. Up-down and down-up partitions. In *Studies in foundations and combinatorics*, volume 1 of *Adv. in Math. Suppl. Stud.*, pages 101–129. Academic Press, New York, 1978.

[41] L. Carlitz and R. Scoville. Up-down sequences. *Duke Math. J.*, 39:583–598, 1972.

[42] L. Carlitz, R. Scoville, and T. Vaughan. Enumeration of pairs of sequences by rises, falls and levels. *Manuscripta Math.*, 19(3):211–243, 1976.

[43] L. Carlitz and T. Vaughan. Enumeration of sequences of given specification according to rises, falls and maxima. *Discrete Math.*, 8:147–167, 1974.

[44] F. Cazals. Monomer-dimer tilings. On INRIA web page, 1997.

[45] M. Cerasoli. Enumeration of compositions with prescribed parts. *J. Math. Anal. Appl.*, 89(2):351–358, 1982.

[46] Ch.A. Charalambides. On the enumeration of certain compositions and related sequences of numbers. *Fibonacci Quart.*, 20(2):132–146, 1982.

[47] W.Y.C. Chen, E.Y.P. Deng, R.R.X. Du, R.P. Stanley, and C.H. Yan. Crossings and nestings of matchings and partitions. *Trans. Amer. Math. Soc.*, 359(4):1555–1575 (electronic), 2007.

[48] P. Chinn, R.P. Grimaldi, and S. Heubach. The frequency of summands of a particular size in palindromic compositions. *Ars Combin.*, 69:65–78, 2003.

[49] P. Chinn, R.P. Grimaldi, and S. Heubach. Tilings with ls and squares. *J. Integer Seq.*, 10(2):Art. 07.2.8, 2007.

[50] P. Chinn and S. Heubach. $(1, k)$-compositions. *Congr. Numer.*, 164:183–194, 2003.

[51] P. Chinn and S. Heubach. Compositions of n with no occurrence of k. *Congr. Numer.*, 164:33–51, 2003.

[52] P. Chinn and S. Heubach. Integer sequences related to compositions without 2's. *J. Integer Seq.*, 6(2):Art. 03.2.3, 2003.

[53] T. Chow and J. West. Forbidden subsequences and Chebyshev polynomials. *Discrete Math.*, 204(1-3):119–128, 1999.

[54] F.R.K. Chung, R.L. Graham, V.E. Hoggatt, Jr., and M. Kleiman. The number of Baxter permutations. *J. Combin. Theory Ser. A*, 24(3):382–394, 1978.

[55] A. Claesson. Generalized pattern avoidance. *European J. Combin.*, 22(7):961–971, 2001.

[56] S. Corteel and C.D. Savage. Partitions and compositions defined by inequalities. *Ramanujan J.*, 8(3):357–381, 2004.

[57] S. Corteel, C.D. Savage, and H.S. Wilf. A note on partitions and compositions defined by inequalities. *Integers*, 5(1):A24, 2005.

[58] A. de Mier. k-noncrossing and k-nonnesting graphs and fillings of Ferrers diagrams. *Electron. Notes Discrete Math.*, 28:3–10, 2007.

[59] R. Diestel. *Graph theory*, volume 173 of *Graduate Texts in Mathematics*. Springer-Verlag, Berlin, third edition, 2005.

[60] J.F. Dillon and D.P. Roselle. Simon Newcomb's problem. *SIAM J. Appl. Math.*, 17:1086–1093, 1969.

[61] J. Dollhopf, I. Goulden, and C. Greene. Words avoiding a reflexive acyclic relation. *Electron. J. Combin.*, 11(2):#R28, 2004/06.

[62] A.A. Evdokimov and V. Nyu. The length of a supersequence for a set of binary words with a given number of units. *Metody Diskret. Anal.*, 52:49–58, 1992.

[63] L. Ferrari, E. Pergola, R. Pinzani, and S. Rinaldi. An algebraic characterization of the set of succession rules. *Theoret. Comput. Sci.*, 281(1-2):351–367, 2002.

[64] G. Firro and T. Mansour. Three-letter-pattern-avoiding permutations and functional equations. *Electron. J. Combin.*, 13(1):#R51, 2006.

[65] G. Firro and T. Mansour. Restricted k-ary words and functional equations. *Discrete Appl. Math.*, 157(4):602–616, 2009.

[66] M.E. Fisher. Statistical mechanics of dimers on a plane lattice. *Phys. Rev. (2)*, 124:1664–1672, 1961.

[67] P. Flajolet, X. Gourdon, and P. Dumas. Mellin transforms and asymptotics: harmonic sums. *Theoret. Comput. Sci.*, 144(1-2):3–58, 1995.

[68] P. Flajolet and R. Sedwick. *Analytic Combinatorics.* Cambridge University Press, 2009.

[69] I. Flores. Direct calculation of k-generalized Fibonacci numbers. *Fibonacci Quart.*, 5:259–266, 1967.

[70] D. Foata and G.-N. Han. Inverses of words. *Sém. Lothar. Combin.*, 39:Art. B39d, 1997.

[71] D. Foata and G.-N. Han. Transformations on words. *J. Algorithms*, 28(1):172–191, 1998.

[72] T. Fort. *Finite Differences and Difference Equations in the Real Domain.* Oxford, at the Clarendon Press, 1948.

[73] W.M.Y. Goh and P. Hitczenko. Average number of distinct part sizes in a random Carlitz composition. *European J. Combin.*, 23(6):647–657, 2002.

[74] H.W. Gould. Binomial coefficients, the bracket function, and compositions with relatively prime summands. *Fibonacci Quart.*, 2:241–260, 1964.

[75] I.P. Goulden and D.M. Jackson. *Combinatorial enumeration.* Dover Publications Inc., Mineola, NY, 2004.

[76] R.P. Grimaldi. Compositions with odd summands. *Congr. Numer.*, 142:113–127, 2000.

[77] R.P. Grimaldi. Compositions without the summand 1. *Congr. Numer.*, 152:33–43, 2001.

[78] R.P. Grimaldi. *Discrete and Combinatorial Mathematics: An Applied Introduction.* Addison Wesley, 2003.

[79] R.P. Grimaldi. Compositions and the alternate Fibonacci numbers. *Congr. Numer.*, 186:81–96, 2007.

[80] G.R. Grimmett and D.R. Stirzaker. *Probability and random processes.* Oxford University Press, New York, third edition, 2001.

[81] E.O. Hare and P. Chinn. Tiling with Cuisenaire rods. In *Applications of Fibonacci numbers, Vol. 6 (Pullman, WA, 1994)*, pages 165–171. Kluwer Acad. Publ., Dordrecht, 1996.

[82] H.A. Helfgott and I.M. Gessel. Enumeration of tilings of diamonds and hexagons with defects. *Electron. J. Combin.*, 6:#R16, 1999.

[83] P. Henrici. *Applied and Computational Complex Analysis.* John Wiley & Sons Inc., New York, 1974-1977.

[84] S. Heubach. Tiling an m-by-n area with squares of size up to k-by-k ($m \leq 5$). *Congr. Numer.*, 140:43–64, 1999.

[85] S. Heubach and P. Chinn. Patterns arising from tiling rectangles with 1-by-1 and 2-by-2 squares. *Congr. Numer.*, 150:173–192, 2001.

[86] S. Heubach, P. Chinn, and P. Callahan. Tiling with trominoes. *Congr. Numer.*, 177:33–44, 2005.

[87] S. Heubach, P. Chinn, and R.P. Grimaldi. Rises, levels, drops and "+" signs in compositions: extensions of a paper by Alladi and Hoggatt. *Fibonacci Quart.*, 41(3):229–239, 2003.

[88] S. Heubach, S. Kitaev, and T. Mansour. Partially ordered patterns in compositions. *Pure Math. Appl.*, 17(1-2):123–134, 2006.

[89] S. Heubach and T. Mansour. Compositions of n with parts in a set. *Congr. Numer.*, 168:127–143, 2004.

[90] S. Heubach and T. Mansour. Counting rises, levels, and drops in compositions. *Integers*, 5(1):A11, 2005.

[91] S. Heubach and T. Mansour. Avoiding patterns of length three in compositions and multiset permutations. *Adv. in Appl. Math.*, 36(2):156–174, 2006.

[92] S. Heubach and T. Mansour. Enumeration of 3-letter patterns in compositions. In *Combinatorial Number Theory*, pages 243–264. deGruyter, 2007.

[93] P. Hitczenko and G. Louchard. Distinctness of compositions of an integer: a probabilistic analysis. *Random Structures Algorithms*, 19(3-4):407–437, 2001.

[94] P. Hitczenko, C. Rousseau, and C.D. Savage. A generatingfunctionology approach to a problem of Wilf. *J. Comput. Appl. Math.*, 142:107–114, 2002.

[95] P. Hitczenko and C.D. Savage. On the probability that a randomly chosen part size in a random composition is unrepeated. In *Paul Erdős and his mathematics (Budapest, 1999)*, pages 108–111. János Bolyai Math. Soc., Budapest, 1999.

[96] P. Hitczenko and C.D. Savage. On the multiplicity of parts in a random composition of a large integer. *SIAM J. Discrete Math.*, 18(2):418–435 (electronic), 2004.

[97] P. Hitczenko and G. Stengle. Expected number of distinct part sizes in a random integer composition. *Combin. Probab. Comput.*, 9(6):519–527, 2000.

[98] V.E. Hoggatt, Jr. and M. Bicknell. Palindromic compositions. *Fibonacci Quart.*, 13(4):350–356, 1975.

[99] V.E. Hoggatt, Jr. and D.A. Lind. Fibonacci and binomial properties of weighted compositions. *J. Combin. Theory Ser. A*, 4:121–124, 1968.

[100] V.E. Hoggatt, Jr. and D.A. Lind. Compositions and Fibonacci numbers. *Fibonacci Quart.*, 7(3):253–266, 1969.

[101] J.E. Hopcroft, R. Motwani, and J.D. Ullman. *Introduction to automata theory, languages and computation.* Addison Wesley, 2001.

[102] H.-K. Hwang and Y.-N. Yeh. Measures of distinctness for random partitions and compositions of an integer. *Adv. in Appl. Math.*, 19(3):378–414, 1997.

[103] B. Jackson and F. Ruskey. Meta-Fibonacci sequences, binary trees and extremal compact codes. *Electron. J. Combin.*, 13(1):#R26, 2006.

[104] V. Jelínek and T. Mansour. On pattern-avoiding partitions. *Electron. J. Combin.*, 15(1):#R39, 2008.

[105] V. Jelínek and T. Mansour. Wilf-equivalence on k-ary words, compositions, and parking functions. *Electron. J. Combin.*, 16(1):#R58, 2009.

[106] C. Kenyon, D. Randall, and A. Sinclair. Approximating the number of monomer-dimer coverings of a lattice. *J. Statist. Phys.*, 83(3-4):637–659, 1996.

[107] R. Kenyon. Local statistics of lattice dimers. *Ann. Inst. H. Poincaré Probab. Statist.*, 33(5):591–618, 1997.

[108] S. Kitaev. Partially ordered generalized patterns. *Discrete Math.*, 298(1-3):212–229, 2005.

[109] S. Kitaev and T. Mansour. Partially ordered generalized patterns and *k*-ary words. *Ann. Comb.*, 7(2):191–200, 2003.

[110] S. Kitaev, T. Mansour, and J.B. Remmel. Counting descents, rises, and levels, with prescribed first element, in words. *Discrete Math. Theor. Comput. Sci.*, 10(3):1–22, 2008.

[111] S. Kitaev, T.B. McAllister, and T.K. Petersen. Enumerating segmented patterns in compositions and encoding by restricted permutations. *Integers*, 6:A34, 2006.

[112] M. Klazar. On *abab*-free and *abba*-free set partitions. *European J. Combin.*, 17(1):53–68, 1996.

[113] M. Klazar. On trees and noncrossing partitions. *Discrete Appl. Math.*, 82(1-3):263–269, 1998.

[114] M. Kline. *Mathematical Thought from Ancient to Modern Times.* Oxford University Press, 1972.

[115] A. Knopfmacher and T. Mansour. Record statistics in a random composition. Preprint, 2009.

[116] A. Knopfmacher, T. Mansour, A. Munagi, and H. Prodinger. Smooth words and Chebyshev polynomials. Preprint, 2008.

[117] A. Knopfmacher and M.E. Mays. Compositions with *m* distinct parts. *Ars Combin.*, 53:111–128, 1999.

[118] A. Knopfmacher and M.E. Mays. The sum of distinct parts in compositions and partitions. *Bull. Inst. Combin. Appl.*, 25:66–78, 1999.

[119] A. Knopfmacher and H. Prodinger. On Carlitz compositions. *European J. Combin.*, 19(5):579–589, 1998.

[120] A. Knopfmacher and N. Robbins. On binary and Fibonacci compositions. *Ann. Univ. Sci. Budapest. Sect. Comput.*, 22:193–206, 2003.

[121] A. Knopfmacher and N. Robbins. Compositions with parts constrained by the leading summand. *Ars Combin.*, 76:287–295, 2005.

[122] D.E. Knuth. *The art of computer programming. Volume 3.* Addison-Wesley Publishing Co., Reading, Mass.-London-Don Mills, Ont., 1973.

[123] D.E. Knuth. *The art of computer programming. Volume 1.* Addison-Wesley Publishing Co., Reading, Mass.-London-Amsterdam, second edition, 1975.

[124] D.E. Knuth. The average time for carry propagation. *Nederl. Akad. Wetensch. Indag. Math.*, 40(2):238–242, 1978.

[125] T. Koshy. *Fibonacci and Lucas numbers with applications*. Pure and Applied Mathematics (New York). Wiley-Interscience, New York, 2001.

[126] C. Krattenthaler. Permutations with restricted patterns and Dyck paths. *Adv. in Appl. Math.*, 27(2-3):510–530, 2001.

[127] C. Krattenthaler. Growth diagrams, and increasing and decreasing chains in fillings of Ferrers shapes. *Adv. in Appl. Math.*, 37(3):404–431, 2006.

[128] V. Lakshmibai and B. Sandhya. Criterion for smoothness of Schubert varieties in $\mathrm{Sl}(n)/B$. *Proc. Indian Acad. Sci. Math. Sci.*, 100(1):45–52, 1990.

[129] L. Lipshitz. D-finite power series. *J. Algebra*, 122(2):353–373, 1989.

[130] M. Lothaire. *Combinatorics on Words*. Cambridge Mathematical Library. Cambridge University Press, Cambridge, 1997.

[131] M. Lothaire. *Algebraic combinatorics on words*, volume 90 of *Encyclopedia of Mathematics and its Applications*. Cambridge University Press, Cambridge, 2002.

[132] M. Lothaire. *Applied combinatorics on words*, volume 105 of *Encyclopedia of Mathematics and its Applications*. Cambridge University Press, Cambridge, 2005.

[133] G. Louchard. The number of distinct part sizes of some multiplicity in compositions of an integer: A probabilistic analysis. In *Random Walks*, volume AC of *DMTCS Proceedings*, pages 155–170. Discrete Mathematics and Theoretical Computer Science, 2003.

[134] G. Louchard and H. Prodinger. Probabilistic analysis of Carlitz compositions. *Discrete Math. Theor. Comput. Sci.*, 5(1):71–95, 2002.

[135] I.D. MacDonald. Some group elements defined by commutators. *Math. Sci.*, 4(2):129–131, 1979.

[136] P.A. MacMahon. Memoir on the theory of the compositions of numbers. *Phil. Trans. Royal Society London. A*, 184:835–901, 1893.

[137] P.A. MacMahon. The Indices of Permutations and the Derivation Therefrom of Functions of a Single Variable Associated with the Permutations of any Assemblage of Objects. *Amer. J. Math.*, 35(3):281–322, 1913.

[138] P.A. MacMahon. *Combinatory analysis. Vol. I, II (bound in one volume)*. Dover Phoenix Editions. Dover Publications Inc., Mineola, NY, 2004. Reprint of *An introduction to combinatory analysis* (1920) and *Combinatory analysis. Vol. I, II* (1915, 1916).

[139] T. Mansour. http://math.haifa.ac.il/toufik/automaton01.html. Automaton program, 2003.

[140] T. Mansour. http://math.haifa.ac.il/toufik/table4auto.html. Table of values computed with automaton program, 2003.

[141] T. Mansour. Restricted 132-avoiding k-ary words, Chebyshev polynomials, and continued fractions. *Adv. in Appl. Math.*, 36(2):175–193, 2006.

[142] T. Mansour. Longest alternating subsequences in pattern-restricted k-ary words. *Online J. Analytic Combin.*, 3:#5, 2008.

[143] T. Mansour. Longest alternating subsequences of k-ary words. *Discrete Appl. Math.*, 156(1):119–124, 2008.

[144] T. Mansour. Smooth partitions and Chebyshev polynomials. Preprint, 2008.

[145] T. Mansour and B.O. Sirhan. Counting l-letter subwords in compositions. *Discrete Math. Theor. Comput. Sci.*, 8(1):285–297, 2006.

[146] T. Mansour and A. Vainshtein. Avoiding maximal parabolic subgroups of S_k. *Discrete Math. Theor. Comput. Sci.*, 4(1):67–75, 2000.

[147] T. Mansour and A. Vainshtein. Restricted 132-avoiding permutations. *Adv. in Appl. Math.*, 26(3):258–269, 2001.

[148] T. Mansour and A. Vainshtein. Restricted permutations and Chebyshev polynomials. *Sém. Lothar. Combin.*, 47:Art. B47c, 2001/02.

[149] T. Mansour and A. Vainshtein. Counting occurrences of 132 in a permutation. *Adv. in Appl. Math.*, 28(2):185–195, 2002.

[150] J.E. Marsden and M.J. Hoffman. *Basic complex analysis*. W. H. Freeman and Company, New York, second edition, 1987.

[151] R.B. McQuistan and S.J. Lichtman. Exact recursion relation for 2 x n arrays of dumbbells. *J. Math. Phy.*, 11(10):3095–3099, 1970.

[152] D. Merlini, R. Sprugnoli, and M.C. Verri. Strip tiling and regular grammars. *Theoret. Comput. Sci.*, 242(1-2):109–124, 2000.

[153] D. Merlini, R. Sprugnoli, and M.C. Verri. A strip-like tiling algorithm. *Theoret. Comput. Sci.*, 282(2):337–352, 2002.

[154] D. Merlini, F. Uncini, and M.C. Verri. A unified approach to the study of general and palindromic compositions. *Integers*, 4:A23, 2004.

[155] S.G. Mohanty. Some properties of compositions and their application to the ballot problem. *Canad. Math. Bull.*, 8:359–372, 1965.

[156] S.G. Mohanty. Restricted compositions. *Fibonacci Quart.*, 5:223–234, 1967.

[157] A.N. Myers. Pattern avoidance in multiset permutations: bijective proof. *Ann. Comb.*, 11(3-4):507–517, 2007.

[158] A.N. Myers and H.S. Wilf. Left-to-right maxima in words and multiset permutations. *Israel J. Math.*, 166:167–183, 2008.

[159] G. Narang and A.K. Agarwal. n-colour self-inverse compositions. *Proc. Indian Acad. Sci. Math. Sci.*, 116(3):257–266, 2006.

[160] G. Narang and A.K. Agarwal. Lattice paths and n-colour compositions. *Discrete Math.*, 308(9):1732–1740, 2008.

[161] T.V. Narayana. Sur les treillis formés par les partitions d'un entier et leurs applications à la théorie des probabilités. *C. R. Acad. Sci. Paris*, 240:1188–1189, 1955.

[162] T.V. Narayana and G.E. Fulton. A note on the compositions of an integer. *Canad. Math. Bull.*, 1:169–173, 1958.

[163] J. Noonan and D. Zeilberger. The enumeration of permutations with a prescribed number of "forbidden" patterns. *Adv. in Appl. Math.*, 17(4):381–407, 1996.

[164] A.M. Odlyzko. Asymptotic enumeration methods. In *Handbook of combinatorics, Vol. 1, 2*, pages 1063–1229. Elsevier, Amsterdam, 1995.

[165] H. Prodinger. Combinatorics of geometrically distributed random variables: left-to-right maxima. *Discrete Math.*, 153:253–270, 1996.

[166] H. Prodinger. Records in geometrically distributed words: sum of positions. *Appl. Anal. Discrete Math.*, 2(2):234–240, 2008.

[167] J. Propp. Enumeration of matchings: problems and progress. In L. J. Billera, A. Björner, C. Greene, R. E. Simion, and R. P. Stanley, editors, *New perspectives in algebraic combinatorics*, volume 38 of *Mathematical Sciences Research Institute Publications*, pages 255–291. Cambridge University Press, 1999.

[168] D. Rawlings. Restricted words by adjacencies. *Discrete Math.*, 220:183–200, 2000.

[169] A. Regev. Asymptotics of the number of k-words with an l-descent. *Electron. J. Combin.*, 5:#R15, 1998.

[170] M. Régnier and W. Szpankowski. On the approximate pattern occurrences in a text, in Compression and Complexity of Sequences 1997. *IEEE Computer Society*, pages 253–264, 1998.

[171] B. Richmond and A. Knopfmacher. Compositions with distinct parts. *Aequationes Math.*, 49:86–97, 1995.

[172] A. Riese. qMultiSum—a package for proving q-hypergeometric multiple summation identities. *J. Symbolic Comput.*, 35(3):349–376, 2003.

[173] Th. Rivlin. *Chebyshev polynomials*. Pure and Applied Mathematics (New York). John Wiley & Sons Inc., New York, second edition, 1990.

[174] A. Robertson, H.S. Wilf, and D. Zeilberger. Permutation patterns and continued fractions. *Electron. J. Combin.*, 6:#R38, 1999.

[175] B.E. Sagan. Proper partitions of a polygon and k-Catalan numbers. *Ars Combin.*, 88:109–124, 2008.

[176] C.D. Savage and H.S. Wilf. Pattern avoidance in compositions and multiset permutations. *Adv. in Appl. Math.*, 36(2):194–201, 2006.

[177] C. Schensted. Longest increasing and decreasing subsequences. *Canad. J. Math.*, 13:179–191, 1961.

[178] R. Servedio and Y.-N. Yeh. A bijective proof on circular compositions. *Bull. Inst. Math. Acad. Sinica*, 23(4):283–293, 1995.

[179] R. Simion and F.W. Schmidt. Restricted permutations. *European J. Combin.*, 6:383–406, 1985.

[180] N.J.A. Sloane. *The on-line encyclopedia of integer sequences.* published electronically at http://www.research.att.com/~njas/sequences/, 2009.

[181] K. Srinivase Rao and A.K. Agarwal. On a generalized composition function. *J. Indian Math. Soc. (N.S.)*, 67(1-4):99–106, 2000.

[182] R.P. Stanley. *Enumerative combinatorics. Vol. 1*, volume 49 of *Cambridge Studies in Advanced Mathematics*. Cambridge University Press, Cambridge, 1997.

[183] R.P. Stanley. *Enumerative combinatorics. Vol. 2*, volume 62 of *Cambridge Studies in Advanced Mathematics*. Cambridge University Press, Cambridge, 1999.

[184] Axel Thue. Über unendliche Zahlenreihen. In Trygve Nagell, Atle Selberg, Sigmund Selberg, and Knut Thalberg, editors, *Selected Mathematical papers of Axel Thue*, pages 139–158. Universitetsforlaget, 1977. Originally published in Kra. Vidensk. Selsl. Skrifter. I. Mat. Nat. Kl. 1906. No. 7. Kra. 1906.

[185] Axel Thue. Über die gegenseitige Lage gleicher Teile gewisser Zeichenreihen. In Trygve Nagell, Atle Selberg, Sigmund Selberg, and Knut Thalberg, editors, *Selected Mathematical papers of Axel Thue*, pages 413–478. Universitetsforlaget, 1977. Originally published in Kra. Vidensk. Selsl. Skrifter. I. Mat. Nat. Kl. 1912. No. 1. Kra. 1912.

[186] Axel Thue. Probleme über Veränderungen von Zeichenreihen nach gegebenen Regeln. In Trygve Nagell, Atle Selberg, Sigmund Selberg, and Knut Thalberg, editors, *Selected Mathematical papers of Axel Thue*, pages 493–524. Universitetsforlaget, 1977. Originally published in Kra. Vidensk. Selsl. Skrifter. I. Mat. Nat. Kl. 1914. No. 10. Kra. 1914.

[187] C.A. Tracy and H. Widom. On the distributions of the lengths of the longest monotone subsequences in random words. *Probab. Theory Related Fields*, 119(3):350–380, 2001.

[188] A.J. van Zanten. The ranking problem of a Gray code for compositions. *Ars Combin.*, 41:257–268, 1995.

[189] N.Y. Vilenkin. *Combinatorics*. Academic Press, New York-London, 1971.

[190] M. Wachs and D. White. p, q-Stirling numbers and set partition statistics. *J. Combin. Theory Ser. A*, 56(1):27–46, 1991.

[191] J. West. Permutations with forbidden subsequences and stack sortable permutations. Ph.D. thesis, M.I.T., 1990.

[192] J. West. Generating trees and forbidden subsequences. In *Proceedings of the 6th Conference on Formal Power Series and Algebraic Combinatorics (New Brunswick, NJ, 1994)*, volume 157, pages 363–374, 1996.

[193] H.S. Wilf. *Generatingfunctionology, 2nd Edition*. Academic Press, Inc., San Diego, 1994.

[194] J. Wimp and D. Zeilberger. Resurrecting the asymptotics of linear recurrences. *J. Math. Anal. Appl.*, 111(1):162–176, 1985.

Index

A

Printed and bound by CPI Group (UK) Ltd, Croydon, CR0 4YY

21/10/2024

01777083-0019